Automotive Emissions Regulations and Exhaust Aftertreatment Systems

Automotive Emissions Regulations and Exhaust Aftertreatment Systems

INTERNATIONAL

Warrendale, Pennsylvania, USA

400 Commonwealth Drive
Warrendale, PA 15096-0001 USA
E-mail: CustomerService@sae.org
Phone: 877-606-7323 (inside USA and Canada)
724-776-4970 (outside USA)
FAX 724-776-0790

Library of Congress Catalog Number 2020936340
http://dx.doi.org/10.4271/9780768099560

Information contained in this work has been obtained by SAE International from sources believed to be reliable. However, neither SAE International nor its authors guarantee the accuracy or completeness of any information published herein and neither SAE International nor its authors shall be responsible for any errors, omissions, or damages arising out of use of this information. This work is published with the understanding that SAE International and its authors are supplying information but are not attempting to render engineering or other professional services. If such services are required, the assistance of an appropriate professional should be sought.

ISBN-Print 978-0-7680-9955-3

To purchase bulk quantities, please contact: SAE Customer Service

E-mail: CustomerService@sae.org
Phone: 877-606-7323 (inside USA and Canada)
724-776-4970 (outside USA)
Fax: 724-776-0790

Visit the SAE International Bookstore at books.sae.org

Chief Product Officer
Frank Menchaca

Publisher
Sherry Dickinson Nigam

Director of Content Management
Kelli Zilko

Production Associate
Erin Mendicino

Manufacturing Associate
Adam Goebel

contents

CHAPTER 3
Emissions Control Regulations   61

CHAPTER 8

Oxidation Catalysts   251

Over six decades of study on exhaust emissions from internal combustion engines and their control have been done to date. New developments improve upon many generations of both low-emissions engines and exhaust aftertreatment devices that have been developed, as both concern for the environment and governmental regulations have grown increasingly stronger. Worldwide, there has been strong payoff for these efforts—visible impacts in major cities such as Los Angeles, Beijing, London, and New York are a tangible result of these engineering efforts. Improvement continues as modern engineers build on this work and have further impact. So, we must be aware of what has been done, evaluate what can be improved upon, and become increasingly sophisticated in our solutions.

Consistent with the above, the objective of this book is to present a fundamental development of the science and engineering underlying the design of exhaust aftertreatment systems for automotive (internal combustion engine containing) systems. Our emphasis is on gasoline and diesel, spark-ignition and compression-ignition combustion engines, though the same fundamentals may be applied to other fuels or combustion types. Notably, considerable introductory material is included on aspects that are related to emissions, including regulatory standards, gasoline and diesel engine basics, fuels, lubricants, combustion, instrumentation, and formation of pollutants. Several causes of emissions and pollutants are intertwined throughout the various topics and the control and treatment of specific emissions species are discussed by device type. Intentionally, the emphasis is on the concepts and mechanisms of operation of the device, rather than mathematical modeling, which can be found in the scientific literature. No prerequisite knowledge of the field is required: our objective is to acquaint the reader, whom we expect to be new to the field of emissions control, with the underlying principles, control methods, common problems, and fuel effects on catalytic exhaust aftertreatment devices. We do this in hope that they can better understand the previous and current generations of emissions control and improve upon them. This book is designed for the engineer, researcher, designer, student, or any combination of those, who is concerned with the control of automotive exhaust emissions. It includes discussion of theory and fundamentals applicable to hardware development.

Chapter 1 is a general introduction to exhaust emissions and air pollution and the role that vehicles play. It also introduces several general concepts used throughout the book.

Chapter 2 details the formation of carbon monoxide, carbon dioxide, hydrocarbons, nitrogen oxides (NOx), and particulate emissions in the combustion cylinder by combustion type. In addition, there is discussion of sulfur oxides and the generation of ash, which are also important to the operation and control of the aftertreatment devices.

Chapter 3 describes the motivation for emission control, focusing on regulatory requirements in the United States, California, the European Union, Japan, the People's

Republic of China, and India, with some discussion of requirements in other nations or regions. There is also discussion of unregulated emissions and ambient air quality.

Chapter 4 presents the emission testing protocols used to evaluate the efficacy of the exhaust aftertreatment system in the vehicle. This scope includes both light-duty and heavy-duty engines and vehicles in on-road and off-road applications. Test protocols for criteria pollutants and for greenhouse gases or fuel economy are described here.

Chapter 5 discusses the emissions measurement equipment used to evaluate engine and vehicle performance during the test protocols described in Chapter 4. Care is taken to distinguish between stationary and mobile detection methods and the equipment used in each.

Chapter 6 contains a discussion of sensors and on-board diagnostics for exhaust aftertreatment systems. This includes the temperature probes and pressure transducers, the NOx, particulate matter, and hydrocarbon sensors, as well as model-based controls and feedback controls that make use of the data they monitor.

Chapter 7 gives an overview of exhaust aftertreatment systems and serves as an introduction to the devices, how they work, and their interdependency within the overall system.

Chapter 8 deals with the oxidation catalysts, the close sibling of the three-way catalyst that is used for the oxidation of hydrocarbons, and carbon monoxide in diesel compression-ignition and other fuel-lean combustion strategies. The discussion includes the functions, catalyst formulations, washcoats, substrates, performance, and failure modes.

Chapter 9 describes the functions, catalyst formulations, washcoats, substrates, performance, and failure modes of three-way catalysts, which are used for the oxidation of hydrocarbons and carbon monoxide and reduction of nitrogen oxides in gasoline spark-ignition engine exhaust systems and other stoichiometric air–fuel combustion applications.

Chapter 10 details the functions of the lean NOx trap in exhaust aftertreatment systems. Although the original intent for the lean NOx trap was to make it a competitive technology to remove nitrogen oxides, the formation of ammonia over the catalyst was problematic for deployment. However, in a perfect example of turning a problem into a solution, the lean NOx trap has found a new purpose as an ammonia generator for use with selective catalytic reduction (SCR) systems. The functions, catalyst formulations, washcoats, substrates, performance, and failure modes of the lean NOx trap device are detailed in this chapter.

Chapter 11 describes particulate filters for both diesel-fueled and gasoline-fueled applications. Discussion of the functionality includes filtration, and regeneration, as well as ash accumulation. The catalyst formulations, washcoats, substrates, performance, and failure modes are described.

Chapter 12 is focused on the SCR catalyst using the eutectic urea-water solution known as diesel exhaust fluid or AdBlue. SCR functions, catalyst formulations, washcoats, substrates, performance, and failure modes are discussed. The processing of the urea reductant, from the injection system and dosing strategy, to the decomposition and hydrolysis of the urea–water solution into ammonia, to mixers and the flow distribution is discussed. The problems with improper dosing and mixing, which result in urea crystals or other deposits, are also described. Additional discussion is included on alternative reductants such as ammonia and hydrocarbons.

Chapter 13 gives details on the ammonia slip catalyst, which serves to clean up any residual ammonia "slipping" through the exhaust aftertreatment system. The functions, catalyst formulations, washcoats, substrates, performance, and failure modes are included in the discussion.

Chapter 14 describes passive aftertreatment systems, such as the passive NOx adsorber and hydrocarbon trap, which are designed to trap emissions during the start-up or cold-start period, before the catalysts in the aftertreatment system are hot enough to be active. Trap materials, washcoats, and supports are described for each.

Chapter 15 deals with combined devices, in which the functions of several separate devices are combined into one, such as the SCR or three-way catalyst on particulate filter. This chapter also looks at systems in which two devices work together for a synergistic function, such as the combination of lean NOx trap or three-way catalyst with downstream SCR device.

Chapter 16 discusses several alternative fuels and their effects on emissions and aftertreatment. It motivates the consideration for petroleum replacement fuels such as Fischer–Tropsch fuels, fatty-acid methyl ester biodiesels, hydro-treated bio-oils called "renewable diesel," dimethyl ether, as well as alcohols such as ethanol and butanol. Additional discussion is included on natural gas and liquefied petroleum gas as stand-alone fuels and co-fuels.

This volume is intended as an overview of critical topics and devices in exhaust aftertreatment systems, but not a thorough review of the very latest research, which can be found in the annual "Aftertreatment Year in Review" publications by SAE International.

acknowledgments

A book of this magnitude comes together with help from many corners, and we appreciate the supportive comments and feedback from our professional colleagues as we have been developing the book for you. The authors express their deep gratitude and appreciation to the publishers of the technical publications cited for permission to reproduce selected figures, and particularly to SAE International. The authors especially wish to thank several people who had a direct hand in making this book better, including the following:

Mr. Tom Howell for his support of the book project within AVL Powertrain Engineering; Mr. Larry Buch, Mr. Gary Hunter, and Mr. Russ Truemner of AVL for their heroic efforts in reviewing our chapters and their incredibly insightful feedback; and Mr. Rod Harris for his help with images.

Dr. Matthew Keenan of Ricardo UK for early reviews of and feedback on the outline and key chapters.

Dr. Christine Lambert of Ford Motor Co., Ms. Kelley Gawlik and Ms. Birgitta Gustavson of Scania CV AB, Dr. Elana Chapman and Dr. Rahul Mital of General Motors, Mr. Anders Vilhelmsson and Mr. Per-Martin Johansson of Volvo Trucks, Mr. James Robinson of Honda Motor Co., Dr. Holmes Ahari of Fiat Chrysler Automobiles, and Dr. Alessandro Cozzolini of Daimler Trucks North America for all of their help in securing the engine and aftertreatment system images used in Chapter 7.

Professors Mike Dudukovic and P.A. Ramachandran of Washington University in St. Louis and Professors Charlie Hill, Thatcher Root, Chris Rutland, and Dave Foster of the University of Wisconsin–Madison for their formative influences on our interests in emissions control technology and chemical reactor design.

We especially want to recognize the influences on this book from some key mentors who are no longer with us, including the late Dr. Richard Blint of N2Kinetics, the late Dr. C. Stuart Daw of Oak Ridge National Laboratory, and the late Professor Emeritus Warren E. Stewart of the University of Wisconsin–Madison.

In addition, we would be remiss to not acknowledge the love and support of our families and their encouragement during this process. In addition, their understanding of the time devoted to writing this book is much appreciated.

Lisa Henn and Carl—thank you for your support throughout the writing process, especially at crunch time. Dale and Jim Kasab—thank you for your support of my nerdy interests over the years that helped lead to this book.

Mike and Rosemary Strzelec—thank you for being supportive cheerleaders in this process. You followed the progress of my word counts with interest that only parents could have—especially when I told you early on that the butler didn't do it. I dedicate this book to my late sister Susie (Strzelec) Buda who brought me into this field and supported me intensely, in this project and all others - while lovingly calling me an "EngiNerd."

1

Introduction

It is up to us as professional engineers to see that our engines are controlled to produce "the minimum emissions feasible employing the maximum technical capabilities."

—Phil S. Myers, UW-Madison Engine Research Center and 1969
SAE President in SAE Technical Paper *700182 [1-1]*

Today, the internal combustion engine (ICE) is ubiquitous and an integral part of our lives, driving progress, people, and goods. ICEs were initially developed and demonstrated in the late 1800s by Nikolaus Otto, who created the spark-ignition (SI) engine in 1876, and Rudolf Diesel, who invented the compression-ignition (CI) engine in 1892. It is also noteworthy to recognize the contribution of Alphonse Beau de Rochas, who patented the idea for the four-stroke engine in 1861, though Otto was the first to build a working four-stroke. Since their invention nearly 150 years ago, ICE have undergone constant development, as engineers' understanding of engine combustion and thermodynamic processes has improved, as new technologies have been developed, as fuels have changed, and as environmental concerns have grown.

In the decades since the passage of laws promoting improved air quality through reductions in mobile sources, as discussed in Chapter 3, Emissions Control Regulations, engineers have studied the exhaust emissions from ICEs and developed methods and devices with which to control them. This effort has generated an immense technical literature spanning engine development, combustion strategies, emissions mitigation strategies, aftertreatment catalysts, and the like.

This book has been written as a professional reference to present a fundamental development of the science and engineering underlying the design of exhaust aftertreatment systems (EAS) for automotive and related applications. The book has two main areas of focus. The first is the regulations that require the use of emissions control devices and the related compliance issues such as testing, emissions measurements, and on-board diagnostics. The second is the individual exhaust aftertreatment devices or catalytic converters, their chemistry and mechanisms, their constituent materials, and their general operation.

Introductory material is presented for the following additional related topics: SI and CI engine fundamentals, combustion and emissions formation, and alternative fuels.

This text is designed to be an introductory reference. Several key concepts and terms that are used throughout this book are therefore defined in this chapter, which starts with a basic description of engine-operating fundamentals in both SI and CI engines, and in Chapter 7, Exhaust Aftertreatment Systems Overview, which provides a basic description of relevant concepts in chemistry and transport phenomena.

1.1 Engine-Operation Fundamentals

The primary use of engines is to convert the chemical energy stored in hydrocarbon (HC) fuel into useful, mechanical work. Specifically, for ICEs, this energy released from the chemical reactions of combusting the fuel, forming water (H_2O) and carbon dioxide (CO_2), occurs inside of the engine. The ICE does not follow an air-standard Otto or Diesel cycle because the working fluid is not constant over the cycle. The air-fuel mixture that is introduced into or created in the engine and the products of combustion are the working fluids involved over this mechanical cycle. This is why an ICE is NOT a thermodynamic cycle, but rather a thermodyanmic process that operates on a mechanical cycle.

Power production using a piston-in-cylinder expander was done for 150 years before ICEs were developed [1-2]. These devices, like ICEs, are reciprocating engines. The combustion of the fuel releases energy within the cylinder, causing the temperature to increase, and the resultant expanding product gases push the piston to produce mechanical work. The two types of ICEs that are of interest for the emissions and aftertreatment focus of this book are SI and CI engines, which are typically fueled with gasoline and diesel, respectively. Engines represent the vast majority of the transportation applications for land and sea. They also have a presence in stationary power generation, and in moving products through oil and natural gas pipelines [1-3]. As this book is focused on automotive applications, we have assumed that the reader will primarily be working with four-stroke, reciprocating ICEs, as described below. First, we will define some key engine components and performance parameters. These terms, along with the basics of engine operation, are necessary as a foundation for the discussion of EAS, but these descriptions are not nearly as detailed as those found in a book dedicated to ICEs. The reader is directed to the works of Heywood [1-2], Stone [1-4], and Ferguson [1-5] for more detailed information on ICE topics.

1.1.1 Engine Components

A cutaway drawing of a multicylinder engine is shown in **Figure 1.1**, with the major components labeled. A brief description of the major components follows, in alphabetical order, with the component name italicized. The *block* is the body of the engine. In the image, the orientation of the cylinders within the engine block can be seen. The block also contains channels for coolant, which are cast into the block. It is often made of cast aluminum for gasoline engines and cast iron for diesels. The *camshaft* is the rotating shaft that is used to push down on the valves, either directly or through a mechanical linkage such as a pushrod, rocker arm, or tappets. The *combustion chamb*er is the portion of the

FIGURE 1.1 Cutaway drawing of an engine, with major components labeled.

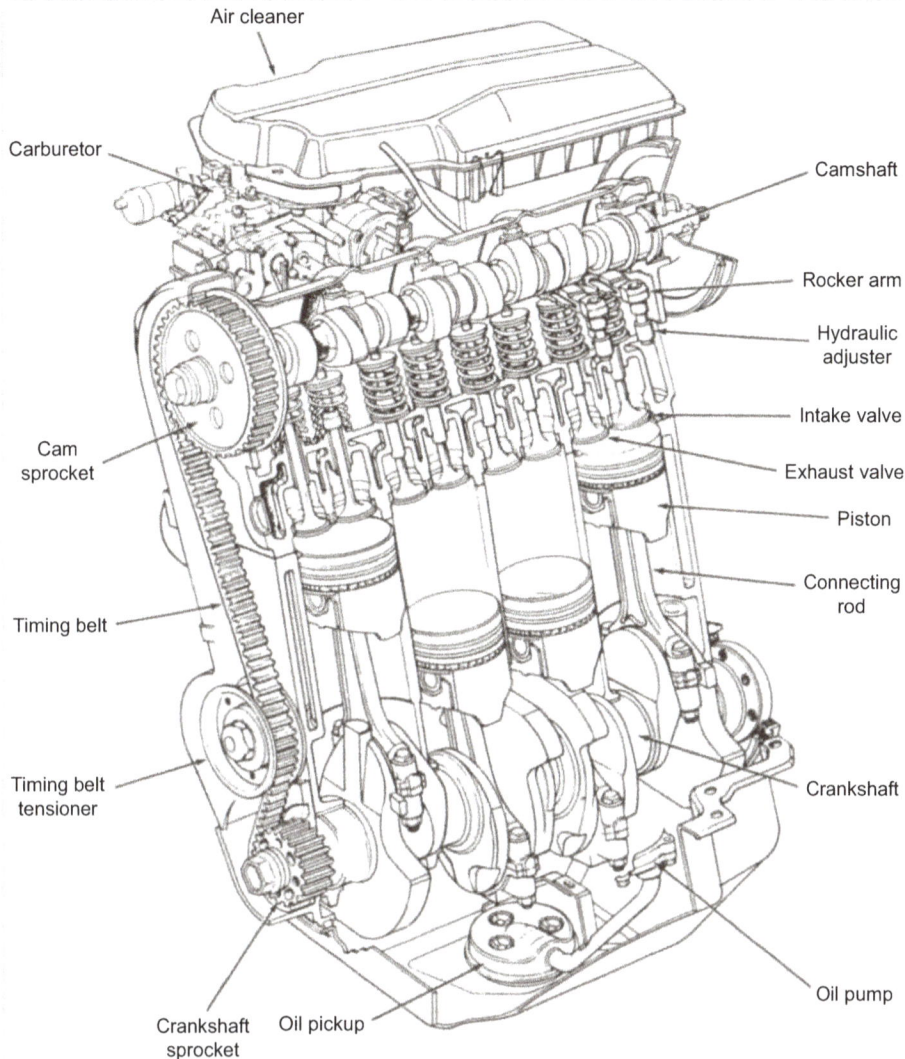

cylinder volume between the head and the top face of the piston, where combustion occurs. *Connecting rods* are the mechanical linkages between the piston and the crankshaft. The *crankcase* is the part of the engine block that surrounds the rotating *crankshaft*, through which the engine work is transferred to the external systems. It is connected to the engine block by the main bearings. The crankshaft is rotated due to the reciprocating motion of the pistons (connected via the *connecting rods*). The *cylinders* are the circular channels within the block through which the pistons move up and down in a reciprocating manner. The exhaust manifold is the piping system that carries the exhaust gases away from the cylinders and into the *exhaust system*, which transmits the spent combustion gases to the tailpipe, before releasing them to the environment. The exhaust aftertreatment components, which are the focus of this book, are contained within the exhaust system. Most engines have an engine-driven *fan* that is used to increase the airflow to the engine compartment and *radiator* for cooling purposes. The *flywheel*, connected to the crankshaft, is a rotating mass with a large moment of inertia. Its purpose is to store energy and deliver the angular momentum required to keep the engine rotating smoothly between power strokes. The *fuel injector* is a pressurized nozzle for introducing fuel into the incoming air for conventional SI engines, or into the cylinder for CI and SI direct injection (DI) engines. In classical SI engines, the fuel injector is located in the intake port, which is why they can also be referred to as port-fuel injected (PFI) engines. The fuel is supplied to the fuel injectors by the *fuel pump*, through the *fuel lines*. A *glow plug* is a small, electric resistance heater that can be mounted inside the combustion chamber of a CI engine to preheat the chamber for cold start. The *head* of the engine is what closes off the top end of the cylinders, defining the clearance volume when the piston is at its top-most position. The valves, fuel injectors, and camshaft are generally located within the head. The *head gasket** is the gasket seal between the block and the head. The *intake manifold* is the system of piping that delivers the intake air into the cylinders. The *main bearings* are connected to the engine block, in which the crankshaft rotates. The number of bearings is generally the number of pistons plus one, or the number of pistons plus three, depending on the bearing size used. The *oil pan* is the reservoir at the bottom of the engine block, often integral to the crankcase. The *oil pump* is used to distribute oil to the required lubrication points within the engine. The *oil sump* is the reservoir for the lubricating oil system, and often part of the crankcase. *Pistons* are the cylindrically shaped masses that reciprocate within the cylinder to transmit the pressure forces developed in the combustion chamber to the rotating crankshaft. The top of the piston is referred to as its *crown* and the sides as its *skirt*. *Piston rings* are the metal rings that fit into the circumferential grooves in the piston skirt to provide the sliding surface against the cylinder walls. There are generally two compression rings and one oil ring per piston. The *radiator* is a liquid-to-air heat exchanger used to dump heat from the engine coolant. It is usually mounted in front of the engine, in the flow of air. A *spark plug* is an electrical device used to initiate combustion in SI engines by creating a high-voltage discharge spark across an electrode gap. The *starter* is a small electric motor geared to the engine flywheel whose energy is supplied by the battery. Its purpose is to overturn or "crank" the engine to start it. A *supercharger* is a mechanical compressor, sometimes used on SI engines, powered off the crankshaft that can be used to compress the intake air. The *throttle* is a butterfly valve mounted upstream of the intake system that is used to control the amount of air introduced into an SI engine. A *turbocharger* is the name for a turbine–compressor combination used to compress the intake air before it enters the engine. The compressor (in the intake system) is powered by the exhaust flow moving through the turbine (in the exhaust system) and therefore only represents a small loss from the engine as compared to the supercharger, which is parasitically powered off the crankshaft. The *valves* allow the flow into (intake valves) and out (exhaust valves) of the cylinder at the proper times during the cycle. The surface against which the valves are closed is called the *valve seats*. The *water jacket* is the system of liquid flow channels that surround the cylinders in the head through which the coolant flows to keep the engine from overheating. The *water pump* is used to circulate the engine coolant through the water jacket and radiator. The *wrist pin* fastens the piston to the connecting rod.

---

* The head gasket is a particularly expensive item to replace on an engine, due to its location and the amount of effort needed for disassembly and reassembly.

Figure 1.2 shows a cutaway view of the slider crank components of a reciprocating engine. This diagram can be consulted to view the locations of the piston at top dead center (TDC), when the connecting rod is fully extended and the piston in its top-most point, closest to the head of the engine. The volume that remains between the head and the face of the piston is called the *clearance volume*, V_c. This volume at TDC, along with the surface area, is a key factor in the efficiency of the engine. When the piston is in its bottom-most point, at bottom dead center (BDC), the volume between the head and the face of the piston is called the total volume of the cylinder. The difference between the total cylinder volume and the clearance volume is the *displacement volume*, V_d, that is swept by the motion of the reciprocating piston. We can define the *compression ratio* (CR) of the engine as the ratio of the engine volume at BDC ($V_d + V_c$) to the volume at TDC (V_c).

$$\text{Compression ratio} = \frac{V_d + V_c}{V_c} \tag{1.1}$$

The reliance on a controlled spark in SI engines historically resulted in lower CRs to manage the temperature history in order to avoid "knock," an undesirable autoignition

FIGURE 1.2 Components of a piston/cylinder reciprocating engine.

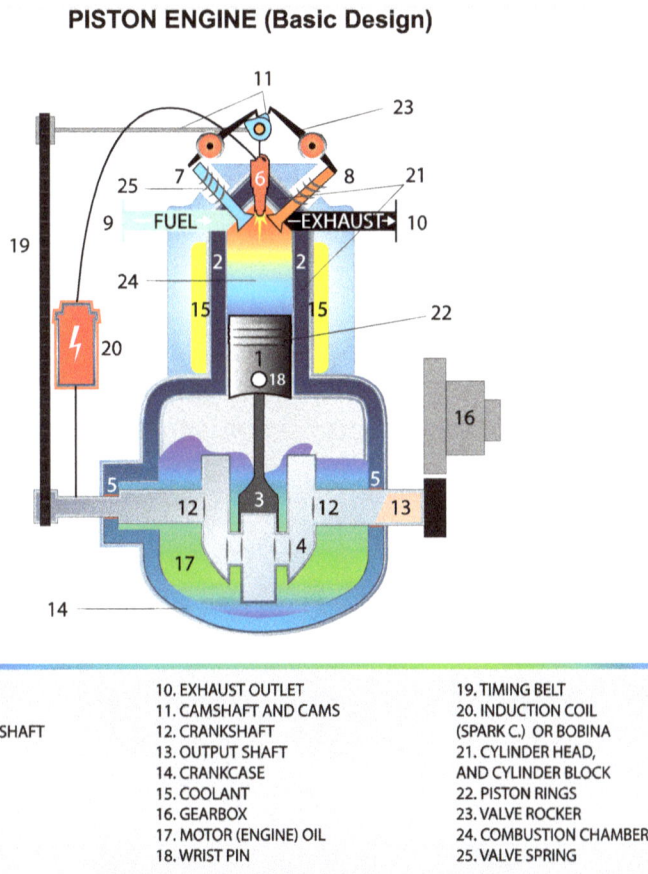

PISTON ENGINE (Basic Design)

1. PISTON
2. CYLINDER
3. CONNECTING ROD SHAFT
4. CRANK
5. MAIN BEARINGS
6. SPARK PLUG
7. INTAKE VALVE
8. EXHAUST VALVE
9. FUEL-AIR MIXTURE
10. EXHAUST OUTLET
11. CAMSHAFT AND CAMS
12. CRANKSHAFT
13. OUTPUT SHAFT
14. CRANKCASE
15. COOLANT
16. GEARBOX
17. MOTOR (ENGINE) OIL
18. WRIST PIN
19. TIMING BELT
20. INDUCTION COIL (SPARK C.) OR BOBINA
21. CYLINDER HEAD, AND CYLINDER BLOCK
22. PISTON RINGS
23. VALVE ROCKER
24. COMBUSTION CHAMBER
25. VALVE SPRING

© Shutterstock

phenomenon that is described further in Section 1.2.1. A lower CR reduces the expansion ratio, and therefore thermodynamic efficiency, which is defined in Section 1.1.4. CI engines employ compression heating of the intake charge to create temperatures high enough to auto-ignite the fuel. Higher CRs are required to achieve the required temperature, which improves the thermodynamic efficiency, but requires heavier materials of construction and therefore adds cost.

1.1.2 The Four-Stroke Cycle

Both SI and CI engines are reciprocating ICE that can operate on a four-stroke cycle. The classic definition of this sequence is composed of the following steps, which are also illustrated in **Figure 1.3**:

1. An *intake stroke*, which draws the fresh charge into the cylinder. Before the intake stroke starts, the intake valve opens to enable the air or air/fuel mixture to enter. Then, the piston moves downward from TDC to BDC, drawing the charge into the cylinder. The intake valve closes shortly before the piston reaches BDC.

2. A *compression stroke*, with the valves closed, and the charge trapped inside the cylinder, the piston travels upward from BDC to TDC, compressing the gas from the maximum cylinder volume down to the clearance volume. Near the end of this stroke, as the piston approaches TDC, combustion starts, which increases the in-cylinder pressure rapidly.

3. A *power stroke*, where the expansion of the gases, due to the high pressure and temperature resulting from combustion, pushes the piston down from TDC to BDC, with more force than was used to compress the charge, thus recovering

FIGURE 1.3   Illustrations of the four-stroke cycle.

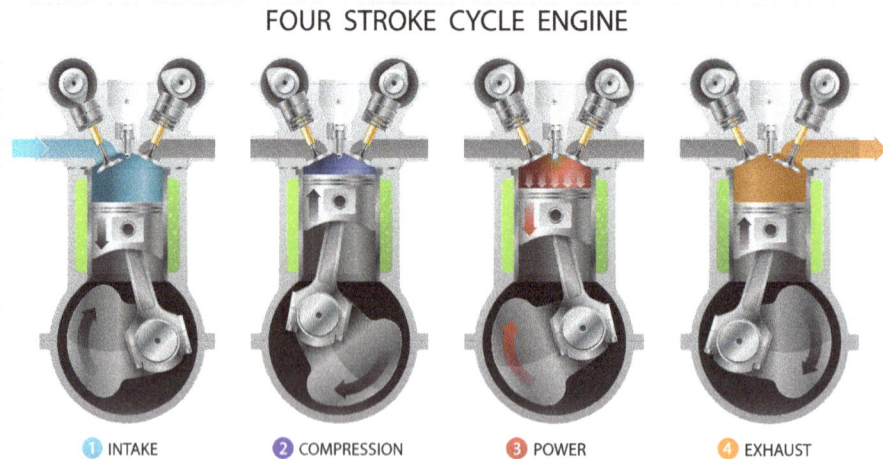

FOUR STROKE CYCLE ENGINE

① INTAKE   ② COMPRESSION   ③ POWER   ④ EXHAUST

© Shutterstock

the chemical energy released from the fuel via combustion. As the piston approaches BDC, the exhaust valve opens to begin the exhaust process.

4. *An exhaust stroke*, when the spent product gases from combustion are pushed out of the cylinder. The driving force for the exhaust to leave the cylinder has two parts—first, the pressure differential between the combustion products in the cylinder and the exhaust manifold, and second, the piston sweeping the cylinder volume as it moves from BDC back to TDC. The exhaust valve closes just after the piston reaches TDC, setting up the cycle to begin again.

These four strokes of the piston result in two complete revolutions of the crankshaft. Together, this makes a single power stroke.

1.1.3 The Two-Stroke Cycle

The four-stroke process, described above, requires each piston to make two revolutions of the crankshaft for each power stroke. The two-stroke cycle was developed to deliver higher power from a given cylinder size, while allowing for a simplified valve design. Like the four-stroke cycle, the two-stroke cycle is applicable for both CI and SI engines. The two-stroke process is only briefly described for completeness. The reader can find further details regarding their use and emission considerations in reference texts on ICEs [1-2, 1-4, 1-5].

Figure 1.4 depicts a simple two-stroke engine, with ports in the liner that are opened and closed by the motion of the piston. These ports control the intake and exhaust flows into and from the cylinder when the piston is near BDC. The processes involved are as follows:

1. *A compression stroke*, which begins upon the closing of the intake-transfer then the exhaust ports. With the cylinder sealed, its contents are compressed as the piston moves up toward TDC, also drawing a fresh charge into the crankcase through the inlet reed spring valve.

2. *A power-expansion stroke*, which is similar in nature to the four-stroke cycle up until the point where the piston nears BDC and first the exhaust, then the inlet-transfer ports open as they are uncovered by the motion of the piston. The majority of the burnt gases leave the cylinder during the blowdown process. When the inlet-transfer ports are uncovered, the fresh charge that was compressed in the crankcase is allowed to flow into the cylinder.

FIGURE 1.4 Illustration of the two-stroke cycle for spark-ignition engines.

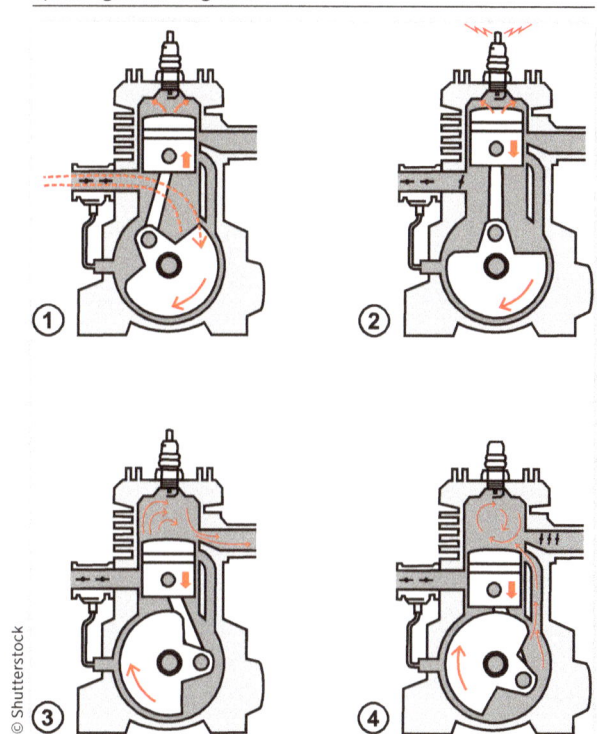

© Shutterstock

The shape of the piston is used to prevent the incoming charge from directly flowing out of the exhaust ports and to promote the scavenging of any residual burned gases remaining in the cylinder by the fresh charge air.

In contrast to the four-stroke cycle, each crankshaft revolution results in a power stroke, giving the two-stroke a higher power density for a given engine size. However, there are several drawbacks to the two-stroke cycle. It is very difficult to accomplish a complete cylinder volume change, and some of the unburnt fuel containing fresh charge can flow directly out of the cylinder during scavenging.

1.1.4 Performance Parameters

A reciprocating ICE produces power from the fuel energy released as heat during combustion in the closed volume of the combustion chamber. The oxygen (O_2) contained in the intake air provides the oxidizer for the fuel, and the released heat increases the temperature within the cylinder, which results in an increase in pressure by the Ideal Gas Law. The increased gas pressure pushes on the moveable piston, producing *power*, a measure of the rate at which work is done. Power is the product of shaft work and *torque*, which is the work done by a spinning shaft, in this case, the crankshaft. The relationship for power and torque is given in Eq. 1.2, where power is the work (force on the piston over the distance it travels) per unit time, torque is force times length, and N is the engine speed in revolutions per unit time.

$$\text{Power} = \text{Torque} \times N_{rev} \tag{1.2}$$

The pressure rise from combustion must be coordinated with the piston motion to extract the maximum work from the in-cylinder energy release, while not exceeding the physical limits of the engine. This process is rapidly repeated over and over, with a fresh charge of air and fuel introduced into the closed volume. The mechanical efficiency of the engine is affected by friction, which depends on several factors, including engine speed, engine load, engine size, accessories, and type of lubrication. Friction is proportional to the square of the engine speed. Friction is roughly constant with increasing load, but at very high loads it increases as the loadings on the bearings, valve trains, and pistons increase. Larger engines have correspondingly larger magnitudes of friction, though on a volume-normalized basis, they have lower specific friction. All of the belt-driven accessories represent losses; the work used to drive them cannot be used for motive power.

The most commonly used measure of energy conversion in ICEs is the *brake-specific fuel consumption (BSFC)*, which is calculated by dividing the mass flow rate of the fuel by the power.

$$\text{BSFC} = \frac{\dot{m}_{fuel}}{\text{Power}} \tag{1.3}$$

Thermal efficiency is another measure of energy conversion in the engine, defined as the ratio of the power generated to the energy content of the fuel consumed. The lower heating value (LHV) of the fuel represents the energy content that can be extracted by completely combusting the fuel to CO_2 and water vapor. The higher heating value of the fuel considers the additional energy that could be extracted by condensing the water vapor to liquid. LHV is used for automotive engine applications, since the water remains in the vapor form in the exhaust.

$$\eta_{brake} = \frac{Power_{brake}}{\left(\dot{m}_{fuel}\right)(LHV)} \quad (1.4)$$

In the equation, η_{brake} is the dimensionless brake thermal efficiency (BTE), Power$_{brake}$ is the brake power delivered by the engine at the flywheel, $\dot{m}_{fuel}$ is the fuel mass flow rate, and LHV is the lower heating value of the fuel. The BSFC and thermal efficiency are related by

$$BSFC = \frac{1}{\eta_b LHV}. \quad (1.5)$$

The volumetric efficiency, η_{vol}, is a measure of the engine's ability to draw air into the cylinder during the intake processad is defined by

$$\eta_{vol} = \frac{\dot{m}_{actual}}{\dot{m}_{ideal}} = \frac{\dot{m}_{actual}}{\rho_{ref}\left(\dfrac{DN\,rpm}{n_r}\right)}, \quad (1.6)$$

where η_{vol} is the volumetric efficiency, $\dot{m}_{actual}$ is the measured flow rate through the engine (mass/time), ρ_{ref} is the density of the charge air being supplied to the engine (mass/volume), and n_r is the number of revolutions per event.

The work done by the piston moving in the cylinder bore of a reciprocating engine is pressure–volume (P-V) work. The relationship of these two parameters over the four-stroke cycle described in Section 1.1.2 is often plotted in a P–V diagram, as shown in **Figure 1.5**. This P–V trace allows for the calculation of work over the cycle as the integral of the pressure with respect to volume, since both are changing simultaneously. Overall, the network produced as a result of the combination of the compression and expansion processes can be found by

FIGURE 1.5 P-V diagram for a four-stroke engine.

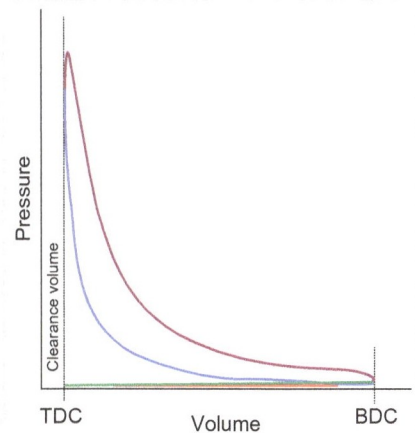

$$W = \int_{V_2}^{V_1} PdV. \quad (1.7)$$

In considering the piston work balance, there are friction losses that must be accounted for as they reduce the amount of work transferred. This includes the friction of the piston rings contacting the cylinder liner, as well as the friction between the connecting rod and flywheel. Normally, the issue of the crankcase pressure can be ignored, because if it is assumed to be approximately constant it drops out of the analysis.

Torque is an effective measure of the engine's ability to do work but scaled by the engine displacement. *Mean effective pressure* (MEP) is a relative performance measure, calculated by dividing the work developed per cycle by the volume displaced per cycle:

$$\text{MEP} = \frac{P \, n_\text{R}}{V_\text{d} N},$$

(1.8)

where P is the power and n_R is the number of revolutions per each power stroke "event." The *indicated mean effective pressure* (IMEP) is the constant pressure which would produce the same indicated work when acting over the same displacement volume. The *brake mean effective pressure* (BMEP) is the average constant pressure that would deliver the same measured (or brake) power output over the same displacement volume if imposed on the pistons. The difference between the IMEP and BMEP for the engine is the *friction mean effective pressure* (FMEP), which is the fraction of the IMEP that is used to overcome the friction of the piston moving in the cylinder. Similarly, the *pumping mean effect pressure* (PMEP) is the portion of the *gross mean effective pressure* (GMEP) that must be used to move the air in and out of the cylinder. The MEP values are related as follows:

$$\text{GMEP} = \text{PMEP} + \text{IMEP} = \text{PMEP} + \left(\text{BMEP} + \text{FMEP} \right)$$

(1.9)

Note that the PMEP may be a negative or positive value. As the engine displacement increases, the BMEP tends to increase because the FMEP and heat transfer losses.

When an engine is tested on a dynamometer, the fuel consumption is measured as a flow rate, which was used in the earlier equations, but a more useful term is the *specific fuel consumption* (SFC), which is a measure of the fuel used per unit power developed. This normalized parameter is independent of engine size and gives an indication of how effectively the engine is using the fuel to provide work at a given operating condition.

$$\text{SFC} = \frac{\dot{m}_\text{f}}{P}$$

(1.10)

Low values for SFC are desirable, as it means that it takes less fuel to deliver the power from the engine. The SFC has units of mass per power, but there is significant fundamental value in a dimensionless factor that relates the engine power to the amount of fuel necessary to deliver it, giving a measure of the engine's efficiency. The *fuel conversion efficiency*, η_f, is such a value, as it is the ratio of power produced in one cycle to the fuel energy released by combustion in that cycle. The fuel energy released by the combustion process is a function of the fuel flow rate, $\dot{m}_\text{fuel}$, supplied to the engine multiplied by the (lower) heating value of the fuel, Q_HV.

$$\eta_f = \frac{P_\text{cycle}}{\dot{m}_\text{fuel} Q_\text{HV}}.$$

(1.11)

From comparison of Eqs. 1.10 and 1.11, it can be seen that SFC is inversely proportional to fuel conversion efficiency, shown in Eq. 1.12, for the types of HC fuels used in engines. The typical LHV for these fuels are on the order of 42–44 MJ/kg (18,000–19,000 Btu/lb$_m$).

$$\eta_f = \frac{1}{sfc\, Q_{HV}} \tag{1.12}$$

It is worth noting that the fuel energy delivered per cycle is not fully released as thermal energy in the combustion process, because the combustion process does not proceed to completion. However, when there is sufficient air present in the cylinder to react with the fuel completely, which is not always the case, approximately 95% (or higher) of the fuel energy supplied can be transferred into the combustion product gases, which is the working fluid for the cycle. If there is not enough air present to oxidize all of the fuel, the incomplete reaction inhibits the release of the supplied fuel energy. Therefore, it is imperative to know what the ratio of air and fuel is in cylinder, as it has a strong effect on both engine performance and emissions.

1.1.5 Air–Fuel and Fuel–Air Ratios

In the test cell, the mass flow rates of both air ($\dot{m}_{air}$) and fuel ($\dot{m}_{fuel}$) are routinely measured as the engine is operated on the dynamometer. The ratio of these flow rates is useful to know, to define the engine-operating condition, and both versions of this flow rate ratio are widely used in practice. They are simply inverse of one another.

$$\text{Air} - \text{fuel ratio}\,(\text{AFR}) = \frac{\dot{m}_{air}}{\dot{m}_{fuel}} \tag{1.13}$$

$$\text{Fuel} - \text{air ratio}\,(\text{FAR}) = \frac{\dot{m}_{fuel}}{\dot{m}_{air}} = \frac{1}{\text{AFR}} \tag{1.14}$$

Since the mass flow rate of air is always greater than the mass flow rate of fuel, it is often easy to tell which ratio is being presented by the value. A conventional SI engine fueled with gasoline, in a normal operating range, has an AFR between 12 and 15 (FAR between 0.067 and 0.083); a CI engine fueled with diesel has an AFR between 20 and 80 (FAR between 0.013 and 0.05) [1-2].

1.1.6 Combustion and Reaction Stoichiometry

The rapid oxidation between the HC fuel and oxygen generates both the desirable thermal energy release into the working fluid and the emissions species that require conversion in the EAS. Combustion transforms the potential energy stored in chemical bonds into thermal energy that can be transferred to the working fluid and extracted as expansion work [1-6]. This definition emphasizes the intrinsic importance of the chemical reaction nature of combustion.

Since the fuel and oxygen react to become combustion products over the course of the reaction, a balanced chemical equation represents an atomic level mass balance on the elements involved. If enough oxygen is present, and the reactants are sufficiently mixed, a HC fuel may undergo *complete combustion* to CO_2 and H_2O. The relationship between the relative quantities of reactants taking part in the reaction, and products formed, is the reaction *stoichiometry*. The reaction stoichiometry for the oxidation of octane (C_8H_{18}) is shown in generalized form as an example.

$$C_8H_{18} + xO_2 \rightarrow yCO_2 + zH_2O \qquad (1.15)$$

As mass is conserved, the number of each type of atom must be the same on each side of the arrow. Octane has eight carbon atoms, so the value of y must be 8. Likewise, octane has 18 hydrogen atoms, and since there are 2 hydrogen atoms per water molecule, the value of z must be 9. Knowing that there are two $y + z$ oxygen atoms on the right side of the arrow, and that molecular oxygen has two oxygen atoms, the value of x is calculated to be 25/2. Putting these values into Eq. 1.15 gives the balanced reaction:

$$C_8H_{18} + \frac{25}{2}O_2 \rightarrow 8CO_2 + 9H_2O \qquad (1.16)$$

The reaction stoichiometry only relates the quantities of elements in the reactant and product species. It does not give any information on the mechanism or rate of reaction, both of which are much more complex.

If the combustion reactions that occurred within engines were as simple and complete as the one represented in Eq. 1.16, there would be no need for emissions control or this book. But, alas, that is not the case. The combustion reaction does not occur in a pure oxygen environment, but in air, which also contains nitrogen (N_2). Since the composition of dry air is approximately 79% nitrogen and 21% oxygen by volume, the ratio of N_2-to-O_2 in air is 3.76:1. Therefore, if the O_2 in Eq. 1.16 is replaced with air, N_2 is introduced into the reaction balance. When the temperature of the reaction is below the threshold of Zel'dovich kinetics, described in Section 2.4, the nitrogen is an observer molecule, rather than a participant in

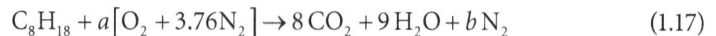

$$C_8H_{18} + a[O_2 + 3.76N_2] \rightarrow 8CO_2 + 9H_2O + bN_2 \qquad (1.17)$$

Using the ratio of oxygen-to-nitrogen, it is possible to keep Eq. 1.17 on the same oxygen basis as the reaction shown in Eq. 1.16. Therefore, a equals $\frac{25}{2}$ and b equals $\frac{25}{2}(3.76) = 47.0$. The final balanced reaction with air is

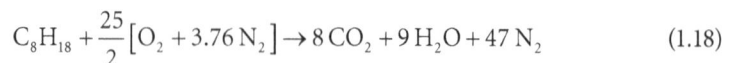

$$C_8H_{18} + \frac{25}{2}[O_2 + 3.76N_2] \rightarrow 8CO_2 + 9H_2O + 47N_2 \qquad (1.18)$$

The reaction shown above defines the stoichiometric proportions of air and fuel necessary to completely combust all of the fuel. This information allows for the stoichiometric AFR or FAR to be calculated. Using the molecular weights of carbon (12.01), hydrogen (1.01), oxygen (16.00), and nitrogen (14.01), the molecular weight for air is 29 and the stoichiometric AFR for octane is calculated as

$$\text{AFR}_{\text{stoich}} = \left(\text{FAR}_{\text{stoich}}\right)^{-1} =$$
$$\left[\left(\frac{25}{2} \times 2 \times 16.00 + \frac{25}{2} \times 3.76 \times 2 \times 14.01\right) \Big/ \left(8 \times 12.01 + 18 \times 1.01\right)\right] = 15.03. \tag{1.19}$$

As long as the AFR is equal to or greater than the $\text{AFR}_{\text{stoich}}$, the mixture can generally be combusted unless it becomes too dilute. However, if the AFR is less than the $\text{AFR}_{\text{stoich}}$, there is not enough oxygen present to completely combust the mixture to CO_2 and H_2O, and the product gases will also contain CO and unburnt HCs. The composition of the product species cannot be determined from the stoichiometric balance alone—it must be measured or additional assumptions must be made. Because the product composition will be significantly different depending on whether the in-cylinder gas mixture is fuel-lean ($\text{AFR}_{\text{actual}} > \text{AFR}_{\text{stoich}}$) or fuel-rich ($\text{AFR}_{\text{actual}} < \text{AFR}_{\text{stoich}}$), the ratio of the actual AFR to the stoichiometric AFR (and the similar ratio of FARs) is an informative parameter for defining the mixture composition. This ratio, shown in Eq. 1.20, is the relative – AFR, λ. The related ratio of FARs is the equivalence ratio, ϕ. The parameters λ and ϕ are the inverse of one another and both are used extensively.

$$\lambda = \frac{\text{AFR}_{\text{actual}}}{\text{AFR}_{\text{stoich}}} = \phi^{-1} \tag{1.20}$$

$$\phi = \frac{\text{FAR}_{\text{actual}}}{\text{FAR}_{\text{stoich}}} \tag{1.21}$$

These parameters are used to define the air-fuel mixture regimes:
Fuel-lean mixtures: $\lambda > 1$, $\phi < 1$
Stoichiometric mixtures: $\lambda = 1 = \phi$
Fuel-rich mixtures: $\lambda < 1$, $\phi > 1$

1.2 Types of Automotive Engines

Automotive engines have traditionally been classified by fuel type (gasoline or diesel) or ignition mechanism (spark or compression), somewhat interchangeably. However, due to advances in engine materials, combustion strategies, injectors, and controls, these criteria are now too narrow. Now, engines require more information to distinguish them by combustion process and mechanical configuration. A gasoline engine, which traditionally had a homogenous air–fuel charge that was ignited by a spark, may now be directly injected (gasoline direct injection [GDI]), with a stratified charge, or gasoline compression ignition (GCI). Likewise, a diesel engine, which traditionally had DI of fuel into compressed air that was auto-ignited during further compression, can now have a homogeneous or premixed charge. Leading to homogeneous charge CI (HCCI) or premixed charge CI (PCCI), respectively.

1.2.1 Spark Ignition Engines

SI engines in vehicles follow the four-stroke cycle described in the previous section. One major differentiating factor is that in *classic* SI engines, the intake charge is a mixture of air and vaporized fuel, premixed before it enters the cylinder. Then, they employ a deliberate mechanism (the spark) to ignite this fuel-air mixture. The most common fuel for SI is gasoline, and fuel injection into the intake manifold is the most common method for premixing the charge in modern engines, having replaced carburation in the 1990s. The nearly homogeneous fuel–air mixture is then compressed, before being ignited by a spark that occurs slightly before TDC. The spark discharge creates a flame kernel in the center of the cylinder, which leads to the development of a turbulent flame that propagates outward, toward the cylinder walls. The flame consumes the air–fuel mixture as it travels, before being extinguished by contact with the cylinder wall. The duration of the burn varies with engine design and operation, and spark timing can be optimized for a given air and fuel charge mass, for maximum torque. The exhaust valve opens late into the expansion stroke (at around 70% of the expansion), and blowdown, the escape of the hot gasses from the cylinder, occurs. The exhaust valve stays open until just after TDC, and as the intake valve opens slightly before TDC, there is a period where both valves are open, referred to as valve overlap. Then, the cycle begins again.

The previous paragraph describes normal operation and combustion within the SI engine, where the flame, once ignited by the spark, proceeds across the combustion chamber until the fuel–air mixture is completely consumed. However, many factors can disrupt the intended operation including combustion chamber design, fuel chemistry, in-cylinder deposits, and even engine-operating conditions. The most important and most concerning type of abnormal combustion that can occur is referred to as *knock*. The name for this unwanted autoignition phenomena comes from the noise that results. Knock occurs when the fuel-air mixture outside of the propagating flame, called the *end gas*, ignites spontaneously due to the temperature and pressure conditions in the cylinder. This autoignition of the end gas can burn up to 25 times faster than normal combustion and result in high-frequency pressure oscillations that produce a sharp noise that sounds like a sharp metallic rap, as the shockwave collapses against the cylinder wall. Repeated knocking can lead to erosion of the metal from the hydraulic cavitation. Knock will not occur if the propagating flame front reaches the end gas before it autoignites, and there are several techniques that can be employed to prevent it, including using higher octane rated fuel, enriching the AFR, reducing the peak cylinder pressure, reducing the engine load, retarding the ignition timing, and by controlling exhaust gas recirculation (EGR). Reduction of the peak cylinder pressure can be achieved by altering the CR or by decreasing the manifold or boost pressure. Other abnormal combustion issues include surface ignition and misfire, which can interrupt the flame propagation in cylinder. Unfortunately, pollutant species formation results from both normal and abnormal combustion, which will be described in detail in Chapter 2, Emissions Formation.

A discussion of SI DI engines, also called GDI engines, is included, as this blending of the two dominant engine types is becoming an increasingly larger part of the fleet [1-7, 1-8, 1-9].

FIGURE 1.6 Spray-guided, wall-guided, and air-guided DI fuel systems for SI engines.

Giannadakis, E, D Papoulias, A Theodorakakos, and M Gavaises. "Simulation of Cavitation in Outward-Opening Piezo-Type Pintle Injector Nozzles." Proceedings of the Institution of Mechanical Engineers, Part D: Journal of Automobile Engineering 222, no. 10, copyright © 2008 by SAGE Publications, Ltd. Reprinted by Permission of SAGE Publications, Ltd.

The desire to hybridize the two ignition concepts to combine the best features of each has existed since the 1920s. Injecting the fuel directly into each of the cylinders allows for better control of the fuel and its vaporization and mixing. This allows the SI DI engine to behave more like a CI DI engine, with increased CRs, reduced throttling losses as compared to PFI and improved dynamic performance [1-2, 1-10, 1-11]. Some of this improvement may be due to the charge densifying cooling effect of the fuel vaporization [1-12].

CI engines have higher brake thermal efficiency than PFI SI engines because they operate fuel-lean, at a near-optimum CR, and achieve the desired work output by the amount of fuel injected per cycle. In comparison, to achieve similar advantages in SI engines, which are otherwise charged with a nearly homogeneous fuel-air mixture, charge stratification needs to be employed. Direct injection of the fuel into the cylinder creates a high degree of turbulence and rotational mixing, called swirl, and improved speed of mixing and combustion realizes this goal [1-2].

SI DI engines are often split into three categories: spray-guided, wall-guided, and air-guided [1-13, 1-14, 1-15, 1-16], as shown in **Figure 1.6**. These categories are named based on how the fuel is introduced into the cylinder. In spray-guided systems, the injector and spark plug are in close proximity and some of the fuel is directed toward the spark plug. Wall-guided systems rely on the fuel impingement on the piston bowl surface and the subsequent transport of fuel toward the spark plug by the charge motion and flow field. The Texaco Controlled Combustion System is an example of a DI system where there is both air-guided and spray-guided mixing [1-17]. The Ford PROCO engine is an example of a system that uses spray-guided charge stratification [1-18], and the MAN FM Combustion system [1-19] is an example of wall-guided mixing. In all cases, DI can reduce the engine's tendency to experience knock, as the evaporation of the fuel spray in cylinder also results in some cooling of the charge air.

1.2.2 Compression Ignition Diesel Engines

CI engines also can follow the four-stroke cycle described earlier. In contrast to SI engines where the fuel and air are premixed before they enter the cylinder, in CI engines, air, with or without recirculated exhaust gas (EGR), is inducted into the cylinder. The fuel is introduced separately by being sprayed directly into the cylinder, which initiates combustion. Modern diesel fuel injection systems deliver the fuel at up to 2,600 bar (37,710 psi) through the injector. The high-pressure liquid fuel atomizes into small droplets as it is sprayed by the fuel injector, and the fuel vaporizes as it mixes with the high-pressure, high-temperature air in cylinder. Since the pressure and temperature in cylinder is above the fuel's ignition point, the fuel–air mixture combusts. Interestingly, the autoignition process which is avoided in SI engines is required in CI engines. The increase in the in-cylinder pressure resulting from the heat release of combustion reduces the evaporation time of the remaining liquid fuel and shortens the ignition delay time. Injection continues until the desired fuel mass has been delivered to the cylinder, and the combustion continues until essentially all of the fuel is consumed. Mixing of the excess air and the spent combustion product gases continues throughout the expansion stroke.

1.2.3 Comparison of SI and CI Engine Operation

SI and CI engines have key differences in their design and operation based on the combustion. CI engines have no knock limit, because the fuel injection events initiate the combustion process. Therefore, higher CRs can be used in CI engines, resulting in improved fuel conversion and indicated thermal efficiencies, defined in Section 1.1.2. Additionally, since the fuel injection timing controls the combustion timing, it is necessary to keep the delay between the start of injection and start of combustion short and reproducible between cycles, so that the maximum cylinder pressure, or the peak firing pressure, stays within the engine's design limits. Another key difference between the engine platforms is in the pumping work requirements that represent an energy cost in the engine operation. In CI engines, since the intake airflow is nearly constant, the engine torque is controlled by metering the amount of fuel injected per cycle and therefore the engine can operate unthrottled, with lower pumping work than in SI engines. Both types can use either homogenous (once thought ideal for SI) or stratified (once thought the only possibility for diesel) air–fuel mixtures. Stratification, enabled by GDI, can be used to enable higher CRs by providing less time and some evaporative cooling to control pre-knock reactions. Diesels can also take advantage of "pre-mixing" through pilot and/or longer ignition delays to mitigate soot. There is still some level of stratification though, owing to the spray-mixing process. True premixed fuel/air mixtures can be auto-ignited in HCCI, if the temperature history can be controlled well enough. Generally, both SI (homogenous or stratified) and diffusion-controlled CI are flame-based combustion systems. The energy to ignite the next zone to burn is derived from what was just burned. The alternative (exemplified by HCCI) is bulk reactions where zones are ignited spontaneously. Related to the previous point regarding efficiency, since diesel CI operated with excess air (lean FARs), the fuel conversion efficiency is higher for a given expansion ratio. Finally, there is one significant and detrimental difference with regard to CI engines, and the lack of sufficient and rapid mixing, which leads to the creation of a fourth criteria pollutant, soot. As the amount of fuel that is

TABLE 1.1 Comparison of spark-ignition and compression-ignition features.

	Spark ignition	Compression ignition
Fuel	Resists autoignition	Supports autoignition
	For example, gasoline	For example, diesel
Compression ratio	9:1 to 12:1	14:1 to 22:1
Mixture formation	Conventional: homogeneous	Conventional: stratified
	DI: homogeneous or stratified	HCCI: homogeneous
Ignition	Deliberate mechanism (positive ignition)	Result of heating due to gas compression
Combustion	Flame front from spark	Diffusion flame

© SAE International

injected in-cylinder increases, there are deficiencies in the mixing of the fuel and air, leading to locally fuel-rich zones, where soot is formed. As the amount of soot formation increases, it quickly outpaces the rate of in-cylinder oxidation, leading to increased particulate concentrations in the exhaust, and the black smoke that is associated with diesel exhaust. The formation of particulate matter (PM), or soot, as well as carbon monoxide (CO), HCs, and NOx is further discussed in Chapter 2, Emissions Formation. The key differences are summarized in Table 1.1.

1.3 Criteria Pollutants

Criteria pollutants cover the set of chemical species that are commonly regulated for air quality, as they are known to cause atmospheric and health hazards such as smog and acid rain. The six criteria pollutants, first defined in 40 CFR part 50 [1-20], are typically the products of combustion or byproducts of industrial processes including sulfur oxides, PM, CO, ground-level ozone (O_3), NOx, and lead (Pb) [1-21]. In addition to these compounds, other criteria pollutants regulated at the tailpipe exit include non-methane hydrocarbons or non-methane organic gases, ammonia (NH_3), and formaldehyde (H_2CO). NOx encompass both nitrogen dioxide (NO_2) and nitrogen monoxide (NO) and may also include nitrous oxide (N_2O). In addition to PM, some jurisdictions are regulating particulate number (PN), as described in Sections 3.2 and 3.3 in Chapter 3, Emissions Control Regulations.

1.3.1 Sulfur Oxides

Sulfur oxides (SOx) are highly reactive gases, and as a category, it usually encompasses sulfur dioxide (SO_2) and sulfur trioxide (SO_3), as they are the most prevalent SOx species. SO_2 is toxic in large amounts, as it can block nerve signal transduction and impede cardiac function [1-22]. SO_3 is a strong and corrosive oxidizing agent, which can damage lung tissue [1-23]. Both species contribute to the formation of acid rain [1-24].

Within the control volume of the engine and EAS, SOx are also known to impair the function of catalysts in EAS. Thus, the sulfur content of both fuels and lubricants has been significantly reduced from the mid-2000s, including the introduction of ultralow sulfur diesel fuels in various markets around the world.

1.3.2 Particulate Matter

PM comprises microscopic solids and liquid droplets suspended as aerosols. Particulates are typically broken into three subgroups: ultrafine particles ($PM_{2.5}$), which have a diameter of 2.5 μm or less; inhalable coarse particulates (PM_{10}), which are those with diameters between 2.5 and 10 μm; and soot, which are the combustible solids in engine exhaust. Numerous studies have shown that particulates are carcinogenic to humans [1-25, 1-26, 1-27, 1-28] and both direct and indirect radiative effects that impact the earth's atmosphere by changing the radiative accounting [1-29, 1-30]. Related to particulate, is Ash, the noncombustible, non-regenerable PM that is derived mainly from metals in the lube oil. Ash is discussed in detail in Section 11.6 of Chapter 11, Particulate Filters.

1.3.3 Carbon Monoxide

CO originates from oxygen-deficient or partial combustion, as may occur in ICEs. CO emissions are a more significant issue in combustion modes that operate with a fuel-rich or stoichiometric AFR. Fuel-lean combustion modes, such as in diesel-fueled CI engines, only generate trace amounts of CO since there is excess air, and therefore oxygen, present during combustion. Although there are atmospheric processes that can convert CO to its fully oxidized state, CO_2, the localized CO concentration in tailpipe exhaust is of great concern due to the toxic nature of the colorless and odorless gas. CO poisoning can start at concentrations as low as 200 ppm, including symptoms such as headache, dizziness, nausea, impaired thinking, and, potentially, death.

1.3.4 Ozone

Ozone (O_3) is a naturally found gaseous constituent of the stratosphere, where it creates a protective layer that absorbs much of the potentially damaging ultraviolet radiation from the sun. In the surface-level troposphere, though, O_3 naturally occurs at only trace levels of approximately 20 ppb. In contrast to ozone in the stratosphere, tropospheric O_3 concentrations are of great concern because O_3 has been shown to have significant human health and environmental effects. Ozone is a powerful oxidizing agent that is known to attack lung tissue and thereby cause irritation leading to asthma and other pulmonary diseases [1-31, 1-32].

1.3.5 Nitrogen Oxides

Nitrogen oxides (NOx) include the species NO_2, NO, and N_2O. Of these three, NO_2 is the most hazardous, because it readily forms nitric acid (HNO_3) with water, has negative health effects on lung tissue [1-33], and contributes to tropospheric ozone formation [1-34]. NO_2 is also the main visible constituent in photochemical smog, with a distinctive orange-brown color.

1.3.6 **Lead**

Lead (Pb) was first defined as a criteria pollutant in the 1970s that was emitted from ICEs because tetraethyl lead ($(C_2H_5)_4Pb$) was used as a knock reduction agent in gasoline-fueled engines. Lead use in fuels was phased out because it poisons the catalysts in three-way catalysts and similar devices. Though lead is a relatively unreactive element, it accumulates in soft tissue, interferes with biological enzymes, and is toxic to the human nervous system, especially in children. Note that there is no safe exposure level for lead in humans, especially children. Lead is therefore extremely problematic at airborne levels of just 100 mg/m^3 (approximately 82 ppm). Inhaled lead can easily dissolve into the blood and cause severe, even fatal damage to the brain and kidneys [1-35, 1-36, 1-37]. Airborne lead can settle into the soil and become an ingestion hazard. Lead has also been shown to have environmental toxicity in plants and soils [1-38].

1.4 **Greenhouse Gases**

Greenhouse gases (GHG) are chemical species that present a long-term threat to human health because they absorb and trap thermal infrared radiation, causing an effect on climate change [1-39]. The overwhelming majority of GHG emissions on Earth are in the form of carbon dioxide (CO_2), which is a primary product of the combustion of HC fuels. Other GHG include water vapor ($H_2O_{(g)}$), methane (CH_4), N_2O, O_3, chlorofluorocarbons (CFC), and hydrofluorocarbons (HCFC). Of these, water vapor is the least concerning since it does not accumulate in the atmosphere over time the way other GHG species do. CFC and HCFC are used as refrigerants for on-vehicle air conditioning systems, and minimizing their use can provide compliance credits, as described in Section 3.4.

Some GHG have been assigned a so-called global warming potential (GWP) that is a multiple of the effect of CO_2. Two GHG species of particular interest with ICEs include methane and N_2O. The main source of methane emissions from ICEs is incomplete combustion in the engine fueled by natural gas, but smaller amounts can also be formed during the combustion of heavier HC fuels. Each methane molecule has a GWP factor equivalent to 25 CO_2 molecules. N_2O is formed as a byproduct during the removal of NOx species in the exhaust system, and it has a GWP factor of 300.

References

1-1.   Myers, P.S., "Automobile Emissions-A Study in Environmental Benefits versus Technological Costs." SAE Technical Paper 700182, 1970, https://doi.org/10.4271/700182.

1-2.   Heywood, J., *Internal Combustion Engine Fundamentals*, 2nd ed. (McGraw-Hill Education, 2018).

1-3.   Kirchgessner, D.A., Lott, R.A., Cowgill, R.M., Harrison, M.R., and Shires, T.M., "Estimate of Methane Emissions from the US Natural Gas Industry." *Chemosphere* 35, no. 6 (1997): 1365–1390.

1-4.   Stone, R., *Introduction to Internal Combustion Engines* (London: Palgrave, 1999), https://doi.org/10.1007/978-1-349-15079-3.

1-5.   Ferguson, C.R. and Kirkpatrick, A.T. 2015. *Internal Combustion Engines: Applied Thermosciences* (New York: John Wiley & Sons).

1-6.   Turns, S.R., *An Introduction to Combustion*, Vol. **499** (New York: McGraw-Hill, 1996).

1-7.   Simmons, R.A., Shaver, G.M., Tyner, W.E., Garimella, S.V., "A Benefit-Cost Assessment of New Vehicle Technologies and Fuel Economy in the U.S. Market." *Applied Energy* **157**, no. 11 (2015): 940–952.

1-8.   Storey, J.M.E., Barone, T.L., Thomas, J.F., Huff, S.P., "Exhaust Particle Characterization for Lean and Stoichiometric DI Vehicles Operating on Ethanol-Gasoline Blends." SAE Technical Paper 2012-01-0437, 2012, https://doi.org/10.4271/2012-01-0437

1-9.   Ke, W., Zhang, S., He, X., Wu, Y., Hao, J., "Well-to-Wheels Energy Consumption and Emissions of Electric Vehicles: Mid-Term Implications from Real-World Features and Air Pollution Control Progress." *Applied Energy* (**188**): 367–377, 2017.

1-10.   Zhao, F., David L. Harrington, and Ming-Chia D. Lai "Automotive Gasoline Direct-Injection Engines." SAE Technical Paper 2002-05-15, 2002, https://doi.org/10.4271/2002-05-15.

1-11.   Iwamoto, Y., Noma, K., Nakayama, O., Yamauchi, T., and Ando, H., "Development of Gasoline Direct Injection Engine." SAE Technical Paper 970541, 1997, https://doi.org/10.4271/970541.

1-12.   Marriott, C.D., Wiles, M.A., Gwidt, J.M., and Parrish, S.E., "Development of a Naturally Aspirated Spark Ignition Direct-Injection Flex-Fuel Engine." *SAE Int. J. Engines* **1**, no. 1 (2009): 267–295, https://doi.org/10.4271/2008-01-0319.

1-13.   Williams, B., Ewart, P., Stone, R., Ma, H., Walmsley, H., Cracknell, R., Stevens, R., Richardson, D., Qiao, J., and Wallace, S., "Multi-Component Quantitative PLIF: Robust Engineering Measurements of Cyclic Variation in a Firing Spray-Guided Gasoline Direct Injection Engine." SAE Technical Paper 2008-01-1073, 2008, https://doi.org/10.4271/2008-01-1073.

1-14.   Drake, M.C., Fansler, T.D., and Lippert, A.M., "Stratified-Charge Combustion: Modeling and Imaging of a Spray-Guided Direct-Injection Spark-Ignition Engine." *Proceedings of the Combustion Institute* **30**, no. 2 (2005): 2683–2691.

1-15.   Fischer, J., Kettner, M., Nauwerck, A., Pfeil, J., and Spicher, U., "Influence of an Adjustable Tumble-System on In-Cylinder Air Motion and Stratification in a Gasoline Direct Injection Engine." SAE Technical Paper 2002-01-1645, 2002, https://doi.org/10.4271/2002-01-1645.

1-16.   Alkidas, A.C., "Combustion Advancements in Gasoline Engines." *Energy Conversion Management* **48**, no. 11 (2007): 2751–2761.

1-17.   Alperstein, M., Schafer, G.H. and F.J. Villforth., "Texaco's Stratified Charge Engine-Multifuel Efficient, Clean, and Practical." SAE Technical Paper 740563, 1974, https://doi.org/10.4271/740563.

1-18.   Scussel, A.J., A.O. Simko, and W.R. Wade., "The Ford PROCO Engine Update." SAE Technical Paper 780699, 1978, https://doi.org/10.4271/780699.

1-19.   Meurer, J.S., and A.C. Urlaub., "Development and Operational Results of the MAN FM Combustion System." SAE Technical Paper 690255, 1969, https://doi.org/10.4271/690255.

1-20.   Environmental Protection Agency, "Part 50-National Primary and Secondary Ambient Air Quality Standards," Code of Federal Regulations, Title 40, Vol. **2** (2018): 5–169, https://www.govinfo.gov/content/pkg/CFR-2018-title40-vol2/pdf/CFR-2018-title40-vol2-chapI-subchapC.pdf.

1-21.   Environmental Protection Agency, "NAAQS Table," https://www.epa.gov/criteria-air-pollutants/naaqs-table. Updated 20 Dec. 2016, accessed 14 Nov. 2018.

1-22.   Rall, D.P., "Review of the Health Effects of Sulfur Oxides." *Environmental Health Perspectives* **8** (1974): 97–121.

1-23.   Ferris Jr, B., Speizer, F., Spengler, J., Dockery, D., Bishop, Y., Wolfson, M., and Humble, C., "Effects of Sulfur Oxides and Respirable Particles on Human Health: Methodology and Demography of Populations in Study." *American Review of Respiratory Disease* **120**, no. 4 (1979): 767–779.

1-24.   Wark, K. and Warner, C.F. , *Air Pollution: Its Origin and Control* (New York: Harper & Row, 1981).

1-25.   Anderson, J.O., Thundiyil, J.G., and Stolbach, A., "Clearing the Air: A Review of the Effects of Particulate Matter Air Pollution on Human Health." *Journal of Medical Toxicology* **8**, no. 2 (2012): 166–175.

1-26.   World Health Organization, "Health Effects of Black Carbon," 2012.

1-27.   Benbrahim-Tallaa, L., Baan, R.A., Grosse, Y., Lauby-Secretan, B., El Ghissassi, F., Bouvard, V., Guha, N., Loomis, D. and Straif, K., "Carcinogenicity of Diesel-Engine and Gasoline-Engine Exhausts and Some Nitroarenes." *The Lancet Oncology* **13**, no. 7 (2012): 663–664.

1-28.   Harrison, R.M. and Yin, J., "Particulate Matter in the Atmosphere: Which Particle Properties are Important for Its Effects on Health?" *Science of the Total Environment* 249(1-3): 85–101, 2000.

1-29.   Grantz, D., Garner, J., and Johnson, D., "Ecological Effects of Particulate Matter." *Environment International* 29(2-3): 213–239, 2003.

1-30.   Jacobson, M.Z., "Control of Fossil-Fuel Particulate Black Carbon and Organic Matter, Possibly the Most Effective Method of Slowing Global Warming," *Journal of Geophysical Research: Atmospheres* **107**(D19): ACH 16-11–ACH 16-22, 2002.

1-31.   Lippmann, M., "Health Effects of Ozone a Critical Review." *Japca* **39**(5): 672–695, 1989.

1-32.   Krupnick, A.J., Harrington, W., and Ostro, B., "Ambient Ozone and Acute Health Effects: Evidence from Daily Data." *Journal of Environmental Economics Management* **18**(1): 1–18, 1990.

1-33.   Kampa, M. and Castanas, E., "Human Health Effects of Air Pollution." *Environmental Pollution* **151**(2): 362–367, 2008.

1-34.   Crutzen, P.J., "The Influence of Nitrogen Oxides on the ATM Ospheric Ozone Content." *Quarterly Journal of the Royal Meteorological Society* **96**(408): 320–325, 1970.

1-35.   Goyer, R.A., "Lead Toxicity: Current Concerns." *Environmental Health Perspectives* **100**: 177–187, 1993.

1-36.   Flora, G., Gupta, D., and Tiwari, A., "Toxicity of Lead: A Review with Recent Updates." *Interdisciplinary Toxicology* **5**(2): 47–58, 2012.

1-37.   National Institute for Occupational Health and Safety (NIOSH), "Lead," https://www.cdc.gov/niosh/npg/npgd0368.html. Accessed November 29, 2018.

1-38.   Sharma, P. and Dubey, R.S., "Lead Toxicity in Plants." *Journal of Plant Physiology* **17**(1): 35–52, 2005.

1-39.   Environmental Protection Agency, "Overview of Greenhouse Gases," https://www.epa.gov/ghgemissions/overview-greenhouse-gases. Updated 9 Oct. 2018, Accessed 30 Oct. 2018.

CHAPTER 2

Emissions Formation

Air-borne pollutants are certainly not restricted to our modern age; for down the centuries people have always breathed smoke-laden air derived from the combustion of wood, peat and coal.

—Peter Eastwood, Critical Topics in Exhaust Gas Aftertreatment [2-1]

The emissions from internal combustion engines (ICEs) consist largely of carbon monoxide (CO), unburnt hydrocarbons (HC), nitrogen oxides (NOx), and particulate matter (PM). Both spark-ignition (SI) and compression-ignition (CI) engines are major sources of these pollutants, but the relative amounts of these species in the engine-out exhaust heavily depend on the engine design and operating conditions. Regulators, environmental advocates, public health professionals, engine manufacturers, and suppliers all spend vast amounts of time and resources on identification, measurement, and abatement of these species. The purpose of this chapter is to discuss how exhaust emissions are formed before addressing the catalytic converters that remove those emissions in the exhaust aftertreatment system (EAS).

The exhaust concentrations of the pollutant species are not representative of chemical equilibrium values; therefore, the kinetics and mechanisms of formation are important to understand. Some species, such as CO, HC, and PM, are strongly affected by combustion reactions, when these species are both formed—or reformed as in the case of hydrocarbons—and consumed. Therefore, a need exists to understand the detailed combustion conditions, including temperature, pressure, and composition, and the chemical kinetics. The formation of other species, such as NOx and sulfur oxides (SOx), do not directly depend on the chemistry of the fuel being combusted. Instead, their formation depends on local conditions within the cylinder, especially temperature, oxygen (O_2) concentration, and nitrogen (N_2) concentrations, all of which are affected by the combustion.

2.1 Emissions Formation in Internal Combustion Engines

The formation of emissions species in ICEs is a problem for both SI and CI engines. Both engine types can generate significant amounts of NOx and HC emissions, particularly in the crevice regions, or locations of wall-wetting, as shown in **Figure 2.1**. CI engines typically have lower HC emissions than SI engines, whereas SI engines have substantial CO emissions. By contrast, CI and other direct injection (DI) stratified combustion strategies form PM, or soot, emissions. Using oxygenated fuels in ICEs can significantly increase aldehyde emissions, contributing to photochemical smog. In addition, both gasoline and diesel fuels typically contain some amount of sulfur (S). The limit for sulfur depends on the fuel type, application, region of the world, and regulatory standards. When sulfur is oxidized, it produces sulfur dioxide (SO_2) and to a lesser extent sulfur trioxide (SO_3), which can combine with water (H_2O) to form sulfuric acid (H_2SO_4). SOx are particularly problematic for catalytic aftertreatment devices, as it is a common poison, decreasing the effectiveness of the catalyst as discussed in Chapter 7.

Another source of emissions from a vehicle is evaporative emissions, where fuel or lubricant compounds are released to the environment from the fuel tank or the engine, including the piston and crankcase. With the introduction of regulations to limit evaporative emissions, these have largely been eliminated. Modern vehicles recycle the blowby gases and the crankcase vapors to the intake manifold, where they can re-enter the cylinder and undergo combustion. Burning the lubricant does release some combustion products of the additive package into the exhaust, which can poison the catalysts in the EAS. The evaporative emissions from the fuel tank are absorbed with an added carbon canister, which is regenerated by purging it into the intake air during the course of normal engine operation. The HC species released during purge then become part of the engine intake charge and undergo combustion.

2.2 CI Combustion Emissions

Designed in 1892 by Rudolf Diesel [2-2], the CI engine was originally developed to burn coal dust, which required a high compression ratio to facilitate ignition. Even today, a CI combustion system will burn a wide range of fuels if they can be supplied to the cylinder.

In a modern CI engine, rapid compression of air raises the gas temperature. When the high-pressure liquid fuel is injected into the cylinder, it atomizes and vaporizes as it mixes with the air. When the pressure and temperature in cylinder are above the fuel's ignition point, spontaneous ignition, called autoignition, of the fuel—air mixture occurs. The rise in the in-cylinder pressure to the peak firing pressure (PFP) resulting from the heat release of combustion reduces the evaporation time of the remaining liquid fuel and shortens the ignition delay time. Injection continues until the desired fuel mass has been delivered to the cylinder, and the combustion continues until essentially all of the fuel is consumed. Mixing of the excess air and the spent combustion product gases continues throughout the expansion stroke.

FIGURE 2.1 Locations where emissions are formed in SI and CI engines include the crevice volumes and locations of cylinder wall-wetting.

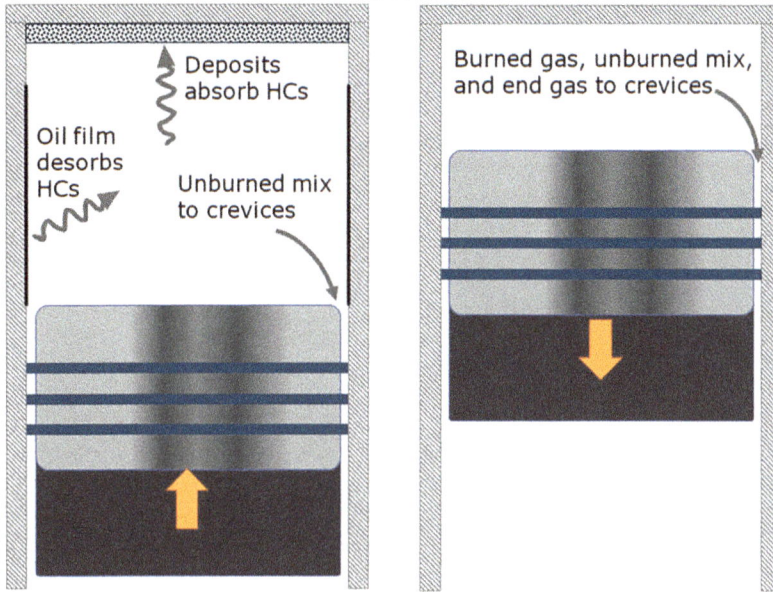

(a) Compression   (b) Combustion

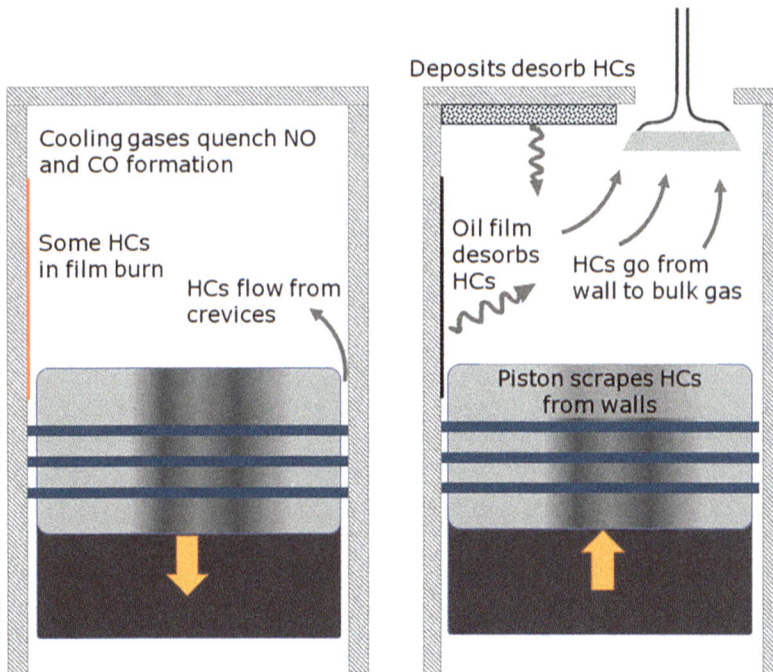

(c) Expansion   (b) Exhaust

© SAE International

CI engines are constructed of heavier materials to accommodate higher compression ratios and PFP. These design features, in turn, result in improved fuel conversion and indicated thermal efficiency. Additionally, the fuel injection timing is used to control the combustion timing. Thus, it is necessary to keep the delay between the start of injection and start of combustion (SOC) short and reproducible between cycles, in order to keep the PFP within the engine's limits. In order to achieve this, the cetane number of the diesel fuel used, which is a representation of how easily the fuel autoignites, must be above a specified minimum.

CI engines are generally operated with excess air, so the engine torque is primarily controlled by metering the amount of fuel injected per cycle. The CI engine is usually operated unthrottled, with lower pumping work than in SI engines, but the turbocharger creates a potential flow restriction. CI engines commonly have a boosted charge—usually from a turbocharger—that can force more air into the cylinder and thereby allow more fuel to be delivered while maintaining excess air.

Related to the previous point regarding efficiency, since CI engines operate with excess air (fuel-lean), the fuel conversion efficiency is higher for a given expansion ratio. NOx emissions are comparable to those from SI engines, HC emissions levels are typically lower, though still requiring control, and CO levels are minor.

Finally, one notable difference with regard to CI engines is the lack of sufficient and rapid mixing of the fuel and air, which leads to the creation of a fourth criteria pollutant, soot. As the amount of fuel that is injected into the cylinder increases, the deficiencies in the mixing lead to local fuel-rich zones, where particulates are formed.

2.3 **SI Combustion Emissions**

In SI engines, the intake charge is ideally a homogeneous mixture of air and vaporized fuel, premixed before it enters the cylinder. The most common fuel for SI engines is gasoline, and fuel injection into the intake ports is a common method for premixing the charge in modern engines, having replaced carburation in the 1980s. The fuel–air mixture is compressed and then ignited by a spark that occurs slightly before top dead center (TDC). The spark discharge creates a flame kernel that leads to the development of a turbulent flame that propagates outward from the spark site. The flame consumes the air–fuel mixture before being extinguished by contact with the relatively cold cylinder wall. The duration of the burn varies with engine design and operation, and the spark timing can be optimized for a given air and fuel charge mass to provide maximum torque.

Many factors can disrupt the intended operation, including combustion chamber design, fuel chemistry, in-cylinder deposits, and engine operating conditions. The most important and most concerning type of abnormal combustion that can occur is referred to as *knock*, which is described in detail in Chapter 1, Introduction. Repeated knock events can lead to erosion of the metal from the creation of local surface temperatures that melt the piston material. Knock will not occur if the propagating flame front reaches the end gas before autoignition. Other abnormal combustion issues include surface ignition or preignition, which can cause unintended ignition of the fuel air mixture, and misfire, which can interrupt the flame propagation in cylinder. Unfortunately, pollutant species formation results from both normal and abnormal combustion—though to a greater degree in the latter case.

2.4 **Nitrogen Oxides**

NOx includes both nitrogen monoxide (NO, a.k.a. nitric oxide) and nitrogen dioxide (NO$_2$) species. Of these, NO is the predominant species produced during combustion. The origin of these species is from molecular nitrogen (N$_2$) from the air, which is nearly 80% N$_2$. Although N$_2$ is usually inert, it becomes reactive at the high temperatures achieved by combustion. The oxidation process of N$_2$ to NO, first described in 1946 by Zel'dovich [2-3], has been extensively studied in ICEs since the 1970s [2-4, 2-5, 2-6]. The Zel'dovich mechanism was notably extended by Lavoie [2-5], and is often referred to as the "Extended Zel'dovich Mechanism," and comprises just three reactions for near-stoichiometric fuel–air mixtures as follows:

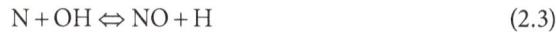

$$N_2 + O \Leftrightarrow NO + N \qquad (2.1)$$

$$N + O_2 \Leftrightarrow NO + O \qquad (2.2)$$

$$N + OH \Leftrightarrow NO + H \qquad (2.3)$$

The double-ended arrow in each reaction indicates that the reaction can go in either the forward or reverse direction. The forward and reverse reaction rate constants, $k_{f,i}$ and $k_{r,i}$, compiled from the literature [2-7], are given in **Table 2.1**. These rate constants can be used to find the rate of change of NO concentration with respect to time using

$$\frac{d[NO]}{dt} = k_{f,1}[N_2][O] - k_{r,1}[NO][N] + k_{f,2}[N][O_2] - k_{r,2}[NO][O]$$
$$+ k_{f,3}[N][OH] - k_{r,3}[NO][H], \qquad (2.4)$$

where $[j]$ is the molar concentration (in g/mL) of species j. The equilibrium rate constants, $K_{eq,i}$, can be calculated from the ratio of the forward and reverse rates for each of the three reactions using

$$K_{eq,i} = k_{f,i} / k_{r,i}. \qquad (2.5)$$

The forward rate constant from the first reaction in Eq. 2.1 and the reverse rate constants from the second and third reactions in Eqs. 2.2 and 2.3 have large activation energies. The large activation energies indicate a very strong temperature dependence for these reactions, which is known to be the case for in-cylinder NOx formation during combustion. Due to this temperature dependence, NO can form on the fuel-lean side of the combustion in a CI engine cylinder or in the flame front in an SI engine cylinder, as well as the post-flame gases. The flame zone itself, due to the high-pressure condition in

TABLE 2.1 Kinetic parameters for the extended Zel'dovich mechanism reactions.

Reaction	Forward rate constant (cm³/mol·s)	Temperature range (K)	Reverse rate constant (cm³/mol·s)	Temperature range (K)	K_{eq} at 2,000 K
(1) N$_2$ + O ⇔ NO + N	7.6×10^{13} exp(−38,000/T)	2,000–5,000	1.6×10^{13}	300–5,000	2.66×10^{-8}
(2) N + O$_2$ ⇔ NO + O	6.4×10^9 T exp(−3,150/T)	300–3,000	1.5×10^9 T exp(−19,500/T)	1,000–3,000	1.52×10^4
(3) N + OH ⇔ NO + H	4.1×10^{13}	300–2,500	2×10^{14} exp(−23,650/T)	2,200–4,500	2.80×10^4

cylinder, is very thin, and the gas residence time is very short. However, because the pressure in cylinder continues to rise throughout the combustion process, the burned gases that were produced early in the process get compressed to a higher temperature than they had reached during combustion. Therefore, it is the post-flame formation process for NO that typically dominates its production. This is the major reason why it is often assumed that the chemical reactions of combustion and the NO formation processes are decoupled. Specifically, the first reaction requires temperatures in excess of 2,000 K to become active, and this reaction enables the other two by providing the nitrogen. If the temperature remains below 2,000 K, little NO is formed; however, the OH-producing (for chain branching) reactions are inhibited below 2,200 K, creating a trade-off.

NO is the more common product, though the complete oxidation product is NO_2. However, at the reaction conditions of the combustion or post-combustion processes, the chemical kinetics for the NO_2 reaction are negligible. This keeps the concentration of NO_2 formed small relative to NO. In CI engines, the NO_2 fraction can be up to 30% of the engine-out NOx emissions, but in SI engines, the percentage is much smaller. NO_2 is formed by NO being further oxidized following the forward reaction

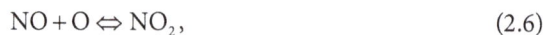

$$NO + O \Leftrightarrow NO_2, \tag{2.6}$$

although the reverse reaction, the thermal decomposition of NO_2 to NO, is just as likely to happen in the flame zone conditions unless the reaction is quenched by cooling. This is why the highest engine-out NO_2:NO ratio occurs at light load in diesel engines, as there will be cooler regions in cylinder where quenching of the thermal decomposition reaction can occur.

2.4.1 CI Engines

While the chemistry of NOx formation is independent of the engine platform, one key difference when assessing the NOx formation mechanisms in CI engines is the nonuniformity of the burned gas temperature and composition throughout the cylinder. Since the DI of the fuel occurs just before the SOC, the fuel–air mixture in cylinder is stratified and nonhomogeneous. Therefore, the combustion products are similarly nonhomogeneous, with respect to temperature, composition, and concentration.

Just after the ignition delay, in the so-called premixed combustion process, the rich fuel–air mixture of the fuel spray autoignites. Following this, during mixing-controlled combustion, the burning mixture of the turbulent diffusion flame entrains air and leans out, causing the AFR to become closer to stoichiometric. Throughout the combustion process, mixing between the atomized fuel vapor, air, and burned gases continues, changing the composition and concentrations of the gases. This mixing also changes the temperature, as do the compression and expansion processes, all of which affect NOx formation. The critical AFR for the formation of NO is close to stoichiometric. For CI engines, the majority of the NO is formed during the mixing-controlled, diffusion flame combustion process. The diffusion flame is what surrounds each fuel spray jet as the fuel is introduced into the cylinder. The fuel-rich, partially burned fuel–air mixture diffuses into the flame from the interior of the spray jet and the air diffuses into the flame from the surroundings of the cylinder. The combustion reaction occurs where the mixture is close to stoichiometric, producing a very high rate of NO formation. Then, the burned gases and NO formed then diffuse away from the flame.

The NO formation is maximized when the gas temperatures are highest, which is close to the time of peak cylinder pressure. The gas temperature rises as the air is compressed, then rises further as the combustion reaction proceeds. Once the peak cylinder pressure has been achieved, the temperature of the gases will decrease during expansion, first slowing the formation of NO and then quenching the reaction altogether. The NO formation reaction is also quenched by the high-temperature combustion gases mixing with the excess air. This effect is responsible for the NO formation reaction quenching much more rapidly in CI engines than in SI engines, and allowing less decomposition of NO back to N_2.

Akinyemi and Cheng [2-8] described the NO formation process in CI engines, as shown in **Figure 2.2**. They described how the profiles of gas temperature, NO concentration, oxidizer concentration, and NO formation rate change with distance across the flame sheet, as the gases move outward from the center of the diffusion flame surrounding each diesel fuel spray injected into the cylinder. The peak rate of NO formation occurs just lean of stoichiometry. This description of the NO formation process is supported by experimental measurements of gaseous species concentrations [2-9, 2-10], as well as from the total amount of NO within the cylinder during the NO formation [2-9] in normally operating DI diesel engines.

NO concentrations from CI engines have been shown to decrease with decreasing overall equivalence ratio [2-11, 2-12], though less rapidly than in SI engines. In CI combustion, the zones where NO formation is highest are nearly stoichiometric. Even though the amount of fuel injected decreases as the overall equivalence ratio decreases, the fuel still burns very close to the stoichiometric ratio. Therefore, the NO emissions are proportional to the mass of fuel injected in a general sense.

Diluting the intake air with *recycled exhaust gas*, a process referred to as exhaust gas recirculation (EGR), is effective at reducing the rate of NO formation in cylinder and the resulting NOx emissions at the tailpipe [2-13]. A schematic of EGR flow is shown in **Figure 2.3**. EGR dilutes the O_2 concentration in the cylinder, which reduces the resulting combustion temperature. The combustion temperature is further lowered because the inert combustion products have a higher heat capacity than N_2 does. For example, the heat capacity of CO_2 is nearly double than that of N_2. Since NO formation is a strong function of temperature, less NO is generally formed when EGR is used. An exception to this effect occurs at low load, since the recirculated gas is predominantly air, and there is consequently little effect on NO formation.

FIGURE 2.2 Change in NO concentration and production with respect to distance across the flame sheet (Figure 2 in [2-8]).

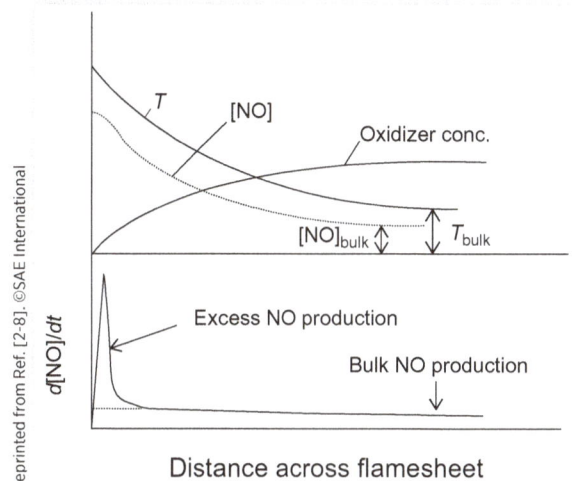

Reprinted from Ref. [2-8]. ©SAE International

Schematic of mass flows in engine with EGR (Figure 1 in [2-14]).

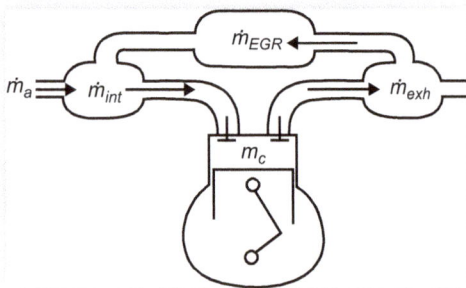

Reprinted from Ref. [2-14]. © SAE International

2.4.2 SI Engines

In port-fuel-injected (PFI) SI engines, the air and fuel are mixed in the intake port, before the charge enters the cylinder. This intake charge is further mixed with any residual gases remaining inside the cylinder. Therefore, the AFR and diluent concentration from the EGR and residual gases is ideally nearly homogeneous throughout the cylinder. With a uniform composition, the NO formation process is a direct application of the Zel'dovich kinetics described earlier with the gas temperature distribution and in-cylinder pressure during the combustion and expansion processes.

The fraction of the mixture that has been combusted earlier in the process is compressed to yet higher temperatures as the cylinder pressures continue to rise during the combustion event. The fraction of the mixture that burns later in the combustion process was compressed in its mostly unburnt state and ends up at a lower temperature in comparison. Therefore, the earlier burning portions of the fuel contribute much more to the formation of NO emissions than do the later burning fractions, with the highest NO concentrations occurring near the spark plug. The temperature-dependent NO chemistry quenches in the early part of the expansion stroke. There is substantial evidence for this from experimental studies in the literature [2-5, 2-15, 2-16, 2-17].

As we saw earlier, for CI combustion the AFR has the single largest effect on NO formation [2-18]. This is not solely dependent on the intake fuel–air charge, but also on the burnt gas fraction and the spark timing. The burnt gas fraction is a combination of the EGR and the residuals from the previous cycle. Fuel properties also have an effect on the burnt gas conditions, though the variations from gasoline are small. All of these factors change the temperature and oxygen concentrations during the combustion process and the initial part of the expansion stroke. Spark timing and burn rate affect NO concentrations significantly. If the spark timing is advanced in a manner such that combustion occurs earlier in the cycle, the result is increased cylinder pressure, as more fuel is burned before the piston reaches TDC, when the volume is the smallest. If the spark is retarded, the peak cylinder pressure achieved is lower, because more of the fuel is burned after TDC, when the cylinder volume is increasing due to the downward piston movement. As discussed previously for CI engines, higher peak cylinder pressures result in higher temperatures and therefore higher NO formation rates.

2.4.3 Exhaust Gas Recirculation

EGR is an effective technique to reduce the formation of NOx in-cylinder for both SI and CI engines [2-13, 2-19, 2-20]. The volume of recirculated exhaust gases displaces a portion of the fresh intake air and thereby changes both the concentration of O_2 present and the heat capacity of the intake charge.

There are two key mechanisms by which EGR reduces NOx emissions. First, by displacing some of the O_2, it changes the concentration of O_2 available to react with the fuel and the resulting rate of the combustion reaction. The second mechanism is using the higher heat capacity of the exhaust gas components, particularly CO_2 and water, to lower the in-cylinder gas temperature, thereby stunting the Zel'dovich kinetics and the resulting NOx formation [2-21, 2-22]. EGR has been used extensively in gasoline and light-duty diesel engines (LDDEs) for many years but was only applied to heavy-duty diesel engines (HDDEs) when the NOx limits were tightened to 2.0 g/bhp·h (2.7 g/kW·h) [2-23, 2-24].

Exhaust gas is primarily composed of water ($\hat{C}_p$ = 4.18 kJ/kg·K) and carbon dioxide (CO_2, $\hat{C}_p$ = 0.85 kJ/kg·K), where as a fresh intake charge is largely composed of N_2 ($\hat{C}_p$ = 1.04 kJ/kg·K). EGR is measured in terms of volume, and **Figure 2.4** shows that increasing the volume of exhaust gas diluent reduces the formation of NO during the combustion process. It is also worthwhile to note that while diluting the intake charge with any species is shown to decrease the NO formation, the largest effect is seen when the diluting species has the highest heat capacity.

There are two common routes for EGR in an engine: short and long. Short-route EGR is also known as high-pressure EGR, because it is drawn from where the exhaust manifold enters into the turbine on a boosted engine. The EGR is then introduced to the intake charge in the intake manifold. The advantages of high-pressure EGR are that it is compact and that it responds quickly to changes in demand. The disadvantages are that an EGR cooler may be needed to manage the intake charge temperature and that the turbine may need to be adjusted so that there is a pressure gradient between the exhaust manifold and intake manifold to drive the flow.

Long-route EGR is also known as low-pressure EGR, because it is drawn from downstream of the turbine and is introduced to the intake air-flow upstream of the compressor. The advantage of low-pressure EGR is that it can be drawn from after the EAS components to clean the recirculated exhaust gas. The main disadvantages are packaging the EGR route efficiently and managing transients in EGR demand, since the transport delays are appreciable.

There is a limit to the amount of EGR that can be used. If too much of the intake charge is replaced with inert species, the combustion can be quenched or inconsistent.

FIGURE 2.4   Effect of diluent volume and composition on the reduction of NO formed in-cylinder (data from [2-25]).

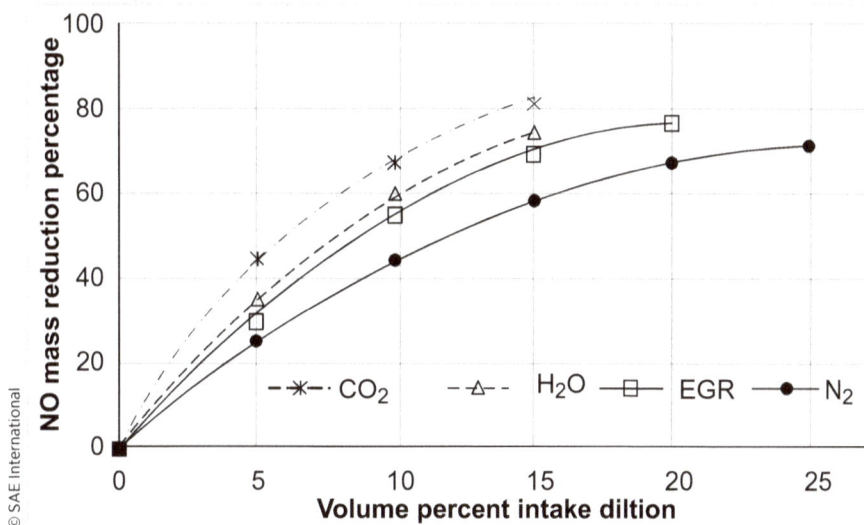

There is also a diminishing return on NO decrease in the engine-out exhaust as the EGR level is increased past 25%. In addition, there can be an unintended consequence of increasing PM emissions when using EGR. This well-known phenomenon is often referred to as the "soot-NOx tradeoff."

2.5 **Carbon Monoxide**

Similar to NO formation rates, CO emissions are predominantly controlled by the fuel–air mixture in cylinder [2-26], independent of fuel type. When the AFR is enriched, there is more fuel than the stoichiometric amount that can react with the air in the cylinder. Therefore, not all of the fuel can be completely combusted, and CO emissions increase. When the AFR is lean and there is excess air, the likelihood of complete combustion in increased and the CO emissions decrease.

CI engines always operate on the fuel-lean side of stoichiometric, with excess air present to enable complete combustion. Because of this, CO emissions from diesels are generally considered negligible. However, SI engines have significant CO emissions that must be controlled. This is because their operation usually occurs with nearly stoichiometric AFR at part-load conditions and can be fuel-rich at certain high-load conditions, which is an ideal condition for formation of CO. For the premixed SI engines, the concentration of CO increases due to rapid formation in the flame zone, reaching a higher level than for adiabatic combustion of the same fuel–air conditions. The CO concentrations formed in cylinder decrease as the gases exit the cylinder. Even then, the concentrations are higher than the equilibrium concentrations would be for the exhaust temperature, which indicates that the CO formation and destruction processes are kinetically controlled [2-27]. CO formation is one of the key reaction steps in the HC combustion reaction described by

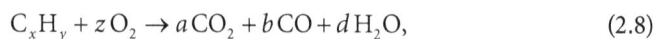

$$C_xH_y + z\,O_2 \rightarrow a\,CO_2 + b\,CO + d\,H_2O, \tag{2.8}$$

where $x = a + b$, $y = 2d$, and $z = a + b/2 + d/2$. The CO formed by Eq. 2.8 can be further oxidized to CO_2 by one of the following reactions:

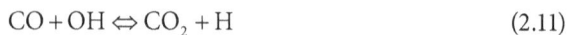

$$2\,CO + O_2 \rightarrow 2\,CO_2 \tag{2.9}$$

$$CO + H_2O \Leftrightarrow CO_2 + H_2 \tag{2.10}$$

$$CO + OH \Leftrightarrow CO_2 + H \tag{2.11}$$

Eq. 2.9 represents the further oxidation of CO to its complete combustion product CO_2 by any O_2 remaining in the cylinder. Eq. 2.10 is the so-called water–gas shift reaction, which continues as the gases cool in the exhaust process. Eq. 2.11 represents the oxidation of CO by hydroxyl radicals, which is the dominant CO oxidation mechanism in HC flames. Detailed studies on engine-generated CO formation [2-28] showed that the CO concentrations in the exhaust for fuel-rich mixtures are nearly equal to the equilibrium concentration during the expansion stroke. Conversely, as the gases cool and expand further in the expansion and exhaust strokes, the CO oxidation process may not remain in equilibrium depending on how the local temperature changes [2-28]. Concentrations of CO in the exhaust from fuel-rich combustion mixtures remain close to equilibrium during the expansion process. For combustion mixtures near the stoichiometric AFR, the CO concentrations in the exhaust are much higher than the equilibrium concentration should be. Fuel-lean mixtures result in exhaust CO concentrations that are markedly higher than expected from predictions of kinetically controlled chemistry. It is postulated that in the fuel-lean mixture, only partial oxidation of the CO occurs as HC previously stored in the crevice volumes burn.

Overall, the literature documents that CO emissions are determined by the relative AFR of the combustion mixture. The extent to which engine developers have been able to control the CO formation within the engine has predominantly originated from improvements in the intake mixture uniformity and the ability to lean-out this mixture while sustaining combustion. Because the CO concentration quickly increases as the inlet mixture becomes richer than stoichiometric, cylinder-to-cylinder variations in fuel–air equivalence ratio are very important. A nonuniform distribution increases the average emissions for the same overall mean AFR. Relatedly, enrichment of the fuel–air mixture is necessary during cold start, which is why the CO concentrations are higher at this time than during normal, fully warmed-up operation. Another time that CO concentrations may be higher due to fuel enrichment is during acceleration and deceleration, which can be mitigated by efficient fuel metering control. Nevertheless, the most important point is that oxidation of CO in the exhaust does not occur to any significant extent without the use of catalytic converter, despite the large quantities of oxygen present, because the exhaust temperature is too low. Therefore, the CO kinetics for both formation and oxidation are effectively quenched as the exhaust gases move out of the cylinder and into the exhaust system.

2.6 Hydrocarbons

In 1952, Haagen-Smit, who would later go on to be the first chair of the California Air Resources Board (CARB), identified HC as a significant contributor to photochemical smog [2-29]. HC emissions in engine exhaust have several different sources but primarily originate from fuel HC that are partially burned, recombined, or unburnt. A smaller fraction comes from the lubricating oil, which may have been partially oxidized or volatilized. The HC in the exhaust are usually reported as total hydrocarbons (THC), independent of their source. The THC concentration in the exhaust can be a convenient way to quantify the combustion efficiency, as higher THC concentrations mean less efficient combustion. Nevertheless, THC concentration may not be a very good indicator for pollutant emissions, as HC species represent a wide range of carbon numbers. Some of these HC species are essentially inert in the environment with respect to smog; some have no implications for human health and physiology; and others have strong reactivity and ozone-forming potential or are carcinogenic.

The *Incremental Reactivity Scale* is commonly used as a resource for estimating the reactivity of HC mixtures in the environment. Recent data [2-30] from a University of

TABLE 2.2 Incremental reactivity of hydrocarbons [2-30].

Hydrocarbon species	Incremental reactivity in gO_3/gNMOG
C_1–C_4 alkanes	0.25–1
Benzene	0.4
Methanol	0.6
>C_5 alkanes	1–2
Ethanol	1.3
Cyclic alkanes	1–2.5
Toluene	2.7
Olefins	4–10
Acetaldehyde	5.5
Aromatics	6–10
Formaldehyde	7.2
Ethylene	7.3
Diolefins	10

California-Riverside report to CARB are given in Table 2.2. All of the HC species, with the exception of methane, react in the course of time. Therefore, in some cases, it is useful to further split this concentration into methane (CH_4) and non-methane organics (NMOGs).

The specific components in the fuel blend significantly impact the speciation and concentrations of the THC emissions, especially in SI engines. Fuels with high olefin (alkene) content or aromatic fractions have correspondingly higher HC emissions, especially of reactive HC. Many of the HC species found in the exhaust gas are not present in the fuel or lubricating oil, which indicates that they are formed via synthesis or pyrolytic processes during combustion.

One particularly problematic class of species, in terms of reactivity for production of photochemical smog, is oxygenates. This class includes aldehydes such as formaldehyde (HCHO) and acetaldehyde (CH_3CHO), which are oxygenate species that are particularly noxious or irritating to the respiratory system. In addition, using alcohols and esters as fuels or in blend stocks increases oxygenated HC emissions, particularly methanol (CH_3OH), formaldehyde, and acetaldehyde.

2.6.1 Incomplete Combustion—Quenching and Crevices

Incomplete combustion of fuel HC can occur due to flame quenching or radical extinction at the SI combustion chamber wall. The walls of the combustion chamber are provided by the piston crown, the cylinder head, and the engine block. The latter two are actively cooled to maintain the structural integrity of the head and block. Therefore, the walls are a thermal sink for the energy released during combustion, as well as a solid-body interactor for extinction of the radical species generated within the flame. Flame quenching is essentially a two-step process: the first step is the extinction of the flame as it gets close to the cold wall. The distance from the wall at which this happens depends on the ratio of the conduction from the hot reaction zone to the wall to the thermal energy released in the reaction zone. The second step is diffusion and oxidation of the post-quench HC. This second step occurs very quickly, on the order of milliseconds, and is the process that is responsible for reducing the mass of wall quench HC.

Flame quenching can be the result of flame propagation into the wall, or into a crevice. In either case, when the flame is quenched, it leaves unburnt HC. Whether these result in exhaust HC emissions or not depends on whether further oxidation occurs in the expansion or exhaust. The mechanism with the single largest contribution to the engine-out THC emissions is the crevice volumes. The largest quenching crevice is formed by the gap between the piston crown and the cylinder liner, creating a corner of sorts for the flame to propagate into, with two surfaces that can accomplish the flame cooling. These crevices are filled with unburnt HC during the compression and combustion strokes, when the high pressures in-cylinder can cause the fuel–air mixture to penetrate into the crevices within the combustion chamber. Any unburnt fuel in these volumes evades combustion as the flame cannot penetrate into these tight areas, then during the expansion process, these gases are then able to leave the crevices and escape the cylinder without being oxidized.

Flame quenching, though originally thought to be a significant source of HC emissions from SI engines, has since been shown to be only a small fraction. The quench layer is where the flame is extinguished a short distance from the cooled combustion chamber wall, leaving a layer of unburnt or partially burnt fuel compound. HC from the quench layer diffuse into the hot combustion products and oxidize during the expansion process [2-31]. Therefore, flame quenching only contributes a small amount to efficiency loss and HC emissions. Conversely, when the flame is extinguished before the flame quenching at the wall, also known as bulk quenching, there is a larger contribution to THCs. This quenching can occur when the engine is operating close to its lean operating limit, or with high levels of EGR. Modern engines, operating under normal conditions, with fast-burn combustion systems are not likely to create significant HC emissions. Nevertheless, misfire, engine transients, cycles with high levels of EGR, or retarded spark timings may generate enough HC emissions to merit attention.

Another important mechanism for HC emissions formation is fuel absorption and desorption from the oil layers on the cylinder liners, or walls of the combustion chamber, which increases the residual HC in the gas phase post-combustion [2-32, 2-33, 2-34, 2-35]. These HC are primarily composed of fuel molecules and their concentrations depend on their solubility of the fuel in the oil. The absorption or dissolution of the fuel into the oil occurs prior to the ignition event; then the HC are later desorbed into the hot product gases after the combustion event has completed.

The largest sources of HC emissions in the engine come from the so-called crevice volume during stoichiometric or lean-burn operation. During the compression and combustion strokes, the high pressures in cylinder can force gases into the crevices within the combustion chamber including the very narrow volumes between the piston and cylinder wall. Any unburnt fuel in these volumes evades combustion as the flame cannot penetrate into these tight areas. During the expansion process, these gases are then able to escape the crevices.

2.6.2 Hydrocarbon Emissions from CI Engines

Overall, the engine-out HC emissions from a CI engine are relatively low over most of the operating range. The exception is at low load conditions, where THC emissions can be higher because of the very lean AFR or because of fuel injector control stability. Diesel fuel contains higher molecular weight HC species with correspondingly higher boiling points than gasoline fuel. In addition, due to the nature of combustion in the diffusion flame, the conditions are favorable for significant pyrolysis of the fuel within the spray. This results in a large distribution of HC species over a large molecular size range. There are two main processes by which HC can remain unburnt during the combustion process in a CI engine: the fuel–air mixture can be made too lean to support autoignition of the fuel or flame propagation, or made too rich to ignite. In either case, the fuel *could* later be consumed by slow, thermal oxidation processes in the expansion process, but in reality, the majority of these HC remain unburnt due to quenching. **Figure 2.5** illustrates how HC can originate from incomplete combustion during the phases of ignition delay and the premixed burn and during the mixing-controlled burn.

FIGURE 2.5   Illustrations of the hydrocarbon formation processes in CI combustion.

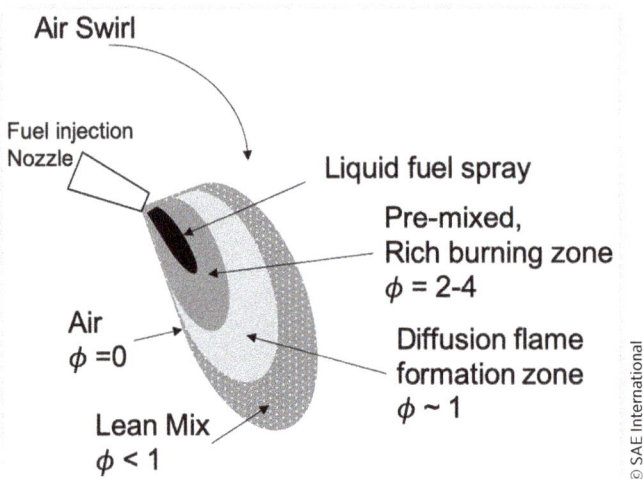

Fuel that is injected during the ignition delay vaporizes and mixes with air that is entrained into the fuel spray, resulting in a rich fuel–air mixture in the bulk of the spray. Due to turbulence at the edges of the spray, there will be regions where the fuel will overmix with air, resulting in a locally lean fuel–air mixture that is difficult to autoignite and burn. This is considered to be the most significant mechanism for HC formation [2-13]. Some of this mixture may later burn if it mixes with high-temperature burned gases early in the expansion stroke. After the ignition delay, during the premixed burn, the ignition occurs in the bulk of the spray where the rich fuel–air mixture is hot enough to autoignite. Whether or not this overrich mixture achieves complete combustion is dependent on further mixing with air in the diffusion flame that surrounds each spray jet or is later able to mix with the hot burned gases before expansion and the associated cooling occurs.

During mixing-controlled burning, there is rapid oxidation of the partially burned fuel and fuel pyrolysis reaction products as they further mix with air in the diffusion flame surrounding each spray jet, resulting in complete combustion. If the mixing of the air and fuel and pyrolysis products is slow, it can result in an over-rich mixture or quenching of the combustion reactions. Such a scenario would result in the HC emissions consisting of species from incomplete fuel combustion, pyrolysis and unburnt fuel [2-13].

The HC concentrations in CI engine exhaust vary widely with operating condition, depending on which of the various HC formation mechanisms dominates. Light-load and idling conditions produce substantially higher HC emission concentrations than full-load operation. Additionally, HC emissions increase significantly when the engine is over-fueled. The cylinder wall temperatures can lead to flame quenching and increased HC emissions, and under-mixing, which results in an over-rich fuel–air mixture, can also occur when the engine is over-fueled.

During light-load operation, over-mixing, also called over-leaning, can be a note-worthy source of HC emissions. When the fuel is injected into the cylinder, a distribution in the fuel–air equivalence ratio quickly develops across each fuel spray, and the amount of fuel that has been mixed leaner than the lean combustion limit increases nearly quickly with time. In the core of each spray, the mixture is substantially richer than stoichiometric, while close to the boundary has already mixed beyond the lean limit for combustion and will not autoignite. In this zone, unburnt fuel, fuel decomposition products, and partial oxidation products will coexist, some of which will escape the combustion chamber without being burnt. The extent of unburnt HC from the over-lean zones will depend on the amount of fuel that was injected during the ignition delay, the mixing rate, and the cylinder conditions—and whether or not they will support autoignition. This explains the (nearly exponential) correlation between HC emissions and ignition delay duration described in the literature [2-36].

Fuel that leaves the injector nozzle at low velocity and excess fuel that enters the cylinder in an over-fueling condition are two causes for under-mixing. The small volume formed in the tip of the injector after the needle seats is called the injector sac volume, and it remains filled with fuel at the end of the injection process. As the engine progresses through the combustion and expansion strokes, this fuel volume is heated and vaporizes, entering the cylinder at low velocity, when it mixes slowly with air and is at risk of escaping combustion. In DI engines, particulate formation limits the fuel–air equivalence ratio (ϕ) at full-load to approximately 0.7, where ϕ is the ratio of the stoichiometric air–fuel ratio (AFR) to the current AFR. However, during engine transients, such as acceleration, over-fueling can occur resulting in local rich zones, even though the overall equivalence ratio remains lean. Generally, THC concentrations do not change with increasing equivalence ratio until a threshold is reached ($\phi \approx 0.9$), when the levels then begin increasing significantly. Under normal operating conditions, this mechanism is not dominant, but it has a more significant contribution to the THCs when over-fueling occurs, such as during heavy acceleration.

Quenching and misfire are two additional sources for HC emissions in a CI engine. In that regard, THCs have been shown to depend strongly on oil and coolant temperature [2-13]. Keeping ignition delay, and therefore over-mixing, constant and changing the oil and coolant temperatures from 40 to 90°C was shown to reduce THC by up to 30%. Therefore, both injected fuel temperature and spray impingement wall quenching of the flame affect THC emissions. Cycle-to-cycle variability in CI engines is generally low, and the risk of the complete lack of combustion, also referred to as *misfire*, is relatively low in a CI engine's normal operating range.

Overall, the engine-out THC emissions from a CI engine are relatively low over most of the operating range. The exception is at low load conditions, where the very lean nature of the fuel–air mixture in the cylinder can increase THC emissions. New developments in CI engine combustion, such as fuel injection strategies that use multiple fuel injections per cycle and homogeneous charge compression ignition strategies, are also likely to complicate the relationship with THC emissions. However, it is unlikely to be problematic, as will later be described in Chapter 8, Oxidation Catalysts.

2.6.3 Hydrocarbon Emissions from SI Engines

HC concentrations in the engine-out exhaust of SI engines are significantly higher than from CI engines. The relative magnitudes are shown in Table 2.3. They can be approximately 1,000–3,000 ppm (0.1%–0.3%).

TABLE 2.3 Sources and magnitudes of hydrocarbon emissions data from [2-7].

Source	Percentage of fuel escaping combustion	Percentage of engine-out HC emissions
Crevices	5.2	38
Deposits	1	16
Exhaust valve leaks	0.1	5
Liquid fuel	1.2	20
Oil layers	1	16
Quenching	0.5	5
Total	9.0	100

As described in Section 2.6.2, THC emissions increase with the enrichment of the fuel–air mixture past stoichiometric and as the fuel–air mixture is leaned out. A number of mechanisms for HC emissions have been proposed for SI engines, including the following [2-7]:

- Fuel–air mixture compressed into the crevice volumes
- Fuel species absorbed into the oil layer on the cylinder liner walls
- Fuel molecules absorbed by in-cylinder deposits on the piston crown and cylinder head
- Quenching on the cylinder walls, flame extinction, liquid fuel that did not evaporate in cylinder, and leakage from the exhaust valve

All these processes combine to result in 5%–10% of the fuel surviving the combustion process, which in addition to increasing exhaust THC emissions, also represent a loss of engine efficiency [2-37, 2-38, 2-39, 2-40]. All of these mechanisms, except misfire, create a situation where the unburnt HC are in close proximity to the cylinder walls, rather than in the bulk gas, creating a nonuniform distribution of THC within the cylinder. The processes by which the unburnt HC can escape oxidation and leave the cylinder include the following [2-7]:

- Outflow of unburnt fuel that was previously trapped in the crevice volumes
- Vaporization of HC from the oil layers and deposits
- Wall and bulk quench gases being mixed into the burned gases
- Unburnt HC escaping during the exhaust blow-down

While these HC could oxidize in the exhaust port, they are more likely to contribute to engine-out THC emissions. Because of this, the HC concentration in the exhaust gases fluctuates substantially during the exhaust process.

The crevices in the SI engine combustion chamber are the largest source of THC emissions from unburnt HC. The largest crevice region in a SI engine is the volume that exists between the piston, piston rings, and cylinder wall. Additional crevice volumes in SI engine cylinders include the volume surrounding the spark plug threads, the volume around the spark plug electrode, the volumes around the intake and exhaust valves, and the head gasket crevice volume. These crevice volumes are illustrated in **Figure 2.6**.

The crevice volumes together typically represent approximately 2% of the total clearance volume [2-38]. The crevice volumes are filled when the increasing pressure of

FIGURE 2.6 Illustration of the crevice volumes and locations of pollutant formation in a spark ignition engine.

SI Pollutant Formation Mechanisms

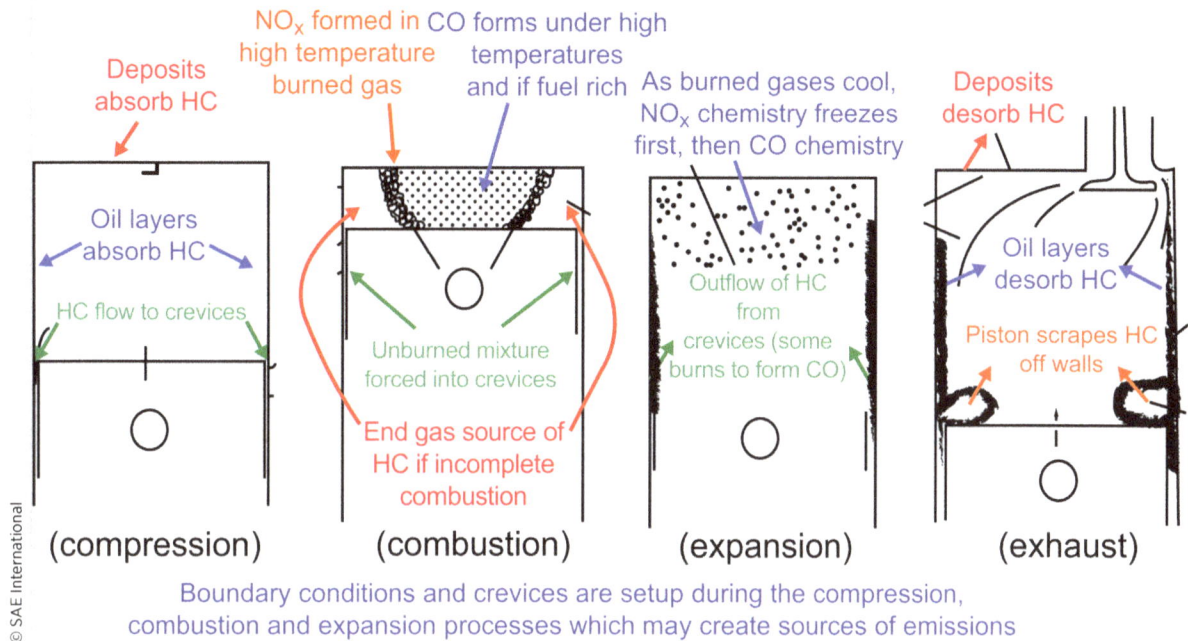

the in-cylinder gases forces the unburnt fuel–air mixture into them. The large surface-to-volume ratio of these small volumes results in the cooling of the gases due to the significant heat transfer to the walls. Since the pressure continues to rise during the combustion process, the gases continue to enter the crevice volumes and be trapped. As the flame propagates to each crevice, depending on the size, the flame can either enter the crevice and burn the stored fuel–air mixture, or if the space is too small or not geometrically accessible, the flame is quenched at the crevice volume entrance, allowing the stored fuel–air mixture to remain unburnt. Once the PFP has been reached, the pressure begins to decrease as the transition to the expansion stroke starts and the gases are able to return to the cylinder from the crevice volumes.

Another important source of HC emissions is from *blowby*, which is the gas that flows from the combustion chamber, past the piston rings, and into the crankcase. Blowby can be thought of as the leakage from the high-pressure conditions in-cylinder through the volume where the piston, piston ring, and cylinder liner meet during the compression, combustion, and expansion strokes. This leakage includes the gases that were previously stored in the crevice volumes, which will then not have a chance to return to the cylinder and react. The blowby gases from the crankcase were previously vented to the atmosphere and represented a substantial source of HC emissions from the vehicle. In 1958, General Motors Research identified that the road draft tube from the crankcase was responsible for nearly half of the HC emissions and the Positive Crankcase Ventilation (PVC) systems became the first emission control device deployed, with the majority of vehicles being equipped with PVC systems by 1964. In modern vehicles, the PVC system routes the vented HC to the intake system, recycling the blowby gases to where they can be burned.

HC emissions can also originate from fuel species absorption into and desorption from the lubricating oil that coats the cylinder liner. This oil layer is revealed and exposed to the intake charge fuel–air mixture as the piston travels downward during the intake stroke. During the time between the intake and compression strokes, when the oil layer is in contact with the unburnt fuel vapor and residual gas mixture, some of the components from the bulk gas can be transferred to the oil layer, where they can absorb. The solubility of the gaseous HC components is proportional to the temperatures of the gas and oil, as described by Henry's law, therefore the highest solubility occurs at higher gas pressures and lower oil temperatures. Later, during the expansion stroke, the oil layer is once again uncovered due to the downward motion of the piston. During this time, the oil has higher concentrations of the fuel components than the burned gases in the cylinder and therefore the HC desorb from the oil layer and diffuse into the burnt gases in-cylinder. It is less likely that they will oxidize as the gas temperature drops during the expansion process, and if they do not oxidize, they can exit the cylinder during the exhaust stroke and contribute to the HC emissions. It has been experimentally determined that this absorption and desorption process is responsible for up to 25% of the THC emissions [2-37].

Another source of HC emissions from SI engines is the deposits on the intake valves, piston crown, and combustion chamber walls that build up with time or mileage. Intake valve deposits effect the THC emissions primarily during cold-start, warm-up, and during changes in load. The porous deposits that build up on the backside of the intake valves have the opportunity to absorb some of the liquid fuel as it is sprayed into the intake port in PFI engines. This results in the in-cylinder fuel–air mixture being leaner during accelerations, when the throttle is opening and richer during decelerations, when the throttle is closing. These changes in the fuel–air mixture composition have an impact on both the HC species and concentrations being emitted from the cylinder during the engine transients, or changes in load. In comparison, the deposits that build up in the combustion chamber increase both the steady-state and transient THC emissions. These deposits typically build up and stabilize on the piston crown and chamber walls over 10,000 mi, though they can affect the THC emissions sooner. Engine-out HC emissions experience the fastest increase over the clean engine values in the first 10 h of operation, followed by slower increases over the next 20 h of operation, before stabilizing over the next 20 h of operation, as the deposits continue to grow [2-41, 2-42, 2-43]. Like valve deposits, combustion chamber deposits are porous in structure, allowing them to act similarly to the crevices by harboring some of the HC during combustion.

The final source for HC emissions in PFI SI engines is liquid fuel impingement on the intake port walls [2-44]. In PFI engines, the fuel injector directs the liquid gasoline fuel toward the back of the intake valve and surrounding intake port surface area in order to induce mixing of the fuel with air and to heat the fuel while cooling the air and the valve. Though a small portion of the fuel will directly evaporate from the atomized droplets, the majority of the fuel will impinge on these surfaces. If the fuel is injected onto a closed valve, the backflow of combustion product gases from the cylinder, which occurs when the valve opens, will aid in the fuel vaporization of the surfaces [2-45]. When the engine is operating under fully warmed-up, part-load conditions, nearly all of the fuel will enter the cylinder in the vapor phase. If the fuel injection is done with the valve open, a larger fraction of the fuel will enter the cylinder in the liquid phase as the droplets are entrained in the intake airflow. Most of the liquid will evaporate as the droplets mix with the air and residual gasses in the cylinder, forming a combustible mixture during the compression stroke. Any remaining liquid, especially the heavier, less volatile fuel compounds, is likely to lead to HC emissions, by forming a liquid film or being absorbed in the deposits or crevices. As this fuel would remain unburnt or be only partially burnt, it would contribute to the THCs in the exhaust. The amount of liquid fuel in the cylinder is especially problematic during cold start, when the intake valves and ports, and the intake air are cold, and the conditions are less favorable for fuel evaporation. In order to accomplish fast and efficient startup, an excess of fuel must be injected, to achieve enough vaporization for a combustible mixture, but a significant amount of this injected fuel will not become part of that mixture and is unaccounted for. It is hypothesized that a fraction of this fuel may end up in the oil sump, with the remainder mixing with the charge later in the cycle, where if it is not reacted, it will contribute to the THC emissions.

DI SI engines have the fuel injected into the cylinder instead of into the intake port as in PFI. DI SI engines combine direct fuel injection and SI technologies to simultaneously enable high-fuel efficiency, improved performance, and low emissions. For a DI SI engine, during the compression and power cycle, the stratified fuel–air mixture combusts like a lean-burn CI engine does, which increases the thermal efficiency to about 35%, compared to 28% fuel efficiency of the homogenous fuel–air mixture in PFI engines. Due to the thermal efficiency increase, DI SI engines have seen a rapid adoption by the automotive industry over the past few years and are predicted to increase their market share by model year 2025 [2-46].

Further description of DI SI engines can be found in Chapter 1, Introduction. Directly injecting the fuel changes the behavior of the liquid fuel in the cylinder and has significant ramifications for HC emissions. For example, liquid films do not form on the valves or intake ports, which remove a reservoir of fuel from within the engine. On the other hand, there will be more liquid fuel in the cylinder than in PFI engines, especially during cold start and warm up. The fuel is where it is intended to be, rather than adsorbed in a deposit layer, and the event can be tailored to minimize liner impingement and fuel ending up in the crevice volume. This phenomenon requires the fuel metering calibration for cold start and engine transients to differ from the calibration for steady-state operation. DI has the potential to reduce the requirement for cold-start fuel enrichment since all of the fuel injected enters the cylinder instead of pooling on the cool intake port wall, reducing the likelihood of increased THC emissions during that time. DI SI engines that are stratified at light and medium loads generally have higher HC emissions as compared to homogenous, stoichiometric DI SI engines due to increased flame quenching in the periphery of very lean mixtures, as described above. Higher HC emissions can also result from the lower exhaust temperatures due to the lower expansion gas temperatures and higher efficiency as the injected fuel cools the charge through evaporation and as more energy is extracted as work. These higher THC emissions from lean, stratified DI SI engines remain one of their most significant challenges, keeping them from deployment in the fleet [2-47].

2.7 Particulate Matter

PM or "soot" emissions have been an issue for ICEs since their invention. PM emissions from diesel-fueled CI engines are a more significant problem than with SI engines. On a mass basis, they are 75 times higher than with PFI SI engines and about 7.5 times higher than with DI SI engines.

2.7.1 Particulate Formation and Characterization

Particulate formation takes place within the rich regions of the combusting fuel–air mixture, at temperatures around 1,600 K and pressures of approximately 100 bar, where there is not enough air in the local mixture to completely combust the fuel. This process occurs at the locally rich regions of the fuel spray, as shown in **Figure 2.7**.

FIGURE 2.7 Dec conceptual model of particulate formation [2-48].

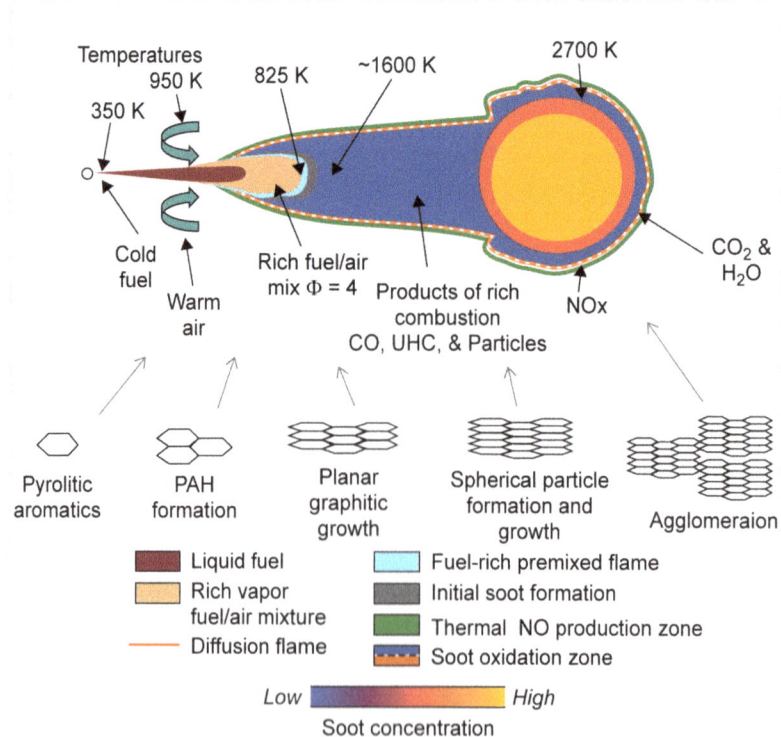

© SAE International

FIGURE 2.8   Primary (a) aggregate, (b) agglomerate, and (c) [2-54] HR-TEM. Images by Randy Vander Wal.

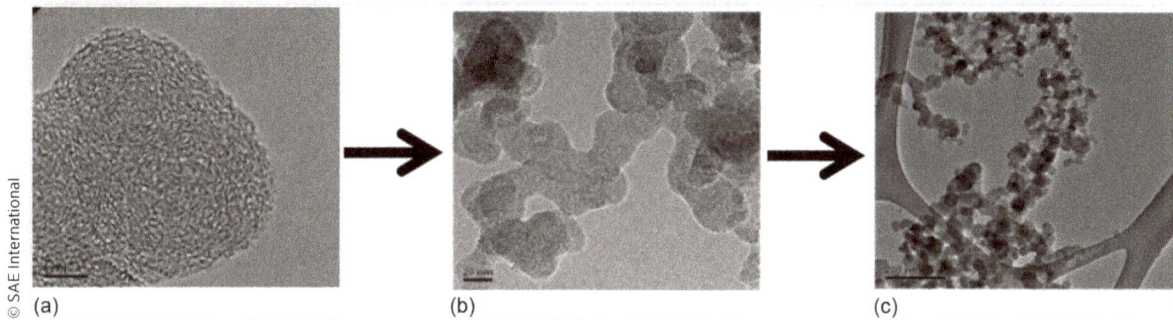

(a)   (b)   (c)

In the fuel-rich regions, incompletely burned aromatics pyrolyze into polyaromatic hydrocarbon precursors in the *inception mode*. These precursors undergo the *nucleation mode*, forming nuclei particles [2-49, 2-50]. The nuclei coalesce and undergo surface growth to become *primary particles* (3–20 nm in diameter) [2-51, 2-52, 2-53], **Figure 2.8a**, which are composed of equidistant hexagonal graphitic crystallites bent around a sphere. In the accumulation mode, primary particles coalesce with one another or other species and absorb mobile HC species, forming complex *aggregate* structures (20–200 nm in diameter), **Figure 2.8b**, with non-homogeneous chemical properties. Finally, the HC-laden aggregates chain together into *agglomerates* (200 nm–1 µm in diameter), **Figure 2.8c**.

These processes are illustrated schematically in **Figure 2.9**.

While these stages of particle generation and growth are occurring, it is possible that, simultaneously, oxidation may be as well, and the ultimate particulate emissions concentration leaving the engine will depend on the balance of these processes. Once the particulates leave the engine, further addition of mass occurs as the exhaust gas cools and adsorption of the HC species present commences. PM formation is a nonequilibrium process, and these processes are not distinct and separate, but rather they can be concurrent and overlapping. In addition, as the gas mixture in cylinder and in the exhaust is nonhomogeneous, not all of the particulates are in the same stage of formation or growth at a particular time. As a result of the competing processes, the engine-out

FIGURE 2.9   Processes leading to the production of particulates [2-55].

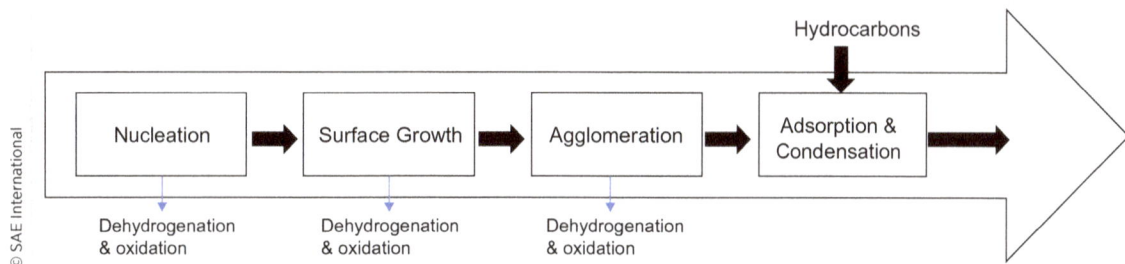

FIGURE 2.10   Example particle size distribution from diesel engine exhaust.

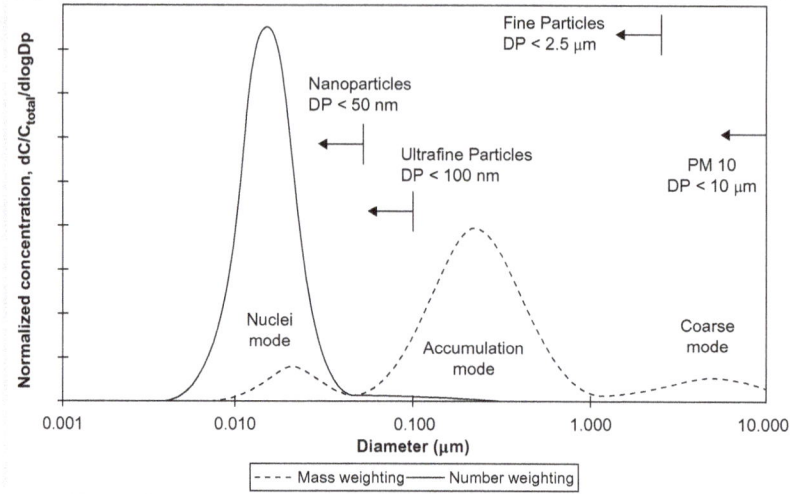

particulate size is not singular, but rather a distribution, with everything from primary particles to agglomerates represented, as seen in **Figure 2.10**.

Diesel particulates tend to be on the large end of the size ranges given, with DI SI particulates being smaller, and PFI gasoline being dominated by nucleation mode particle sizes. It has been shown that the higher pressures (above 3,000 bar) lead to reduced particle mass and number; however, it is suspected that these pressures reduced the diameters below the detection limit [2-56]. A comparison of the size and emissions rates of particulates from different light-duty engine combustion types is shown in **Figure 2.11**.

FIGURE 2.11   (a, b) Comparison plots of PM number versus mass emissions from different light-duty engine combustion strategies [2-88].

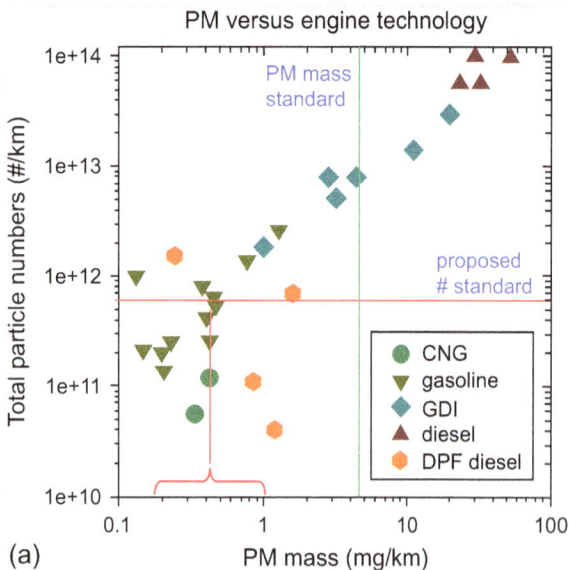

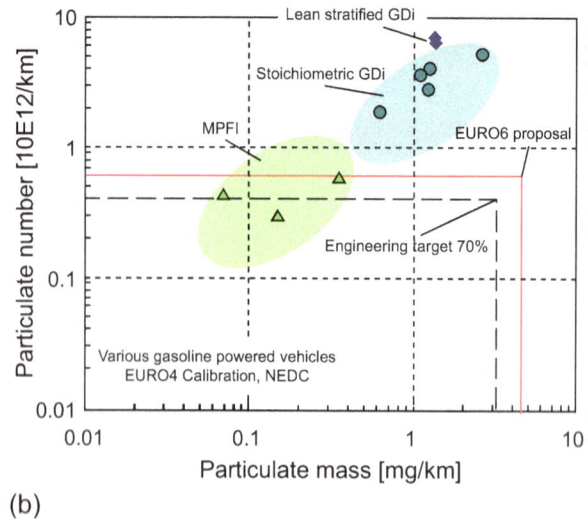

For both CI engine and DI SI engine exhaust, nucleation mode particles can also form in the end region of the tailpipe, or just after the exhaust exits the tailpipe. While these particles represent only a small fraction (single-digit percentage) of the particulate mass, they comprise the majority (90+%) of the particle number measurement. Unlike the carbonaceous particles formed in-cylinder, these tend to consist mainly of condensed HC and therefore tend to be volatile. Measurement of the concentration of these volatile particles is strongly influenced by the sampling methodology, including dilution ratio, dilution temperature, and residence time in the dilution system. Additionally, a small fraction of other non-carbon solids can be found in the nucleation mode particles, such as metallic ash originating from the lubricating oil additives. In cases where the fuel contains significant amounts of sulfur, the nucleation mode particles can also contain hydrated sulfuric acid condensates, formed from gaseous precursors in the exhaust.

Many researchers have conducted characterization analyses on particulates, by chemical, optical, and physical methods. Chemical analysis has allowed the further subdivision of particulates into mobile fractions, the fixed carbon fraction, and sulfates. The fixed carbon fraction is predominantly formed in-cylinder, whereas the mobile and sulfate liquid fractions tend to be formed when the exhaust gas is cooled, such as upon mixing with ambient air after leaving the tailpipe. The mobile fractions, often referred to as organic fractions, contain the species which can be separated from the fixed carbon backbone and are named by the method of mobilization. The volatile organic fraction (VOF) refers to the portion of the mobile fraction that is separated from the fixed carbon by heating, or volatilizing them off. This primarily consists of partially unburnt HC from the fuel. The soluble organic fraction (SOF) includes the VOF as well as heavier HC species from the fuel and lubricating oil, which can be separated from the fixed carbon by organic extraction techniques including Soxhlet and microwave-assisted reaction. The fixed carbon fraction is composed of the solid-carbon hexagonal graphitic crystallites and is also referred to as the elemental carbon fraction. The sulfate portion is typically composed of sulfuric acid and water and can be physisorbed unto the fixed carbon fraction. The exact composition of the relevant fractions is a function of engine type, size and speed-load point, as well as fuel source as shown in **Figure 2.12**

FIGURE 2.12 Particulate composition.

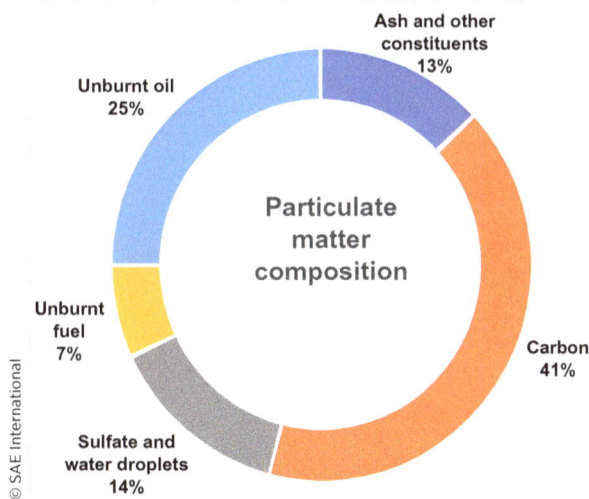

[2-53, 2-57, 2-58, 2-59, 2-60, 2-61]. Additional chemical characterization can be carried out to investigate the reactivity of the particulates for filter regeneration, and those are discussed later, in Chapter 11, Particulate Filters.

Optical analysis of particulates, by high-resolution transmission electron microscopy (HR-TEM), as illustrated in **Figure 2.13** reveals the particle nanostructure—the detailed atomic level structure comprising the carbon lamella. Measures such as size, orientational ordering, relative density, and even separation distance are readily seen within the images and may be quantified by custom image processing. Such analyses permit rigorous definition of similarities and differences between particulates and their sources.

Physical analysis of particulates is possible by Brunauer–Emmett–Teller analysis to measure the specific surface area of the particulates. This analysis, used in conjunction with microscopy and reactivity analysis, can provide forensic-like insight into the in-cylinder formation of particulates, as described by Vander Wal et al. [2-63]. Particulates have a large range of structures associated with them, from the nanostructure associated with the graphene layers forming the primary particles, to the fractal aggregate, composed of agglomerated primary particles [2-64, 2-65]. The nanostructure has long been understood to impact the particulate's physical properties and chemical reactivity, including the reactive surface area and oxidation characteristics [2-64, 2-65, 2-66, 2-67]. Thermal pyrolysis work by Vander Wal and Tomasek [2-68] has shown by HR-TEM that the temperature, pressure, and the rate of temperature rise during formation greatly impact the nanostructure of particulates. At lower temperatures, the particulate structure is more amorphous in nature, consisting of short, randomly oriented, disconnected segments. With increasing temperature, the graphene segments grow in length and are more ordered, or graphitic, in nature.

Vander Wal and collaborators from Cummins Engine Company have also seen significant differences in the structure, size, and behavior of particles from different

FIGURE 2.13   HR-TEM of particulates, defining physical features of the nanostructure [2-62].

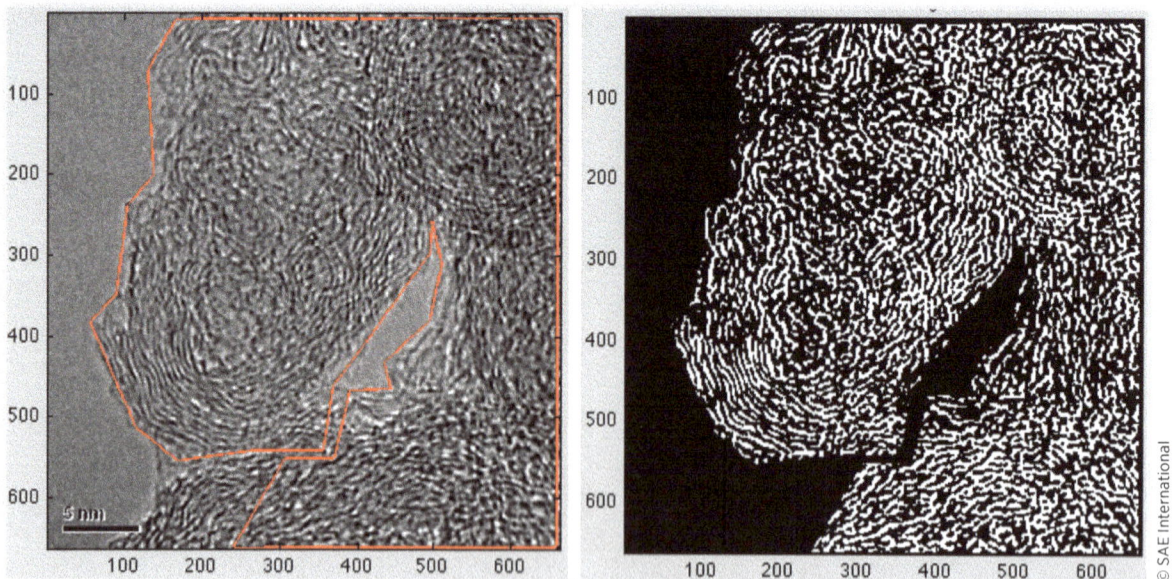

© SAE International

engine load-speed points [2-69]. Braun [2-70, 2-71] examined properties of diesel engine-derived particulates and observed that changes in the nanostructure between idle and load conditions were much greater for oxygenates than conventional diesel.

The oxidative reactivity of a particulate, relevant for DPF regeneration described in Chapter 11, Particulate Filters, has been shown to be dependent on the initial nanostructure; specifically, the more ordered graphitic structure exhibits higher threshold temperatures and is more resistant to oxidation than amorphous carbon [2-72]. Diesel particulates are typically made up of multiple carbon types. This may be the result of several factors, including different effective densities, as well as different particle size distributions. In the Active Site Mechanism theory [2-73], particulate oxidation is said to proceed through dissociation of edge site carbons, and therefore samples with more edge sites, physical defects in the structure, and less tangential arrangement are known to be more reactive, as there are geometrical and electrical availability advantages [2-74, 2-75, 2-76].

The ability to alter particulate nanostructure properties by changing fuel or operating conditions has been clearly demonstrated by previous work [2-68, 2-77, 2-78]; however, many of the studies employed synthesis conditions that were not equivalent to engine combustion or did not isolate the impact of the fuel from the combustion conditions. Later work by McCormick, Song and Boehman, Vander Wal and Strzelec, and others [2-61, 2-63, 2-79, 2-80, 2-81] used conventional combustion conditions and was able to clearly show the impact of fuel type, in an engine setting on the particulate physical morphology and nanostructure.

The chemical characteristics of the mobile carbon fraction (MC) have been studied for conventional and advanced combustion engines [2-82, 2-83, 2-84, 2-85, 2-86]. Proposed low temperature combustion regimes typically have higher organic (MC) content than conventional diesel combustion [2-83]. For conventional combustion, it was shown that higher load points have less mobile content than at lower loads, likely due to the exhaust gas temperature [2-87].

Gasoline-fueled PFI SI engines have very low levels of PM emissions, on the order of only a few milligram per mile. Three types of SI PM emissions have been traditionally defined: inorganic particulates, organic particulates including carbonaceous PM, and sulfates. The inorganic particulates, or ash, were predominantly lead-based until lead lubricity additives were taken out of gasoline fuels. The sulfate particulates depend heavily on the sulfur concentration in the fuel—which can be between 10 and 500 ppm for gasoline—and the operating condition of the engine. Sulfates are discussed further in Section 2.7.2. The organic particulates can consist of black carbon under rich fuel–air mixture conditions, similar to diesel PM, but more generally consist of condensed organic compounds. Compared to PFI engines, DI engines produce higher concentrations of both PM and particulate number (PN) [2-88, 2-89, 2-90, 2-91]. Along with the high fuel efficiency, the lean burn combustion of SI DI engines also produces diesel-like particulate emissions. DI SI engines have approximately 10 times the PM emissions compared to PFI engines and also higher PN emissions (EU standard). Stratified charge DI SI engines have higher numbers of particulates, especially nucleation mode particulates, than homogeneous charge, stoichiometric DI SI engines do because of the presence of local rich zones where particulate emissions are formed.

A more detailed description of DI SI engine operation is given in Section 1.2.1; only a few of the key factors that affect PM formation in DI SI and how they differ from those formed in CI combustion will be discussed [2-92]. The fuel injection system plays a critical role in both the mixing of the air and fuel, and in turn formation of the particulates. As described in Section 1.2.1, there are three common types of fuel injection systems that are broadly classified by how they achieve the charge mixture stratification: air-guided, spray-guided, and wall-guided. Practically, all of three types of mixings occur concurrently in cylinder. All of the systems have issues with mixing that are relevant to PM formation and several engine design parameters must be optimized to achieve low particulate emissions over a wide range of engine speed-load conditions. In spray-guided systems there is less turbulence generated for promoting fuel evaporation and mixing, whereas in wall-guided systems the fuel can impinge on surfaces leading to higher particle emissions [2-89, 2-93]. With the injection of the liquid fuel occurring directly into the cylinder, greater penetration lengths for the fuel droplets, coupled with slower evaporation, give rise to the issues of liquid fuel evaporation and wall wetting of the piston crown, cylinder head, and liner, as described above for HC emissions become relevant for particulate formation as well. The particles formed are broadly in the 10–100-nm size range, and while the emissions may be low on a mass basis, on a number basis the emissions can be exceed the standard by one or two orders of magnitude. A study of Euro 5 vehicles showed that modern SI engines can emit more particulates than CI engines fitted with diesel particulate filters (DPF), making them a target for regulators [2-91, 2-94, 2-95]. In addition, the aromatic content of the gasoline soot was found to be much higher than that from the diesel soot after the DPF. This aromatics content supports increased secondary organic aerosol (SOA) formation from DI SI. While DPF capture any SOA precursors, they are currently emitted to the ambient in the case of SI DI vehicles without filters. The SOA concentrations have been found to be 6.5 times that of primary organic aerosols after atmospheric aging.

Most of the particles emitted on a FTP-75 test cycle are in the 50–90-nm size range; however, the distribution can be very different over the more aggressive US06 cycle [2-96]. The FTP-75 test cycle is described in Section 4.1.1, and the US06 cycle, Section 4.1.3. For the US06 cycle, a bimodal distribution of particulate sizes has been described, with nanoparticles <20 nm completely dominating the distribution, and with a much smaller contribution from the accumulation mode 40–90-nm particles. A bimodal distribution was also observed in a study of two DI SI vehicles over the LA92 and US06 test cycles. The LA92 cycle is described in Section 4.1.6. They found that 40–120 nm accumulation mode particles were dominant over the LA92 cycle, because most of the emissions occurred during the cold start portion of LA92. This result is consistent with the belief that the nucleation mode particulates come from the higher molecular weight HC in the fuel, which condense at lower temperature conditions, such as those that exist during cold start, or late during the expansion stroke, or during the exhaust process.

The average DI SI primary and aggregate particle sizes have been shown to be gradually increasing with advancing injection timing, implying that, like for diesel, fuel–air mixing influences particle size distributions. The compactness of the aggregates from DI SI engines is between those from LDDE and HDDE. Additionally, particles from the

DI SI engine have been shown to have less-ordered graphitic structures than from typical diesel engines, implying they may be more easily oxidized.

2.7.2 Sulfur and Sulfur Oxides

Sulfur is a natural component in crude oil present in distillate gasoline and diesel fuels, unless it is actively removed. There are two distinct effects of sulfur, as follows:

1. Sulfur compounds poison the catalysts in the EAS
2. SOx and sulfur deposits contribute to particulate matter

Sulfur emissions originate from the fuel and to a lesser extent from the lubricating oil and have implications on multiple emission species. Sulfur that is oxidized to SO_2 during combustion can then be further oxidized to SO_3. The extent of this conversion depends on the presence of a catalyst and the exhaust temperature, as increasing temperatures favor SO_3 formation. SO_2 is the dominant species in engine exhaust, and its emissions are important because they are a precursor to the emissions of very small particulates, also known as PM 2.5 [2-97]. SO_3 is of interest because it can further react with water at ambient temperatures to become sulfuric acid (H_2SO_4). SOx compounds can be chemically reduced in the EAS to form hydrogen sulfide (H_2S) under fuel-rich conditions.

Since the concentration of SO_2 emissions in the exhaust directly correlates to the sulfur content of the fuel, reductions in the maximum fuel sulfur level have greatly reduced SOx emissions from engines. For example, the sulfur level of gasoline has decreased from 30 ppm in 2006 to 10 ppm in 2017, and of diesel, from 500 ppm in 1993 to 15 ppm in 2006.

2.7.3 Ash

Ash is noncombustible mineral material that is derived primarily from metallic additives in the lubricating oil. Ash can also originate from engine wear, engine corrosion, and trace metals in the fuel. It consists of metals such as calcium, zinc, and magnesium, often combined with phosphate or sulfate ions [2-98, 2-99, 2-100]. These metals can vaporize in a reduced form into the bulk gas due to the extreme temperature and pressure conditions in-cylinder. Then, they subsequently condense as the gas temperature decreases, forming submicron ash particles that consist of the matter that is bound to the metallic elements in the oil. Alternatively, ash may result as a residue from combustion of the oil. Similar to carbonaceous particulates, ash particles grow by condensation and collision, or coalescence mechanisms at elevated temperatures. The growth continues until the temperature decreases causing the ash particles to agglomerate into chain or cluster aggregates. The amount of ash that is produced depends on the nature and quantities of the metallic compounds in the additive package. The sources of ash in diesel engine oil are anti-wear agents, such as zinc dialkyldithiolphosphate (ZDDP), calcium and magnesium-based detergents, dispersants, antioxidants, viscosity improvers, and friction modifiers [2-101]. Their relative contributions to ash formation are shown in Table 2.4, adapted from Ref. [2-102]. According to

TABLE 2.4 Component Contributions to Ash, Data from Ref. [2-102].

Component	Ash contribution wt.% of oil
Detergent	0.6–1.3
ZDDP	0.15
Other	0.0–0.15
Antioxidants	
Viscosity improvers	
Friction modifiers	
Total	0.75–1.6

the ASTM D876 specification, lubricant sulfated ash is defined as the material remaining after the oil has been treated with sulfuric acid and heated until the weight of the residual material remains constant [2-103]. However, there is variability in the test method, as it is affected by magnesium-based detergents and boron-based dispersants, as well as various phosphorous compounds [2-104, 2-105, 2-106]. Not all of the additive compounds are consumed or emitted from the engine at the same rate, as differences in volatility affect their relative emission rates. Data on the contributions from the individual lubricating oil additives to ash emissions are given in Table 2.4. Further discussion on ash and especially its effects on particulate filters can be found in Chapter 11, Particulate Filters, in a special section contributed by Dr. Justin Kamp.

2.8 Aldehydes and Ketones

Aldehydes and aliphatic ketones are a class of HC known as carbonyls. Aldehydes have the carbonyl functional group at the end of the HC chain (with an H bound to the C=O), while ketones contain the C=O functional group within the molecule, as shown in **Figure 2.14**. Both forms are found in engine exhaust and are under increased scrutiny in recent years and beginning to be subject to regulation due to their ozone-forming potential and carcinogenic impacts on human health.

Aldehydes in particular have very high incremental reactivity, as shown in Table 2.2, in the earlier section on HC, Section 2.6.3. The three most common aldehyde species are formaldehyde (HCHO), acetaldehyde (CH_3CHO), and acrolein (CH_2CHCHO), schematics of which are shown in **Figure 2.15**. Of these three, formaldehyde is a volatile and reactive molecule that tends to form ozone from photochemical oxidation. Since neither aldehydes nor ketones are a significant fraction of the HC species present in the fuel, any aldehydes or ketones found in the engine exhaust gases were formed in the engine or EAS. The basic aldehyde formation reactions are driven by radical chain reactions created during combustion, with temperature having a very strong effect on aldehyde formation. The formation is higher at lower temperatures, such as those experienced during cold start [2-107]. In addition, the blending of alcohols into fuel significantly increases the levels of aldehyde emissions in both CI and SI engines. Further discussion of the fuel effects of oxygenates will occur in Chapter 16, Alternative Fuels.

FIGURE 2.14 Molecular schematics for (a) aldehydes and (b) ketones.

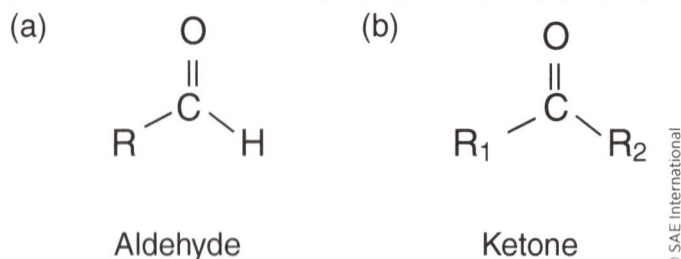

© SAE International

FIGURE 2.15 Molecular schematics for (a) formaldehyde, (b) acetaldehyde, and (c) acrolein.

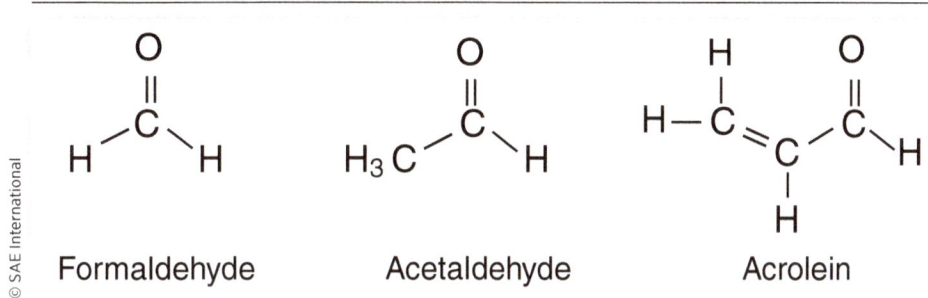

Formaldehyde   Acetaldehyde   Acrolein

© SAE International

2.8.1 CI Engines

Carbonyls account for approximately 5%–10% of the HC emissions from light duty diesels. In particular, the species found in diesel exhaust are formaldehyde, acrolein, isobutyral, benzaldehyde, and acetone [2-108, 2-109, 2-110, 2-111]. Diesel fuel formulation has a strong impact on carbonyl emissions. Aldehyde emissions from CI engines were shown to decrease with the increasing cetane number of the diesel fuel, which corresponds to increasing fuel density and decreasing aromatic content. As oxygenate content increases, or when using biofuels, such as biodiesel, aldehyde emissions increase. In addition, aldehyde emissions from turbocharged engines were found to be slightly lower for low cetane number fuels and higher for high cetane number fuels [2-107]. It is thought that this effect is caused by the higher oxygen content due to the large amount of excess air present in diesel combustion and exhaust. The fuel effects of oxygenates are discussed further in Chapter 16, Alternative Fuels.

2.8.2 SI Engines

The carbonyl species present in SI engine exhaust include formaldehyde (44%), acrolein and acetone (29%), acetaldehyde (10%), and benzaldehyde (5%) [2-107]. Just as for CI engines, the fuel formulation for SI engines has a large effect on the carbonyl emissions. It has been well documented that acetaldehyde concentrations increase when ethanol (CH_3CHOH) is blended with gasoline [2-112]. The concentration of exhaust aldehydes has been strongly linked with exhaust temperature and fuel hydrogen-to-carbon ratio. Nevertheless, the concentration of the acetaldehyde and acetone species presents complex behavior, as they increase in some cases and decrease in others [2-113].

2.9 Carbon Dioxide

CO_2 is not a regulated criteria pollutant as of writing, but it is increasingly regulated as a greenhouse gas (GHG). CO_2 thus remains an important topic for regulators and manufacturers alike. CO_2 is an important trace-level component of Earth's atmosphere, as it is an essential part of the carbon cycle. However, its increasing concentration in the atmosphere, from approximately 280 ppm at the time of the Industrial Revolution to 410 ppm in 2018 [2-114, 2-115], as well as its GHG-defining ability to absorb and emit infrared radiation at 4.26 and 14.99 μm [2-116, 2-117], is a concern. CO_2 is formed from the combustion of HC fuels in an engine, the decomposition of urea, and the conversion of carbon-containing criteria pollutants in the EAS.

References

2-1.   Eastwood, P., *Critical Topics in Exhaust Gas Aftertreatment* (Baldock, Hertfordshire, UK: Research Studies Press Ltd., 2000).

2-2.   Diesel, R., "Internal Combustion Engine," US608845A, Aug. 9, 1898.

2-3.   Zel'dovich, Y.B., "The Oxidation of Nitrogen in Combustion Explosions," *Acta Physicochimica U.S.S.R* **21** (1946): 577-628.

2-4.   Hanson, R.K. and Salimian, S., "Survey of Rate Constants in the N/H/O System," *Combustion Chemistry* (Springer, 1984), 361-421.

2-5.   Lavoie, G.A., Heywood, J.B., and Keck, J.C., "Experimental and Theoretical Study of Nitric Oxide Formation in Internal Combustion Engines," *Combustion Science & Technology* **1**, no. 4 (1970): 313-326.

2-6.   Williams, F.A. *Combustion Theory, (1985)* (Cummings Publ. Co, 1985).

2-7.   Heywood, J., *Internal Combustion Engine Fundamentals,* 2nd ed. (McGraw-Hill Education, 2018).

2-8.   Akinyemi, O.C. and Cheng, W.K., "A Flame Sheet Model for NO Production in Diesel Combustion Simulation," SAE Technical Paper 982586, 1998, https://doi.org/10.4271/982586.

2-9.   Voiculescu, I.A. and Borman, G.L., "An Experimental Study of Diesel Engine Cylinder-Averaged NOx Histories," SAE Technical Paper 780228, 1978, https://doi.org/10.4271/780228.

2-10. Aoyagi, Y., Kamimoto, T., Matsui, Y., and Matsuoka, S.J.G.S., "A Gas Sampling Study on the Formation Processes of Soot and NO in a DI Diesel Engine," SAE Technical Paper 800254, 1980, https://doi.org/10.4271/800254.

2-11. Ahmad, T. and Plee, S.L., "Application of Flame Temperature Correlations to Emissions from a Direct-Injection Diesel Engine," SAE Technical Paper 831734, 1983, https://doi.org/10.4271/831734.

2-12. Turns, S.R., "Understanding NOx Formation in Nonpremixed Flames: Experiments and Modeling," *Progress in Energy Combustion Science & Technology* **21**, no. 5 (1995): 361-385, https://doi.org/10.1016/0360-1285(94)00006-9.

2-13. Yu, R.C. and Shahed, S.M., "Effects of Injection Timing and Exhaust Gas Recirculation on Emissions from a D.I. Diesel Engine," SAE Technical Paper 811234, 1981, https://doi.org/10.4271/811234.

2-14. Lapuerta, M., Salavert, J.M., and Doménech, C., "Modelling and Experimental Study About the Effect of Exhaust Gas Recirculation on Diesel Engine Combustion and Emissions," SAE Technical Paper 950216, 1995, https://doi.org/10.4271/950216.

2-15. Lavoie, G.A., "Spectroscopic Measurements of Nitric Oxide in Spark Ignition Engines," *Combustion & Flame* **15**, no. 2 (1970): 97-108.

2-16. Alperstein, M. and Bradow, R.L., "Exhaust Emissions Related to Engine Combustion Reactions," SAE Technical Paper 660781, 1966, https://doi.org/10.4271/660781.

2-17.   Starkman, E.S., Stewart, H.E., and Zvonow, V.A., "Investigation into Formation and Modification of Exhaust Emission Precursors," SAE Technical Paper 690020, 1969, https://doi.org/10.4271/690020.

2-18.   Blumberg, P. and Kummer, J., "Prediction of NO Formation in Spark-Ignited Engines—An Analysis of Methods of Control," *Combustion Science & Technology* **4**, no. 1 (1971): 73-95.

2-19.   Stumpp, G. and Banzhaf, W., "An Exhaust Gas Recirculation System for Diesel Engines," SAE Technical Paper 780222, 1978, https://doi.org/10.4271/780222.

2-20.   Alger, T., Gingrich, J., Roberts, C., and Mangold, B., "Cooled Exhaust-Gas Recirculation for Fuel Economy and Emissions Improvement in Gasoline Engines," *International Journal of Engine Research* **12**, no. 3 (2011): 252-264.

2-21.   Komiyama, K. and Heywood, J.B., "Predicting NOx Emissions and Effects of Exhaust Gas Recirculation in Spark-Ignition Engines," SAE Technical Paper 730475, 1973, https://doi.org/10.4271/730475.

2-22.   Abd-Alla, G.H., "Using Exhaust Gas Recirculation in Internal Combustion Engines: A Review," *Energy Conversion and Management* **43**, no. 8 (2002): 1027-1042.

2-23.   Majewski, W.A. and Khair, M.K., *Diesel Emissions and Their Control* (Warrendale, PA: SAE International, 2006).

2-24.   Zheng, M., Reader, G.T., and Hawley, J.G., "Diesel Engine Exhaust Gas Recirculation--A Review on Advanced and Novel Concepts," *Energy Conversion and Management* **45**, no. 6 (2004): 883-900.

2-25.   Majewski, W. and Khair, M., *Diesel Emissions and Their Control* (Warrendale, PA: SAE International, 2006), ISBN-13:978-970.

2-26.   Harrington, J.A. and Shishu, R.C., "A Single-Cylinder Engine Study of the Effects of Fuel Type, Fuel Stoichiometry, and Hydrogen to Carbon Ratio and CO, NOx and HC Exhaust Emissions," SAE Technical Paper 730476, 1973, https://doi.org/10.4271/730476.

2-27.   Pitts, W.M., "The Global Equivalence Ratio Concept and the Formation Mechanisms of Carbon Monoxide in Enclosure Fires," *Progress in Energy and Combustion Science* **21**, no. 3 (1995): 197-237.

2-28.   Newhall, H.K., "Kinetics of Engine-Generated Nitrogen Oxides and Carbon Monoxide," in *Symposium (International) on Combustion*, Elsevier, 1969.

2-29.   Haagen-Smit, A.J., "Chemistry and Physiology of Los Angeles smog," *Industrial & Engineering Chemistry* **44**, no. 6 (1952): 1342-1346, https://doi.org/10.1021/ie50510a045.

2-30.   Carter, W.P.L., "Updated Maximum Incremental Reactivity Scale and Hydrocarbon Bin Reactivities for Regulatory Applications," University of California-Riverside Report No. Report to California Air Resources Board (Contract 07-339), 2010.

2-31.   LoRusso, J., Kaiser, E., and Lavoie, G., "Quench Layer Contribution to Exhaust Hydrocarbons from a Spark-Ignited Engine," SAE Technical Paper 800045, 1980, https://doi.org/10.4271/800045.

2-32.   Lavoie, G., Adamczyk, A., Kaiser, E., Cooper, J., and Rothschild, W., "Engine HC Emissions Modeling: Partial Burn Effects," *Combustion Science & Technology* **49**, no. 1-2 (1986): 99-105.

2-33.   Kaiser, E., LoRusso, J., Lavoie, G., and Adamczyk, A., "The Effect of Oil Layers on the Hydrocarbon Emissions from Spark-Ignited Engines," *Combustion Science & Technology* **28**, no. 1-2 (1982): 69-73.

2-34.   Kaiser, E., Adamczyk, A., and Lavoie, G., "The Effect of Oil Layers on the Hydrocarbon Emissions Generated during Closed Vessel Combustion," in *Symposium (International) on Combustion*, Elsevier, 1981.

2-35. Adamczyk, A.A., Kaiser, E., Cavolowsky, J., and Lavoie, G., "An Experimental Study of Hydrocarbon Emissions from Closed Vessel Explosions," in *Symposium (International) on Combustion*, Elsevier, 1981.

2-36. Greeves, G., Khan, I., Wang, C., and Fenne, I., "Origins of Hydrocarbon Emissions from Diesel Engines," SAE Technical Paper 770259, 1977, https://doi.org/10.4271/770259.

2-37. Cheng, W.K., Hamrin, D., Heywood, J.B., Hochgreb, S., Min, K., and Norris, M., "An Overview of Hydrocarbon Emissions Mechanisms in Spark-Ignition Engines," SAE Technical Paper 932708, 1993, https://doi.org/10.4271/932708.

2-38. Smith, P.M., "Crevice Volume Effect on Spark Ignition Engine Efficiency," *Mechanical Engineering* (Massachusetts Institute of Technology, 2013).

2-39. Haidar, H.A. and Heywood, J.B., "Combustion Chamber Deposit Effects on Hydrocarbon Emissions from a Spark-Ignition Engine," SAE Technical Paper 972887, 1997, https://doi.org/10.4271/972887.

2-40. Min, K., Cheng, W.K., and Heywood, J.B., "The Effects of Crevices on the Engine-Out Hydrocarbon Emissions in SI Engines," SAE Technical Paper 940306, 1994, https://doi.org/10.4271/940306.

2-41. Bitting, W.H., Firmstone, G.P., and Keller, C.T., "Effects of Combustion Chamber Deposits on Tailpipe Emissions," SAE Technical Paper 940345, 1994, https://doi.org/10.4271/940345.

2-42. Kalghatgi, G., "Combustion Chamber Deposits in Spark-Ignition Engines: A Literature Review," SAE Technical Paper 952443, 1995, https://doi.org/10.4271/952443.

2-43. Valtadoros, T.H., Wong, V.W., and Heywood, J.B. "Fuel Additive Effects on Deposit Build-Up and Engine Operating Characteristics," *Abstracts of Papers of the American Chemical Society*, Amer Chemical Soc, 1991.

2-44. Shin, Y., Cheng, W.K., and Heywood, J.B.J.S.T., "Liquid gasoline behavior in the engine cylinder of a SI engine," SAE Technical Paper 941872, 1994, https://doi.org/10.4271/941872.

2-45. Cheng, C.-O., Cheng, W.K., Heywood, J.B., Maroteaux, D., and Collings, N., "Intake Port Phenomena in a Spark-Ignition Engine at Part Load," SAE Technical Paper 912401, 1991, https://doi.org/10.4271/912401.

2-46. Gladstein, N., "Ultrafine Particulate Matter and the Benefits of Reducing Particle Numbers in the United States," A Report to the Manufacturers of Emissions Controls Association (MECA), 2013.

2-47. Zhao, F., Harrington, D.L., and Lai, M.-C.D., *Automotive Gasoline Direct-Injection Engines* (Warrendale, PA: Society of Automotive Engineers, 2002).

2-48. Dec, J.E., "A Conceptual Model of Dl Diesel Combustion Based on Laser-Sheet Imaging," SAE Technical Paper 970873, 1997, https://doi.org/10.4271/970873.

2-49. Harris, S.J. and Maricq, M.M., "Signature Size Distributions for Diesel and Gasoline Engine Exhaust Particulate Matter," *Journal of Aerosol Science* **32**, no. 6 (2001): 749-764.

2-50. Harris, S. and Kennedy, I., "The Coagulation of Soot Particles with van der Waals Forces," *Combustion Science and Technology* **59** (1988): 443-454.

2-51. Bukowiecki, N., Kittelson, D.B., Watts, W.F., Burtscher, H., Weingartner, E., and Baltensperger, U., "Real-Time Characterization of Ultrafine and Accumulation Mode Particles in Ambient Combustion Aerosols," *Journal of Aerosol Science* **33**, no. 8 (2002): 1139-1154.

2-52. Kittelson, D.B., McMurry, P., Park, K., Sakurai, H., Tobias, H., and Ziemann, P., "Chemical and Physical Characteristics of Diesel Aerosol," *6th International ETH Conference on Nanoparticle Measurement*, Zurich, 2002.

2-53.   Kittelson, D.B., "Engines and Nanoparticles: A Review," *Journal of Aerosol Science* **29**, no. 5-6 (1998): 575-588.

2-54.   Strzelec, A., *Kinetic Model Development for the Combustion of Particulate Matter from Conventional and Soy Methyl Ester Diesel Fuels* (Madison, Wisconsin: University of Wisconsin-Madison, 2009), 168.

2-55.   Amann, C.A. and Siegla, D.C., "Diesel Particulates-What they are and Why," *Journal of Aerosol Science Technology* **1**, no. 1 (1981): 73-101, https://doi.org/10.1080/02786828208958580.

2-56.   Agarwal, A.K., Dhar, A., Srivastava, D.K., Maurya, R.K., and Singh, A.P., "Effect of Fuel Injection Pressure on Diesel Particulate Size and Number Distribution in a CRDI Single Cylinder Research Engine," *Fuel* **107** (2013): 84-89.

2-57.   Strzelec, A., Storey, J.M.E., Lewis, S.A., Daw, C.S., Foster, D.E., and Rutland, C.J., "Effect of Biodiesel Blending on the Speciation of Soluble Organic Fraction from a Light Duty Diesel Engine," SAE Technical Paper 2010-01-1273, 2010, https://doi.org/10.4271/2010-01-1273.

2-58.   Graboski, M., Mccormick, R., Alleman, T., and Herring, A., "The Effect of Biodiesel Composition on Engine Emissions from a DDC Series 60 Diesel Engine: Final Report; Report 2 in a Series of 6," Report:1-91, 2003.

2-59.   Environmental Protection Agency, "A Comparative Analysis of Biodiesel Impacts on Exhaust Emissions," US Environmental Protection Agency Report No. 420-P-02-001, 2002.

2-60.   Boehman, A., Song, J., and Alam, M., "Impact of Biodiesel Blending on Diesel Soot and the Regeneration of Particulate Filters," *Energy & Fuels* 19 (2005): 1857-1864.

2-61.   Song, J., Alam, M., Boehman, A., and Kim, U., "Examination of the Oxidation Behavior of Biodiesel Soot," *Combustion and Flame* **146**, no. 4 (2006): 589-604.

2-62.   Strzelec, A., Vander Wal, R.L., Lewis, S.A., Toops, T.J., and Daw, C.S., "Nanostructure and Burning Mode of Light-Duty Diesel Particulate with Conventional Diesel, Biodiesel, and Intermediate Blends," *International Journal of Engine Research* **18**, no. 5-6 (2017): 520-531, https://doi.org/10.1177/1468087416674414.

2-63.   Vander Wal, R.L., Strzelec, A., Toops, T.J., Stuart Daw, C., and Genzale, C.L., "Forensics of Soot: C5-Related Nanostructure as a Diagnostic of In-Cylinder Chemistry," *Fuel* **113** (2013): 522-526, https://doi.org/10.1016/j.fuel.2013.05.104.

2-64.   Stasio, S., "Electron Microscopy Evidence of Aggregation Under Three Different Size Scales for Soot Nanoparticles in Flames," *Carbon* **39** (2001).

2-65.   Lahaye, J. and Pardo, G., "Morphology and Internal Strucutre of Soot and Carbon Blacks," in *Particulate Carbon Formation During Combustion*, Siegla, Editor (New York: Plenum Press, 1981), 33-55.

2-66.   Vander Wal, R. and Tomasek, A., "Soot Oxidation Dependence upon Initial Nanostructure,"1 *Combustion and Flame* **134**, no. 1-2 (2003): 1-9.

2-67.   Palmer, H. and Cullis, C., *The Chemistry and Physics of Carbon*, Walker, P. Jr, Editor (New York: Dekker, 1965), 265.

2-68.   Vander Wal, R. and Tomasek, A., "Soot Nanostructure: Dependence upon Synthesis Conditions," *Combustion and Flame* **136**, no. 1-2 (2004): 129-140.

2-69.   Yezerets, A., Currier, N., Epling, W., Kim, D., Peden, C., Muntean, G., Wang, C., Burton, S., and Vander Wal, R., *Towards Fuel Efficient DPF Systems: Understanding the Soot Oxidation Process* (DEER, 2005).

2-70.   Braun, A., Huggins, F., Seifert, S., Ilavsky, J., Shah, N., Kelly, K., Sarofim, A., and Huffman, G., "Size-Range Analysis of Diesel Soot with Ultra-Small Angle X-Ray Scattering," *Combustion and Flame* **137**, no. 1-2 (2004): 63-72.

2-71. Braun, A., Shah, N., Huggins, F., Huffman, G., Wiricj, S., Jacobsen, C., Kelly, K., and Sarofim, A., "A Study of Diesel PM with X-Ray Microscopy," *Fuel* **83**, no. 7 (2004): 997-1000.

2-72. Su, D., Jentoft, R., Müller, J., Rothe, D., Jacob, E., Simpson, C., Tomović, Ž., Müllen, K., Messerer, A., and Pöschl, U., "Microstructure and Oxidation Behaviour of Euro IV Diesel Engine Soot: A Comparative Study with Synthetic Model Soot Substances," *Catalysis Today* **90**, no. 1-2 (2004): 127-132.

2-73. Marsh, H., *Kinetics and Catalysis of Carbon Gassification, in Introduction to Carbon Science* (Butterworths: London, 1989).

2-74. Smith, W.R. and Polley, M.H., "The Oxidation of Graphitized Carbon Black," *Journal of Electrochemistry* **20** (1956): 689-691.

2-75. Rosner, D. and Allendorf, H., "Comparitive Studies of the Attack of Pyrolytic and Isotropic Graphite by Atomic and Molecular Oxygen at High Temperature," *AIAA Journal* **6**, no. 4 (1968): 650-654.

2-76. Hippo, E., Murdie, N., and Hyjazie, A., "The Role of Active Sites in the Inhibition of Gas-Carbon Reactions," *Carbon* **27**, no. 5 (1989): 689-695.

2-77. Vander Wal, R., Tomasek, A., Berger, G., Street, K., Hull, D., and Thompson, W., "Soot Nanostructure: Physical and Chemical Characterization," *ORNL Fuels, Engines and Emissions Research Group Presentation*, 2007.

2-78. Vander Wal, R. and Mueller, C., "Initial Investigation of Effects of Fuel Oxygenation on Nanostructure of Soot from a Direct-Injection Diesel Engine," *Energy Fuels* **20**, no. 6 (2006): 2364-2369.

2-79. Strzelec, A., Toops, T.J., and Daw, C.S., "Oxygen Reactivity of Devolatilized Diesel Engine Particulates from Conventional and Biodiesel Fuels," *Energy & Fuels* **27**, no. 7 (2013): 3944-3951.

2-80. McCormick, R.L., Alleman, T., Barnitt, R., Clark, W., Hayes, B., Ireland, J., Proc, K., Ratcliff, M., Thornton, M., Whitacre, S., and Williams, A., "Biodiesel R&D at NREL," Report No. NREL/PR-540-39538, 2006.

2-81. McCormick, R.L., Williams, A., Ireland, J., Brimhall, M., and Hayes, R.R., "Effects of Biodiesel Blends on Vehicle Emissions: Fiscal Year 2006 Annual Operating Plan Milestone 10.4," Report No. NREL/MP-540-40554, 2006.

2-82. Higgins, K., Jung, H., Kittelson, D., Roberts, J., and Zachariah, M., "Size-Selected Nanoparticle Chemistry: Kinetics of Soot Oxidation," *Journal of Physical Chemistry A* **106**, no. 1 (2002): 96-103.

2-83. Sluder, C., Wagner, R., Lewis, S., and Storey, J., "Fuel Property Effects on Emissions from High Efficiency Clean Combustion in a Diesel Engine," SAE Technical Paper 2006-01-0080, 2006, https://doi.org/10.4271/2006-01-0080.

2-84. Sluder, C., Wagner, R., Lewis, S., and Storey, J., "Exhaust Chemistry of Low-NOX, Low-PM Diesel Combustion," SAE Technical Paper 2004-01-0114, 2004, https://doi.org/10.4271/2004-01-0114.

2-85. Sluder, S., Wagner, R., Storey, J., and Lewis, S., "Implications of Particulate and Precursor Compounds Formed During High-Efficiency Clean Combustion in a Diesel Engine," SAE Technical Paper 2005-01-3844, 2005, https://doi.org/10.4271/2005-01-3844.

2-86. Lewis, S., Storey, J.M.E., Bunting, B., and Szybist, J., "Partial Oxidation Products and Other Hydrocarbon Species in Diesel HCCI Exhaust," SAE Technical Paper 2005-01-3737, 2005, https://doi.org/10.4271/2005-01-3737.

2-87. Kolodziej, C., "Comprehensive Characterization of Particulate Emissions from Advanced Diesel Combustion," SAE Technical Paper 2017-01-1945, 2007, https://doi.org/10.4271/2017-01-1945.

2-88.   Maricq, M., "How are Emissions of Nuclei Mode Particles Affected by Emission Control," in *HEI Conference*, May 2009.

2-89.   Chen, L., Liang, Z., Zhang, X., and Shuai, S., "Characterizing Particulate Matter Emissions from GDI and PFI Vehicles under Transient and Cold Start Conditions," *Fuel* **189** (2017): 131-140.

2-90.   Saito, C., Nakatani, T., Miyairi, Y., Yuuki, K., Makino, M., Kurachi, H., Heuss, W., Kuki, T., Furuta, Y., and Kattouah, P., "New Particulate Filter Concept to Reduce Particle Number Emissions," SAE Technical Paper 2011-01-0814, 2011, https://doi.org/10.4271/2011-01-0814.

2-91.   Johnson, T. and Joshi, A., "Review of Vehicle Engine Efficiency and Emissions," *SAE Int. J. Engines* **11**, no. 6 (2018): 1307-1330, https://doi.org/10.4271/2018-01-0329.

2-92.   Richter, J.M., Klingmann, R., Spiess, S., and Wong, K.-F., "Application of Catalyzed Gasoline Particulate Filters to GDI Vehicles," *International Journal of Engines* **5**, no. 3 (2012): 1361-1370.

2-93.   Dalla Nora, M., Lanzanova, T.D.M., and Zhao, H., "Effects of Valve Timing, Valve Lift and Exhaust Backpressure on Performance and Gas Exchanging of a Two-Stroke GDI Engine with Overhead Valves," *Energy Conversion and Management* **123** (2016): 71-83.

2-94.   Mayer, A., *Particle Filter Retrofit for All Diesel Engines* (Renningen, Germany: Expert Verlag, 2008), 97.

2-95.   Taylor, A.M., "Science Review of Internal Combustion Engines," *Energy Policy* 36, no. 12 (2008): 4657-4667.

2-96.   Storey, J.M., Barone, T., Norman, K., and Lewis, S., "Ethanol Blend Effects on Direct Injection Spark-Ignition Gasoline Vehicle Particulate Matter Emissions," *SAE Int. J. Fuels Lubr.* **3**, no. 2 (2010): 650-659, https://doi.org/10.4271/2010-01-2129.

2-97.   "Diesel Fuel Sulfur Effects on Particulate Matter Emissions," National Renewable Energy Laboratory, 1999.

2-98.   Manufacturers of Emissions Controls Association (MECA), "Diesel Particulate Filter Maintenance: Current Practices and Experience," 2005.

2-99.   Givens, W.A., Buck, W.H., Jackson, A., Klador, A., Hertzberg, A., Moehrmann, W., Mueller-Lunz, S., Pelz, N., and Wenniger, G., "Lube Formulation Effects on Transfer of Elements to Exhaust After-Treatment System Components," SAE Technical Paper 2003-01-3109, 2003, https://doi.org/10.4271/2003-01-3109.

2-100. Bardasz, E., Cowling, S., Panesar, A., Durham, J., and Tadrous, T., "Effects of Lubricant Derived Chemistries on Performance of the Catalyzed Diesel Particulate Filters," SAE Technical Paper 2005-01-2168, 2005, https://doi.org/10.4271/2005-01-2168.

2-101. Sappok, A.G., "The Nature of Lubricant-Derived Ash-Related Emissions and Their Impact on Diesel Aftertreatment System Performance," *Mechanical Engineering*, Massachusetts Institute of Technology, 2009, 306.

2-102. Watson, S.A.G., "Lubricant-Derived Ash: In-Engine Sources and Opportunities for Reduction," *Mechanical Engineering*, Massachusetts Institute of Technology, 2010, 235.

2-103. ASTM, "D 874-06: Standard Test Method for Sulfated Ash from Lubrication Oils and Additives."

2-104. Bodek, K.M. and Wong, V.V., "The Effects of Sulfated Ash, Phosphorus and Sulfur on Diesel Aftertreatment Systems-A Review," SAE Technical Paper 2007-01-1922, 2007, https://doi.org/10.4271/2007-01-1922.

2-105. Haycock, R., Caines, A., and Hiller, J. *Automotive Lubricants Reference Book* (Society of Automotive Engineers, 1996).

2-106. Totten, G., *Handbook of Lubrication and Tribology: Applications and Maintenance* (Taylor & Francis, 2006).

2-107. Wagner, T. and Wyszyński, M.L., "Aldehydes and Ketones in Engine Exhaust Emissions-A Review," *Proceedings of the Institution of Mechanical Engineers, Part D: Journal of Automobile Engineering* **210**, no. 2 (1996): 109-122.

2-108. Dietzmann, H.E. and Lee, T.P., "Emissions Characterization of Diesel Forklift Engines," SAE Technical Paper 841396, 1984, https://doi.org/10.4271/841396.

2-109. Li, L., Ge, Y., Wang, M., Peng, Z., Song, Y., Zhang, L., and Yuan, W., "Exhaust and Evaporative Emissions from Motorcycles Fueled with Ethanol Gasoline Blends," *Science of the Total Environment* **502** (2015): 627-631.

2-110. Grosjean, D., Grosjean, E., and Gertler, A.W., "On-Road Emissions of Carbonyls from Light-Duty and Heavy-Duty Vehicles," *Environmental Science & Technology* **35**, no. 1 (2001): 45-53.

2-111. Correa, S.M. and Arbilla, G., "Carbonyl Emissions in Diesel and Biodiesel Exhaust," *Atmospheric Environment* **42**, no. 4 (2008): 769-775.

2-112. Reuter, R.M., Benson, J.D., Burns, V., Gorse Jr., R.A., Hochhauser, A.M., Koehl, W.J., Painter, L.J., Rippos, B.H., and Ruterford, J.A., "Effects of Oxygenated Fuels and RVP on Automotive Emissions - Auto/Oil Air Quality Improvement Program," SAE Technical Paper 920326, 1992, https://doi.org/10.4271/920326.

2-113. Zervas, E., Montagne, X., and Lahaye, J., "Emission of Alcohols and Carbonyl Compounds from a Spark Ignition Engine: Influence of Fuel and Air/Fuel Equivalence Ratio," *Environmental Science & Technology* **36**, no. 11 (2002): 2414-2421.

2-114. Foster, G.L., Royer, D.L., and Lunt, D.J., "Future Climate Forcing Potentially without Precedent in the Last 420 Million Years," *Nature Communications* **8** (2017): 14845.

2-115. Eggleton, T., *A Short Introduction to Climate Change* (Cambridge University Press, 2013).

2-116. Dlugokencky, E., *Annual Mean Carbon Dioxide Data* (Earth System Research Laboratory, 2016).

2-117. Petty, G.W., A First Course in Atmospheric Radiation," *EOS Transactions* **85**, no. 36 (2004): 229-251, https://doi.org/10.1029/2004EO360007.

CHAPTER 3

Emissions Control Regulations

Any society has to delegate the responsibility to maintain a certain kind of order. Enforcing regulations, making sure people stop at stoplights. We can't function as a society without rules and regulations, and the enforcement mechanism of those rules and regulations. It sounds abstract, but it's central to how any advanced society or economy operates.

—Joseph Stiglitz [3-1]

Government regulation has historically been the primary motivation for emissions control technology improvements in global markets including internal combustion engines (ICEs) and vehicles. In the United Kingdom (UK), air quality regulations date back to the proclamation by King Edward I in 1307 that coal should not be burned in the City of London or nearby when his consort Queen Margaret was resident in the Tower of London [3-2]. Starting in the late nineteenth century, general air quality regulations, such as the 1875 Public Health Act in the UK were implemented in reaction to smoke emissions from power plants and other industrial facilities. The UK introduced Clean Air Acts (CAA) in 1956 and 1968 to manage industrial pollutants, and by the 1970s these expanded to cover mobile sources. The European Union (EU) and its member states also implemented rules to improve air quality starting in the early 1970s [3-3]. Similar legislative trends occurred in other countries. For example, within the United States (US), the State of California implemented the California Air Pollution Control Act in 1947 and established the California Motor Vehicle Pollution Control Board in 1960 and the California Air Resources Board (ARB or CARB) in 1967. At the US Federal level, Congress passed the 1963 and 1970 CAA and their subsequent amendments in 1977 and 1990 [3-4], and the US Environmental Protection Agency (EPA) was founded in 1970. The CAA is a comprehensive law regulating emissions from stationary and mobile sources of air pollution and authorized EPA to establish National

Ambient Air Quality Standards (NAAQS) to regulate the emissions of specific hazardous pollutants in the interest of human health and welfare, as defined in Title 40, Part 50 of the *Code of Federal Regulations* (40 CFR part 50) [3-5, 3-6].

Since California had enacted its laws before the Federal CAA came into force, California has maintained its ability to set standards that enhance the Federal requirements. Other states may also choose to adopt the California standards.

The regulatory emphasis is concerned with improving ambient air quality by reducing criteria pollutants from the atmosphere, especially locally. Criteria pollutant regulations are designed to minimize the release of compounds known or suspected to be harmful to human health, usually acutely so. These compounds are described further in Sections 1.3 and 3.1. Recent regulations also focus on reducing greenhouse gas (GHG) emissions.

The global focus on fuel economy started in the 1970s with the sudden volatility in oil prices caused by the cartel behavior of oil-producing countries to restrict supplies. For example, in the US the Corporate Average Fuel Economy (CAFE) standards for light-duty vehicles (LDVs) were promulgated in 1975 and were implemented in 1978. In more recent years, the emphasis for LDVs has shifted from fuel economy only to a more comprehensive approach that includes greenhouse gas emissions from vehicles and engines. Heavy-duty (HD) applications have had a similar trend where the long-standing market interest in reduced fuel consumption has been reinforced by regulations targeting lower GHG emissions. The rules for GHG emissions primarily target carbon dioxide (CO_2), although tailpipe GHG emissions rules may include methane (CH_4) or nitrous oxide (N_2O). Other GHG compounds of interest include those used as refrigerants, especially hydrofluorocarbons (HFCs).

The purpose of this chapter is to provide the reader with an overview of the key regulatory requirements, especially the emissions targets for each application, based on their current state as of writing (August 2019). The most definitive source for regulatory requirements is always the regulations themselves, although they can be difficult reading for the beginner. Fortunately, several useful summaries of the rules and requirements are updated regularly and made available, such as those prepared by AVL [3-7], Delphi Technologies [3-8, 3-9], DieselNet [3-10], and Ricardo [3-11]. These summaries are updated as the regulations change and are a good starting point to understand what may have changed since publication. Evaporative emissions rules are not within the scope of this book, and so are omitted.

This chapter describes both the criteria pollutant regulations that have motivated exhaust emissions control system development over the last several decades in both light-duty and HD applications and the newer and coordinated fuel economy and GHG emissions rules. The rules are discussed by application and emissions type, and then by region. Thus, ambient air quality requirements are discussed in more detail in Sections 3.1, 3.2, and 3.3 describe the criteria pollutant emissions requirements for light-duty and HD applications, respectively, at the vehicle or engine level. Sections 3.4 and 3.5 discuss fuel economy and GHG emissions requirements at the vehicle or engine level for light-duty and HD applications, respectively. Each section includes a discussion of regulations from multiple countries or regions covering a majority of the global economy.

TABLE 3.1 Light-duty and heavy-duty vehicle Class based on gross vehicle weight rating (GVWR), US.

Vehicle Class	1	2a	2b	3	4	5	6	7	8
Minimum GVWR (lb.)	0	6,000	8,500	10,000	14,000	16,000	19,500	26,000	33,000
Minimum GVWR (kg)	0	2,722	3,855	4,536	6,350	7,257	8,845	11,793	14,969
Vehicle Type	LDV		Light HDV				Medium HDV		Heavy HDV
Emissions Certification Type	Light-duty vehicle chassis certification		MDV chassis certification for homologated vehicle Heavy-duty certification for the engine only		Heavy-duty certification for engine				

In general, an engine or vehicle must have the appropriate regulatory body approve the product for sale after the manufacturer has demonstrated that it complies with the applicable emissions rules. LDVs and medium-duty vehicles (MDVs) are typically homologated and certified as complete vehicles including the engine. Here, "homologation" determines the vehicle configuration for evaluation to get regulatory approval for sale. MDVs are defined as light heavy-duty vehicles (LHDVs) with a gross vehicle weight rating (GVWR) of 8,500 lb (3,856 kg) to 14,000 lb (6,350 kg) that are sold as complete vehicles and certified on a chassis dynamometer. By contrast, HD engines are certified separately from their final on-road or non-road applications. A full list of vehicle classes for the US concerning GVWR—including LDVs, MDVs, and heavy-duty vehicles (HDVs)—is shown in Table 3.1, along with the most common regulatory approaches for those classes. In the EU, the three relevant commercial vehicle classes for regulatory purposes are N_1, with a GVWR under 3,500 kg; N_2, with a GVWR between 3,500 and 12,000 kg; and N_3, with a GVWR over 12,000 kg. In all cases, the manufacturer needs to work closely with its suppliers to ensure that the product will comply out to its full useful life (FUL) as defined by the applicable regulations.

3.1 Ambient Air Quality

The goal of legislation and regulations for cleaner air is to improve ambient air quality, especially for criteria pollutants. These criteria pollutants usually present an acute risk to human health but some may also increase health risks after chronic exposure. In the US, for example, the CAA requires EPA to set NAAQS which are defined in 40 CFR part 50 [3-5]. EPA has defined NAAQS for the following chemical species from 40 CFR §§50.4–50.13, which are described in more detail in Section 1.3:

- Sulfur oxides (SOx), especially sulfur dioxide (SO_2)
- Particulates with an aerodynamic diameter less than or equal to a nominal 10 μm and greater than 2.5 μm (PM_{10})
- Particulates with an aerodynamic diameter less than or equal to a nominal 2.5 μm ($PM_{2.5}$)
- Carbon monoxide (CO)
- Ozone (O_3)
- Nitrogen oxides (NOx), especially nitrogen dioxide (NO_2)
- Lead (Pb)

Several of these criteria pollutants are generated by combustion in ICEs. PM from combustion sources, which includes soot, has been identified by the World Health Organization (WHO) as a Class 1 carcinogen [3-12]. In addition, ultrafine particles, particularly those below 2.5 nm ($PM_{2.5}$) have been shown to have harmful effects on the cardiovascular system, including causing heart attacks and strokes [3-13, 3-14, 3-15], because these small particles can cross through the lung membranes into the blood. CO is a colorless, odorless, and tasteless gas that displaces the oxygen dissolved in the blood with detrimental health effects. It is toxic in concentrations above 35 ppm in hemoglobic species, such as humans and other animals, both vertebrate and invertebrate [3-16, 3-17, 3-18]. The NOx species, which include nitrogen monoxide (NO) and NO_2 can inflame respiratory tissues and aggravate diseases such as asthma. In particular, NO_2 is a highly active oxidizer that can damage mucus membranes of the eyes, nose, and lungs [3-19, 3-20].

Section 112 of the CAA specifically addresses the emissions of hazardous airborne pollutants, which the 2007 Supreme Court decision in *Massachusetts v. EPA* affirmed included GHG. Before the 1990 amendment, there had been a risk-based program, under which few standards were developed. Section 112 was revised to require technology-based standards to be issued. On-road or highway vehicles, as well as other non-road mobile sources, represent a significant fraction of the national emissions profile. As shown in **Figure 3.1**, vehicles are strong contributors to CO, NOx, volatile organic carbon (VOCs), and PM emissions.

Numerous scientific studies have linked the pollutant species identified in Figure 3.1 to health effects in people. In particular, some populations are at greater risk, including children, older adults, and those that are suffering from cardio or pulmonary diseases.

FIGURE 3.1   National emissions by source category [3-21].

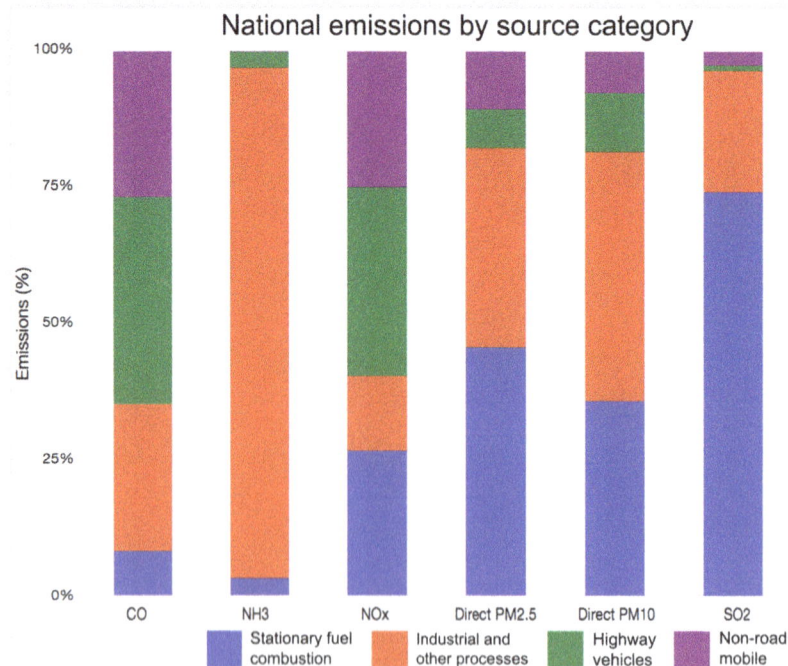

Reprinted from Ref. [3-21]. Environmental Protection Agency

These same pollutant species have been shown to have serious environmental conse-quences. For example, CO emissions contribute to the formation of CO_2, NO_2, and ozone (O_3). These species are considered GHG that have a warming effect on the atmosphere, as shown in **Figure 3.2**. NOx and sulfur oxides (SOx) form acids in the atmosphere and contribute to the acidification of soils and surface water. NOx species can react with VOCs to form O_3 or react with ammonia (NH_3) and other species to form particulates. SOx species can affect plants in both aquatic and terrestrial ecosystems, as well as form sulfate particles. PM emissions are related to visibility-reducing haze, staining deposits on stone and other surfaces, and increased thermal absorption by surfaces they coat. Additionally, when NOx and HC species co-mingle in the environment, in the presence of UV light, there is strong potential for the formation of photochemical smog, the brown haze associated with air pollution in major cities [3-22, 3-23].

It is for these health and environmental concerns that regulatory bodies have promulgated rules to control and remediate exhaust emissions species, as discussed in this chapter. It is the characterization, regulation, control, and remediation of these species that are the focus of this text.

The air quality standards set by EPA and CARB have significantly improved air quality since they first started being implemented in the 1970s, both by the US Federal government and through state implementation plans (SIP) that apply to the particular industrial sources within each state. Likewise, Gemmer and Bo note that in the EU air

FIGURE 3.2   US GHG inventory by type [3-24].

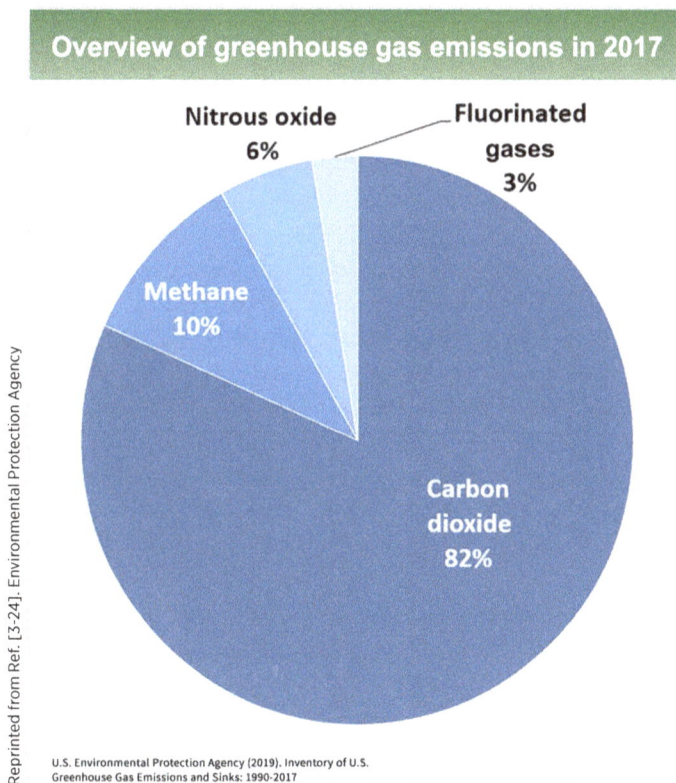

Reprinted from Ref. [3-24]. Environmental Protection Agency

U.S. Environmental Protection Agency (2019). Inventory of U.S. Greenhouse Gas Emissions and Sinks: 1990-2017

quality has significantly improved since the 1990s, with reductions not just with air pollutants such as "sulfur dioxide (SO_2), lead (Pb), nitrogen oxides (NOx), carbon monoxide (CO) and benzene (C_6H_6)," but also contributors to smog, such as PM_{10}, $PM_{2.5}$, and O_3.

A significant amount of research in recent years has indicated that smaller sized particles have a greater potential for adverse health effects [3-25, 3-26, 3-27, 3-28, 3-29, 3-30, 3-31]. Epidemiological studies have shown that inhalation of particles under 100 nm can cause cancer and increase the incidences of cardiovascular and respiratory diseases, which can result in challenged breathing, irregular heartbeat and even premature death [3-29]. PM may contribute to an estimated 800,000 annual fatalities [3-25]. Particles with diameters below 100 nm have been shown to have greater toxicity within the human body despite no difference in chemical composition from larger particles. As a result, particle measurements have increasingly focused on particle number (PN) concentration and size, particularly in the EU As discussed in Sections 3.2.3 and 3.3.3, the EU has set limits for PN emissions for particulates greater than 23 nm in diameter [3-32]. This size cutoff of 23 nm was chosen because of the limitations of the measurement systems.

The challenge for regulators is to translate the ambient air quality standards into requirements for mobile and stationary sources, such as vehicles and engines. The compounds covered by the engine and vehicle-level regulations typically include NOx, CO, unburned hydrocarbons (HC), and particulates. Methane is usually excluded from criteria pollutant limits for HC because it is less reactive than heavier HC species and does not contribute to ozone formation, so the regulatory limits usually refer to non-methane hydrocarbons (NMHC) or non-methane organic gases (NMOG).[*] Particulates have been regulated on a particulate mass (PM) basis; however, the EU is also regulating particulates on a PN basis, since the measurement of PN at modern regulatory limits is more robust than the measurement of PM.

One of the challenges with implementing regulations on mobile sources is the aggregated effect can be significant and lead to non-attainment of ambient air quality standards, even if each source complies with its requirements. A second challenge is that mobile source standards generally only apply to new engines or vehicles when sold, whereas the ambient air quality in a given region is influenced by the whole set of engines and vehicles used within that region, also known as the parc, as well as other sources such as power plants, trains, ships, and small engines. For example, in the US the average LDV age is about 12 years, which suggests that a long tail of older LDVs is in the parc since the minimum age is 0 years. Therefore, ambient air quality will be strongly influenced by older LDVs that were approved for sale under earlier, less stringent emissions standards or that may even be beyond the FUL of the product. Similarly, some applications, such as specialty vehicles working at ports, operate over a very long functional life, meaning that engines that are decades old may still be in service. This aggregation of newer and older engines and vehicles contributes to the persistent and continuing issues with air quality in such areas as metropolitan Houston, Texas; California's South Coast Air Quality Management District (SCAQMD); and several core urban areas in the UK.

---

* We have elected to use "NMHC" throughout this book, but recognize that there may be differences in the definitions of NMHC and NMOG in various jurisdictions that can affect approval for sale.

3.2 **Light-Duty Criteria Pollutants**

Criteria pollutant regulations for mobile sources started with LDVs and MDVs. This section describes the regulations for criteria pollutant emissions from LDVs and MDVs in various regions. In general, LDV and MDV criteria pollutants are regulated at the level of a homologated, complete vehicle or vehicle family. The testing and sale of MDVs as a complete homologated vehicle distinguishes them from other LHDVs. Vehicles are tested on a chassis dynamometer over prescribed drive cycles that have vehicle speed set points as a function of time. In the chassis dynamometer facility, the test conditions are controlled and emissions are measured using bench-mounted sensors. In recent years, some jurisdictions have implemented Real Driving Emissions (RDE) tests or similar evaluations on the road to evaluate off-cycle performance. These RDE tests use portable emissions measurement systems (PEMS) mounted to the vehicle to measure tailpipe emissions while the vehicle is driving on the roads. PEMS are described further in Chapter 5, Emissions Measurements. In this section, details of the regulations are reviewed by region, using the current state of the rules as of writing.

3.2.1 **United States (Federal)**

EPA's regulations for LDV criteria pollutants are published in 40 CFR part 86 [3-33]. In its rule-making process, EPA assumes that LDVs and MDVs will comply with the emissions limits at all reasonably accessible parts of the engine operating map during vehicle operation. Thus, the CAA has explicitly prohibited defeat devices since 1972 in 40 CFR §86.1809-01 [3-34], §86.1809-10 [3-35], and §86.1809-12 [3-36]. The LDV or MDV manufacturer "must show to the satisfaction of the Administrator that the vehicle design does not incorporate strategies that unnecessarily reduce emission control effectiveness … when the vehicle is operated under conditions that may reasonably be expected to be encountered in normal operation and use." [3-36] Violations of this prohibition may be penalized with a fine of up to $37,000 per vehicle sold or criminal prosecution or both.

The only exceptions to this prohibition are for declared and approved Auxiliary Emissions Control Devices (AECD), as defined in 40 CFR §86.1803-01 [3-37]. AECD are any elements of design, including software, in which an engine or vehicle parameter is measured and used to alter the operation of any part of the emissions control system, including the exhaust aftertreatment system (EAS). These engine or vehicle parameters include, but are not limited to, temperature, vehicle speed, steering position, and time since key on. The allowed exceptions are as follows:

1. The conditions are substantially included in the applicable test cycles and regulations
2. The need for the AECD to protect the vehicle against damage or accident is justified
3. The AECD does not go beyond the requirements of engine starting

Any AECD must be limited to the minimum intervention necessary. Before an AECD is implemented in a vehicle for sale, it must be thoroughly documented and

TABLE 3.2 Tier 3 certification standards by bin [3-38].

Bin	Limits CO (mg/mi.)	NMHC + NOx (mg/mi.)	HCHO (mg/mi.)	PM (mg/mi.)
Bin 160	4.2	160	4	3
Bin 125	2.1	125	4	3
Bin 70	1.7	70	4	3
Bin 50	1.7	50	4	3
Bin 30	1.0	30	4	3
Bin 20	1.0	20	4	3
Bin 0	0.0	0	0	0

* In MY2017–2020, the PM standard applies only to that segment of a manufacturer's vehicles covered by the percent of sales phase-in for that MY.

justified by the manufacturer, and this documentation must be approved by EPA [3-33]. EPA reserves the right to road test vehicles with PEMS to assess the presence and performance of AECD. An example of a commonly approved AECD is fuel enrichment at high engine load to lower the combustion and exhaust gas temperatures and thereby protect components even though it leads to excess NMHC emissions.

EPA has been setting LDV exhaust emissions limits since the 1970s, with the current set of standards being called Tier 3. The Tier 3 requirements defined in 40 CFR §86.1811-17 [3-38] divide vehicles into various bins based on their emissions performance, and each bin is named after its combined NMHC + NOx emissions limit as evaluated on the applicable drive cycle. For example, Tier 3 bin 50 has a combined NMHC + NOx limit of 50 mg/mi.

The EPA Tier 3 limits shown in Table 3.2 are for an LDV at FUL, which is 15 years or 150,000 mi., whichever comes first [3-38]. Vehicles have their emissions performance evaluated using the FTP-75 cycle, which is described in Section 4.1.1.

In addition to testing vehicles on FTP-75, vehicles are also evaluated on the Supplemental Federal Test Protocol (SFTP) which includes criteria pollutant emissions results measured on the US06 and SC03 drive cycles. These drive cycles are described in Sections 4.1.3 and 4.1.4, respectively. The composite SFTP criteria pollutant result is the following weighted average of the results from the three drive cycles:

$$SFTP = 0.35 \times FTP - 75 + 0.28 \times US06 + 0.37 \times SC03. \quad (3.1)$$

Also, by MY2025, all LDVs must not exceed 6 mg/mi. PM on US06. In addition to the categorization for each vehicle, each manufacturer must meet a fleet average rating of NMHC + NOx that ramps down from 72 mg/mi. for passenger cars and 83 mg/mi. for light trucks in model year (MY) 2019 to 30 mg/mi. for all vehicles by MY2025. Therefore, from MY2025 onward the fleet average rating will be the equivalent of Tier 3, bin 30. This sales-weighted fleet average assumes that some vehicles will emit pollutants at a higher Tier 3 bin as long as their emissions are offset by others that emit at the lower levels of Tier 3, bin 20, or Tier 3, bin 0.

The limits for MDVs are presented with the other US (Federal) regulations for on-road HD applications in Section 3.3.1.1.

3.2.2 California

Criteria pollutant emissions are regulated in California by CARB under its California Clear Air Act authority in parallel with the Federal US regulations described in Section 3.2.1. Like EPA, CARB assumes that a vehicle being certified for sale complies with its regulatory requirements at all reasonable operating conditions. Likewise, any necessary AECDs must be documented and justified by the manufacturer and then approved by CARB before the vehicle can be sold.

CARB has set a series of Low Emissions Vehicle (LEV) standards to regulate criteria pollutants from LDVs and MDVs, with the current standards being LEV III [3-39]. These LEV III standards, shown in **Table 3.3**, apply at the vehicle's FUL of 150,000 mi. (241,400 km) or 10 years, whichever comes first. The LEV III standards cover both LDVs and MDVs, as both are sold as complete vehicles and certified on a chassis dynamometer. By MY2025, the sales-weighted fleet average for LDVs must be 30 mg/mi. for NMHC + NOx, which is equivalent to the SULEV30 level.

TABLE 3.3 LEV III criteria pollutant standards by bin [3-39].

Vehicle Type	Vehicle Emission Category	Limits				
		CO (mg/mi.)	NMHC + NOx (mg/mi.)	HCHO (mg/mi.)	PM* (mg/mi.)	PM* (mg/mi.) MY2028
All LDV with	LEV160	4.2	160	4	3	1
GVWR ≤ 8,500 lb.	ULEV125	2.1	125	4	3	1
(3,856 kg)	ULEV70	1.7	70	4	3	1
	ULEV50	1.7	50	4	3	1
	SULEV30	1.0	30	4	3	1
	SULEV20	1.0	20	4	3	1
MDV with	LEV395†	6.4	395	6	120	—
GVWR ≥ 8,501 lb	ULEV340†	6.4	340	6	60	—
(3,856 kg) and	ULEV250	6.4	250	6	60	8
GVWR ≤ 10,000 lb	ULEV200	4.2	200	6	60	8
(4,536 kg)	SULEV170	4.2	170	6	60	8
	SULEV150	3.2	150	6	60	8
MDV with	LEV630†	7.3	630	6	120	—
GVWR ≥ 10,001 lb	ULEV570†	7.3	570	6	60	—
(4,536 kg)	ULEV400	7.3	400	6	60	10
and	ULEV270	4.2	270	6	60	10
GVWR ≤ 14,000 lb	SULEV230	4.2	230	6	60	10
(6,350 kg)	SULEV200	3.7	200	6	60	10

* In MY2024–2028, the PM standards for each vehicle category transition from the initial PM limit shown to the MY2028 PM limit.

† Category is discontinued after MY2021.

3.2.3 European Union

The EU implemented its Euro 6 standard for LDVs in 2007 [3-40], and has extended the standard to Euro 6(d) as the EU has transitioned vehicle testing from the New European Drive Cycle (NEDC) to the World-Harmonized Light-Duty Transient Cycle (WLTC). These drive cycles are described in Sections 4.1.6 and 4.1.8, respectively. The PN limit of 6×10^{11} #/km in the Euro 6 standards applies not just to compression ignition (CI) engines but also to spark ignition (SI) engines with direct injection (DI) fuel systems [3-41]. The Euro 6 standards using the WLTC are in transition as of publication. The Euro 6(b) and 6(c) emissions standards using the Euro 6-1 on-board diagnostics (OBD) standard were phased out on 31 August 2018, and as of 1 September 2019, the Euro 6(c) emissions standard using the Euro 6-2 OBD standard applies to all new vehicles. The Euro 6(d) standards with Euro 6-2 OBD standard started phasing in as of 1 September 2018 for new vehicle type approvals, and all new vehicles sold must comply by 1 January 2022.

In general, the Not-to-Exceed (NTE) limits are defined by multiplying the on-cycle Euro 6 pollutant limit for a given criteria pollutant shown in **Table 3.4** by a conformity factor (CF). The limit modified by the CF is then used during actual use, including the RDE test protocols described in Section 4.1.9. The Euro 6 CF for RDE are $1+\varepsilon_{NOx}$ for NOx and $1+\varepsilon_{PN}$ for PN, where ε_{NOx} and ε_{PN} are margin parameters that account for the measurement uncertainties from the PEMS equipment for NOx and PN, respectively. As of publication, the values of both margin parameters are set to 0.5. CO emissions need to be measured and reported during RDE tests but there is neither an NTE limit nor a CF for CO.

3.2.4 Japan

Japan has long had criteria pollutant standards in place for LDVs, including for passenger cars and light commercial vehicles (LCVs). As of 1 October 2019, all LDV new type approvals need to comply with the regulatory limits of the Post Post New Long-Term Regulation, shown in **Table 3.5**. LDVs in continuing production have through 1 October 2021 to comply with these limits. LDVs are evaluated against the limits in Table 3.5 using

TABLE 3.4 Euro 6 Type 1 criteria pollutant emission standards.

Vehicle Category	Level	Reference Mass (RM) (kg)	CO (mg/km) SI	CO (mg/km) CI	NMHC (mg/km) SI	NMHC (mg/km) CI	THC (mg/km) SI	THC (mg/km) CI	NOx (mg/km) SI	NOx (mg/km) CI	THC+NOx (mg/km) SI	THC+NOx (mg/km) CI	PM (mg/km) SI*	PM (mg/km) CI	PN (#/km) SI*,†	PN (#/km) CI
M	—	All	1,000	500	68	—	100	—	60	80	—	170	4.5		6.0×10^{11}	
N1	I	RM ≤ 1,305	1,000	500	68	—	100	—	60	80	—	170	4.5		6.0×10^{11}	
	II	1,305 < RM ≤ 1760	1,810	630	90	—	130	—	75	105	—	195	4.5		6.0×10^{11}	
	III	1,760 < RM	2,270	740	108	—	160	—	82	125	—	215	4.5		6.0×10^{11}	
N2	—	All	2,270	740	108	—	160	—	82	125	—	215	4.5		6.0×10^{11}	

* The PM and PN limits for SI engines only apply when there is a direct injection fuel system.

† SI engines have a transitional limit of 6×10^{12}/km.

TABLE 3.5 Japanese Post Post New Long-Term criteria pollutant emission standards [3-42].

Motor Vehicle Category	CO (g/km) SI	CI	NMHC (g/km) SI	CI	NOx (g/km) SI	CI	PM (g/km) SI*	CI
1. Ordinary-sized, small-sized, or mini-sized motor vehicles for carrying up to 10 passengers.	1.15	0.63	0.10	0.024	0.05	0.15	0.005	0.005
2. Ordinary-sized or small-sized motor vehicles with GVW ≤ 1,700 kg not in category 1.	1.15	0.63	0.10	0.024	0.05	0.15	0.005	0.005
3. Ordinary-sized or small-sized motor vehicles with GVW ≤ 3,500 kg not in category 1 or 2.	2.55	0.63	0.15	0.024	0.07	0.24	0.007	0.007
4. Mini-sized motor vehicles not in category 1.	4.02	—	0.10	—	0.05	—	0.005	—

* The PM limits for SI engines only apply when there is a direct injection fuel system.

the WLTC, which is described further in Section 4.1.8. During the phase-in period, vehicles will use the JC08 combined cycle, which is described further in Section 4.1.7.

In the wake of the Volkswagen emissions cheating revelations in September 2015,[†] Japan also adopted an explicit ban on defeat devices in 2017. This ban includes emissions control strategies that deactivate the EAS after a certain time elapses after the vehicle starts (after "key on") or when the vehicle detects that it is away from known test centers. To confirm that defeat devices are not present on LDVs, Japan has adopted RDE testing as part of the vehicle certification process, and these protocols are described in Section 4.1.9. The Japanese RDE has a CF of 2 for NOx that applies as of 1 October 2022 for new type approvals, and 1 October 2024 for continuing production approvals.

3.2.5 People's Republic of China

Emission legislation in the People's Republic of China (PRC) historically followed those set by the EU, although with delays in implementation dates for urban and rural areas. For criteria pollutants from LDVs, the standard in force as of publication is the China Emission Standard Phase V ("China 5") [3-44], which has applied nationwide to all LDVs since 1 January 2018. The China Emission Standard Phase 6a ("China 6a") [3-45] takes effect as of 1 July 2020, and China 6b will take effect from 1 July 2023. China 5, China 6a, and China 6b standards all apply to LDVs of types M_1, M_2, and N_1 where the vehicle mass does not exceed 3,500 kg. The China 5 standards [3-44], shown in Table 3.6, are designed to use the NEDC, which is described in Section 4.1.6. Note that Vehicle Category I maps to the EU's vehicle category M, and that Vehicle Category II maps to the EU's vehicle category N_1. The introduction of the China 6a standards, shown in Table 3.7, included a shift to using the WLTC for evaluation [3-45]. The WLTC is

[†] In the Volkswagen "Dieselgate" diesel LDV emissions cheating scandal, Volkswagen admitted to implementing a defeat device that used on-board sensors to determine when the vehicle was being tested on a chassis dynamometer. When it was, the engine control unit changed the active engine and EAS calibration to a dynamometer-test calibration that allowed the vehicle to pass the on-cycle criteria pollutant tests. When the vehicle was on the road being driven normally, an on-road calibration was used that improved performance and fuel economy at the expense of increasing in-use NOx emissions by up to 40 times the NTE limit [3-43]. It is expected that other regulatory bodies will explicitly ban defeat devices and emphasize real-world driving conditions for official emissions testing to prevent another manufacturer from attempting something similar.

TABLE 3.6 China 5 criteria pollutant emission standards [3-44].

Vehicle Category	Level	Reference Mass (RM) (kg)	Limits CO (mg/km) SI	CI	NMHC (mg/km) SI	CI	THC (mg/km) SI	CI	NOx (mg/km) SI	CI	THC+NOx (mg/km) SI	CI	PM (mg/km) SI*	CI	PN (#/km) SI	CI
I	—	All	1,000	500	68	—	100	—	60	180	—	230	4.5	4.5	—	6.0 × 10¹¹
II	I	RM ≤ 1305	1,000	500	68	—	100	—	60	180	—	230	4.5	4.5	—	6.0 × 10¹¹
	II	1305 < RM ≤ 1760	1,810	630	90	—	130	—	75	235	—	295	4.5	4.5	—	6.0 × 10¹¹
	III	1760 < RM	2,270	740	108	—	160	—	82	280	—	350	4.5	4.5	—	6.0 × 10¹¹

* The PM limit for SI engines only applies when there is a direct injection fuel system.

TABLE 3.7 China 6a Type I test criteria pollutant emission standards [3-45].

Vehicle Category	Level	Test Mass (TM) (kg)	Limits CO (mg/km)	THC (mg/km)	NMHC (mg/km)	NOx (mg/km)	PM (mg/km)	PN* (#/km)	N_2O (mg/km)
I	—	All	700	100	68	60	4.5	6.0 × 10¹¹	20
II	I	TM ≤ 1,305	700	100	68	60	4.5	6.0 × 10¹¹	20
	II	1,305 < TM ≤ 1,760	880	130	90	75	4.5	6.0 × 10¹¹	25
	III	1,760 < TM	1,000	160	108	82	4.5	6.0 × 10¹¹	30

* Before 1 July 2020 SI engines have a transition limit of 6.0 × 10¹²/km.

TABLE 3.8 China 6b Type I test criteria pollutant emission standards [3-45].

Vehicle Category	Level	Test Mass (TM) (kg)	Limits CO (mg/km)	THC (mg/km)	NMHC (mg/km)	NOx (mg/km)	PM (mg/km)	PN (#/km)	N_2O (mg/km)
I	—	All	500	50	35	35	3.0	6.0 × 10¹¹	20
II	I	TM ≤ 1,305	500	50	35	35	3.0	6.0 × 10¹¹	20
	II	1,305 < TM ≤ 1760	630	65	45	45	3.0	6.0 × 10¹¹	25
	III	1,760 < TM	740	80	55	50	3.0	6.0 × 10¹¹	30

described in Section 4.1.8. The China 6b standards include both the on-cycle limits shown in Table 3.8 and introduce RDE standards. The China 6b RDE standards use a CF of 2.1 for NOx and 2.1 for PN.

In addition to the typical Type I test on WLTC, China 6a and 6b also require Type II or RDE testing and Type VI or cold ambient testing on the first two parts of the WLTC [3-45]. The Type VI test is described in Section 4.1.8. It only monitors CO, THC, and NOx and the limits are shown in Table 3.9.

3.2.6 India

In general, criteria pollutant standards in India have mirrored the EU's standards. For example, the Bharat VI (BS-VI) standards [3-46] use the NEDC with a 90 km/h maximum

TABLE 3.9 China 6b Type VI test criteria pollutant emission standards [3-45].

Vehicle Category	Level	Test Mass (TM) (kg)	Limits		
			CO (g/km)	THC (g/km)	NOx (g/km)
I	—	All	10.0	1.20	0.25
II	I	TM ≤ 1,305	10.0	1.20	0.25
	II	1,305 < TM ≤ 1,760	16.0	1.80	0.50
	III	1,760 < TM	20.0	2.10	0.80

speed and are the same as the Euro 6 standards shown in Table 3.4. They apply to new vehicles built from 1 April 2020. RDE measurements using PEMS will also be required from 1 April 2023. Besides, BS-VI defines OBD threshold values for LDVs with SI or CI engines that are manufactured after 1 April 2020 for BS VI-1 OBD, or 1 April 2023 for BS VI-2 OBD. These OBD threshold values are described further in Section 6.3.2.8.

3.2.7 Brazil

The currently applicable standard for LDV criteria pollutants is PROCONVE L6 [3-47], which uses FTP-75 as its main test cycle. The FTP-75 cycle, also known as the NBR 6601 cycle in Brazil, is described in Section 4.1.1. PROCONVE L6 has applied to all LDVs since 1 January 2015, and the regulatory limits are shown in Table 3.10.

LCVs or LHDVs with an RM over 1700 kg have looser standards for criteria pollutants than those shown in Table 3.10, notably including a 2.0 g/km limit for CO, a 0.25 g/km limit for NOx, and a 0.2 g/km limit for PM.

In late 2018, Brazil approved the PROCONVE L7 and L8 standards [3-48]. The PROCONVE L7 standards will take effect as of 1 January 2022, and the PROCONVE L8 standards, 1 January 2025. The main pollutant limits for PROCONVE L7 are shown in Table 3.11, and as with PROCONVE L6, are measured on FTP-75 (a.k.a. NBR 6601).

TABLE 3.10 Brazil PROCONVE L6 emissions limits for LDVs [3-47].

Fuel	Limits						
	CO (g/km)	Total HC (g/km)	NMHC (g/km)	NOx (g/km)	HCHO (g/km)	PM (g/km)	CO at idle (%)
Gasoline	1.3	—	0.05	0.08	0.02	—	0.2
Ethanol	1.3	—	0.05	0.08	—	—	0.2
Diesel	1.3	—	0.05	0.08	—	0.025	—
CNG	1.3	0.3	0.05	0.08	0.02	—	0.2

TABLE 3.11 Brazil PROCONVE L7 emissions limits for LDVs [3-48].

Category	Limits			
	CO (mg/km)	NMHC + NOx (mg/km)	HCHO (mg/km)	PM (mg/km)
Light passenger	1,000	80	15	6
Light commercial (SI)	1,000	140	—	6
Light commercial (CI)	1,000	320	—	20

Manufacturers must declare the ammonia emissions measured, although there is no specified upper limit. The formaldehyde (HCHO) limit applies only to light passenger vehicles with SI engines; otherwise, the standard covers vehicles with either SI or CI engines.

Likewise, the main pollutant limits for PROCONVE L8 are shown in Table 3.12. These regulatory levels apply to different types of LDVs, both passenger vehicles and LCVs. Levels 0 through 320 apply to LCVs with diesel-fueled CI engines; levels 0 through 140 apply to LCVs with an RM over 1,700 kg and levels 0 through 80 apply to light passenger vehicles and LCVs with an RM up to 1,700 kg. All of the levels shown in Table 3.12 also include a 10 ppm limit on ammonia, except for level 0, which has a 0 mg/km limit.

In addition to the lower vehicular emissions standards, PROCONVE L8 introduces corporate fleet-average emissions limits, shown in Table 3.13, for light passenger vehicles and for LCVs that decrease over time.

TABLE 3.12 Brazil PROCONVE L8 emissions limits for LDVs [3-48].

| Level | Limits | | | |
	CO (mg/mi.)	NMHC + NOx (mg/mi.)	HCHO (mg/mi.)	PM (mg/mi.)
320	1,000	320	—	20
280	1,000	280	—	20
250	1,000	250	—	20
220	1,000	220	—	10
200	1,000	200	—	10
170	1,000	170	—	9
140	1,000	140	15	6
110	1,000	110	15	6
80	1,000	80	15	6
70	600	70	10	4
60	600	60	10	4
50	600	50	10	4
40	500	40	10	4
30	500	30	8	3
20	400	20	8	2
0	0	0	0	0

TABLE 3.13 Brazil PROCONVE L8 corporate-average emissions limits for LDVs [3-48].

| Implementation Date | Corporate Fleet-Averaged Level for NMHC + NOx (mg/km) | |
	Light passenger vehicles	Light commercial vehicles
1 Jan. 2025	140	50
1 Jan. 2027	110	40
1 Jan. 2029	50	30
1 Jan. 2031	30	30

3.2.8 **Republic of Korea**

The exhaust emissions regulations for LDVs in the Republic of Korea (South Korea) are split between limits for SI engines using gasoline or gaseous fuels and limits for CI engines [3-49]. The regulations reference the US test cycles, including FTP-75, US06, and SC03, although the FTP-75 test is called CVS-75 in the South Korean regulations. These test cycles are described in Chapter 4, Emissions Testing Protocols. The South Korean emissions regulations for LDVs with SI engines have applied since 1 January 2016, and are shown in Table 3.14.

The limits for LDVs with CI engines have applied since 1 October 2017, and are shown in Table 3.15. Note that these limits are equivalent to the Euro 6 Type 1 test limits and that the vehicle performance is evaluated on WLTC.

TABLE 3.14 South Korean emissions limits for LDVs with SI engines [3-49].

Standard	Test Cycle	Limits CO (g/km)	NMHC + NOx (g/km)	HCHO* (g/km)	PM† (g/km)
1	CVS-75	2.61	0.100	0.0025	0.002
	US06	5.97	0.087	—	0.006
	SC03	2.0	0.062	—	—
2	CVS-75	1.31	0.078	0.0025	0.002
	US06	5.97	0.075	—	0.006
	SC03	2.0	0.044	—	—
3	CVS-75	1.06	0.044	0.0025	0.002
	US06	5.97	0.075	—	0.006
	SC03	2.0	0.044	—	—
4	CVS-75	1.06	0.031	0.0025	0.002
	US06	5.97	0.075	—	0.006
	SC03	2.0	0.044	—	—
5	CVS-75	0.625	0.019	0.0025	0.002
	US06	5.97	0.031	—	0.006
	SC03	2.0	0.012	—	—
6	CVS-75	0.625	0.0125	0.0025	0.002
	US06	5.97	0.031	—	0.006
	SC03	2.0	0.012	—	—
7	CVS-75	0	0	0	0

* HCHO limits only apply to vehicles fueled by alcohol only or alcohol bi-fuel.

† The PM limit on CVS-75 is 0.004 g/km, and on US06 is 0.011 g/km, for LDVs certified through 13 December 2020 and delivered within 3 years of certification.

TABLE 3.15 South Korean emissions limits for LDVs with CI engines [3-49].

Vehicle Category	Level	Reference Mass (RM) (kg)	Limits CO (mg/km)	NOx (mg/km)	HC+NOx (mg/km)	PM (mg/km)	PN (#/km)
M	—	All	500	80	170	4.5	6.0×10^{11}
N1	I	RM ≤ 1,305	500	105	170	4.5	6.0×10^{11}
	II	1,305 < RM ≤ 1,760	630	125	195	4.5	6.0×10^{11}
	III	1,760 < RM	740	125	215	4.5	6.0×10^{11}

3.2.9 Other Countries

Other countries also have criteria pollutant standards in place for LDVs, although most are modeled after the US or EU standards. Please refer to the appropriate local regulations to see how they are defined.

3.3 Heavy-Duty Criteria Pollutants

Criteria pollutants from HD sources have historically been regulated at the engine level rather than at the full vehicle level. This approach accommodates traditional divisions in the markets for commercial vehicle and non-road applications, where the manufacture of final products is not vertically integrated as it is with LDVs. Commercial vehicles have a broad range of vehicle and powertrain configurations that are offered for sale for a given vehicle model. This proliferation is especially true for vocational vehicles, which are straight trucks between Class 2b and Class 7 or in the EU N_2 category. For example, one vehicle from a manufacturer might use the engine, gearbox, and driveline made by the vehicle manufacturer, and another vehicle of the same model might use external suppliers for the powertrain components. Overall, HDVs do not neatly fall into vehicle families the way LDVs typically do.

Likewise, non-road applications have a proliferation of configuration options that defy easy categorization. Non-road engines may be used in stationary or mobile applications and with a wide variety of potential duty cycles in real use. Besides, a given model of non-road machines from one manufacturer may be offered with a choice of engines from both the machine manufacturer and external suppliers. Therefore, for criteria pollutants, most regulatory bodies focus on measurements from the engine and EAS for emissions certification and not on measurements from a homologated vehicle or application. Note that the standards also reflect the working nature of these applications, so the pollutants are per unit of work done.

This section covers criteria pollutant emissions regulations for on-road and non-road applications in several jurisdictions. As elsewhere in this chapter, the regulatory limits presented are current as of publication.

3.3.1 United States (Federal)

Similar to the approach for LDVs described in Section 3.2.1, the US requires HD engines for both on-road and off-road applications to comply with the appropriate regulations, not just on-cycle, but in the NTE operating range. AECD that lead to deviations from this performance need to be justified and well-documented to receive regulatory approval. Note that the regulatory body, such as EPA or CARB, decides at their discretion whether an AECD is approved, not the manufacturer.

3.3.1.1 **On-Road Applications:** The current criteria pollutant emissions standard for on-road heavy-duty spark-ignition (HDSI) engines and heavy-duty diesel engines

TABLE 3.16 US heavy-duty engine emissions limits [3-50].

Engine Type	Limits				
	CO (g/bhp·h)	NMHC (g/bhp·h)	NOx (g/bhp·h)	NMHC + NOx (g/bhp·h)	PM (g/bhp·h)
HD SI	14.4	0.14	0.20	—	0.01
HD CI	15.5	0.14	0.20	2.4	0.01

TABLE 3.17 US Class 2b medium-duty vehicle emissions limits [3-52].

	Limits				
	NMHC+NOx (g/mi.)		CO (g/mi.)		PM (g/mi.)
Bin	FTP-75	HD-SFTP	FTP-75	HD-SFTP	FTP-75
250	0.250	0.800	6.4	22.0	0.01
200	0.200	0.800	4.2	22.0	0.01
170	0.170	0.450	4.2	12.0	0.01
150	0.150	0.450	3.2	12.0	0.01
0	0.000	0.000	0.0	0.0	0.0

(HDDEs) was set by EPA in 40 CFR §86.007-11 [3-50], and has applied to all new engines since 2010. Additional on-board diagnostics (OBD) requirements defined by 40 CFR §86.010-18 [3-51] started in 2010, and have strengthened since 2013. Those requirements are described in more detail in Section 6.3.2.2. The limits for criteria pollutants, shown in Table 3.16 are for the engine and EAS only and apply at the end of FUL.

LHDVs that are MDVs as defined in Section 3.2.1 may be approved for sale using a vehicle-based chassis certification for emissions that complies with the rules that govern LHDVs in 40 CFR 86.1816-18 [3-52] instead of using the HD engine certification process. The limits for Class 2b MDVs—covering the GVWR range of 8,500 lb. (3,856 kg) to 10,000 lb. (4,536 kg)—are shown in Table 3.17. Likewise, the limits for Class 3 MDVs— covering the GVWR range of 10,001 lb. (4,536 kg) to 14,000 lb. (6,350 kg)—are shown in Table 3.18. The HD-SFTP includes testing on the FTP-75, the SC03, and either the US06 for Class 2b MDVs or the Hot LA-92 cycle for Class 3 MDVs. The results from the three cycles are averaged as shown in Eq. 3.1, although substituting the Hot LA-92 cycle result for the US06 when evaluating a Class 3 MDVs. All of the full-vehicle drive cycles for certification are described in more detail in Section 4.1.

TABLE 3.18 US Class 3 medium-duty vehicle emissions limits [3-52].

	Limits				
	NMHC+NOx (g/mi.)		CO (g/mi.)		PM (g/mi.)
Bin	FTP-75	HD-SFTP	FTP-75	HD-SFTP	FTP-75
400	0.400	0.550	7.3	6.0	0.01
270	0.270	0.550	4.2	6.0	0.01
230	0.230	0.350	4.2	4.0	0.01
200	0.200	0.350	3.7	4.0	0.01
0	0.000	0.000	0.0	0.0	0.0

TABLE 3.19 US heavy-duty engine emissions control durability or full useful life (FUL) periods.

Engine Category	Durability Period	
HD Spark-Ignition	110,000 mi.	10 years
Light HDDE		
Medium HDDE	185,000 mi.	10 years
Heavy HDDE	435,000 mi.	10 years

By MY2022, the fleet averaged NMHC+NOx value on FTP-75 is expected to be 0.178 g/mi. for Class 2b MDVs, and 0.247 g/mi. for Class 3 MDVs. Unlike with LDVs, neither of these values correspond neatly to one of the Tier 3 bins for NMHC+NOx shown in Tables 3.17 and 3.18.

The limits for HDDEs shown in Table 3.16 are expected to be met through the durability periods listed in Table 3.19, which apply to HDSI engines and HDDEs. Note that heavy HDDEs (HHDDEs) also have a FUL of 22,000 operating hours, which may be used instead of the mileage or time use periods if it is reached first.

In November 2018, EPA announced [3-53] that it started work on new rules for lower tailpipe NOx standards for HDDEs known as the ultra-low NOx standard. This rule-making process started to ensure that there is a national standard for NOx emissions from HDDEs that is harmonized with the California ultra-low NOx standard described in Section 3.3.2 that started development in 2017 [3-54]. The new, ultra-low NOx limit is expected to be one of 0.100, 0.080, 0.050, or 0.020 (g/bhp·h), and is expected to come into effect in MY2027. The lowest of these levels would represent a 90% reduction in tailpipe NOx emissions from the EPA 2010 standard.

3.3.1.2 Non-Road Applications:
For non-road applications in the US, the applicable regulatory standards are the Tier 4 final standards that are defined in 40 CFR part 1039 [3-55] and shown in Table 3.20. In general, the range of power levels shown for each set of pollutant limits includes the minimum of the range but excludes the maximum. The emissions on non-road engines are evaluated using the Non-Road Transient Cycle (NRTC) and the Non-Road Steady-State Cycles (NRSC) described in Sections 4.2.7 and 4.2.8, respectively.

TABLE 3.20 US Tier 4 final non-road engine emissions limits [3-56].

Power range (kW) Minimum	Maximum	Engine ignition type	CO (g/kW·h)	HC (g/kW·h)	NOx (g/kW·h)	PM (g/kW·h)	FUL (h)
0	8	All	8.00	HC + NOx ≤ 7.50		0.40	3,000
8	19	All	6.60	HC + NOx ≤ 7.50		0.40	3,000
19	37	All	5.50	HC + NOx ≤ 4.70		0.03	5,000*
37	56	All	5.00	HC + NOx ≤ 4.70		0.03	8,000
56	130	All	5.00	0.19	0.40	0.02	8,000
130	560	All	3.50	0.19	0.40	0.02	8,000
560	—	All	3.50	0.19	3.50	0.04	8,000

* FUL is 3,000 h for constant-speed applications.

These standards were intended to be harmonized with the EU Stage IV limits described in Section 3.3.3.2. Nevertheless, there are some differences between the two, including the PM standards and the weighting of cold start and hot start results from the NRTC.

3.3.2 California

For both on-road and non-road applications that will be sold within the State of California, a manufacturer must receive approval for sale from both EPA and California. The criteria pollutant rules in California are the same as those promulgated by EPA for the US, although California's rules only apply to new engines or vehicles sold within California. California has started a rule-making process to implement an ultra-low NOx standard for on-road HDDEs, to have the new standard apply from MY2024. In early 2019, CARB indicated that the new levels under strong consideration include 0.080 (g NOx/bhp·h) or 0.050 (g NOx/bhp·h), possibly with more stringent limits for MY2027 and beyond to follow [3-54]. These targets represent a 60% to 75% reduction in tailpipe NOx levels from the current standard shown in Table 3.16. The new rules may also include a longer FUL, a lower PM target of 0.005 (g/bhp·h), a new Low Load Cycle (LLC), and new OBD requirements [3-54].

3.3.3 European Union

The EU has also developed a series of standards for both on-road HD engines and non-road engines. In addition to applying within the EU, several other jurisdictions around the world have adopted rules that are based on EU standards. Note that the EU standards refer to positive ignition (PI) engines; SI has been used here for consistency with the rest of the text.

3.3.3.1 On-Road Applications: The current EU standard for on-road HDSI engines and HDDEs is Euro VI, which has been fully in effect since 31 December 2013 [3-57]. The emissions standards [3-58], shown in Table 3.21, use the World-Harmonized Heavy-Duty Transient Cycle (WHTC) and the World-Harmonized Heavy-Duty Steady-State Cycle (WHSC) for evaluation, which is described in Sections 4.2.2 and 4.2.3, respectively. The two sets of WHSC targets shown are for CI and SI engines, respectively. These limits are expected to be met through the durability periods listed in Table 3.22.

Under the Euro VI scheme, commercial vehicles also have an NTE limit for CO, NOx, and either THC or NMHC and CH_4 that is based on a CF of 1.50×. Euro VI also includes a set of OBD requirements, which is described in Section 6.3.2.4.

TABLE 3.21 Euro VI criteria pollutant emission standards [3-57, 3-58].

| Test | Limits | | | | | | | |
	CO (mg/kW·h)	THC (mg/kW·h)	NMHC (mg/kW·h)	CH_4 (mg/kW·h)	NOx (mg/kW·h)	NH_3 (ppm)	PM (mg/kW·h)	PN (#/kW·h)
WHSC (CI)	1,500	130	—	—	400	10	10	8.0×10^{11}
WHTC (CI)	4,000	160	—	—	460	10	10	6.0×10^{11}
WHTC (SI)	4,000	—	160	500	460	10	10	—

TABLE 3.22 Euro VI emissions control durability periods.

Commercial Vehicle Category	Durability Period	
M₁, N₁, and M₂	160,000 km	5 years
N₂, N₃ with GVW up to 16 tonnes and M₃, up to 7.5 tonnes.	300,000 km	6 years
N₃ with GVW over 16 tonnes and M₃, over 7.5 tonnes	700,000 km	7 years

3.3.3.2 Non-Road Applications: The EU adopted the Stage IV standard in Paragraph 4.1.2.6 of [3-59], which was harmonized with the US Tier 4 final standard for non-road engines, as part of a program to harmonize the non-road emissions requirements from Stage III through to Stage IV across the globe. These Stage IV limits are shown in Table 3.23 [3-59]. The Stage IV standards also include a 25 ppm limit for ammonia (NH_3), which must not be exceeded over the test cycle.

From 1 January 2017, the EU has been using the Stage V standard for non-road engines, which was adopted in 2016 [3-60]. These limits are shown in Table 3.24, and include a tighter limit for PM and the first-ever limit for PN.

TABLE 3.23 EU Stage IV non-road engine emissions limits [3-59].

Power range (kW) Minimum	Maximum	Engine ignition type	CO (g/ kW·h)	HC (g/ kW·h)	NOx (g/ kW·h)	PM (g/ kW·h)	FUL (h)
0	19	All	—	—		—	—
19	37†	All	5.5	HC + NOx ≤ 7.50		0.6	5,000*
37	56‡	All	5.0	HC + NOx ≤ 4.70		0.025	8,000
56	130	All	5.0	0.19	0.40	0.025	8,000
130	560	All	3.5	0.19	0.40	0.025	8,000
560	—	All	—	—	—	—	—

* FUL is 3,000 h for constant-speed applications.
† Defined in Stage IIIA regulations, Para. 4.1.2.4 of [3-59], and carried forward to Stage IV.
‡ Defined in Stage IIIB regulations, Para. 4.1.2.5 of [3-59], and carried forward to Stage IV.

TABLE 3.24 EU Stage V non-road engine emissions limits [3-60].

Power range (kW) Minimum	Maximum	Engine ignition type	CO (g/kW·h)	HC (g/kW·h)	NOx (g/kW·h)	PM (g/kW·h)	PN (#/kW·h)	FUL (h)
0	8	CI	8.00	HC + NOx ≤ 7.50		0.4	—	3,000
8	19	CI	6.60	HC + NOx ≤ 7.50		0.4	—	3,000
19	37	CI	5.00	HC + NOx ≤ 4.70		0.015	1×10^{12}	5,000*
37	56	CI	5.00	HC + NOx ≤ 4.70		0.015	1×10^{12}	8,000
56	130	All	5.00	0.19	0.40	0.015	1×10^{12}	8,000
130	560	All	3.50	0.19	0.40	0.015	1×10^{12}	8,000
560	—	All	3.50	0.19	3.50	0.045	—	8,000

* 3,000 h FUL for constant-speed applications.

For engines that are partially or fully run on gaseous fuel, a so-called A factor is used to calculate an alternate limit for HC emissions as follows:

$$HC = 0.19 + (1.5 \times A \times GER), \qquad (3.2)$$

where HC is the hydrocarbon limit in (g/kW·h), A is the scaling factor, and GER is the average gas energy ratio over the drive cycle, preferably the hot-start NRTC. The A factor for these engines is 1.10, except for engines with a rated power over 560 kW, for which the A factor is 6.0.

3.3.4 Japan

For HD on-road applications, Japan has been developing its own emissions standards since the 2000s, as described in Section 3.3.4.1. For non-road applications, Japan is harmonized with other jurisdictions, as described in Section 3.3.4.2.

3.3.4.1 On-Road Applications: Japan requires HDVs with a GVWR of 2.5 Mg (2.5 tonnes) or more to comply with the full-vehicle emissions standards from the Post New Long-Term Regulations 2009 [3-61], shown in Table 3.25. HDV emissions are evaluated on the JE05 test cycle, which is described in Section 4.3.4. HDVs in Japan are expected to meet these regulatory limits out to the durability periods listed in Table 3.26.

From 1 October 2016 through 1 October 2019, Japan phased in the HDDE standards shown in Table 3.27 known as the Post Post New Long-Term Regulations. For each of the steady-state (WHSC) and transient (WHTC) cycles, the rules state the average and maximum emissions upper limits allowed. Advanced OBD requirements, described in Section 6.3.2.7, phase in from 1 October 2018 through 1 October 2021. In both cases, the limits apply first to the heaviest HDVs with GVW over 7.5 Mg, and then to successively lighter HDVs.

TABLE 3.25 Japan Post New Long-Term Regulations heavy-duty engine emissions limits [3-61].

Vehicle	Limits			
	CO (g/kW·h)	NMHC (g/kW·h)	NOx (g/kW·h)	PM (g/kW·h)
Gasoline and LPG vehicles	16.0	0.23	0.7	0.01
Diesel vehicles	2.22	0.17	0.7	0.01
Comparison table for other vehicles	16.0	0.23	0.07	0.01

TABLE 3.26 Japan emissions control durability periods [3-61].

Commercial Vehicle GVW (Mg)	Durability Period
3.5–8.0	250,000 km
8–12	450,000 km
Over 12	650,000 km

TABLE 3.27 Japan Post Post New Long-Term Regulations HD engine emissions limits [3-61].

Test	Limits CO (g/kW·h)	NMHC (g/kW·h)	NOx (g/kW·h)	PM (g/kW·h)
WHSC (avg.)	2.22	0.17	0.4	0.010
WHSC (NTE)	2.95	0.23	0.7	0.013
WHTC (avg.)	2.22	0.17	0.4	0.010
WHTC (NTE)	2.95	0.23	0.7	0.013

3.3.4.2 **Non-Road Applications:** Non-road standards in Japan are at the Stage 4 level as of 2014, and these are mostly harmonized with the US Tier 4 final and EU Stage IV non-road standards as shown in Table 3.28. The exception is that Japan splits the NMHC and NOx limits for engines between 19 and 56 kW, whereas the US and EU lump the two together into a combined target.

3.3.5 People's Republic of China

Historically, the PRC has followed the EU standards, but with implementation dates that lag the EU by a few years, and with urban areas leading rural ones. Increasingly, the PRC's standards are more closely following the EU's and with shorter lag times. In some areas, such as in the forthcoming China Phase VI-b rules [3-62], the PRC's standards have started to exceed what the EU standards require.

3.3.5.1 **On-road Applications:** The criteria pollutant standards for the on-road China Phase VI regulations [3-62] are the same as for Euro VI, as shown in Table 3.21. There are some subtle differences in the requirements, including the longer useful life period for category M_1, N_1, and M_2 LHDVs that is shown in Table 3.29. The useful life periods for heavier vehicles are the same as for the Euro VI standards.

TABLE 3.28 Japan Stage 4 non-road engine emissions limits.

Power range (kW) Minimum	Maximum	Engine ignition type	CO (g/kW·h)	NMHC (g/kW·h)	NOx (g/kW·h)	PM (g/kW·h)	FUL (h)
19	37	All	5.0	0.7	4.0	0.03	5,000
37	56	All	5.0	0.70	4.0	0.025	8,000
56	130	All	5.0	0.19	0.4	0.02	8,000
130	560	All	3.5	0.19	0.4	0.02	8,000

TABLE 3.29 China VI-b emissions control useful life periods [3-62].

Commercial Vehicle Category	Durability Period	
M_1, N_1, and M_2	200,000 km	5 years
N_2, N_3 with GVW up to 16 Mg and M_3, up to 7.5 Mg.	300,000 km	6 years
N_3 with GVW over 16 Mg and M_3, over 7.5 Mg	700,000 km	7 years

TABLE 3.30 Implementation dates for China Stage VI standards.

Standard Stage	Vehicle Type	Implementation Date
Stage VI-a	Gas vehicles	1 July 2019
	Urban vehicles	1 July 2020
	All vehicles	1 July 2021
Stage VI-b	Gas vehicles	1 July 2021
	All vehicles	1 July 2023

The China VI standards are being phased in with a slight lag to the corresponding EU standards. The phase-in schedule is shown in **Table 3.30**, with the full implementation for China VI-b expected by 1 July 2023.

In addition to the tailpipe emissions standards based on engine testing, the China VI-b standards also introduce remote monitoring of vehicle exhaust. These remote monitoring standards rely on the OBD systems to monitor key emissions performance metrics and then report them to a central system every 10 s. The OBD system also needs to be tamper-resistant, as discussed in Section 6.3.2.6. These requirements add a level of stringency to the China VI-b standards that are not in the EU VI standards. These differing requirements also reflect strong cultural differences regarding data privacy between the two jurisdictions.

3.3.5.2 Non-Road Applications: China Stage III [3-63] is currently in place for off-road applications and mirrors the EU Stage III-A standards. China Stage IV is expected to take effect from 1 January 2020 and is based on the EU Stage IV limits shown in Table 3.23. One difference between the China and EU standards is that all engines with rated power below 37 kW, not just those between 19 and 37 kW, need to conform to the following limits for both China Stage III and Stage IV:

- CO limit of 5.5 (g/kW·h)
- HC+NOx limit of 7.5 (g/kW·h)
- PM limit of 0.6 (g/kW·h)

China Stage IV also includes additional requirements over the test cycle, including the following:

- NH_3 limit of 25 ppm
- PN limit of 5×10^{12} (#/kW·h)

All engines rated up to 560 kW are to be tested on the NRTC, which is described in Section 4.2.7. The net NRTC result is to be a weighted average of cold and hot start results, using a 10% weight for the cold start and 90% for the hot start. Engines rated over 560 kW use the NRSC instead, which is described in Section 4.2.8. OEM is also encouraged to develop and produce non-road engines that conform to EU Stage V standards, as shown in Table 3.24.

3.3.6 India

In India, the Bharat emissions standards cover both on-road and non-road applications. In general, they follow the corresponding EU standards described in Section 3.3.3.

3.3.6.1 On-Road Applications: In India, the Bharat Stage VI rules [3-46] apply as of 1 April 2020 for all on-road HD engines. The emissions test cycles and limits are the same as the Euro VI levels shown in Table 3.21. The deterioration factors for various criteria pollutants at FUL are specified in Table 3.31. The Bharat Stage VI rules specifically note that the rules apply to engines designed for alternative fuels, including the following:

- E5 ethanol-gasoline blend
- E85 ethanol-gasoline blend
- Natural gas, including biomethane, biogas, LNG, and CNG
- Hydrogen and CNG blends
- LPG
- B7 fatty-acid methyl ester biodiesel blend
- ED95 ethanol-biodiesel blend
- Diesel-natural gas dual fuel

ED95 is a 95% ethanol, 5% biodiesel fuel blend that can be used in a CI engine with some adaptation. The manufacturer is expected to use a reference fuel that meets the regulatory requirements to certify the engine on an alternative fuel. Several of these alternative fuels and their effects on emissions are described in Chapter 16, Alternative Fuels.

3.3.6.2 Non-Road Applications: For non-road applications, the Bharat (CEV/TREM) IV standards [3-64] that cover agricultural tractors, construction equipment vehicles, and combine harvesters, apply from 1 October 2020. The Bharat IV standards mirror the EU Stage IV standards shown in Table 3.23. Starting on 1 April 2024, India will move to the Bharat (CEV/TREM) V standards, which mirror the EU Stage V standards shown in Table 3.24, including that the average NH_3 emissions over the NRTC must not exceed 25 ppm.

The rules published to date do not fully define in-use conformity and similar requirements that are well-defined in the EU Stage IV and Stage V rules. It is expected that the forthcoming updates to Automotive Industry Standard No. 137 (AIS 137) will address these remaining issues in the rules.

3.3.7 Other Countries or Regions

Brazil currently uses the PROCONVE P-7 standards [3-65] for on-road HD engines, which are the equivalent of the Euro V standards [3-66, 3-67], shown in Table 3.32. Brazil will move to the PROCONVE P-8 standards [3-68] starting on 1 January 2022 for new type approvals and 1 January 2023 for all new sales. The PROCONVE P-8 standards are equivalent to the Euro VI standards described earlier in Section 3.3.3.1.

TABLE 3.31 Bharat VI criteria pollutant deterioration factors [3-46].

Test Cycle	CO	THC	NMHC	CH_4	NOx	NH_3	PM	PN
WHSC	1.3	1.3	1.4	1.4	1.15	1.0	1.05	1.0
WHTC	1.3	1.3	—	—	1.15	1.0	1.05	1.0

TABLE 3.32 Euro V criteria pollutant emission standards [3-66, 3-67].

Test	CO (g/kW·h)	THC (g/kW·h)	NMHC (g/kW·h)	CH$_4$ (g/kW·h)	NOx (g/kW·h)	PM (g/kW·h)
	Limits					
ESC (CI)	1.5	0.46	—	—	2.0	0.02
ETC (CI)	4.0	—	0.55	—	2.0	0.03
ETC (gas)	4.0	—	0.55	1.1	2.0	0.03

For HD engines and vehicles, most other countries have regulatory standards that are equivalent to the EU regulations, with a few exceptions such as Canada and Mexico that follow the US regulations.

For non-road equipment, Brazil uses the PROCONVE MAR-I regulatory standards [3-69], which apply to all applications from 1 January 2019, and are equivalent to the EPA Tier 4 final or EU Stage IV limits shown in Tables 3.20 and 3.23. Most other countries are likewise following these two standards if they have not yet moved to the EU Stage V limits described in Section 3.3.3.2.

3.4 Light-Duty Vehicle Greenhouse Gases

The interest in using regulations to improve fuel economy started during the oil crisis of the 1970s when fuel prices were artificially and dramatically raised by the action of the Organization of the Petroleum Exporting Countries (OPEC). Initially, many regulatory schemes, especially those in what is now the EU, used taxes on fuel or taxes based on engine displacement or power to promote the adoption and sale of more efficient LDVs. In the US direct action on fuel economy started in the 1970s with an emphasis on improving vehicle-level fuel economy. In the 2000s, the regulatory emphasis started including GHG emissions, which for typical fuels correlate closely with fuel consumption.

3.4.1 United States (Federal)

The original regulatory scheme in the US, the CAFE standard, covers fuel economy only. CAFE is regulated by the National Highway Traffic Safety Administration (NHTSA), which is a part of the US Department of Transportation (DOT), using the framework that is described in Section 3.4.1.1. A GHG emissions standard for LDVs was added to the CAFE standards from MY2012. The LDVs GHG standards are administered by EPA and are described in Section 3.4.1.2. Because over 95% of new LDVs sold in the US use gasoline-fueled SI engines, the GHG standards are harmonized with CAFE standards for gasoline-fueled LDVs. Thus, for LDVs with light-duty diesel engines (LDDEs) and some LDVs with gasoline engines, it is possible for the vehicle to comply with CAFE and not the GHG standards or vice versa.

3.4.1.1 Fuel Economy Standards: The CAFE standards in the US are regulated by NHTSA. The statutory authority for the standards was established by the Energy

Policy and Conservation Act (EPCA) of 1975 [3-70] and was later amended by the Energy Independence and Security Act (EISA) of 2007 [3-71]. By statute, manufacturers evaluate the fuel consumption of their vehicles over the city (FTP-75) and highway (HFET) drive cycles, which are described in Sections 4.1.1 and 4.1.2, respectively. A weighted average is then calculated as follows:

$$FE_{combined} = \frac{0.55}{FE_{city}} + \frac{0.45}{FE_{hwy}}. \tag{3.3}$$

For a given LDV model, the combined fuel economy for CAFE is calculated using the city and highway fuel economy test results as directly measured on the test protocol. The overall CAFE fuel economy for a given manufacturer is calculated using a sales-weighted average of the fuel economy values for each vehicle model. By contrast, the Monroney label fuel economy results, described in Section 3.4.1.3, use a weighted average and scaling factors to translate the on-cycle test results to an estimate of in-use fuel economy for the specific vehicle model.

For MY2017 through MY2025, NHTSA has set the CAFE standards for LDVs. These standards separate passenger cars and light trucks into their bins for averaging, banking, and trading of credits. The fuel economy standards also scale concerning vehicle foot-print, which is the product of the vehicle wheelbase by the vehicle track width, the lateral distance between the centers of the tires. The passenger car targets are shown in **Figure 3.3** below, and the light truck targets are shown in **Figure 3.4** [3-72].

FIGURE 3.3   CAFE targets for passenger cars concerning vehicle footprint, by model year [3-72].

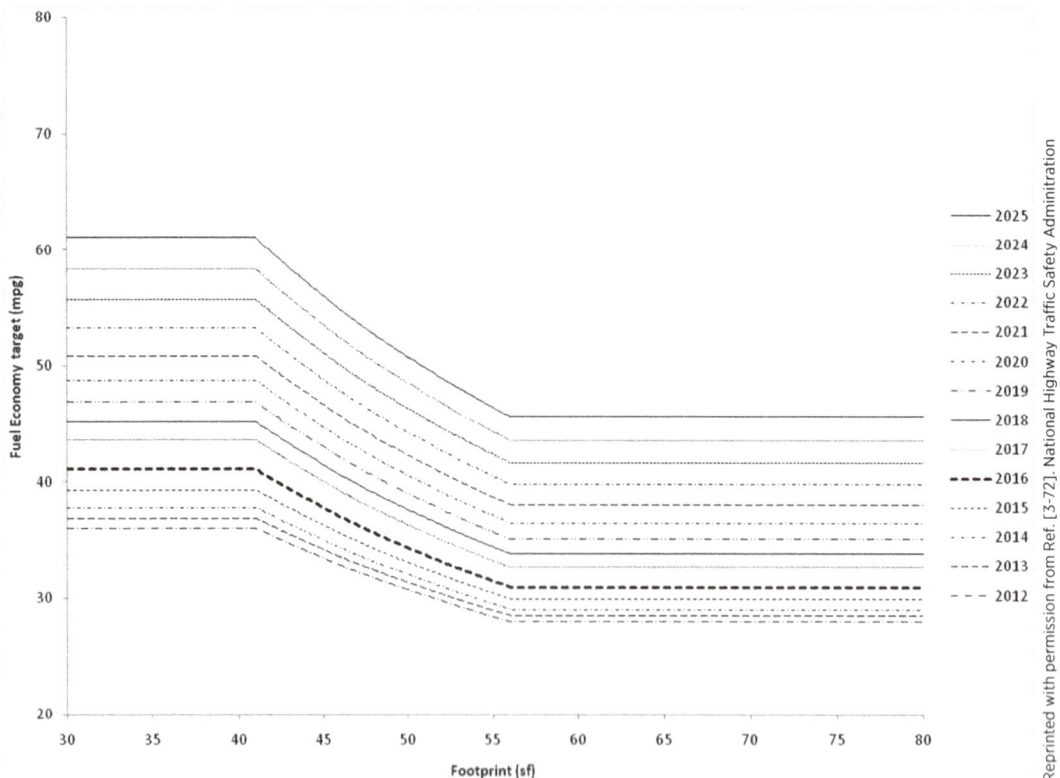

FIGURE 3.4 CAFE targets for light trucks concerning vehicle footprint, by model year [3-72].

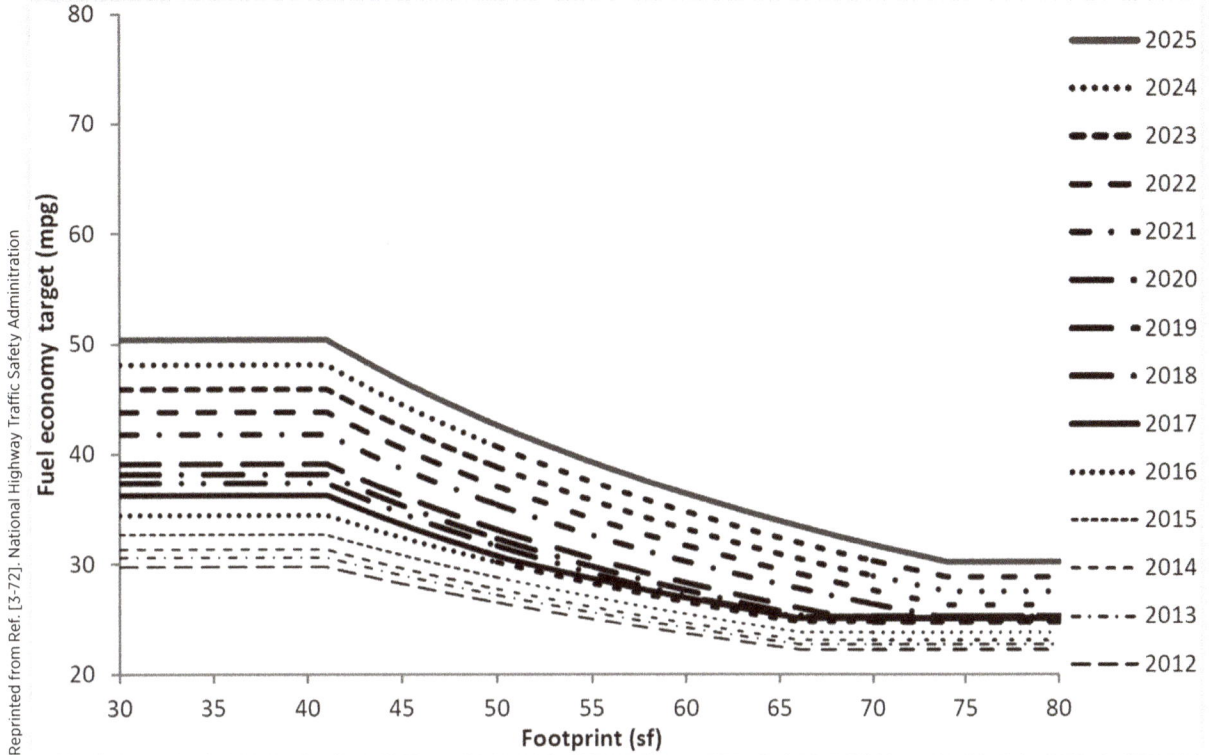

On 31 March 2020, NHTSA and EPA issued the Safer Affordable Fuel-Efficient (SAFE) Vehicles Rule [3-73], which will increase the CAFE target by 1.5% per year from MY2021 through MY2026.

3.4.1.2 Greenhouse Gas Standards:

The GHG standards for LDVs that correspond to the CAFE rules described in Section 3.4.1.1 are promulgated by EPA under their CAA authority. These standards are designed to be harmonized with the fuel economy standards from NHTSA, although there are a few areas in which the standards do not overlap completely. For example, EPA is primarily interested in GHG reduction, thus the LDV GHG standards offer credits for off-cycle reductions that are not explicitly included in the fuel economy standards, such as mitigating the GHG emissions effects of cabin air conditioning system refrigerants and other off-cycle credits.

The LDV GHG standards started taking effect in MY2012, and are currently defined by 40 CFR §86.1818-12 [3-74] through MY2025. The footprint-based GHG standards for passenger cars and light trucks are presented as curves in **Figures 3.5** and **3.6**, respectively. On 31 March 2020, NHTSA and EPA issued the Safer Affordable Fuel-Efficient (SAFE) Vehicles Rule [3-73], which will decrease the GHG target by 1.5% per year from MY2021 through MY2026.

FIGURE 3.5   Greenhouse gas targets for passenger cars concerning vehicle footprint, by model year [3-74].

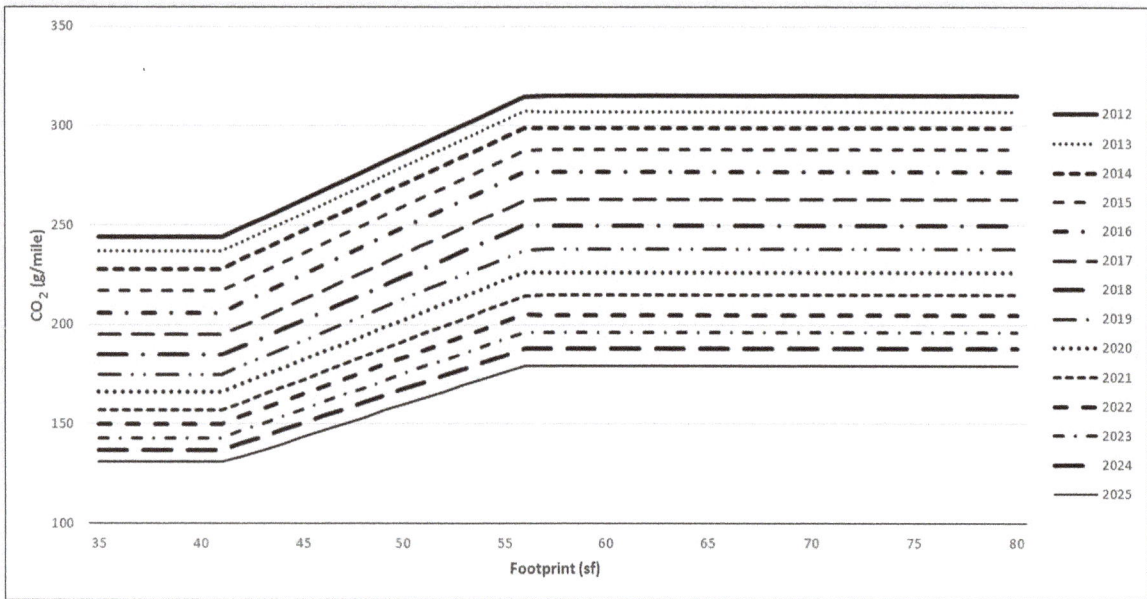

FIGURE 3.6   Greenhouse gas targets for light trucks concerning vehicle footprint, by model year [3-74].

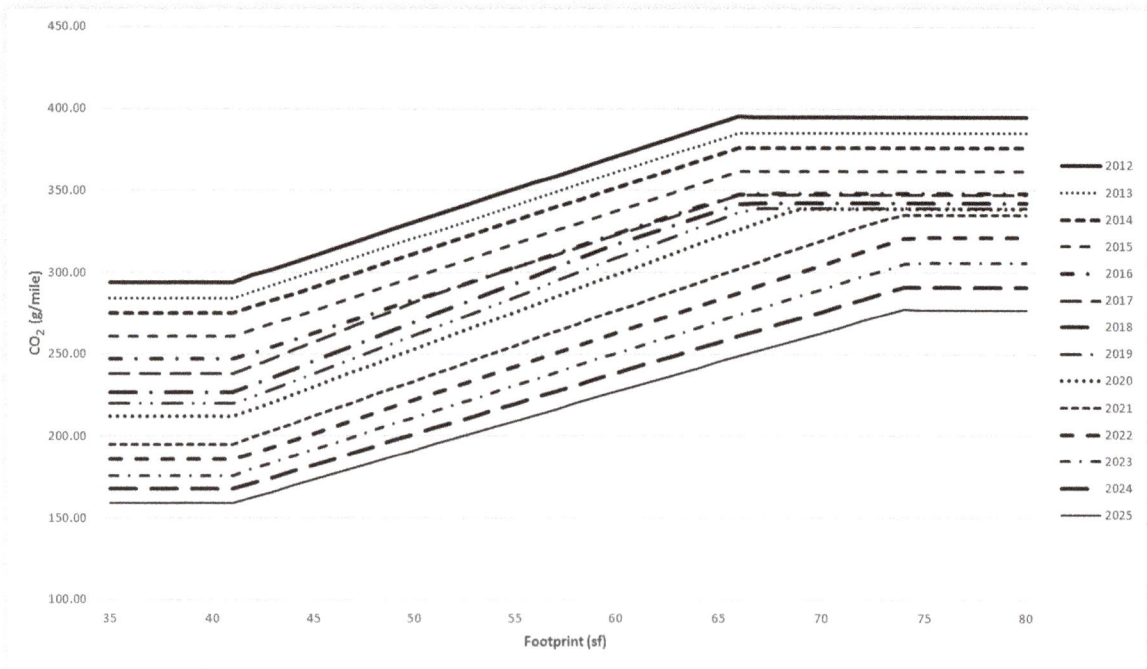

3.4.1.3 Monroney Fuel Economy Label:

The Monroney fuel economy label, or sticker, must be included on a new vehicle for sale in the US and is intended to provide an estimate of real-world fuel economy and GHG emissions. The label, shown in **Figure 3.7**, includes an estimate of fuel economy (in miles per US gallon, mpg) for city driving, highway driving, and combined driving.

The current calculation method [3-76, 3-77] uses a weighted average of the test results from the following five drive cycles, each of which is described further in Section 4.1:

- FTP-75, which represents urban stop-and-go driving from the 1970s
- HFET, which represents rural driving
- US06, which represents aggressive highway driving
- SC03, which represents hot-weather operation with the air conditioning active
- FTP-75 with cold ambient (a.k.a. FTP-20), which represents cold-weather operation

The 5-cycle weighted average for the city fuel economy FE_{City} is calculated as follows [3-76, 3-77]:

$$FE_{City} = 0.905 \left[FC_{CS} + FC_{CR} \right]^{-1}, \tag{3.4}$$

where the starting city-driving fuel consumption FC_{CS} (in gal./mi.) is

$$FC_{CS} = 0.330 \frac{\left(0.76\, SF_{75} + 0.24\, SF_{20} \right)}{4.1}, \tag{3.5}$$

SF_x is the starting fuel consumed (in gal.) for the FTP-75 (x = "75") or FTP-20 (x = "20"),

$$SF_x = 3.6 \left[\frac{1}{FE_{x,Bag1}} - \frac{1}{FE_{x,Bag3}} \right], \tag{3.6}$$

FIGURE 3.7 Sample Monroney label for gasoline-fueled LDV [3-75].

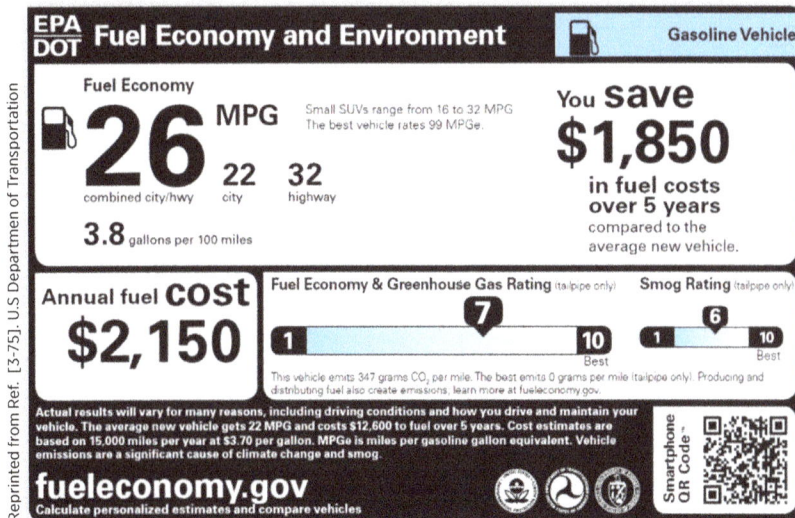

and $FC_{x,\,Bag\,n}$ is the fuel consumption from Bag n of the FTP-75 or FTP-20 drive cycle. The running city-driving fuel consumption FC_{CR} (in gal./mi.) is calculated as follows:

$$FC_{CR} = 0.82\left[\frac{0.48}{FE_{75,Bag\,2}} + \frac{0.41}{FE_{75,Bag\,3}} + \frac{0.11}{FE_{US06,city}}\right] + 0.18\left[\frac{0.5}{FE_{20,Bag\,2}} + \frac{0.5}{FE_{20,Bag\,3}}\right]$$
$$+ 0.133\times1.083\left[\frac{0.5}{FE_{SC03}} - \left(\frac{0.39}{FE_{75,Bag\,2}} + \frac{0.61}{FE_{75,Bag\,3}}\right)\right], \qquad (3.7)$$

Likewise, the 5-cycle weighted average for the highway fuel economy FE_{Hwy} is calculated as follows [3-76, 3-77]:

$$FE_{Hwy} = 0.905\left[FC_{HS} + FC_{HR}\right]^{-1}, \qquad (3.8)$$

where the starting highway-driving fuel consumption FC_{HS} (in gal./mi.) is

$$FC_{HS} = 0.330\frac{(0.76\,SF_{75} + 0.24\,SF_{20})}{60}, \qquad (3.9)$$

and the running highway-driving fuel consumption FC_{HR} (in gal./mi.) is calculated as follows:

$$FC_{HR} = 1.007\left[\frac{0.79}{FE_{US06,hwy}} + \frac{0.21}{FE_{HFET}}\right] + 0.133\times0.377\left[\frac{1}{FE_{SC03}} - \left(\frac{0.39}{FE_{75,Bag\,2}} + \frac{0.61}{FE_{75,Bag\,3}}\right)\right], \qquad (3.10)$$

To calculate the city and highway fuel economy, the city portion of the US06 cycle [3-78] combines the segment from 1 to 130 s and the segment from 495 to 596 s. The highway portion of US06 is, therefore, the segment from 130 to 495 s [3-78].

The combined fuel economy for the Monroney label is then

$$FE_{Combined} = \left[\frac{0.43}{FE_{City}} + \frac{0.57}{FE_{Hwy}}\right]^{-1}. \qquad (3.11)$$

EPA manages the database of Monroney label fuel economy values, and they are documented online at FuelEconomy.gov (https://www.fueleconomy.gov/) [3-75], a website maintained by EPA and the US Department of Energy (DOE) Office of Energy Efficiency and Renewable Energy (EERE). All of the data on the Monroney labels, such as in Figure 3.7, is available on the website.

3.4.2 California

CARB has also set standards for GHG emissions from LDVs under the state's CAA since the mid-2000s that are in addition to the Federal US requirements. California's AB 32, the California Global Warming Solutions Act of 2006, set the base year for GHG emissions in California to 1990 and the target to return to those base year emissions by 2020. This baseline emissions level of 431 Tg (475×10^6 US ton) of CO_2 or equivalent (Million metric tonnes of CO_2 equivalent, or MMTCO$_2$e). In addition, the 2050 target for overall

GHG emissions was set to 86.2 $MMTCO_2e$ (95.0×10^6 ton), which is 20% of the base year emissions value. Subsequent regulations have also set an intermediate target for 2030 of 258.6 $MMTCO_2e$ (285.1×10^6 ton), or 60% of base year emissions. CARB uses a scaling factor of 296 for N_2O and 23 for CH_4.

CARB documents the California GHG inventory each year [3-79], and as of writing, the latest GHG emissions values available were for 2016 and were 429.4 $MMTCO_2e$ (473.3×10^6 ton), which is just under the 2020 target. Of the reported 2016 emissions, 41% come from transportation sources.

CARB's rules for GHG emissions from LDVs are aligned with the EPA program described in Section 2.3.1. California has also implemented a Zero Emissions Vehicle (ZEV) mandate, which requires a certain proportion of sales for these GHG-free vehicle options. The mandate is intended to incentivize the development of battery-electric vehicles (BEV), fuel cell electric vehicles (FCEV), plug-in hybrid vehicles (PHEV), and other options. As with the criteria pollutant rules, other states may also adopt the ZEV mandate, and to date, the following states have: Connecticut, Maine, Maryland, Massachusetts, New Jersey, New York, Oregon, Rhode Island, and Vermont [3-80]. Together with California, these states represent about 30% of the total US market for LDVs.

3.4.3 European Union

As of MY2015, 100% of new passenger cars sold in the EU must meet the 130 (g CO_2/km) standard. In MY2021 the target decreases further to 95 (g CO_2/km), although with an adjustment to account for the shift in-vehicle test cycle from the NEDC to the WLTC. These cycles are described in detail in Sections 4.1.7 and 4.1.9, respectively. Likewise, for LCVs the GHG limits are 175 (g CO_2/km) as of MY2017, and 147 (g CO_2/km) as of MY2020. The shift to the WLTC is intended to narrow the gap between as-tested results and real-world results and keeps LDV certification harmonized with the criteria pollutant standards described in Section 3.1.3. In addition, the RDE protocol described in Section 4.1.10 is used to monitor CO_2 emissions but there is no RDE regulatory standard for CO_2 yet.

These targets for passenger cars scale with the vehicle mass, as follows [3-81]:

$$Ems. = NEDC_{nom} + 0.0333 \times \left(\bar{M} - M_0 \right), \tag{3.12}$$

where $NEDC_{nom}$ is the nominal GHG target in (g CO_2/km), $\bar{M}$ is the sales-weighted average vehicle mass for a manufacturer, and M_0 is a median mass set by regulators in the EU For example, for passenger cars from MY2020, $NEDC_{nom}$ = 95 (g CO_2/km). M_0 is expected to be updated on 31 October 2020 and 31 October 2022, and will be the sales-weighted average mass of all new passenger cars sold in the previous three calendar years. For example, the value of M_0 determined by 31 October 2020 will use data from 2017, 2018, and 2019. As of publication, the current value of M_0 is 1379.88 kg, and is expected to hold at that value through MY2024 [3-81]. The formula is shown in Eq. 3.12, however, assumes the use of the NEDC. With the EU using the WLTC as its regulatory cycle, EU Regulation 2019/631 [3-81] includes a correction factor calculated as follows for a given manufacturer in MY2021:

$$WLTP_{tgt} = WLTP_{CO_2} \times \left(\frac{NEDC_{tgt}}{NEDC_{CO_2}} \right), \tag{3.13}$$

where $WLTP_{CO_2}$ is the sales-weighted average specific CO_2 emissions (in g/km) from testing on the WLTC, $NEDC_{tgt}$ is the sales-weighted CO_2 emissions target, and $NEDC_{CO_2}$ is the sales-weighted average specific CO_2 emissions (in g/km) from testing on the NEDC. Once the WLTP target is calculated per Eq. 3.13, for 2021 through 2024, a manufacturer's specific CO_2 emissions target becomes the following [3-81]:

$$\text{Ems.} = WLTP_{tgt} + 0.0333 \times \left[\left(\bar{M} - M_0 \right) - \left(\bar{M}_{2020} - M_{0,2020} \right) \right], \qquad (3.14)$$

where $\bar{M}$ and M_0 are as for Eq. 3.12, $\bar{M}_{2020}$ is a manufacturer's sales-weighted vehicle mass in 2020, and $M_{0,2020}$ is the reference mass used in 2020, which is 1379.88 kg. Whether using the NEDC or WLTC to evaluate vehicles, each manufacturer is liable for a penalty of €95 per vehicle per (g CO_2/km) over the limit. This penalty is intended to incentivize investment in engine and vehicle technologies to meet the target over simply paying a fee for business as usual. The EU has also set fleet-wide targets in EU Regulation 2019/631 [3-81] as part of its overall GHG targets from EU Regulation 2017/1151 [3-82].

3.4.4 Japan

In 2007, Japan introduced fuel economy standards (in km/L) for LDVs that started with MY2015 [3-83]. These standards are based on vehicle mass, and thus have a lower fuel economy target for heavier vehicles. The fuel economy value for a given vehicle is evaluated on the JC08 cycle, which is described further in Section 4.1.8.

For a given manufacturer, the overall fleet fuel economy standards across all weight classes—currently 16 in total—are shown in **Table 3.33**. Manufacturers can bank and trade across weight classes, though. For MY2020, Japan will further increase the fuel economy target for passenger cars to 20.3 (km/L), which represents another 24.1% improvement from the MY2015 levels [3-84].

In 2015, Japan introduced new fuel economy standards that take effect with MY2022 vehicles [3-85]. These standards, shown in **Table 3.34**, separate vehicles by transmission type—manual or automatic—and by engine type—SI or CI. The vehicles are then categorized by a curb mass (CM) bin for a total of 38 distinct bins. These fuel economy targets use a weighted average of 25% cold-start JC08 results and 75% hot-start JC08 results [3-86].

3.4.5 People's Republic of China

As of 1 January 2016, all passenger cars in China must follow the Stage 4 fuel consumption rules [3-87], which distinguish between vehicles with fewer than three rows of seats and those with three or more rows. By contrast, LCVs must follow the standards outlined

TABLE 3.33 Japan light-duty vehicle fuel economy targets [3-83].

Vehicle Type (Across all weight classes)	MY2015 target (km/L)	MY2004 baseline (km/L)	Improvement (%)
Passenger cars	16.8	13.6	+23.5%
LCV up to 3.5 tonnes	15.2	13.5	+12.6%
Small Buses	8.9	8.3	+7.2%

TABLE 3.34 Japan's light-duty vehicle fuel economy targets from MY2022 [3-85].

Vehicle Type	Curb mass (kg) Minimum	Maximum	Transmission	MY2022 target (km/L) SI (Gasoline)	CI (Diesel)
Passenger cars	—	740	MT or AT	28.1	28.1
Passenger cars	741	855	MT or AT	25.0	25.0
Passenger cars	856	970	MT or AT	22.7	22.7
Passenger cars	971	1,080	MT or AT	20.8	20.8
Passenger cars	1,081	1,195	MT or AT	18.5	18.5
Passenger cars	1,196	—	MT or AT	16.9	16.9
Van or Truck	—	740	MT	21.0	21.0
			AT	20.4	20.4
Van or Truck	741	855	MT	20.4	20.4
			AT	19.8	19.8
Van or Truck	856	970	MT	19.9	19.9
			AT	19.2	19.2
Van or Truck	971	1,080	MT	19.4	19.4
			AT	18.7	18.7
Van or Truck	1,081	1,195	MT	16.7	16.7
			AT	16.3	16.3
Van or Truck	1,196	1,310	MT	15.1	15.1
			AT	14.7	14.7
Van or Truck	1,311	1,420	MT	13.9	13.9
			AT	13.5	13.5
Van or Truck	1,421	1,530	MT	12.9	12.9
			AT	12.5	12.5
Van or Truck	1,531	1,650	MT	12.1	12.1
			AT	11.7	11.7
Van or Truck	1,651	1,760	MT	11.5	16.8
			AT	11.1	14.0
Van or Truck	1,761	1,870	MT	11.0*	15.9
			AT	10.6	13.7
Van or Truck	1,871	1,990	MT	11.0*	15.2
			AT	10.2**	13.5
Van or Truck	1,991	2,100	MT	11.0*	14.6†
			AT	10.2**	13.3
Van or Truck	2,101	—	MT	11.0*	14.6†
			AT	10.2**	13.0

* Van or Truck with manual transmission (MT) and SI (gasoline) engine from 1761 kg and up is all one bin.

** Van or Truck with automatic transmission (AT) and SI (gasoline) engine from 1871 kg and up is all one bin.

† Van or Truck with manual transmission (MT) and CI (diesel) engine from 1991 kg and up is all one bin.

TABLE 3.35 China Stage 4 light-duty vehicle fuel consumption targets [3-87].

Complete vehicle curb mass (kg)		Target (L/100 km)					
Minimum	Maximum	PC 1 or 2 seat rows	PC 3 or more seat rows	N1 gasoline	N1 diesel	M2 gasoline	M2 diesel
—	750	4.3	4.5	5.5	5.0	5.0	4.7
750	865	4.3	4.5	5.8	5.2	5.4	5.0
865	980	4.3	4.5	6.1	5.5	5.8	5.3
980	1,090	4.5	4.7	6.4	5.8	6.2	5.6
1,090	1,205	4.7	4.9	6.7	6.1	6.6	5.9
1,205	1,320	4.9	5.1	7.1	6.4	7.0	6.2
1,320	1,430	5.1	5.3	7.5	6.7	7.4	6.5
1,430	1,540	5.3	5.5	7.9	7.0	7.8	6.8
1,540	1,660	5.5	5.7	8.3	7.3	8.2	7.1
1,660	1,770	5.7	5.9	8.7	7.6	8.6	7.4
1,770	1,880	5.9	6.1	9.1	7.9	9.0	7.7
1,880	2,000	6.2	6.4	9.6	8.3	9.5	8.0
2,000	2,110	6.4	6.6	10.1	8.7	10.0	8.4
2,110	2,280	6.6	6.8	10.6	9.1	10.5	8.8
2,280	2,510	7.0	7.2	11.1	9.5	11.0	9.2
2,510	3,500*	7.3	7.5	11.7	10.0	11.5	9.6

* 3500 kg upper limit on curb mass (CM) applies to Category M_2 LCV; other vehicle types are nominally unbounded.

in GB 20997–2015 [3-88]. Per China standard GB/T 19233–2008 [3-89], the NEDC is used for certifying Stage 4 fuel consumption and CO_2 emissions results of LDVs. NEDC is described further in Section 4.1.8. Fuel consumption targets (in L/100 km) for passenger cars and LCVs are shown in Table 3.35.

As of writing, the following China Stage 5 standards for passenger cars are still in development, although the latest drafts dated January 2019 propose using WLTC as the test cycle for certification and using a mass-based fuel consumption target. There is no proposal yet for tighter fuel consumption rules for LCVs. WLTC is described in more detail in Section 4.1.9. The proposed fuel consumption targets are as follows:

$$\text{If CM} \leq 1090 \text{ kg}, FC_{tgt} = 4.02,$$

$$\text{If } 1090 < \text{CM} \leq 2510, FC_{tgt} = 4.60 + 0.0018 \times (\text{CM} - 1415), \quad (3.15)$$

$$\text{If CM} > 2510 \text{ kg}, FC_{tgt} = 6.57,$$

where CM is the curb mass (in kg), and FC_{tgt} is the fuel consumption target (in L/100 km). FC_{tgt} should be 0.2 (L/100 km) higher for passenger cars with three or more seating rows. To calculate GHG, China assumes that gasoline has 2.37×10^3 (g CO_2/L), and that diesel fuel has 2.60×10^3 (g CO_2/L).

3.4.6 India

India is currently using a fuel consumption standard for LDVs that is adapted from the EU's scheme for GHG emissions. For fiscal year (FY) 2017 through FY2021, the sales-weighted average fuel consumption in liters per 100 km (L/100 km) for a manufacturer in India is as follows [3-90]:

$$\overline{FC} = 0.0024 \times \left(W - 1037 \right) + 5.4922, \tag{3.16}$$

where W is the sales-weighted average vehicle mass sold (in kg). From FY2022 on, the sales-weighted average fuel consumption target changes to the following [3-90]:

$$\overline{FC} = 0.0020 \times \left(W - 1145 \right) + 4.7694. \tag{3.17}$$

India assumes that gasoline has 2.371×10^3 (g CO_2/L), and that diesel fuel has 2.648×10^3 (g CO_2/L).

3.4.7 Other Countries

South Korea has standards in place for LDVs for both fuel economy (in km/L) and GHG emissions (in g CO_2/km). There is a high-level split between LDVs that seat 10 or fewer persons and those that seat 11 to 15, and a floor to the vehicle mass of 1070 kg below which the fuel economy standard does not scale concerning mass.

Brazil has standards that are based on energy efficiency, as a way to normalize the different energy contents of the range of fuels used there. Other countries tend to use rules that are derived from either the EU or US standards, but this is not as guaranteed with fuel economy or GHG emissions as with criteria pollutants.

3.5 Heavy-Duty Vehicle Greenhouse Gases

In recent years, regulatory agencies have begun turning their attention to evaluating and regulating fuel consumption or GHG emissions from HDVs. Japan was one of the first markets to do so, followed by the US and EU. This effort marks a shift in regulatory emphasis since HDVs already have a strong market incentive to keep improving fuel consumption. Fuel consumption costs strongly affect an HDV operator's total cost of ownership and return on investment (ROI).

3.5.1 United States (Federal)

The US EPA and National Highway Traffic Safety Administration (NHTSA) have worked together to develop GHG and fuel economy standards for HDVs that span the range from Class 2b/3 LHDVs through to Class 8 tractors and heavy vocational vehicles. To date, there have been two phases of the rules, which together provide standards for HDVs from MY2014 through MY2027 [3-91, 3-92, 3-93].

As part of the rule-making process, the agencies identified HDV test cycles for evaluating fuel consumption or GHG emissions. These cycles were modified as part of the transition from the Phase 1 GHG standards to the Phase 2 GHG standards. The test cycles for each Phase of the GHG standards have been incorporated into the respective versions of EPA's GHG Emissions Model (GEM) certification tool, and are described in Sections 4.3.1 and 4.3.2, respectively. GEM is the official certification tool for HDVs under both phases of the GHG standards, although GEM only focuses on CO_2 from fuel consumption and not other potential GHG sources.

MDVs, that is, Class 2b and 3 LHDVs that are homologated for sale as complete vehicles, are certified using the chassis dynamometer tests for LDVs that are described in Section 4.1. Unlike LDVs, the fuel consumption or GHG emissions results for MDVs are weighted by cargo mass. If the engine uses a dynamometer certification, though, these LHDVs can be handled like other LHDVs in Classes 4 and 5 and heavier HDVs. Heavier HDVs are certified using a mix of powertrain component data on CO_2 emissions, fuel consumption, driveline efficiency, vehicle data, and GEM. Sections 4.3.1 and 4.3.2 describe the data requirements for the Phase 1 and Phase 2 GHG standards, respectively.

GEM requires several inputs from a manufacturer to assess compliance including the vehicle segment (*e.g.*, light, medium, or heavy HDV); projected use (*e.g.*, multipurpose, regional, or urban); technologies meriting credits; and engine or powertrain characteristics. The key input is the engine fueling map, which may either be the detailed steady-state fueling map from engine testing or a cycle-averaged fueling map that accounts for the full vehicle properties. HDV certificate holders are allowed to use GEM default values for non-engine powertrain components if they choose. It is expected that manufacturers will want to test complete power packs or powertrains to take advantage of the integration of the engine in the rest of the powertrain. The HDV test cycles are incorporated into GEM.

3.5.1.1 Phase 1 Greenhouse Gas Emissions Standard: The Phase 1 GHG
standard was published as a final rule in August 2011 and applies to HDVs from MY2014 through MY2020. In addition to the GHG requirements, there is a corresponding HDV fuel economy measurement and reporting requirement [3-91]. There are three broad categories of HDVs defined by these rules,

1. LHDVs that are homologated and tested as a complete vehicle (*i.e.*, MDVs)
2. Vocational vehicles
3. Tractors above 26,000 lb. GVWR

MDVs have their emissions limits defined in 40 CFR §86.1819-14 [3–94]. For each MDV, a work factor *WF* needs to be calculated using

$$WF = 0.75 \times \left(GVWR - CurbWt + xwd \right) + 0.25 \times \left(GCWR - GVWR \right), \qquad (3.18)$$

where *xwd* is a factor equal to 500 lb. if the vehicle has four-wheel or all-wheel drive, and to 0 lb. for all others; and CW, GVWR, and GCWR have the usual definitions for a given vehicle. This work factor is used in the formulas in **Table 3.36** to calculate the CO_2 target (in g/ton-mi.) for a given MDV in a given model year.

TABLE 3.36 Phase 1 GHG targets for MDV (chassis certified LHDV) by model year [3-94].

Model Year Range	CO$_2$ standard (g/ton-mi.)	
	Spark ignition	Compression ignition
MY2014	0.0482 × WF + 371	0.0478 × WF + 368
MY2015	0.0479 × WF + 369	0.0474 × WF + 366
MY2016	0.0469 × WF + 362	0.0460 × WF + 354
MY2017	0.0460 × WF + 354	0.0445 × WF + 343
MY2018–MY2020	0.0440 × WF + 339	0.0416 × WF + 320

TABLE 3.37 Phase 1 GHG targets for vocational vehicles [3-93].

Vehicle size	CO$_2$ standard (g/ton-mi.)	
	MY2014 through MY2016	MY2017 through MY2020
Light HDV	388	373
Medium HDV	234	225
Heavy HDV	226	222

TABLE 3.38 Phase 1 GHG targets for Class 7 and Class 8 tractors by model year [3-93].

Vehicle Type	CO$_2$ standard (g/ton-mi.)	
	MY2014–MY2016 Phase 1	MY2017–MY2020 Phase 1
Class 7 low roof	107	104
Class 7 mid-roof	119	115
Class 7 high roof	124	120
Class 8 low roof day cab	81	80
Class 8 mid-roof day cab	88	86
Class 8 high roof day cab	92	89
Class 8 low roof sleeper cab	68	66
Class 8 mid-roof sleeper cab	76	73
Class 8 high roof sleeper cab	75	72

The GHG targets for vocational vehicles are shown in Table 3.37, and those for Class 7 and 8 tractors, in Table 3.38. 40 CFR part 1037 [3-93] has a complete definition of the vehicle size classes, but in general, the LHDV size covers Class 2b through Class 5; the Medium HDV (MHDV), Class 6 and 7; and the Heavy HDV (HHDV), Class 8 vocational vehicles.

Tractors are split into nine categories in Phase 1, including low, medium, and high roof versions of Class 7 tractors, Class 8 day cabs, and Class 8 sleeper cabs.

3.5.1.2 **Phase 2 Greenhouse Gas Emissions Standard:** The Phase 2 GHG standard was published in August 2016 and extends the HDV standards from MY2021 through MY2027 [3-92, 3-93]. As with the Phase 1 GHG standard, there is a corresponding HDV fuel economy measurement and reporting requirement [3-91]. The drive cycles and GEM tools were updated to reflect the experience gained by EPA and NHTSA and the regulated community. The Phase 2 GHG standard is considerably more complex than the Phase 1 standard, as there are additional vehicle categories, especially for vocational vehicles. Also, the update in test cycles and input data requirements makes it difficult to compare certification targets and results from Phase 1 to Phase 2.

As with the Phase 1 GHG standards, the Phase 2 standards provide credits for technologies that have benefited either on cycle or off cycle. For example, Class 8 sleeper cab tractors qualify for a 6% credit for including a tamper-resistant battery pack for use as an auxiliary power unit (APU). HDVs also get credits for systems that support engine stop during idle or that monitor or maintain tire pressure.

In the Phase 2 GHG standard, MDVs use the same approach to estimate the CO_2 target for a given vehicle, which starts by using Eq. 3.18 to calculate the work factor *WF*. Table 3.39 shows how the work factor is used to calculate the CO_2 target (in g/ton-mi.) for each MDV in a given model year.

Vocational vehicles are split out into considerably more subcategories in the Phase 2 GHG standards, as shown in Tables 3.40, 3.41, and 3.42. For example, LHDVs and MHDVs are treated separately if they have CI or SI engines, and all types are further split out by expected use: multipurpose, regional, or urban. Tables 3.40 and 3.41 have the transitional values during the Phase 2 implementation, whereas Table 3.42 has the final set of targets.

As for Class 7 and Class 8 tractors in Phase 2 GHG standards, an additional category of the heavy-haul tractor was added to the nine from Phase 1. Table 3.43 shows both the transitional and final targets for these HDVs.

TABLE 3.39 Phase 2 GHG targets for MDV (chassis certified LHDV) by model year [3-94].

Model Year Range	CO_2 standard (g/ton-mi.)	
	Spark ignition	Compression ignition
MY2021	0.0429 × WF + 331	0.0406 × WF + 312
MY2022	0.0418 × WF + 322	0.0395× WF + 304
MY2023	0.0408 × WF + 314	0.0386 × WF + 297
MY2024	0.0398 × WF + 306	0.0376 × WF + 289
MY2025	0.0388 × WF + 299	0.0367 × WF + 282
MY2026	0.0378 × WF + 291	0.0357 × WF + 275
MY2027 and later	0.0369 × WF + 284	0.0348 × WF + 268

TABLE 3.40 Phase 2 GHG targets for MY2021 through MY2023 vocational vehicles [3-93].

Vehicle Type	CO_2 standard (g/ton-mi.)		
	Multipurpose	Regional	Urban
Light HDV, CI	373	311	424
Light HDV, SI	407	335	461
Medium HDV, CI	265	234	296
Medium HDV, SI	293	261	328
Heavy HDV	261	205	308

TABLE 3.41 Phase 2 GHG targets for MY2024 through MY2026 vocational vehicles [3-93].

Vehicle Type	CO$_2$ standard (g/ton-mi.)		
	Multipurpose	Regional	Urban
Light HDV, CI	344	298	385
Light HDV, SI	385	324	432
Medium HDV, CI	246	221	271
Medium HDV, SI	279	251	310
Heavy HDV	242	194	283

TABLE 3.42 Phase 2 GHG targets for MY2027 and later vocational vehicles [3-93].

Vehicle Type	CO$_2$ standard (g/ton-mi.)		
	Multipurpose	Regional	Urban
Light HDV, CI	330	291	367
Light HDV, SI	372	319	413
Medium HDV, CI	235	218	258
Medium HDV, SI	268	247	297
Heavy HDV	230	189	269

TABLE 3.43 Phase 2 GHG targets for Class 7 and Class 8 tractors by model year [3-93].

Vehicle Type	CO$_2$ standard (g/ton-mi.)		
	MY2021–MY2023 Phase 2	MY2024–MY2026 Phase 2	MY2027 and later Phase 2
Class 7 low roof	105.5	99.8	96.2
Class 7 mid-roof	113.2	107.1	103.4
Class 7 high roof	113.5	106.6	100.0
Class 8 low roof day cab	80.5	76.2	73.4
Class 8 mid-roof day cab	85.4	80.9	78.0
Class 8 high roof day cab	85.6	80.4	75.7
Class 8 low roof sleeper cab	72.3	68.0	64.1
Class 8 mid-roof sleeper cab	78.0	73.5	69.6
Class 8 high roof sleeper cab	75.7	70.7	64.3
Heavy-haul tractor	52.4	50.2	48.3

3.5.2 California

As with HD engine criteria pollutant standards, CARB generally follows the US Federal rules for GHG emissions for HD engines and HDVs. Nevertheless, under AB 32 [3-95] California plans to reduce GHG emissions to 60% of 1990 levels by 2030, and 20% of 1990 levels by 2050. Therefore, CARB is taking a multipronged approach to reducing net GHG emissions from the transportation sector given its current contribution to overall GHG emissions in the state.

To support these longer-term GHG targets, California has special rules for urban transit buses and is considering other options to encourage the market penetration of zero-emission HDVs in the state. For example, California rules require 85% of new urban buses to use alternative fuels, that is, fuels that are not distillate diesel. Alternative fuels include compressed natural gas (CNG); propane or liquefied petroleum gas (LPG); methanol; ethanol; hydrogen; electricity; or other advanced technologies that do not use diesel fuel.

3.5.3 European Union

As of writing, the EU has implemented rules for GHG emissions from HDVs that only require manufacturers to report out on their vehicles' emissions. On 17 May 2018, the European Commission [3-96] issued a proposal for new regulations that would transition the EU from a reporting mode to a compliance mode with fuel consumption targets defined on grams of fuel per tonne-km basis. The proposal has the new standards transition in starting from 1 January 2025, with a final standard in place by 1 January 2030.

The current regulatory framework for HDVs is analogous to that in the US, in that the European Commission recognizes that there are a limited number of chassis dynamometers available for full vehicle testing of HDVs and that the fuel consumption results tend to be less precise. Therefore, the EU is using the Vehicle Energy Consumption Calculation Tool (VECTO) to estimate fuel consumption and GHG emissions from HDVs.

As with the US Phase 2 GHG standards, the EU rules require component testing for an HDV's engine, gearbox, axles, and other auxiliary systems. The required engine data include the steady-state fueling map as well as results from the WHTC and WHSC. Other data required include the vehicle mass, tire dimensions, rolling resistance coefficients, aerodynamic drag factors, and moments of inertia from the HDV including a standard body or trailer. These test data are inputs for VECTO simulations used to calculate the official fuel consumption and CO_2 emissions values for a given HDV.

VECTO contains the set of test cycles over which the HDV fuel consumption and CO_2 emissions values are calculated. Which test cycle or cycles from the complete set are used by VECTO depends on the HDV type and expected use. Rexeis *et al.* [3-97], indicated that VECTO includes five different drive cycles, such as the Long Haul 2015, Regional Delivery 2016/2012, and Urban Delivery 2012.

3.5.4 Japan

From 2015, HDVs in Japan must meet fuel consumption targets [3-98]. Certification comes from measuring fuel consumption over two driving cycles and taking a weighted average of the results. The two drive cycles are the JE05 urban cycle and a constant 80 km/h speed intercity highway cycle, both of which are described in more detail in Section 4.3.4. The weighted average is calculated as follows:

$$FC = \left[\frac{1-\alpha}{FC_U} + \frac{\alpha}{FC_H}\right]^{-1}, \tag{3.19}$$

where FC is the averaged fuel consumption in (km/ℓ); FC_U, the urban fuel consumption from JE05; FC_H, the intercity highway fuel consumption; and α, the fraction of time spent in intercity highway driving. HDV fuel economy limits for tractor-trailers are

TABLE 3.44 Fuel economy limits for heavy-duty tractors [3-98, 3-99].

Vehicle classification	Range of GVW (Mg)	FY2015 Standard (km/L)	FY2025 Standard (km/L)
TT1	Up to 20	3.09	3.11
TT2	Over 20	2.01	2.32

TABLE 3.45 Fuel economy limits for heavy-duty trucks [3-99].

Vehicle classification	Range of GVW (Mg) Minimum*	Maximum	Maximum load (Mg) Minimum*	Maximum	FY2015 Standard (km/L)	FY2025 Standard (km/L)
T1	3.5	7.5	0	1.5	10.83	13.45
T2	3.5	7.5	1.5	2	10.35	11.93
T3	3.5	7.5	2	3	9.51	10.59
T4	3.5	7.5	3	—	8.12	9.91
T5	7.5	8	—	—	7.24	8.39
T6	8	10	—	—	6.52	7.46
T7	10	12	—	—	6.00	7.44
T8	12	14	—	—	5.69	6.42
T9	14	16	—	—	4.97	5.89
T10	16	20	—	—	4.15	4.88
T11	20	25	—	—	4.04	4.42

* The minimum value is not included in the range, but the maximum value is.

shown in Table 3.44, and, for trucks, in Table 3.45. Similar limits are in place for buses and may be found in the Japanese regulations. In addition, Japan is introducing limits for HDV fuel economy that start in Japanese FY2025, and these are also shown in Tables 3.44 and 3.45.

3.5.5 People's Republic of China

In the PRC, the Stage 3 fuel consumption limits [3-100] are phasing in as of publication, with all new type approvals following the standard from 1 July 2019, and all approvals following from 1 July 2021. Fuel consumption is measured using the C-WTVC procedure [3-101]. It uses computer simulation over a 900 s urban segment followed by a 468 s highway segment followed by a 432 s expressway segment. The fuel consumption (in L/100 km) is calculated three ways: by a carbon balance method, a gravimetric method, and a volumetric method. These three results are then averaged together over a set of C-WTVC tests and among each other.

Each type of HDV has its mass-based standard to follow. For example, the fuel consumption limits for vocational vehicles are shown in Table 3.46, where the mass range includes the maximum value but excludes the minimum. The corresponding targets for HDV tractors are shown in Table 3.47. The standard [3-100] also includes targets for motor coaches, city buses, and dump trucks separate from those in Table 3.46.

TABLE 3.46 China Stage 3 heavy-duty vocational vehicle fuel consumption targets |[3-100].

GVW (kg)		Target (L/100 km)
Minimum	Maximum	
3,500	4,500	11.5*
4,500	5,500	12.2*
5,500	7,000	13.8*
7,000	8,500	16.3*
8,500	10,500	18.3*
10,500	12,500	21.3*
12,500	16,000	24.0
16,000	20,000	27.0
20,000	25,000	32.5
25,000	31,000	37.5
31,000	-	38.5

* For HDV with SI engines, the limit is 20% higher, with the decimal rounded up.

TABLE 3.47 China Stage 3 heavy-duty vocational vehicle fuel consumption targets [3-100].

GVW (kg)		Target (L/100 km)
Minimum	Maximum	
—	18,000	28.0
18,000	27,000	30.5
27,000	35,000	32.0
35,000	40,000	34.0
40,000	43,000	35.5
43,000	46,000	38.0
46,000	49,000	40.0
49,000	—	40.5

3.5.6 India

India has implemented fuel consumption rules for HDVs [3-102], and they have been in effect since 1 April 2018 in the 1st Phase of implementation. The standards are given as a fuel consumption target (in L/100 km) that is a function of the vehicle's type, GVW, and axle configuration at a constant speed of 40 (km/h) (25 mph) or 60 (km/h) (37 mph) on a 0% grade. The standards for HDVs in the 1st Phase are shown in Table 3.48.

The 2nd Phase standards will start to take effect on 1 April 2021, and they represent a further tightening of the fuel consumption standards. The standards for HDVs in the 2nd Phase are shown in Table 3.49 [3-102].

TABLE 3.48 1st Phase standards (2018) for rigid N_3, M_3, and N_3 tractor HDV [3-102].

Type	GVW (Mg) Minimum	Maximum	Axle configuration	Fuel consumption target (L/100 km) At 40 km/h	At 60 km/h
N_3 rigid	12.0	16.2	4×2	0.362 × GVW + 10.327	0.788 × GVW + 9.003
N_3 rigid	16.2	25.0	6×2	0.603 × GVW + 6.415	0.755 × GVW + 9.546
N_3 rigid	16.2	25.0	6×4	0.723 × GVW + 4.482	1.151 × GVW + 3.122
N_3 rigid	25.0	31.0	8×2	0.527 × GVW + 8.333	0.650 × GVW + 12.160
N_3 rigid	25.0	31.0	8×4	0.928 × GVW – 0.658	0.968 × GVW + 7.692
N_3 rigid	31.0	37.0	10×2	0.960 × GVW – 5.100	0.650 × GVW + 12.160
M_3	12.0	—	4×2 and 6×2	0.509 × GVW + 11.062	0.199 × GVW + 19.342
N_3 tractor	35.2	40.2	4×2	0.968 × GVW – 7.727	0.208 × GVW + 32.198
N_3 tractor	40.2	49.0	6×2	0.628 × GVW + 6.648	0.628 × GVW + 15.298
N_3 tractor	40.2	49.0	6×4	1.255 × GVW – 18.523	1.342 × GVW – 13.390

TABLE 3.49 2nd Phase standards (2021) rigid N_3, M_3, and N_3 tractor HDV [3-102].

Type	GVW (Mg) Minimum	Maximum	Axle configuration	Fuel consumption target (L/100 km) At 40 km/h	At 60 km/h
N_3 rigid	12.0	16.2	4×2	0.329 × GVW + 9.607	0.600 × GVW + 9.890
N_3 rigid	16.2	25.0	6×2	0.523 × GVW + 6.462	0.515 × GVW + 11.271
N_3 rigid	16.2	25.0	6×4	0.673 × GVW + 4.032	0.932 × GVW + 4.515
N_3 rigid	25.0	31.0	8×2	0.430 × GVW + 8.780	0.382 × GVW + 14.598
N_3 rigid	25.0	31.0	8×4	0.732 × GVW + 2.558	1.318 × GVW – 5.148
N_3 rigid	31.0	37.0	10×2	0.963 × GVW – 7.753	1.043 × GVW – 5.913
M_3	12.0	—	4×2 and 6×2	0.659 × GVW + 6.582	0.340 × GVW + 14.300
N_3 tractor	35.2	40.2	4×2	0.826 × GVW – 3.165	0.260 × GVW + 27.888
N_3 tractor	40.2	49.0	6×2	0.630 × GVW + 4.732	0.236 × GVW + 28.838
N_3 tractor	40.2	49.0	6×4	1.008 × GVW – 10.480	0.563 × GVW + 15.728

3.5.7 Other Countries

Other countries are beginning to implement fuel consumption or GHG emissions standards for HDVs, but these rules are still in development. For example, South Korea has rules in effect as of 2019 to monitor measured fuel economy values for HDVs that will transition to fuel economy standards with targets for manufacturers to meet.

3.6 Unregulated Emissions

With the introduction of GHG emissions requirements for light-duty and HD applications, several formerly unregulated gaseous emissions such as CH_4 and N_2O are now covered by standards, since they are substantially more effective greenhouse gases on a per molecule basis than CO_2 is. In addition, NH_3 has become regulated as the criteria pollutant emissions standards have developed. In the US, manufacturers are asked to affirm that they are not aware of increasing any unregulated emissions in the course of mitigating the regulated ones.

References

3-1. Reinhardt, C., "Across the Great Divide: A Conversation with Joseph Stiglitz," https://www.barnesandnoble.com/review/across-the-great-divide-a-conversation-with-joseph-stiglitz. Updated 19 May 2015, Accessed 30 Mar. 2019.

3-2. Brimblecombe, P., "Attitudes and Responses towards Air Pollution in Medieval England," *Journal of the Air Pollution Control Association* **26**, no. 10 (1976): 941–945, https://doi.org/10.1080/00022470.1976.10470341.

3-3. Gemmer, M. and Bo, X., "Air Quality Legislation and Standards in the European Union: Background, Status and Public Participation," *Advances in Climate Change Research* **4**, no. 1 (2013): 50-59, https://doi.org/10.3724/sp.J.1248.2013.050.

3-4. Environmental Protection Agency, "Overview of the Clean Air Act and Air Pollution," https://www.epa.gov/clean-air-act-overview. Updated 28 Aug. 2018, Accessed 14 Nov. 2018.

3-5. Environmental Protection Agency, "Part 50—National Primary and Secondary Ambient Air Quality Standards," *Code of Federal Regulations, Title 40*, Vol. **2** (2018): 5–169, https://www.govinfo.gov/content/pkg/CFR-2018-title40-vol2/pdf/CFR-2018-title40-vol2-chapI-subchapC.pdf.

3-6. Environmental Protection Agency, "NAAQS table," https://www.epa.gov/criteria-air-pollutants/naaqs-table. Updated 20 Dec. 2016, Accessed 14 Nov. 2018.

3-7. AVL, "AVL Emission Reports (Emrep)," https://www.avl.com/emission-reports. Accessed 15 Mar. 2019.

3-8. Delphi Technologies, *Worldwide Emissions Standards: Passenger Cars and Light Duty Vehicles, 2018-2019*, 26th ed (Delphi Technologies plc, 2018).

3-9. Delphi Technologies, *Worldwide Emissions Standards: On and Off-Highway Commercial Vehicles, 2018-2019*, 26th ed (Delphi Technologies plc, 2018).

3-10. DieselNet, "Emission Standards," https://www.dieselnet.com/standards/. Updated 2019, Accessed 20 Mar. 2019.

3-11. Ricardo plc, "Emissions Legislation (EMLEG)," http://www.emleg.com/. Updated 2019, Accessed 20 Mar. 2019.

3-12. World Health Organization, "Health Aspects of Air Pollution with Particulate Matter, Ozone, and Nitrogen Dioxide: Report on a WHO Working Group," 2003.

3-13. Benbrahim-Tallaa, L., Baan, R.A., Grosse, Y., Lauby-Secretan, B., El Ghissassi, F., Bouvard, V., Guha, N., Loomis, D., and Straif, K., "Carcinogenicity of Diesel-Engine and Gasoline-Engine Exhausts and Some Nitroarenes," *The Lancet Oncology* **13**, no. 7 (2012): 663–664.

3-14. Donaldson, K., Duffin, R., Langrish, J.P., Miller, M.R., Mills, N.L., Poland, C.A., Raftis, J., Shah, A., Shaw, C.A., and Newby, D.E., "Nanoparticles and the Cardiovascular System: A Critical Review," *Nanomedicine* **8**, no. 3 (2013): 403–423.

3-15. Heyder, J., "Deposition of Inhaled Particles in the Human Respiratory Tract and Consequences for Regional Targeting in Respiratory Drug Delivery," *Proceedings of the American Thoracic Society* **1**, no. 4 (2004): 315–320.

3-16. Coburn, R.F., "Mechanisms of Carbon Monoxide Toxicity," *Preventive Medicine* **8**, no. 3 (1979): 310–322.

3-17. Omaye, S.T., "Metabolic Modulation of Carbon Monoxide Toxicity," *Toxicology* **180**, no. 2 (2002): 139–150.

3-18. Turino, G.M., "Effect of Carbon Monoxide on the Cardiorespiratory System: Carbon Monoxide Toxicity, Physiology and Biochemistry," *Circulation (United States)* **63**, no. 1 (1981).

3-19.   Morrow, P.E., "An Evaluation of Recent NOx Toxicity Data and an Attempt to Derive an Ambient Air Standard for NOx by Established Toxicological Procedures," *Environmental Research* **10**, no. 1 (1975): 92–112.

3-20.   Book, S.A., "Scaling Toxicity from Laboratory Animals to People: An Example with Nitrogen Dioxide," *Journal of Toxicology and Environmental Health, Part A Current Issues* **9**, no. 5-6 (1982): 719–725.

3-21.   Environmental Protection Agency, "2014 National Emissions Inventory (NEI) Report," 2017.

3-22.   Dickerson, R.R., Kondragunta, S., Stenchikov, G., Civerolo, K.L., Doddridge, B.G., and Holben, B.N., "The Impact of Aerosols on Solar Ultraviolet Radiation and Photochemical Smog," *Science* **278**, no. 5339 (1997): 827–830.

3-23.   Calvert, J.G., "Hydrocarbon Involvement in Photochemical Smog Formation in Los Angeles Atmosphere," *Environmental Science & Technology* **10**, no. 3 (1976): 256–262.

3-24.   Environmental Protection Agency, "Overview of Greenhouse Gases Emissions in 2017," https://www.epa.gov/ghgemissions/inventory-us-greenhouse-gas-emissions-and-sinks-1990-2017. Updated 11 Apr 2019, Accessed 4 Dec 2019.

3-25.   Anderson, J.O., Thundiyil, J.G., and Stolbach, A., "Clearing the Air: A Review of the Effects of Particulate Matter Air Pollution on Human Health," *Journal of Medical Toxicology* **8**, no. 2 (2012): 166–175, https://doi.org/10.1007/s13181-011-0203-1.

3-26.   Attfield, M.D., Schleiff, P.L., Lubin, J.H., Blair, A., Stewart, P.A., Vermeulen, R., Coble, J.B., and Silverman, D.T., "The Diesel Exhaust in Miners Study: A Cohort Mortality Study with Emphasis on Lung Cancer," *Journal of the National Cancer Institute* **104**, no. 11 (2012): 869–883, https://doi.org/10.1093/jnci/djs035.

3-27.   Hoek, G., Krishnan, R.M., Beelen, R., Peters, A., Ostro, B., Brunekreef, B., and Kaufman, J.D., "Long-Term Air Pollution Exposure and Cardio-Respiratory Mortality: A Review," *Environmental Health* **12**, no. 1 (2013): 43, https://doi.org/10.1186/1476-069X-12-43.

3-28.   Dai, L., Zanobetti, A., Koutrakis, P., and Schwartz, J.D., "Associations of Fine Particulate Matter Species with Mortality in the United States: A Multicity Time-Series Analysis," *Environmental Health Perspectives* **122**, no. 8 (2014): 837–842, https://doi.org/10.1289/ehp.1307568.

3-29.   Weinmayr, G., Hennig, F., Fuks, K., Nonnemacher, M., Jakobs, H., Möhlenkamp, S., Erbel, R., Jöckel, K.-H., Hoffmann, B., and Moebus, S., "Long-Term Exposure to Fine Particulate Matter and Incidence of Type 2 Diabetes Mellitus in a Cohort Study: Effects of Total and Traffic-Specific Air Pollution," *Environmental Health* **14**, no. 1 (2015): 53, https://doi.org/10.1186/s12940-015-0031-x.

3-30.   Abdel-Shafy, H.I. and Mansour, M.S.M., "A Review on Polycyclic Aromatic Hydrocarbons: Source, Environmental Impact, Effect on Human Health and Remediation," *Egyptian Journal of Petroleum* **25**, no. 1 (2016): 107-123, https://doi.org/10.1016/j.ejpe.2015.03.011.

3-31.   Kheirbek, I., Haney, J., Douglas, S., Ito, K., and Matte, T., "The Contribution of Motor Vehicle Emissions to Ambient Fine Particulate Matter Public Health Impacts in New York City: A Health Burden Assessment," *Environmental Health* **15**, no. 1 (2016): 89, https://doi.org/10.1186/s12940-016-0172-6.

3-32.   Gladstein, N., "Ultrafine Particulate Matter and the Benefits of Reducing Particle Numbers in the United States," A Report to the Manufacturers of Emissions Controls Association (MECA), 2013.

3-33.   Environmental Protection Agency, "Part 86—Control of Emissions from New and In-Use Highway Vehicles and Engines," Code of Federal Regulations, Title 40, Vol. **21** (2018): 429–1202, https://www.govinfo.gov/app/collection/cfr/2018/title40/chapterI/subchapterC/part86.

3-34. Environmental Protection Agency, "§ 86.1809-01 Prohibition of Defeat Devices," *Code of Federal Regulations, Title 40*, Vol. **21** (2018): 932, https://www.govinfo.gov/app/collection/cfr/2018/title40/chapterI/subchapterC/part86/subpartS.

3-35. Environmental Protection Agency, "§ 86.1809-10 Prohibition of Defeat Devices," *Code of Federal Regulations, Title 40*, Vol. **21** (2018): 933–934, https://www.govinfo.gov/app/collection/cfr/2018/title40/chapterI/subchapterC/part86/subpartS.

3-36. Environmental Protection Agency, "§ 86.1809-12 Prohibition of Defeat Devices," *Code of Federal Regulations, Title 40*, Vol. **21** (2018): 934–935, https://www.govinfo.gov/app/collection/cfr/2018/title40/chapterI/subchapterC/part86/subpartS.

3-37. Environmental Protection Agency, "§ 86.1803-01 Definitions," *Code of Federal Regulations, Title 40*, Vol. **21** (2018): 888–901, https://www.govinfo.gov/app/collection/cfr/2018/title40/chapterI/subchapterC/part86/subpartS.

3-38. Environmental Protection Agency, "§ 86.1811-17 Exhaust Emission Standards for Light-Duty Vehicles, Light-Duty Trucks and Medium-Duty Passenger Vehicles," *Code of Federal Regulations, Title 40*, Vol. **21** (2018): 961–970, https://www.govinfo.gov/app/collection/cfr/2018/title40/chapterI/subchapterC/part86/subpartS.

3-39. State of California, "§1961.2 Exhaust Emission Standards and Test Procedures - 2015 and Subsequent Model Passenger Cars, Light-Duty Trucks, and Medium-Duty Vehicles," *California Code of Regulations*, Title 13 (Mar. 22, 2012): 37–74.

3-40. European Commission, "Regulation (EC) No 715/2007 of the European Parliament and of the Council of 20 June 2007 on type approval of motor vehicles and engines with respect to emissions from light passenger and commercial vehicles (Euro 5 and Euro 6) and on access to vehicle repair and maintenance information," *Official Journal of the European Union*, Vol. **171** (Jun. 29, 2007): 1–16, https://www.eea.europa.eu/policy-documents/regulation-ec-no-715-2007.

3-41. Mamakos, A., Martini, G., Marotta, A., and Manfredi, U., "Assessment of Different Technical Options in Reducing Particle Emissions from Gasoline Direct Injection Vehicles," *Journal of Aerosol Science* **63** (2013): 115–125, https://doi.org/10.1016/j.jaerosci.2013.05.004.

3-42. Road Transport Bureau, Ministry of Land, Infrastructure, Transport, and Tourism (MLIT), "Partial Amendments of the Announcement that prescribes Details of Safety Regulations for Road Transport Vehicles, etc.," 2016.

3-43. National Public Radio, "'It Was Installed For This Purpose', VW's U.S. CEO Tells Congress About Defeat Device," https://www.npr.org/sections/thetwo-way/2015/10/08/446861855/volkswagen-u-s-ceo-faces-questions-on-capitol-hill. Updated 8 Oct. 2015, Accessed 23 Jul. 2019.

3-44. Ministry of Ecology and Environment of PRC, "Limits and Measurement Methods for Emissions from Light-Duty Vehicles (China 5)," GB 18352.5-2013, 2013.

3-45. Ministry of Ecology and Environment of PRC, "Limits and Measurement Methods for Emissions from Light-Duty Vehicles (China 6)," GB 18352.6-2016, 2016.

3-46. Ministry of Road Transport and Highways, "Notification No. G.S.R. 889(E), New Delhi, the 16th September, 2016, regarding mass emission standards for BS-VI," G.S.R. 889(E), 2016, http://egazette.nic.in/WriteReadData/2016/171776.pdf.

3-47. Conselho Nacional do Meio Ambiente (CONAMA), "Resolução CONAMA No. 415/2009 - Dispõe sobre nova fase (PROCONVE L6) de exigências do Programa de Controle da Poluição do Ar por Veículos Automotores-PROCONVE para veículos automotores leves novos de uso rodoviário e dá outras providências," 2009, http://www2.mma.gov.br/port/conama/legiabre.cfm?codlegi=615.

3-48. Conselho Nacional do Meio Ambiente (CONAMA), "Resolução CONAMA No. 492/2018. Estabelece as Fases PROCONVE L7 e PROCONVE L8 de exigências do

Programa de Controle da Poluição do Ar por Veículos Automotores - PROCONVE para veículos automotores leves novos de uso rodoviário, altera a Resolução CONAMA n° 15/1995 e dá outras providências," 2018, http://www.in.gov.br/materia/-/asset_publisher/ Kujrw0TZC2Mb/content/id/56643907.

3-49.   Ministry of Environment, "Enforcement Regulations of Atmospheric Environment Preservation Act," MOE Ord. 714-2017, Oct. 18, 2017.

3-50.   Environmental Protection Agency, "§ 86.007-11 Emission Standards and Supplemental Requirements for 2007 and Later Model Year Diesel Heavy-Duty Engines and Vehicles," Code of Federal Regulations, Title 40, Vol. **21** (2018): 467–471, https://www.govinfo.gov/ app/collection/cfr/2018/title40/chapterI/subchapterC/part86.

3-51.   Environmental Protection Agency, "§ 86.010-18 On-Board Diagnostics for Engines Used in Applications Greater than 14,000 pounds GVWR," *Code of Federal Regulations, Title 40*, Vol. **21** (2018): 483–551, https://www.govinfo.gov/app/collection/cfr/2018/title40/ chapterI/subchapterC/part86.

3-52.   Environmental Protection Agency, "§ 86.1816-18 Emission Standards for Heavy-Duty Vehicles," *Code of Federal Regulations, Title 40*, Vol. **21** (2018): 979–984, https://www. govinfo.gov/app/collection/cfr/2018/title40/chapterI/subchapterC/part86/subpartS.

3-53.   Shepardson, D., "US Plans New Limits on Heavy-Duty Truck Emissions," https://www. hydrocarbonprocessing.com/news/2018/11/us-plans-new-limits-on-heavy-duty-truck-emissions. Updated 13 Nov. 2018, Accessed 14 Nov. 2018.

3-54.   California Air Resources Board, "California Air Resources Board Staff Current Assessment of the Technical Feasibility of Lower NOx Standards and Associated Test Procedures for 2022 and Subsequent Model Year Medium-Duty and Heavy-Duty Diesel Engines," California Air Resources Board, 2019.

3-55.   Environmental Protection Agency, "Part 1039—Control of Emissions from New and In-Use Nonroad Compression-Ignition Engines," *Code of Federal Regulations, Title 40*, Vol. **36** (2018): 291–376, https://www.govinfo.gov/app/collection/cfr/2018/title40/ chapterI/subchapterU/part1039.

3-56.   Environmental Protection Agency, "§ 1039.101 What Exhaust Emission Standards must my Engines Meet after the 2014 Model Year?," Code of Federal Regulations, Title 40, Vol. **36** (2018): 295–298, https://www.govinfo.gov/content/pkg/CFR-2018-title40-vol36/pdf/ CFR-2018-title40-vol36-sec1039-101.pdf.

3-57.   European Commission, "Commission Regulation (EU) No 136/2014 amending Directive 2007/46/EC of the European Parliament and of the Council, Commission Regulation (EC) No 692/2008 as regards emissions from light passenger and commercial vehicles (Euro 5 and Euro 6) and Commission Regulation (EU) No 582/2011 as regards emissions from heavy duty vehicles (Euro VI)," *Official Journal of the European Union*, Vol. **43** (Feb. 11, 2014): 12–35.

3-58.   United Nations, "Agreement concerning the adoption of uniform technical prescriptions for wheeled vehicles, equipment, and parts which can be fitted and/or be used on wheeled vehicles and the conditions for reciprocal recognition of approvals granted on the basis of these prescriptions," Addendum 48: Regulation No. 49, Revision 6 (Mar. 4, 2013): 1–434.

3-59.   European Commission, "Directive 97/68/EC of the European Parliament and of the Council of 16 December 1997 on the approximation of the laws of the Member States relating to measures against the emission of gaseous and particulate pollutants from internal combustion engines to be installed in non-road mobile machinery," *Official Journal of the European Union*, Vol. **59** (1997): 1–85, https://eur-lex.europa.eu/eli/ dir/1997/68/oj.

3-60.   European Commission, "Regulation (EU) 2016/1628 of 14 September 2016 on requiremernts relating to gaseous and particulate pollutant emission limits and type-approval for internal combustion engines for non-road mobile machinery, amending Regulations (EU) No 1024/2012 and (EU) No 167/2013, and amending and repealing Directive 97/68/EC," *Official Journal of the European Union*, Vol. **252** (2016): 53–117.

3-61.   Ministry of Environment, "Post New Long-Term Standards," http://www.env.go.jp/air/car/gas_kisei/kisei.pdf. Accessed Apr. 24, 2019.

3-62.   Ministry of Ecology and Environment of PRC, "Limits and Measurement Methods for Emissions from Diesel-Fueled Heavy-Duty Vehicles (CHINA VI)," GB 17691-2018, 2018.

3-63.   Ministry of Ecology and Environment of PRC, "Limits and Measurement Methods for Exhaust Pollutants from Diesel Engines of Non-Road Mobile Machinery (China III, IV)," GB 20891-2014, 2014.

3-64.   Ministry of Road Transport and Highways, "Notification No. G.S.R. (201)(E) dated 5 March 2018 Regarding Emission Standards for CEV and Agricultural Tractors," Regd. No. D.L.-33004/99, 2018, http://morth.nic.in/showfile.asp?lid=3180.

3-65.   Conselho Nacional do Meio Ambiente (CONAMA), "Resolução CONAMA No. 403, de 11 de novembro de 2008 - Dispõe sobre a nova fase de exigência do Programa de Controle da Poluição do Ar por Veículos Automotores-PROCONVE para veículos pesados novos (Fase P-7) e dá outras providências," Nov. 11, 2008, https://www.legisweb.com.br/legislacao/?id=108775.

3-66.   European Commission, "DIRECTIVE 2005/55/EC OF THE EUROPEAN PARLIAMENT AND OF THE COUNCIL of 28 September 2005 on the approximation of the laws of the Member States relating to the measures to be taken against the emission of gaseous and particulate pollutants from compression-ignition engines for use in vehicles, and the emission of gaseous pollutants from positive-ignition engines fuelled with natural gas or liquefied petroleum gas for use in vehicles," *Official Journal of the European Union*, Vol. **275** (Sep. 28, 2005): 1–163, https://eur-lex.europa.eu/LexUriServ/LexUriServ.do?uri=OJ:L:2005:275:0001:0163:EN:PDF.

3-67.   European Commission, "COMMISSION DIRECTIVE 2008/74/EC of 18 July 2008 amending, as regards the type approval of motor vehicles with respect to emissions from light passenger and commercial vehicles (Euro V and Euro VI) and access to vehicle repair and maintenance information, Directive 2005/55/EC of the European Parliament and of the Council and Directive 2005/78/EC," *Official Journal of the European Union*, Vol. **192** (Jul. 18, 2008): 51–59, https://publications.europa.eu/en/publication-detail/-/publication/d0a3abdc-1501-4627-8256-82570f6bed9b/language-en.

3-68.   Conselho Nacional do Meio Ambiente (CONAMA), "Resolução CONAMA No. 490, de 16 de novembro de 2018 - Estabelece a Fase PROCONVE P8 de exigências do Programa de Controle da Poluição do Ar por Veículos Automotores - PROCONVE para o controle das emissões de gases poluentes e de ruído para veículos automotores pesados novos de uso rodoviário e dá outras providências," (Nov. 16, 2018): 153–156, http://pesquisa.in.gov.br/imprensa/jsp/visualiza/index.jsp?data=21/11/2018&jornal=515&pagina=153.

3-69.   Conselho Nacional do Meio Ambiente (CONAMA), "Resolução CONAMA No. 233, de 13 de julho de 2011 - Dispõe sobre a inclusão no Programa de Controle da Poluição do Ar por Veículos AutomotoresPROCONVE e estabelece limites máximos de emissão de ruídos para máquinas agrícolas e rodoviárias novas.," Jul. 13, 2011: 69-76, http://www2.mma.gov.br/port/conama/legiabre.cfm?codlegi=654.

3-70.   "An Act to increase domestic energy supplies and availability; to restrain energy demand; to prepare for energy emergencies; and for other purposes," Public Law 94-163, U.S. Statutes at Large, Vol. **89** (Dec. 22, 1975): 871–969.

3-71.   "An Act to move the United States toward greater energy independence and security, to increase the production of clean renewable fuels, to protect consumers, to increase the efficiency of products, buildings, and vehicles, to promote research on and deploy greenhouse gas capture and storage options, and to improve the energy performance of the Federal Government, and for other purposes.," Public Law 110-140, U.S. Statutes at Large, Vol. **121** (Dec. 19, 2007): 1492–1801, https://www.govinfo.gov/content/pkg/PLAW-110publ140/pdf/PLAW-110publ140.pdf.

3-72.   National Highway Traffic Safety Administration, "NHTSA and EPA set standards to improve fuel economy and reduce greenhouse gases for passenger cars and light trucks for model years 2017 and beyond," National Highway Traffic Safety Administration, 2011, 1–10.

3-73.   National Highway Traffic Safety Administration, "SAFE - The Safer Affordable Fuel-Efficient 'SAFE' Vehicles Rule," https://www.nhtsa.gov/corporate-average-fuel-economy/safe. Updated 31 Mar. 2020, Accessed 25 May 2020.

3-74.   Environmental Protection Agency, "§ 86.1818-12 Greenhouse gas emission standards for light-duty vehicles, light-duty trucks, and medium-duty passenger vehicles," *Code of Federal Regulations, Title 40*, Vol. **21** (2018): 990-1001, https://www.govinfo.gov/app/collection/cfr/2018/title40/chapterI/subchapterC/part86/subpartS.

3-75.   U.S. Department of Energy and Environmental Protection Agency, "www.FuelEconomy.gov - The official government source for fuel economy information," https://www.fueleconomy.gov/. Updated 25 Mar. 2019, Accessed 25 Mar. 2019.

3-76.   Environmental Protection Agency, "Fuel Economy Labeling of Motor Vehicles: Revisions to Improve Calculation of Fuel Economy Estimates; Final Rule," *Federal Register*, Vol. **71**, No. 248 (Dec. 27, 2006): 77872-77969, https://www.gpo.gov/fdsys/pkg/FR-2006-12-27/pdf/06-9749.pdf.

3-77.   Environmental Protection Agency, "§ 600.114-12 Vehicle-Specific 5-Cycle Fuel Economy and Carbon-Related Exhaust Emission Calculations," *Code of Federal Regulations, Title 40*, Vol. **37** (2011): 901–913, https://www.govinfo.gov/content/pkg/CFR-2018-title40-vol32/pdf/CFR-2018-title40-vol32-part600-subpartB.pdf.

3-78.   Environmental Protection Agency, "§ 1066.831 Exhaust Emission Test Procedures for Aggressive Driving," *Code of Federal Regulations, Title 40*, Vol. **37** (2018): 368–369, https://www.govinfo.gov/app/collection/cfr/2018/title40/chapterI/subchapterU/part1066/subpartI.

3-79.   California Air Resources Board, "California Greenhouse Gas Emission Inventory - 2018 Edition," https://www.arb.ca.gov/cc/inventory/data/data.htm. Updated 11 Jul. 2018, Accessed 29 Apr. 2019.

3-80.   Weissler, P., "California ZEV Law Gets Simpler, More Challenging," *Automotive Engineering* (Dec 2017): 17–19.

3-81.   European Commission, "Regulation (EU) 2019/631 of the European Parliament and of the Council of 17 April 2019 setting CO2 emission performance standards for new passenger cars and for new light commercial vehicles, and repealing Regulations (EC) No 443/2009 and (EU) No 510/2011," *Official Journal of the European Union*, Vol. **111** (2019): 13-53, https://eur-lex.europa.eu/eli/reg/2019/631/oj.

3-82.   European Commission, "Regulation (EU) 2017/1151 of 1 June 2017 supplementing Regulation (EC) No 715/2007 of the European Parliament and of the Council on type-approval of motor vehicles with respect to emissions from light passenger and commercial vehicles (Euro 5 and Euro 6) and on access to vehicle repair and maintenance information, amending Directive 2007/46/EC of the European Parliament and of the Council, Commission Regulation (EC) No 692/2008 and Commission Regulation (EU) No 1230/2012 and repealing Commission Regulation (EC) No 692/2008," *Official Journal of the European Union*, Vol. **175** (2017): 1-643, https://eur-lex.europa.eu/legal-content/EN/TXT/?uri=CELEX%3A32017R1151.

3-83. "Final Report of Joint Meeting between the Automobile Evaluation Standards Subcommittee, Energy Efficiency Standards Subcommittee of the Advisory Committee for Natural Resources and Energy and the Automobile Fuel Efficiency Standards Subcommittee, Automobile Transport Section, Land Transport Division of the Council for Transport Policy," 2007.

3-84. "Final Report of Joint Meeting between the Automobile Evaluation Standards Subcommittee, Energy Efficiency Standards Subcommittee of the Advisory Committee for Natural Resources and Energy and the Automobile Fuel Efficiency Standards Subcommittee, Automobile Section, Land Transport Division of the Council for Transport Policy," 2011.

3-85. Ministry of Land, Infrastructure, Transport, and Tourism (MLIT), "Summary of the Joint Meeting of the Comprehensive Resources and Energy Research Committee, Energy Conservation and New Energy Subcommittee, Energy Conservation Subcommittee Automotive Judgment Working Group and the Transportation Policy Council Land Transport Subcommittee, Automobile Subcommittee, Automobile Fuel Consumption Standards Subcommittee," Japan Ministry of Land, Infrastructure, Transport and Tourism (MLIT), 2015.

3-86. Rutherford, D., "Japan Light Commercial Vehicle Fuel Economy Standards for 2022," The International Council on Clean Transportation, 2018.

3-87. Ministry of Ecology and Environment of PRC, "Fuel Consumption Evaluation Methods and Targets for Passenger Cars," GB 27999-2014, 2014.

3-88. Ministry of Ecology and Environment of PRC, "Limits of Fuel Consumption for Light-Duty Commercial Vehicles," GB20997-2015 2015.

3-89. Ministry of Ecology and Environment of PRC, "Measurement Method of Fuel Consumption for Light-Duty Vehicle," GB/T 19233-2008, 2008.

3-90. Ministry of Power, "Notification No. 837, New Delhi, the 23rd April, 2015," 2015: 1–6.

3-91. National Highway Traffic Safety Administration, "Part 535—Medium- and Heavy-Duty Vehicle Fuel Efficiency Program," Code of Federal Regulations, Title 49, Vol. **6** (2016): 119–179, https://www.govinfo.gov/app/collection/cfr/2018/title49/subtitleB/chapterV/part535.

3-92. Environmental Protection Agency, "Part 1036—Control of Emissions from New and In-Use Heavy-Duty Highway Engines," *Code of Federal Regulations, Title 40*, Vol. **36** (2016): 94–154, https://www.govinfo.gov/app/collection/cfr/2018/title40/chapterI/subchapterU/part1036.

3-93. Environmental Protection Agency, "Part 1037—Control of Emissions from New Heavy-Duty Motor Vehicles," *Code of Federal Regulations, Title 40*, Vol. **36** (2016): 154–291, https://www.govinfo.gov/app/collection/cfr/2018/title40/chapterI/subchapterU/part1037.

3-94. Environmental Protection Agency, "§ 86.1819-14 Greenhouse Gas Emission Standards for Heavy-Duty Vehicles," *Code of Federal Regulations*, Title 40, Vol. **21** (2018): 1001-1010, https://www.govinfo.gov/app/collection/cfr/2018/title40/chapterI/subchapterC/part86/subpartS.

3-95. State of California, "California Global Warming Solutions Act of 2006," Assembly Bill No. 32 (AB 32), Chapter 488 (27 Sep. 2006): 1–13, http://www.leginfo.ca.gov/pub/05-06/bill/asm/ab_0001-0050/ab_32_bill_20060927_chaptered.pdf.

3-96. European Commission, "Proposal for a Regulation of the European Parliament and of the Council Setting CO2 Emission Performance Standards for New Heavy-Duty Vehicles," COM(2018) 284 final (May 17, 2018): 1–11.

3-97. Rexeis, M., Quaritsch, M., Hausberger, S., Silberholz, G., Kies, A., Steven, H., Goschütz, M., and Vermeulen, R., "VECTO Tool Development: Completion of Methodology to

Simulate Heavy Duty Vehicles' Fuel Consumption and CO2 Emissions," European Commission Report No. I 15/17/Rex EM-I 2013/08 1670, 2017.

3-98.   Ministry of Land, Infrastructure, Transport, and Tourism (MLIT), "Criteria for judgment of manufacturers of energy consumption equipment, etc., concerning improvement of energy consumption of lorries," Japan Ministry of Land, Infrastructure, Transport and Tourism (MLIT), 2007.

3-99.   Ministry of Land, Infrastructure, Transport, and Tourism (MLIT), "Integrated Resources and Energy Research Committee Energy Conservation and New Energy Subcommittee, Energy Conservation Subcommittee Automotive Judgment Criteria Working Group, and Transportation Policy Council, Land Transportation Division, Automobile Division, Automobile Fuel Efficiency Standard Subcommittee, Joint meeting Summary (heavy vehicle fuel consumption standard etc.)," 2017.

3-100. Ministry of Industry and Information Technology of the PRC, "Fuel Consumption Limits for Heavy-Duty Commercial Vehicles," GB 30510-2018, 2018.

3-101. Ministry of Ecology and Environment of PRC, "Fuel Consumption Test Methods for Heavy-Duty Commercial Vehicles (C-WTVC)," GB/T 27840-2011, 2011.

3-102. Ministry of Power, "Notification, New Delhi, the 16th August 2017," 2017.

4

Emissions Testing Protocols

It is a capital mistake to theorize before one has data. Insensibly one begins to twist facts to suit theories, instead of theories to suit facts.

—Sherlock Holmes, "A Scandal in Bohemia"

Regulatory bodies use emission-testing protocols to provide a highly repeatable method for manufacturers to demonstrate that their products comply with the regulatory requirements described in Chapter 3, Emissions Control Regulations, for criteria pollutants and for fuel consumption or greenhouse gas (GHG) emissions. These test protocols are written specifically for light-duty vehicles (LDVs), heavy-duty vehicles (HDVs), and on-road and off-road engines and are intended to probe aspects of real world use. The cycles described in this chapter are in use in various countries or regions across the globe to evaluate emissions and fuel consumption characteristics of LDVs, HDVs, on-road engines, and off-road engines.

In general, the emission-testing protocols described in this chapter have the following elements:

- Preparing the test article for the evaluation.
- Testing of the article over a prescribed operating schedule or test cycle.
- Processing the data for comparison of the results against regulatory requirements.

This chapter is intended to provide an overview of the test cycles, with an emphasis on the salient characteristics for vehicle, engine, and exhaust aftertreatment system (EAS) design and development. It is not a comprehensive description of the test protocols, for which the applicable regulations are the most definitive and current source, but key information on the test preparations or test conditions is provided. Details on the systems used to make emissions measurements during testing are provided in Chapter 5, Emissions Measurement.

4.1 **LDV Test Cycles**

The test cycles discussed in this section are used with LDVs to evaluate their criteria pollutant emissions and GHG emissions or fuel economy. Each regulatory agency has a specific protocol for determining which LDVs from a manufacturer need to be tested and the details of how those selected LDVs are prepared and tested. In general, the test vehicle that represents a family of similar vehicles must use a configuration that is available for sale, and it should represent the worst-case scenario for emissions performance.

LDV testing for certification is performed on a chassis dynamometer in a vehicle test facility. A chassis dynamometer is a device that allows a vehicle to be driven over a proscribed vehicle speed schedule while applying the road load expected in real world use. For safety, the vehicle is chained down and cannot be steered. The vehicle test cell is climate controlled and includes test equipment, described in Chapter 5, Emissions Measurement, to measure chemical species of interest in the exhaust gas such as nitrogen oxides (NOx), non-methane hydrocarbons (NMHC), or carbon dioxide (CO_2). For accurate and reproducible chassis dynamometer testing, the vehicle's road load parameters, such as the vehicle mass, aerodynamic drag, and rolling resistance, must be known and entered into the test cell controller so that the dynamometer provides the appropriate resistance to the vehicle during the test cycle. The vehicle mass is measured directly and then adjusted to the appropriate test weight bin. The other road load parameters are measured using a coast-down test, in which the vehicle starts at a target speed and then coasts to a stop. The distance traveled during the coast-down is used to calculate the parameters. For chassis dynamometer testing, so-called A, B, and C coefficients are typically used instead of the physical parameters such as aerodynamic drag or rolling resistance. These A, B, and C road load coefficients are included in the Environmental Protection Agency (EPA) Test Car Lists databases [4-1] for vehicles tested in the United States (US).

This section presents the five test cycles used by the US EPA to measure LDV criteria pollutant emissions, GHG emissions and fuel economy, or both. It also presents the European Union's (EU's) New European Drive Cycle (NEDC), the Japanese JC08 cycle, and the World-Harmonized Light-Duty Transient Cycle (WLTC). Finally, there is a discussion on how various regulatory bodies are assessing real-world emissions from LDVs. The test cycles presented here support the light-duty criteria pollutant rules described in Section 3.1 and the LDV GHG and fuel economy rules described in Section 3.3.

4.1.1 **US Federal Test Protocol FTP-75**

The light-duty Federal Test Protocol test schedule, FTP-75, is the most important drive cycle for LDV regulations in the United States. It is the key test cycle for criteria pollutant emission measurements and is one of two test cycles used to calculate fuel economy and corresponding GHG emissions for the Corporate-Averaged Fuel Economy (CAFE) regulations described in Section 3.4.1. FTP-75 is also one of the five test cycles used to calculate the Monroney sticker fuel economy estimates [4-2] that are maintained by the US Department of Energy and EPA [4-3]. The Monroney sticker fuel economy calculations are described in Section 3.4.1.3.

FIGURE 4.1   Vehicle speed schedule for the full FTP-75 cycle, including the hot soak segment between Bags 2 and 3 [4-5].

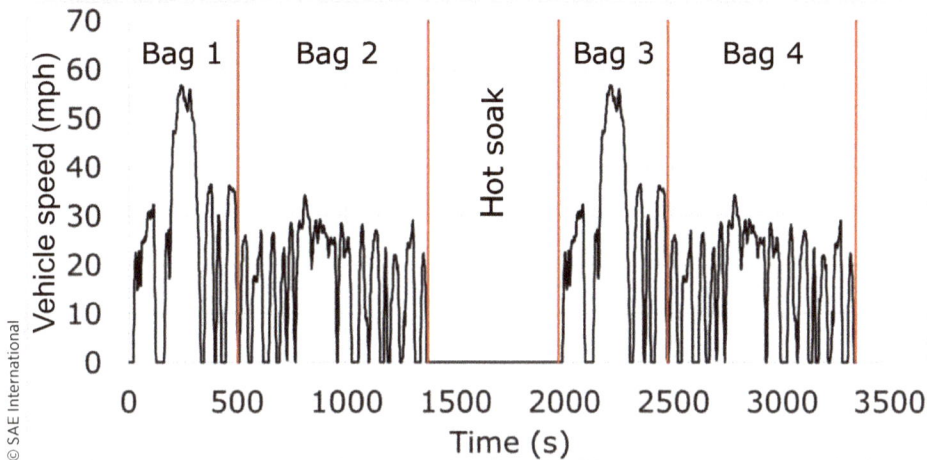

The FTP-75 cycle is built up from EPA's Urban Dynamometer Driving Schedule (UDDS) [4-4], which is also known as the LA-4 cycle or FTP-72. The cycle is formally defined in Appendix I of Title 40, Part 86 of the *Code of Federal Regulations* (40 CFR part 86) [4-5]. As shown in **Figure 4.1**, FTP-75 is divided into four portions that are called "Bags" because the vehicle exhaust sample is literally captured in a bag during each portion of the test cycle. The UDDS covers the pairs of bags—1 and 2 or 3 and 4—that comprise FTP-75, thus the vehicle speed schedule is the same for Bags 1 and 3 and for Bags 2 and 4.

To prepare a test vehicle for FTP-75, the fuel tank needs to be drained and filled with an appropriate amount of certification fuel. The vehicle is run over a few FTP-75 cycles to precondition the vehicle, especially the EAS. Then, the vehicle undergoes a procedure called cold soak, where the vehicle is held for 12–36 h in a preparation room at 20–30°C (68–86°F), with 25°C (77°F) being a common condition. When ready for testing, the vehicle is pushed into the test chamber and secured to the chassis dynamometer. Because FTP-75 is used for both criteria pollutant and fuel economy certification, engineers should plan to measure both fuel consumption and exhaust gas emissions constituents, including NMHC or non-methane organic gases, NOx, particulate matter (PM), carbon monoxide (CO), and CO_2.

The cold soak means that the FTP-75 cycle starts Bag 1 with the complete vehicle equilibrated to the test facility's ambient temperature of 20–30°C (68–86°F). The test vehicle is then driven over the FTP-75 speed schedule and uses its onboard controls to manage the engine and EAS warm-up and subsequent performance. The test vehicle drives through the vehicle speed schedule for Bag 1 and Bag 2 and is then turned off for a hot soak period of about 10 min (540–660 s). The vehicle is restarted and then continues driving over the speed schedule with Bag 3. LDVs with conventional powertrains may stop testing at the end of Bag 3, on the assumption that the typical LDV warm-up profile will yield emissions and fuel consumption results from Bag 4 that would be identical to those from Bag 2. LDVs with hybrid powertrains, especially charge-depleting ones such as plug-in hybrid electric vehicles or pure battery electric vehicles, need to drive through

Bag 4 to the end of the full test cycle. Charge-sustaining hybrid electric vehicles need to start and end the test cycle at the same battery state of charge.

FTP-75 Bags 1 and 3 are each 505 s in duration and 5.78 km (3.59 mi.) in length, and Bags 2 and 4 are each 867 s and 6.21 km (3.86 mi.). The total distance driven over all four bags is 23.98 km (14.90 mi.), whereas the three-bag test distance is 17.77 km (11.04 mi.). With the hot soak, the full four-bag FTP-75 cycle duration is about 3,345 s, as shown in Figure 4.1. The total testing time of Bags 1–3 is 1,874 s, as shown in **Figure 4.2**, although this total test time excludes the 10 min (540-660 s) hot soak between Bags 2 and 3. The average vehicle speed on FTP-75 is 19.5 mph (31.4 km/h), with the average speed on Bag 1 and 3 being slightly higher than that on Bag 2 and 4. The maximum speed on FTP-75 is 56.7 mph (91.2 km/h). Of the time that the vehicle is on—2,745 s over all four bags—the vehicle is stopped and at idle for 520 s, which is 18.9% of the total. For Bags 1–3, the idle time is 358 s of 1,874 s, or 19.1% of the test time.

The criteria pollutant and GHG results from each FTP-75 bag are measured as concentrations from which the species masses are calculated. The total mass for each species of interest over the equivalent of four bags is divided by the total distance traveled on FTP-75 to get an emissions result in grams per mile. Likewise, fuel economy is calculated by taking the total distance traveled over all four bags of FTP-75 by the gallons of fuel consumed. Note that if the test only includes Bags 1–3, the emissions and fuel consumption results from Bag 2 need to be doubled to account for the omitted Bag 4 in order to correctly calculate the averaged results from FTP-75.

Note that EPA requires the driver to stay within 2 mph (3 km/h) of any target speed within 1 s of a given point in the drive cycle [4-6]. This allowable speed range can lead to a large difference in test results, however. A driver who holds precisely to the speed set points with the associated accelerations and decelerations will create stronger transient events in the vehicle and engine than would another driver who takes advantage of the allowable speed range to smooth out the transients. An example of this smoothing is shown in **Figure 4.3**, in which the second "hill" of Bag 1 is shown along with the allowable speed range. The smoothed red curve is an example speed trace that stays

FIGURE 4.2   Vehicle speed schedule for Bags 1–3 of FTP-75, omitting the hot soak segment between Bags 2 and 3 [4-5].

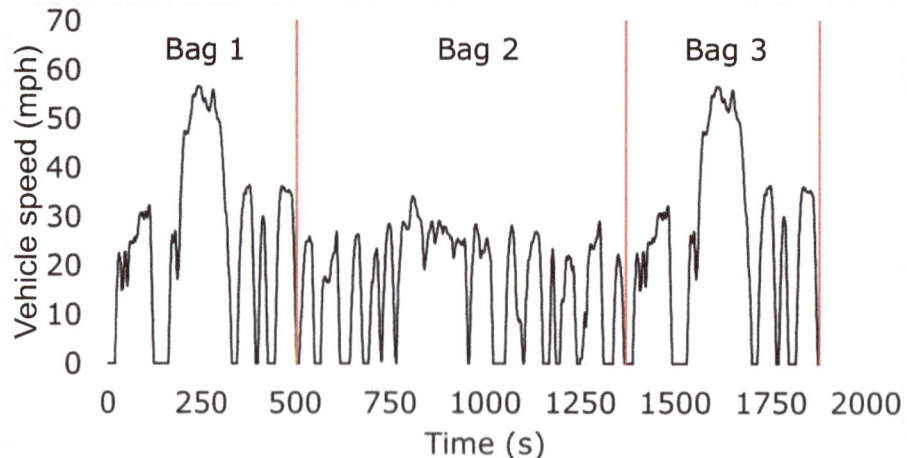

© SAE International

FIGURE 4.3 Vehicle speed schedule for excerpt of FTP-75 Bag 1, showing allowable speed range around the speed schedule and smoothed speed trace.

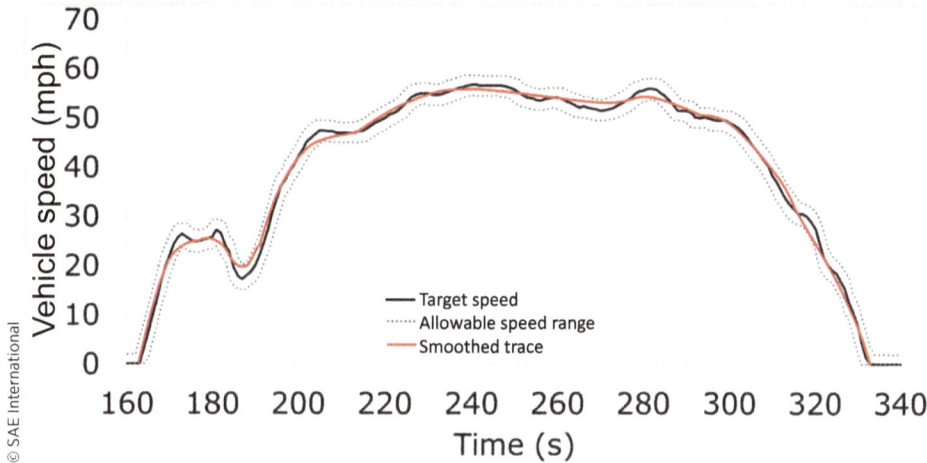

within the allowable speed range but has considerably fewer transient events in it, especially from about 230 s to about 285 s when the smoothed vehicle speed is nearly constant. The smoothed curve is allowed, but a manufacturer runs the risk of a validation or audit test yielding a worse result if a subsequent driver holds closer to the prescribed speed schedule than the original certification test driver did.

4.1.2 US Highway Fuel Economy Test

EPA's Highway Fuel Economy Test (HFET or HWFET) driving schedule is a moderately aggressive highway drive cycle that is defined in Appendix I of 40 CFR part 86 [4-7] and shown in **Figure 4.4**. HFET is the highway drive cycle used to calculate the combined fuel economy for CAFE regulations governed by NHTSA and the corresponding LDV

FIGURE 4.4 EPA HFET vehicle speed schedule [4-7].

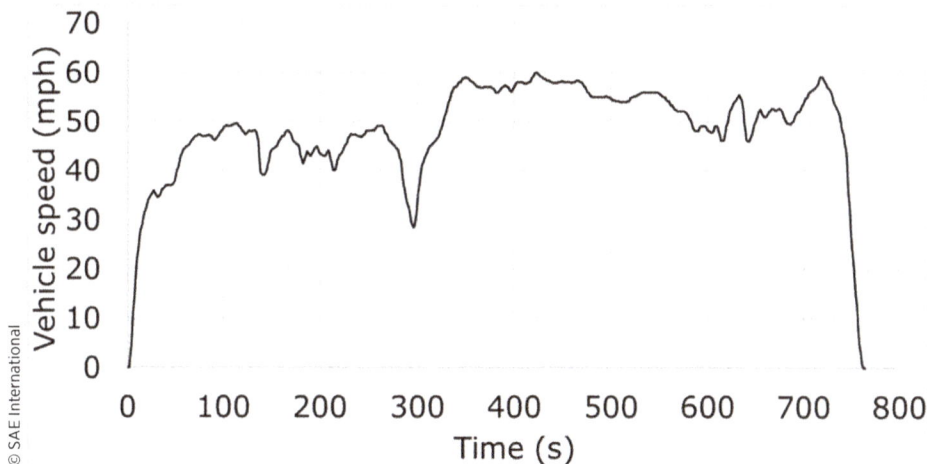

GHG emissions governed by EPA and California Air Resources Board (CARB). Additionally, HFET is one of the five drive cycles used to calculate the EPA's Monroney sticker city and highway fuel economy values [4-2].

HFET is usually run in conjunction with FTP-75, described in Section 4.1.1, for fuel economy testing. HFET is a hot-start cycle that uses the same ambient conditions as FTP-75; therefore, it can immediately follow an FTP-75 test. As shown in Figure 4.4, the HFET has a duration of 765 s, of which only a few seconds at the start and end of the test are at idle. The vehicle travels 10.26 mi. (16.51 km) over the drive cycle at an average speed of 48.3 mph (77.7 km/h). Like FTP-75, the cumulative emissions and fuel consumption are measured over the cycle and then adjusted by the cycle's distance to get emissions in g/mi. or mpg. The maximum speed over HFET is 59.9 mph (96.4 km/h), which is only slightly faster than the maximum speed on FTP-75.

4.1.3 US US06 (Supplemental FTP)

The US06 cycle is an aggressive highway-driving schedule, defined in 40 CFR §1066.831 [4-8], that is part of the EPA Supplemental FTP Driving Schedule (SFTP) for emissions certification. US06 is intended to better represent real-world high-speed driving and is used to evaluate criteria pollutant emissions as part of the SFTP. It is also one of the five drive cycles used to calculate the EPA's Monroney sticker fuel economy values [4-2]. US06 is a hot start test that uses the same ambient conditions as FTP-75 and HFET, and so it can be run immediately following the FTP-75 or HFET tests. Like FTP-75, US06 is used for both criteria pollutant certification and calculating the sticker fuel economy values, and so engineers should plan to measure both fuel consumption and exhaust gas emission constituents, including NMHC, NOx, PM, CO, and CO_2.

As shown in **Figure 4.5**, the duration of the drive cycle is 600 s, of which only 45 s are spent at idle [4-4]. US06 is characterized by high speeds, especially compared to FTP-75 or HFET, and hard acceleration and deceleration events over the 8.01 mi. (12.9 km) traveled. The average vehicle speed is 48.0 mph (77.2 km/h), and the speed is at or above 60 mph (96.6 km/h) for about half of the cycle. The maximum vehicle speed is 80.3 mph (129 km/h).

FIGURE 4.5 EPA US06 vehicle speed schedule [4-8].

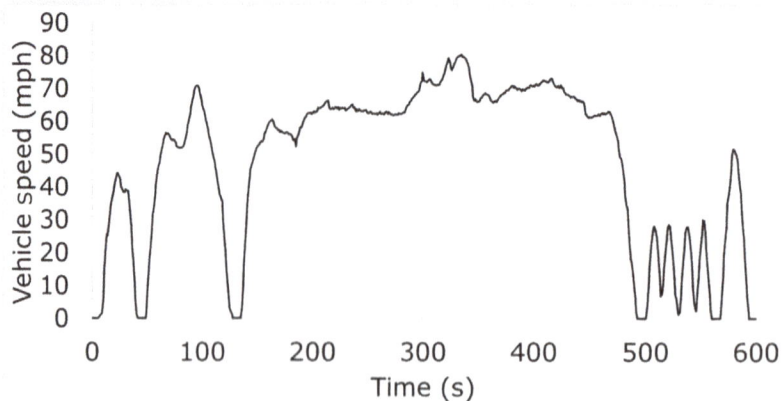

For the purposes of calculating the Monroney sticker fuel economy per the equations presented in Section 3.4.1.3 [4-2], the city portion of the US06 cycle combines the segment from 1 to 130 s and the segment from 495 to 596 s. The highway portion of US06 is, therefore, the segment from 130 to 495 s [4-8].

4.1.4 US SC03 (Supplemental FTP)

The SC03 test cycle is intended to simulate urban driving on a hot summer day and is defined in 40 CFR §1066.835 [4-9]. It is also intended to help identify potential defeat devices that are triggered by running the air conditioning system. SC03 is one of the three cycles in the EPA SFTP, which is used to evaluate criteria pollutant emissions as described in Chapter 3, Emissions Control Regulations. SC03 is also one of the five drive cycles used to calculate the EPA's Monroney sticker city and highway fuel economy values [4-2]. Therefore, engineers should plan to measure both fuel consumption and exhaust gas emission constituents, including NMHC, NOx, PM, CO, and CO_2, like for FTP-75 or US06.

SC03 has several acceleration and deceleration events, as shown in **Figure 4.6** [4-4]. The overall test cycle duration is 600 s, and the vehicle is stopped for 117 s of this total, which is 19.5% of the duration. The vehicle covers a distance of 3.58 mi. (5.76 km) at an average speed of 21.4 mph (34.4 km/h), and the maximum vehicle speed is 54.8 mph (88.2 km/h).

SC03 has several distinct features compared to the other EPA LDV cycles, which are described in 40 CFR §1066.835 [4-9], that reflect the intent of the cycle, including the following:

- 35.0°C ± 3.0°C (95.0°F ± 5.4°F) preconditioning soak temperature
- 35.0°C ± 3.0°C (95.0°F ± 5.4°F) ambient temperature in test cell
- 850 ± 45 W/m² of simulated solar load on vehicle during test, as measured at the base of the windshield and the bottom of the rear window
- Active air conditioning in vehicle during test, set to either 72°F or maximum cooling

The test cell must be equipped to provide the necessary solar load in addition to the hotter ambient temperature. The hotter ambient temperature requirement of 35°C (95°F)

FIGURE 4.6   EPA SC03 vehicle speed schedule [4-9].

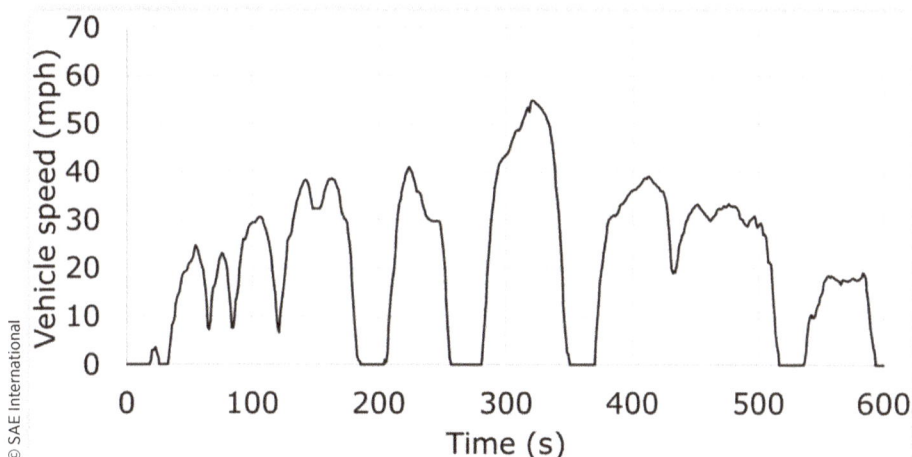

means that SC03 must be run separately from the other US chassis dynamometer tests for LDVs. Note that SC03 is the only US drive cycle that requires the air conditioning system to be active.

4.1.5 US FTP-75 Cold Cycle

The FTP-75 cold cycle, also known as FTP20, is a version of FTP-75 that represents driving on a cold winter day. The FTP-75 cold cycle is one of the five drive cycles used to calculate the EPA's Monroney sticker fuel economy values [4-2]. Its purpose is to evaluate the additional fuel consumption penalty of starting the vehicle and operating it in winter conditions. Therefore, EPA is interested in measurements of CO_2, CO, and NMHC at the cold ambient conditions. The facility may choose to measure other criteria pollutants to assess the EAS warm-up strategy and performance in a cold environment.

Per the procedure in 40 CFR §1066.710 [4-10], the vehicle is soaked to a winter cold temperature of $-7.0°C \pm 1.7°C$ ($20°F \pm 3°F$) and then tested at that temperature. The test cycle is the same as for FTP-75, as shown in Figure 4.1. In addition, during the test the cabin heater must be active and set to defrost the windshield but with the air conditioning compressor turned off. The heater should be set to maximum heat if manually controlled or at least 72°F if automatically controlled.

Depending on how quickly the vehicle warms up, the vehicle may only require testing on Bags 1, 2, and 3 of the cold FTP-75. However, it may be worth running the vehicle through Bag 4 in case the performance during Bag 4 is better than that in Bag 2.

4.1.6 California (US) LA-92

The LA-92 cycle was developed by CARB as the Unified LA-92 Dynamometer Driving Schedule and is based on real LDV driving behavior in the greater Los Angeles area [4-11]. Thus, LA-92 contains a mixture of urban, suburban, and freeway driving that is intended to be more representative of real LDV use than FTP-75. The Hot LA-92, that is, the LA-92 cycle that starts with a warmed up vehicle, is used for chassis certification of criteria pollutant levels from Class 3 light heavy-duty vehicles (LHDVs) as discussed in Section 3.3.1.1. LA-92 is used in place of the US06 cycle for these vehicles. Thus, engineers should plan to measure both fuel consumption and exhaust gas emissions constituents, including NMHC, NOx, PM, CO, and CO_2.

LA-92 is a highly transient cycle that has several acceleration and deceleration events, as shown in **Figure 4.7** [4-4]. The overall test cycle duration is 1,435 s, and the vehicle is stopped for 234 s of this total, which is 16.3% of the duration. The vehicle covers a distance of 9.82 mi. (15.8 km) at an average speed of 24.6 mph (39.6 km/h) including stops, and the maximum vehicle speed is 67.2 mph (108 km/h). Thus, by comparison to FTP-75, a vehicle on LA-92 is stopped for proportionally less time and drives at faster speeds. Compared to the US06 cycle, however, the LA-92 cycle is less aggressive.

4.1.7 EU NEDC

The NEDC is a combined driving cycle used for fuel consumption and emission testing in the EU. For new vehicle registrations, the NEDC is being phased out of use in the EU

FIGURE 4.7 LA-92 "Unified" vehicle speed schedule [4-11].

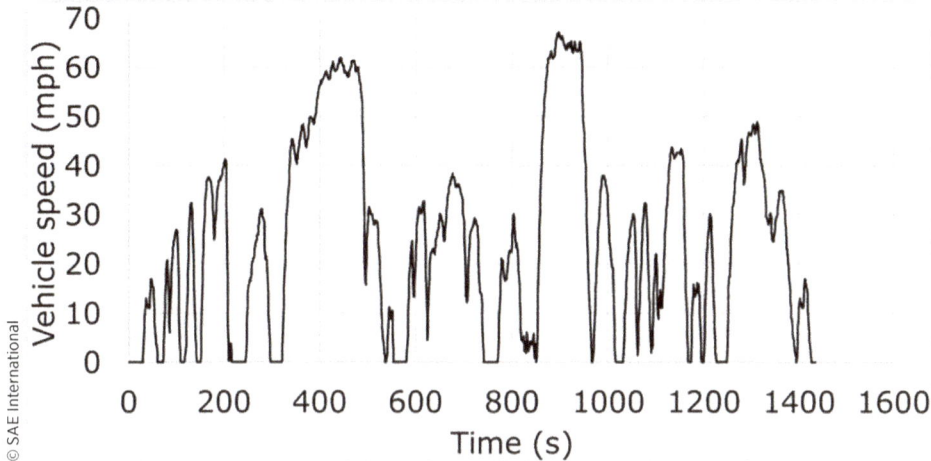

in favor of the WLTC, described in Section 4.1.9, but it is a well-established cycle for which legacy vehicle data are likely available.

NEDC comprises four Elementary Urban Cycle (ECE 15) schedules followed by one Extra Urban Drive Cycle (EUDC) schedule [4-12], as shown in **Figure 4.8**. The NEDC is also known as the Type I test from ECE R83.06 [4-12]. The NEDC is a very gentle drive cycle since it comprises a series of steady speeds and smooth ramped transitions between them. The test cycle takes a total of 1,180 s to complete. Over the test cycle, the vehicle is at a stop for 267 s, which is 22.6% of the total duration. The vehicle travels a total of 10.9314 km (6.79 mi.) over the test cycle.

Official guidance on NEDC is provided in the UN/ECE WP.29 1958 Agreement and its Addenda, with details in Regulation 83 of the Regulations for the Construction of Vehicles [4-12]. The test vehicle needs to have been run in for 3,000 km before testing. The vehicle is preconditioned by running three Part Two cycles, or EUDC. It is then set to soak between 20°C (68°F) and 30°C (86°F) for at least 6 h and not more than 30 h

FIGURE 4.8 NEDC vehicle speed schedule [4-12].

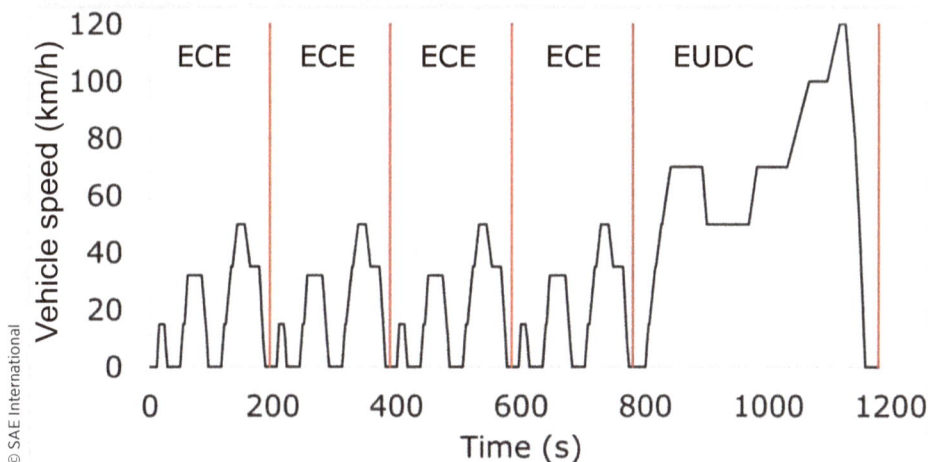

before testing. During the test, the ambient temperature in the test cell needs to be between 20°C (68°F) and 30°C (86°F), and the absolute humidity of the intake air needs to be between 5.5 and 12.2 (g H$_2$O/kg dry air).

4.1.8 Japan JC08

The JC08 cycle was introduced in Japan for use in the so-called Post New Long Term Regulation of 2009 for evaluating criteria pollutants and fuel economy for new type approvals [4-13]. The cycle's use was fully phased in by October 2011 for all LDVs. JC08 is a combined cycle that represents a mix of urban and suburban driving that is typical for Japan. JC08 superseded the previous 10 + 15-mode drive cycle in Japan that had been in use from 1983. As of publication, JC08 is being superseded by the WLTC for approvals.

JC08 is used for both criteria pollutant emissions and fuel consumption measurements, so appropriate measurements of exhaust gas constituents and fuel flow will be needed. The emission measurements are used to calculate a weighted average of the cold-start and hot-start results over the cycle. The weighting is 25% for the cold-start JC08 and 75% for the hot-start JC08, as shown in the following equation:

$$E_{Comb.} = 0.25\,E_{Cold} + 0.75\,E_{Hot} \tag{4.1}$$

As shown in **Figure 4.9**, the duration of the JC08 drive cycle is 1,204 s, including 357 s spent at idle, which represents 29.7% of the total cycle time. JC08 has several low-to-moderate speed segments with stops in between. The drive cycle distance is 8.171 km (5.077 mi.). The average vehicle speed is 24.4 km/h (15.2 mph), and the top speed is 81.6 km/h (50.7 mph).

4.1.9 World-Harmonized Light-Duty Transient Cycle

The WLTC is actually a set of drive cycles for LDVs that was developed as part of the World-Harmonized Light-Duty Transient Protocol (WLTP) [4-14, 4-15]. The WLTP also

FIGURE 4.9 JC08 vehicle speed schedule [4-13].

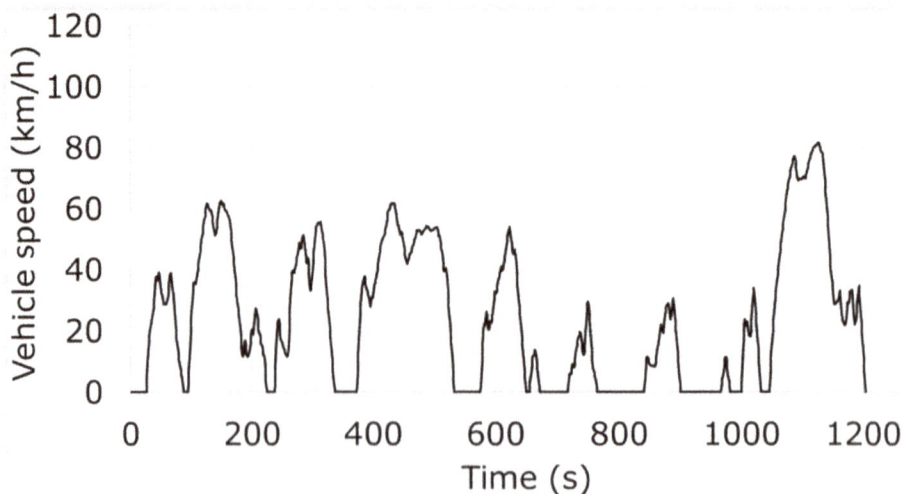

TABLE 4.1 WLTC segments by vehicle class [4-15].

Class	Segment 1	Segment 2	Segment 3	Segment 4
1	Low 1	Medium 1	Low 1	-
2	Low 2	Medium 2	High 2	Extra-high 2
3a	Low 3	Medium 3-1	High 3-1	Extra-high 3
3b	Low 3	Medium 3-2	High 3-2	Extra-high 3

includes other details about the test procedures beyond just the drive cycle, although many use WLTC and WLTP interchangeably. There are four main variants of the WLTC, and which one is used depends on the vehicle power-to-mass ratio (PMR) and top speed. The PMR is calculated by dividing the rated power of the engine (in W) by the vehicle mass (in kg). The four LDV classes for the WLTC are defined as follows:

- Class 1 has PMR $\leq$ 22 W/kg
- Class 2 has PMR > 22 W/kg and $\leq$ 34 W/kg
- Class 3a has PMR > 34 W/kg and v_{max} < 120 km/h (75 mph)
- Class 3b has PMR > 34 W/kg and v_{max} $\geq$ 120 km/h (75 mph)

Class 1 and Class 2 vehicles are representative of relatively low-powered LDVs driven in India and similar markets. Class 2 LDVs may also be found as lower powered LDV variants driven in Japan and the EU. Most LDVs sold in Europe or Japan, though, are Class 3b, with some lower powered variants in Class 3a. The main distinction between Class 3a and Class 3b is the achievable top speed of the vehicle, where Class 3b vehicles have a top speed of 120 km/h (75 mph) or higher.

The WLTC for Class 1 LDVs is a combination of two low-speed segments and one medium-speed segment, as described in **Table 4.1** and shown in **Figure 4.10**. For Class 2, Class 3a, and Class 3b LDVs, the WLTC includes the following four general segments that are intended to represent a mixture of urban, suburban, rural, and freeway driving conditions:

- Low-speed segment with an average speed of about 20 km/h (12 mph) and a top speed of about 50 km/h (31 mph).
- Medium-speed segment with an average speed of about 40 km/h (25 mph) and a top speed of 65–70 km/h (40.4–43.5 mph).
- High-speed segment with an average speed of about 55 km/h (34 mph) and a top speed of 85–95 km/h (53–59 mph).
- Extra-high speed segment with an average speed of about 90 km/h (56 mph) and a top speed of 120–130 km/h (75–81 mph).

The specific segments used in the WLTC depend on the LDV Class as defined above. There are three versions of the low-speed segment, four of the medium-speed segment, three of the high-speed segment, and two of the extra-high speed segment. The specific segments used are shown in Table 4.1. All of the test cycles are plotted in **Figures 4.10– 4.13**. **Figure 4.14** plots the speed schedules together to allow a direct comparison for the Class 2, Class 3a, and Class 3b vehicles. The largest differences are between the Class 2 and Class 3a or 3b test cycles, whereas the Class 3a and Class 3b schedules only have small differences in the second (medium) and third (high) segments.

FIGURE 4.10 WLTC speed schedule for Class 1 light-duty vehicles [4-15].

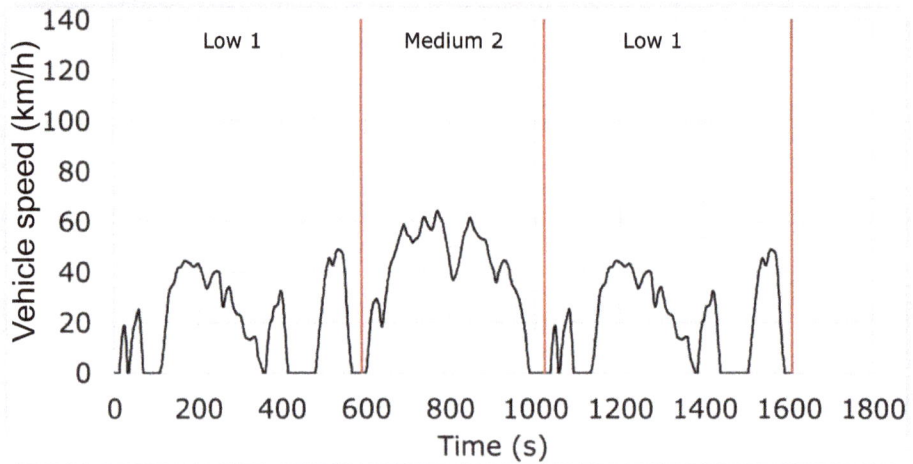

FIGURE 4.11 WLTC speed schedule for Class 2 light-duty vehicles [4-15].

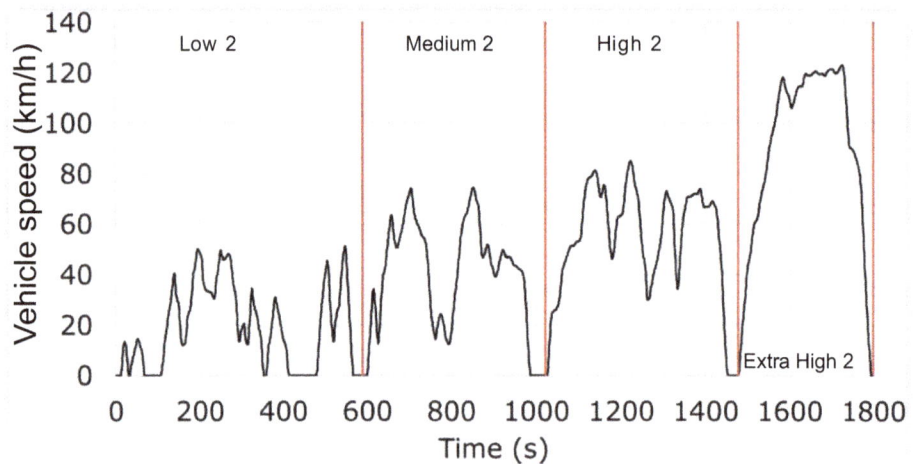

The Type I test where the vehicle starts at 20–30°C (68–86°F), or a nominal 25°C (77°F), is the most common initial condition for the WLTC test. The People's Republic of China (PRC) and other jurisdictions also require a Type VI test, however, where the vehicle starts at −10°C to −4°C (14–25°F), or a nominal −7°C (20°F), and is only exercised over the low- and medium-speed parts of the WLTC. The Type VI test only measures CO, THC, and NOx.

4.1.10 Real Driving Emissions

Regulatory bodies were moving toward evaluating the real-world emission performance of LDVs even before the Volkswagen diesel emission cheating ("Dieselgate") scandal of 2015 emerged.* The scandal accelerated a trend toward tightening up perceived loopholes

_____

* See the explanation in Chapter 3, Emissions Control Regulations.

FIGURE 4.12 WLTC speed schedule for Class 3 light-duty vehicles with top speed under 120 km/h [4-15].

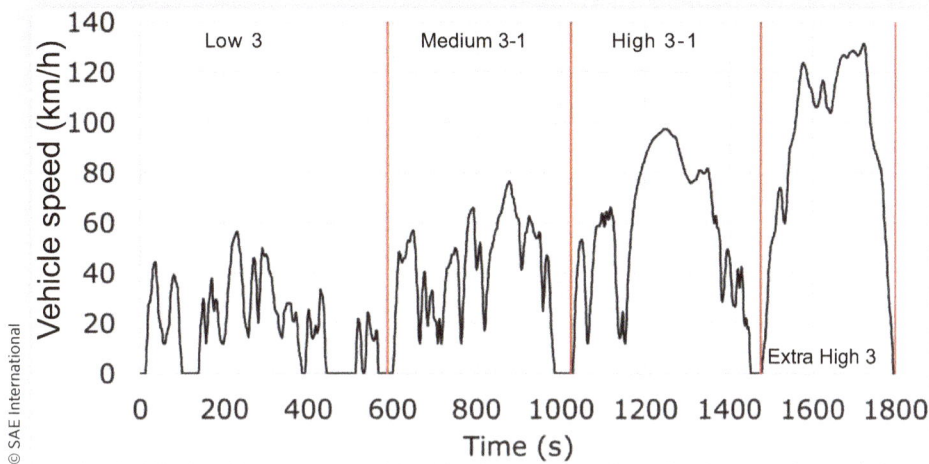

FIGURE 4.13 WLTC speed schedule for Class 3 light-duty vehicles with top speed over 120 km/h [4-15].

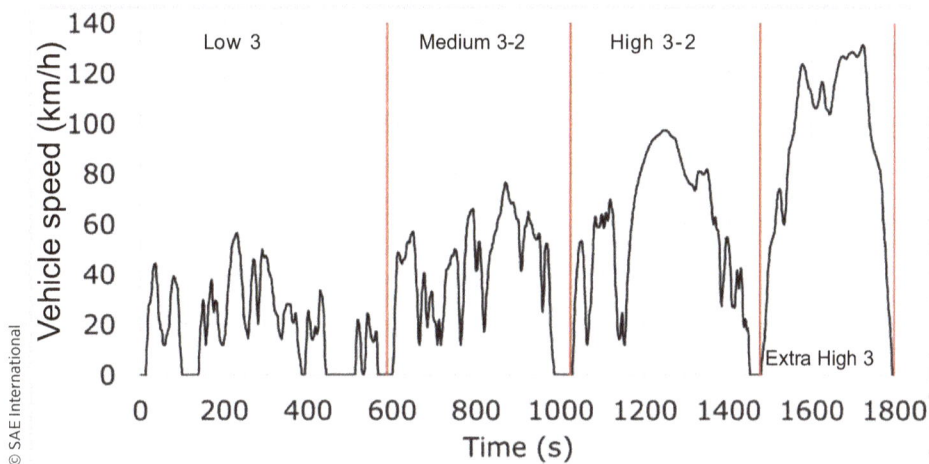

in the test cycles. One approach that has gained regulatory emphasis globally is evaluating the real-world or on-road emissions performance of LDVs and HDVs.

Given the broader regulatory philosophies for emission compliance in the US and California, as described in Sections 3.2.1 and 3.2.2, respectively, EPA and CARB typically fit selected vehicles with portable emissions measurement system (PEMS) equipment and drive them under reasonable driving conditions to assess criteria pollutant emissions and to identify the presence of declared and undeclared auxiliary emissions control devices (AECD). AECD are discussed in Section 3.2.1 and defined in 40 CFR §86.1803-01 [4-16]. EPA and CARB have been deliberately opaque about where and how they are operating the vehicles tested to prevent a recurrence of cheating. EPA and CARB are also known to purchase or rent in-use LDVs for spot

FIGURE 4.14   Comparison of WLTC speed schedules for Class 2 and Class 3 light-duty vehicles.

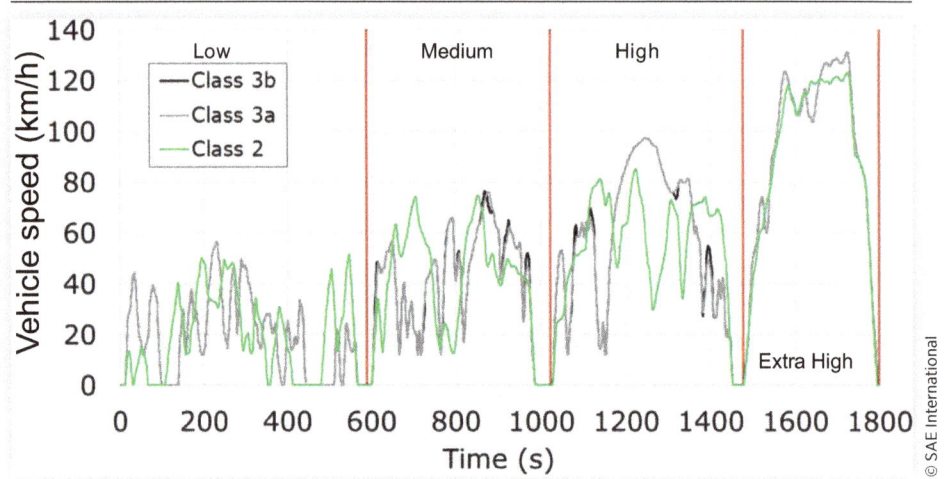

checks of performance, fuel economy, and emissions, especially if they have received complaints from owners.

In the EU and Japan, new regulations require an evaluation of real-world performance using Real Driving Emissions (RDE) test protocols that are described further in this section. Other countries, including PRC, India, Brazil, and the Republic of Korea (South Korea), are also considering introducing RDE testing requirements.

4.1.10.1   RDE in the EU: RDE for the EU Euro 6 standards are defined by the requirements in Annex IIIA to Commission Regulation (EU) No. 2017/1151 [4-15]. In general, the LDV is preconditioned, left to soak, and then tested on the road over three segments: urban, rural, and motorway. Assuming the test complies with the RDE requirements and is valid, the measured results are post-processed to normalize them to a standard trip. These normalized results are compared against the on-cycle targets modified by a multiplicative conformity factor [4-15, 4-17].

The vehicle pre-conditioning consists of fitting the vehicle with PEMS equipment, driving the vehicle for up to 30 min, and then leaving the car parked with doors and hood closed in the preferred ambient temperature range of 0°C (32°F) and 30°C (86°F) for 6–56 h. The PEMS equipment installed must be part of a validated family according to Appendix 7 of Annex IIIA [4-15]. The test vehicle weight is the vehicle's curb weight plus the weight of the driver, an optional witness (or equivalent ballast), and the PEMS equipment. No modifications to the vehicle aerodynamics are allowed beyond the effects of the installed PEMS unit. The fuel and lubricants put into the test vehicle need to be ones that meet the manufacturer specification for the vehicle as sold and should be ones that a customer would normally use for the vehicle.

For the RDE test trip, the ambient temperature is expected to be between 0°C (32°F) and 30°C (86°F) and the altitude of the route should not exceed 700 m (2,297 ft.) above sea level. The ambient conditions may be extended as follows:

- Extending the temperature range down to between −7°C (20°F) and 0°C (32°F).

- Extending the temperature range up to between 30°C (86°F) and 35°C (95°F).
- Increasing the altitude range to between 700 m (2,297 ft.) and 1,000 m (3,281 ft.) above sea level.

The RDE test trip duration shall be 90–120 min without interruption. The overall trip needs to start and end within 100 m (325 ft.) of the same altitude, but the trip route does not need to start and end at the same location. The RDE test trip shall include three segments in the order and proportion shown in **Table 4.2**. Each segment must cover a distance of at least 16 km (9.9 mi.).

As shown in Table 4.2, the urban segment is characterized by vehicle speeds that do not exceed 60 km/h (37.3 mph). The average vehicle speed over the urban segment is expected to be between 15 km/h (9.3 mph) and 40 km/h (24.9 mph). Because it is the first of the three segments, the urban segment includes the vehicle cold start period when the engine, EAS, and vehicle are warming up. The vehicle must start moving within 15 s of the first ignition of the engine, and during the cold start period, the vehicle speed must not exceed 60 km/h (37.3 mph). Vehicle stops during the cold start period must be kept to a minimum and may not exceed 90 s total. For the overall, the urban segment stops may be 6–30% of the time in the urban segment, although stops may not exceed 300 s for a valid test. In addition, the urban segment cannot have a positive grade that exceeds +1.2% (1,200 m/100 km).

The Rural segment is characterized by vehicle speeds between 60 km/h (37.3 mph) and 90 km/h (55.9 mph), unless the LDV is a N_2 category vehicle with a top speed limited to 90 km/h (55.9 mph), in which case the Rural operation top speed is 80 km/h (49.7 mph). The Rural operation segment may be interrupted by short periods of lower speed Urban operation.

The Motorway segment has vehicle speeds that are above 90 km/h (55.9 mph) and are expected to be below 145 km/h (90.1 mph). Motorway segment includes at least 300 s of operation with vehicle speeds above 100 km/h (62.1 mph). The Motorway segment may also be interrupted by short periods of Urban or Rural operation, for example, because of road construction. N_2 category vehicles that are speed-limited to 90 km/h (55.9 mph) will operate in the Motorway segment between 80 km/h (49.7 mph) and 90 km/h (55.9 mph) and do not have the requirement to drive over 100 km/h (62.1 mph).

If the vehicle has a periodically regenerating system, such as a diesel particulate filter (DPF), then the vehicle data will need to be monitored to determine if a periodic regeneration event took place during the test. If so, the measured results will be adjusted by a multiplicative or additive factor consistent with the type of periodically regenerating

TABLE 4.2 Definition of valid segments of EU RDE test trip [4-15].

	Proportion of trip time		Vehicle speed	
Segment	Minimum (%)	Maximum (%)	Minimum	Maximum
1. Urban	29	44	0 km/h (0 mph)	60 km/h (37.3 mph)
2. Rural	23	43	60 km/h (37.3 mph)	90* km/h (55.9 mph)
3. Motorway	23	43	90* km/h (55.9 mph)	145* km/h (90.1 mph)

* N_2 category LDV with a top speed limited to 90 km/h (55.9 mph) should use 80 km/h (49.7 mph) as the boundary between the Rural and Motorway speed ranges.

TABLE 4.3 Definition of valid segments of Japan RDE test trip [4-18].

Segment	Proportion of trip time		Vehicle speed	
	Minimum (%)	Maximum (%)	Minimum	Maximum
1. Low speed	20	35	0 km/h (0 mph)	40 km/h (24.9 mph)
2. Middle speed	20	40	40 km/h (24.9 mph)	60 km/h (37.3 mph)
3. High speed	35	55	60 km/h (37.3 mph)	-*

* No upper bound is given.

system per the procedures in Appendix 1 to Sub-Annex 6 of Annex XXI for type approval. Alternately, the RDE test trip may be repeated, but if there is a periodic regeneration event during the second test, the manufacturer must use the adjusted results from the second test.

4.1.10.2 **RDE in Japan:** The RDE rules adopted in Japan [4-18] are limited in scope by comparison to the rules in the EU and only apply to LDVs with compression-ignition (CI) engines and one of the following attributes: having a GVW at or below 3,500 kg or being a passenger vehicle carrying nine or fewer passengers. The RDE test requirement starts with new type approvals on October 1, 2022, with all vehicles needing approval from October 1, 2024.

The Japanese RDE procedure references the RDE procedure from the Euro 6 regulations [4-15], described in the previous subsection. The Japanese RDE protocol has the following characteristics:

- Measure emissions during normal road and highway driving using PEMS.
- Test trip duration is 90–120 min.
- The speed range segments shown in **Table 4.3** need to be included in the test trip.
- The ambient conditions shall be between −2°C (28.4°F) and +38°C (100.4°F).
- The test route shall not exceed 1,000 m altitude.

If the ambient temperature is between −2°C (28.4°F) and 0°C (32°F) or between +35°C (95°F) and +38°C (100.4°F), the test results shall be divided by 1.6. Likewise, if the test altitude is over 700 m, the test result shall be divided by 1.6. The results are compared to the not-to-exceed (NTE) threshold described in Section 3.1.4.

4.2 Heavy-Duty Engine Test Cycles

As described in Chapter 3, Emissions Control Regulations, most regulatory bodies prefer that the certificate holder evaluates heavy-duty engines (HDEs) separately on an engine dyno-test cycle instead of evaluating the engine and EAS as homologated into a vehicle or other product. To support the appropriate regulations, most jurisdictions require that engines are tested on drive cycles that include both transient operation and steady-state or quasi-steady-state operation.

In this section, several transient and steady-state test cycles for heavy-duty (HD) engines are presented and discussed. These cycles are used for both HD diesel (CI) engines (HDDEs) and HD spark-ignition (HDSI) engines, except as noted. Test cycles for on-road applications are presented first, then ones for non-road applications. For the steady-state tests, the engine and EAS may be run at a steady-state operating point until they stabilize or the engine and EAS may be run over a quasi-steady ramped modal cycle (RMC) that includes the engine's operation at a series of steady-state operating points and also the ramped transitions between those points.

For the engine test cycles presented in this chapter, 0% of normalized speed is the engine's idle speed, not 0 rpm. The 100% normalized speed is usually the engine speed at which the engine generates its rated power and not the advertised rated speed of an engine. Thus, it is possible for an engine to be able to operate above 100% normalized speed. For example, the engine speed at rated power for the Volvo D13 is 1,600 rpm, as shown in Figure 4.15, but it can operate up to 2,100 rpm. Similarly, 100% normalized torque in the test schedule is the maximum torque achievable at a given engine speed. Over the full range of engine speeds, this maximum torque defines the torque curves, such as those shown in **Figures 4.15** and **4.16**. Zero percent torque is zero net torque, where the engine is burning enough fuel to keep itself spinning at the given engine speed, but with no net power flow between the engine and dynamometer. The motoring torque for an engine is also a function of engine speed and represents the torque that must be provided by the dynamometer to spin the engine when the engine has zero fuel flow.

The US test cycles need to be run in an engine test cell that meets the requirements set forth in 40 CFR part 1065 [4-19] for emissions measurements. A 1065-compliant cell also needs to include a fully transient, or motoring, engine dynamometer because of the requirements of the test cycles. 40 CFR part 1065 [4-19] also prescribes the appropriate certification fuels to be used for a valid test.

In other parts of the world, regulatory bodies follow the guidelines for test cells and measurement equipment that are set forth in Global Technical Regulations, which are coordinated by the United Nations (UN). For example, the measurement procedures

FIGURE 4.15   Volvo D13 engine torque and power curves with respect to engine speed.

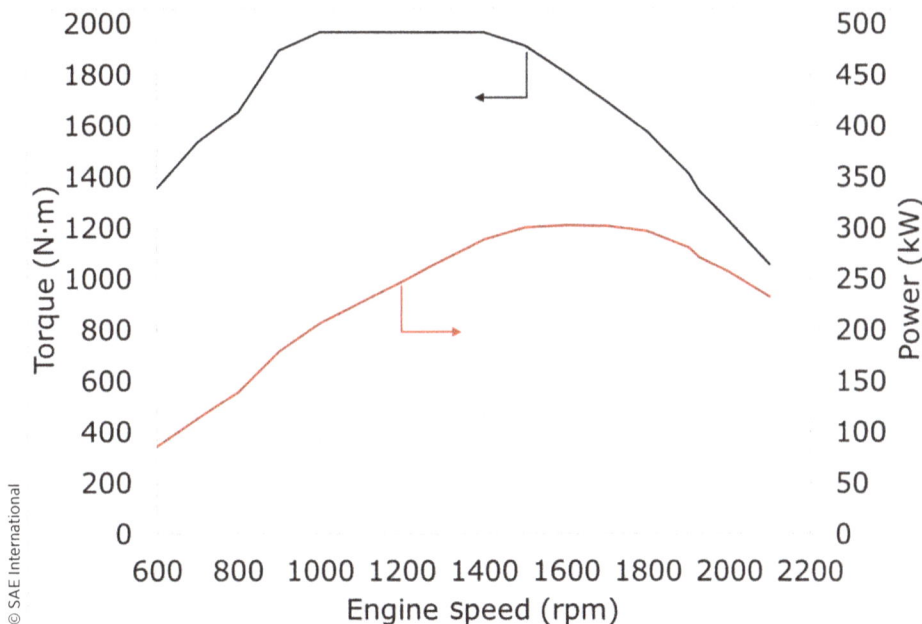

© SAE International

FIGURE 4.16 Scania DC13 085A torque and power curves with respect to engine speed.

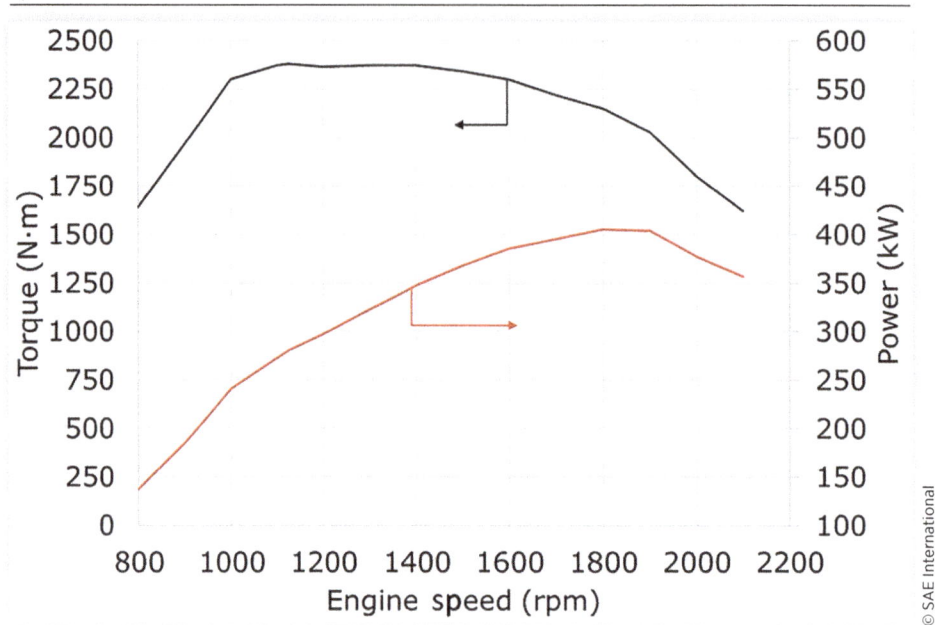

used for the on-road HDE tests described in Sections 4.2.2 and 4.2.3 are defined in a UN Agreement [4-20, 4-21]. Valid measurement procedures are also defined by international standards, including ISO 8178-1:2017 [4-22] for gaseous and particulate emissions on a test bed and ISO 8178-3:2019 [4-23] for smoke measurements.

Measured pollutant levels from these cycles are usually reported on an energy-specific basis, typically grams of pollutant per kilowatt-hour (g/kW·h) or grams of pollutant per brake horsepower-hour (g/bhp·h). It may be useful to think of these results as pollutant mass flow per unit power, that is, (g/h)/kW, although the numerical results are equivalent. The newest regulatory limits are increasingly low enough that they may be stated on a milligram per unit energy basis (mg/kW·h or mg/bhp·h). One exception to this general scheme is particulate number, which is measured on a number flow per unit power basis (#/kW·h).

4.2.1 Example HDEs

To help illustrate the drive cycles in this section, two example HDDEs are used. The first is the D13 engine from Volvo Trucks, which is a typical heavy HDDE used in line-haul tractors and similar commercial vehicle applications. This engine has been chosen to illustrate the drive cycles and operating points for on-road applications. The second engine is the DC13 085A from Scania, which is a typical non-road HDDE used in this section to illustrate the drive cycles and operating points for non-road applications. Note that the torque and power curves shown are taken from public sources and may differ from the performance of the actual engine during certification testing.

The D13 engine from Volvo Trucks is a 12.8 L inline 6 (I-6) HDDE with boost system and direct injection (DI). The version of the engine with the torque and power curves shown in Figure 4.15 has a rated power of 302 kW (405 bhp) at 1,600 rpm, and a peak torque of 1,966 N·m (1,450 lb.-ft.) at 1,000 rpm [4-24]. The engine is certified to the EPA 2010 emissions levels with a rated speed of 1,700 rpm. Volvo Truck has fitted the engine with a diesel oxidation catalyst, DPF, and selective catalytic reduction (SCR) system to manage the criteria pollutants in the EAS. More information on this engine and its EAS may be found in Section 7.5.5.

The Scania DC13 085A is a 12.7 L I-6 non-road HDDE with boost system and DI. The version with the curves shown in Figure 4.16 has a rated power of 405 kW (550 bhp) at 1,800 rpm, and a peak torque of 2,363 N·m (1,755 lb.-ft.) at 1,200 rpm [4-25]. The engine is certified to EU Stage IV and EPA Tier 4 final emissions levels. It uses exhaust gas recirculation and SCR systems to manage the tailpipe-out criteria pollutant levels. More information on this engine and its EAS may be found in Section 7.5.5.

4.2.2 World-Harmonized Heavy-Duty Transient Cycle

The WHTC [4-20, 4-21] is a transient engine cycle that is used by several regulatory bodies globally, including the EU, Japan, and PRC. In the EU, the WHTC supplanted the use of the European Transient Cycle (ETC), which is described further in Section 4.2.9. As defined in Annex I of ECE/TRANS/WP.29/2006/128 [4-20], the test cycle lasts 1,800 s (30 min), and has a series of engine acceleration and deceleration events as shown in **Figure 4.17**. The deceleration events even include operating on the motoring curve,

FIGURE 4.17 WHTC traces showing (a) normalized engine speed and (b) normalized engine torque with respect to time.

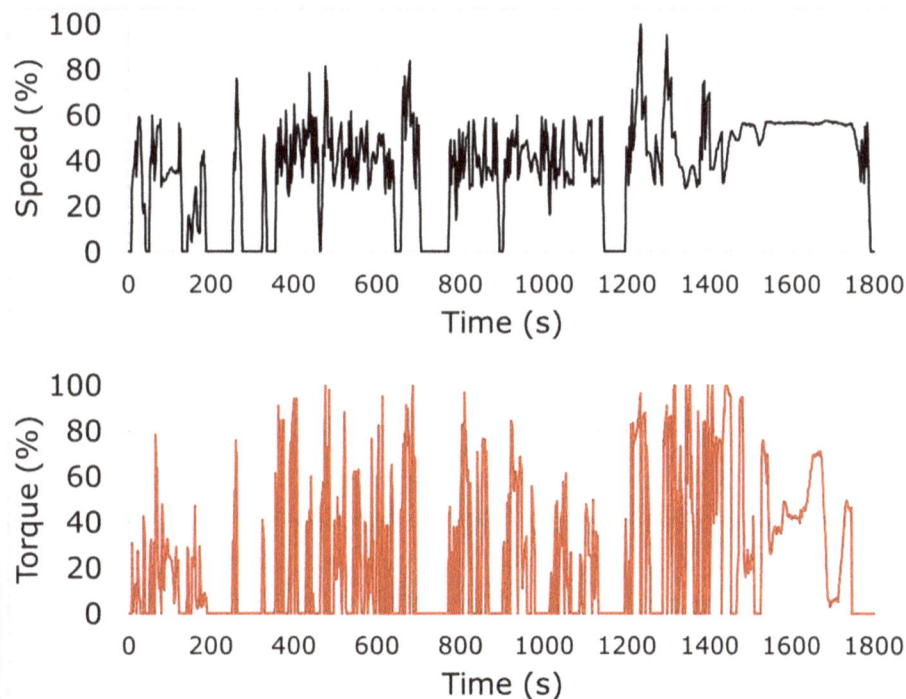

FIGURE 4.18   WHTC test points on normalized speed and load ranges.

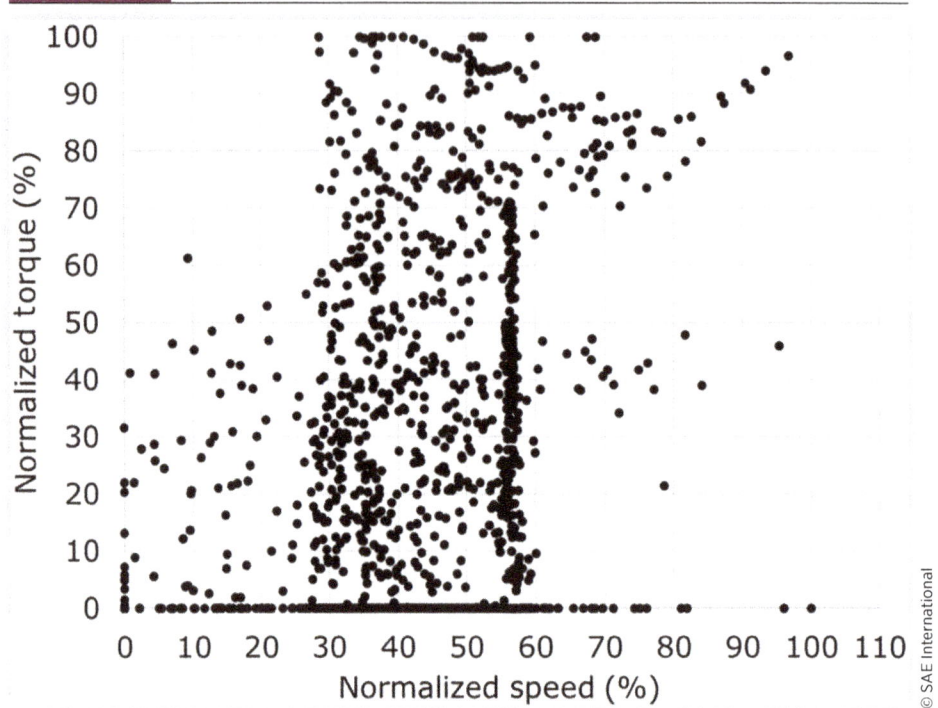

although in Figure 4.17, these motoring points have been rounded up to 0% torque. During the cycle, 401 s are spent motoring the engine and another 391 s are spent at idle torque. Together, these represent 44.0% of the total test cycle.

The WHTC engine operating points are also shown together in **Figure 4.18**, with engine load versus engine speed. As with Figure 4.17, the operating points on the motoring curve have been rounded up to 0% torque on this plot. Note that most of the operation on this cycle occurs between 25% and 60% of the speed range between idle and the rated power speed.

The WHTC operating points are also shown on the Volvo D13 operating map to provide a more concrete example. The blue curve in **Figure 4.19** is the torque curve for the engine, which is also shown in Figure 4.15. The operating points cluster between 1,000 and 1,300 rpm, which is a typical preferred speed range for this type of engine. The motoring curve shown is an approximation used to illustrate the drive cycle.

4.2.3 World-Harmonized Heavy-Duty Steady-State Cycle

The World-Harmonized Heavy-Duty Steady-State Cycle (WHSC) [4-20, 4-21] is a steady-state engine test cycle that was developed so that it could be used broadly around the globe. WHSC has been incorporated into the testing required by several regulatory bodies, including in the EU, Japan, and PRC. WHSC supersedes the 13-mode European Steady-State Cycle (ESC) used in the EU that is the basis for the US Supplemental Engine Test (SET) described in Section 4.2.5 and supersedes the previous 13-mode cycle used in Japan.

FIGURE 4.19 Example of WHTC test points on Volvo D13 operating map.

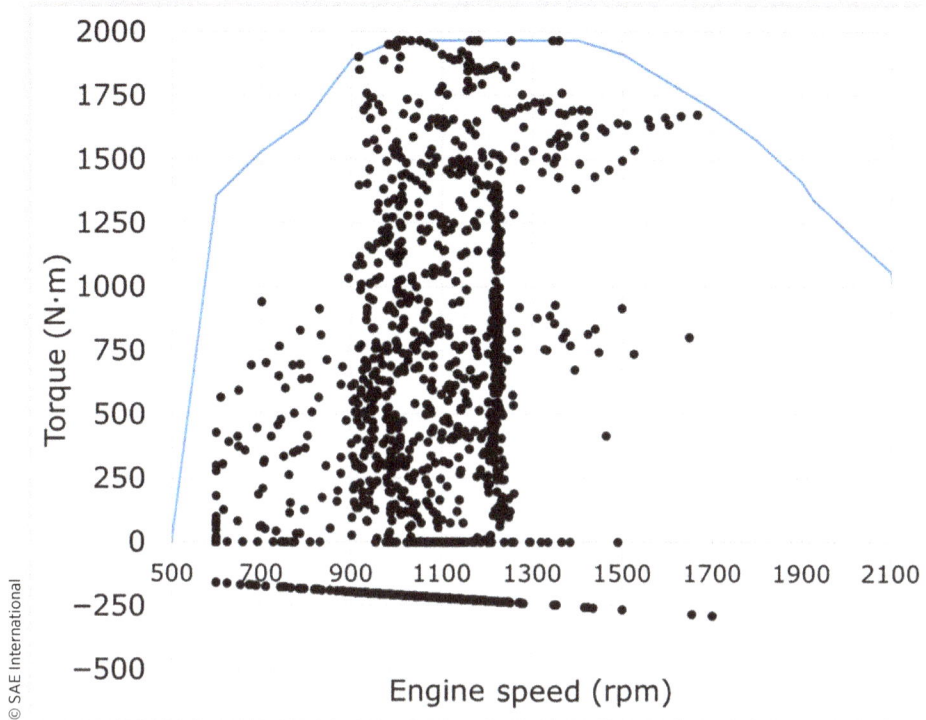

As defined in Annex I of ECE/TRANS/WP.29/2006/128 [4-20], the WHSC comprises 13 operating modes, which includes operating twice at idle, once at the start and once at end of the test, as shown in Table 4.4. The 13 operating modes are also plotted in Figure 4.20, where each mode is marked with its associated weight. The idle condition shows the sum of weights for the two test modes that operate there. For WHSC the 100% normalized speed is the engine speed at which the engine has its rated power, for example, 1,600 rpm for the Volvo D13, as shown in Figure 4.15.

The time indicated for each mode in Table 4.4 also includes a 20-s ramp to that mode from the previous one, with the exception of Mode 1 (initial idle). These ramps are also shown in Figure 4.21. The total test time for all 13 modes is 1,895 s (nearly 32 min). Test results are averaged over the modes to generate a brake-specific emissions value in (g/kW·h). Because each mode is run for a period of time corresponding to its weight, the test results may be averaged over the complete test cycle to get the appropriate cycle-averaged results.

4.2.4 US Heavy-Duty Federal Transient Protocol

The US Heavy-Duty Federal Transient Protocol (HD FTP) is the transient engine cycle used to evaluate both criteria

TABLE 4.4 WHSC operating modes [4-20].

Mode number	Normalized speed (%)	Normalized load (%)	Time in mode (s)	Weight (%)
1	0	0	210	11.1
2	55	100	50	2.6
3	55	25	250	13.2
4	55	70	75	4.0
5	35	100	50	2.6
6	25	25	200	10.6
7	45	70	75	4.0
8	45	25	150	7.9
9	55	50	125	6.6
10	75	100	50	2.6
11	35	50	200	10.6
12	35	25	250	13.2
13	0	0	210	11.1

FIGURE 4.20 WHSC operating points plotted on normalized engine map [4-20].

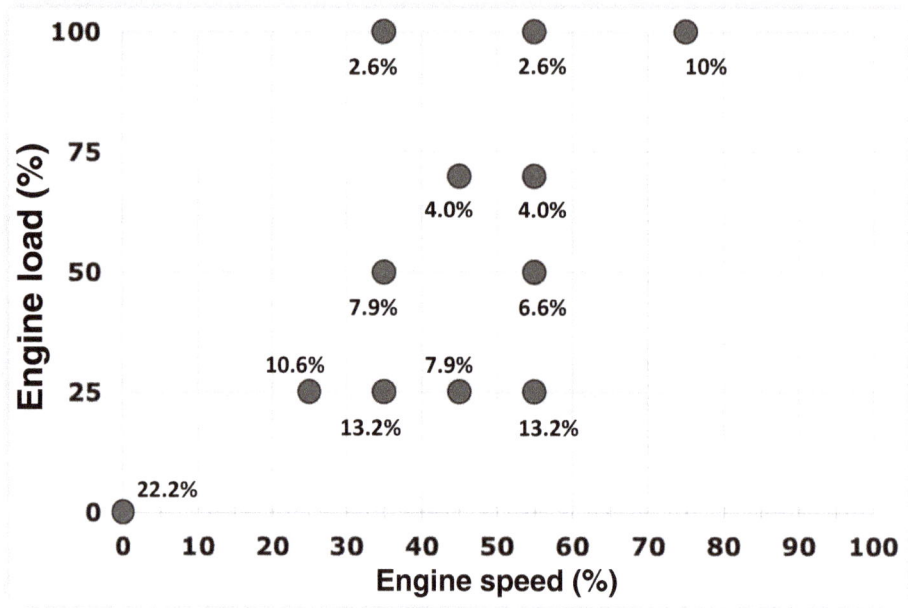

FIGURE 4.21 WHSC ramped modal cycle traces showing (a) normalized engine speed and (b) normalized engine torque with respect to time [4-20].

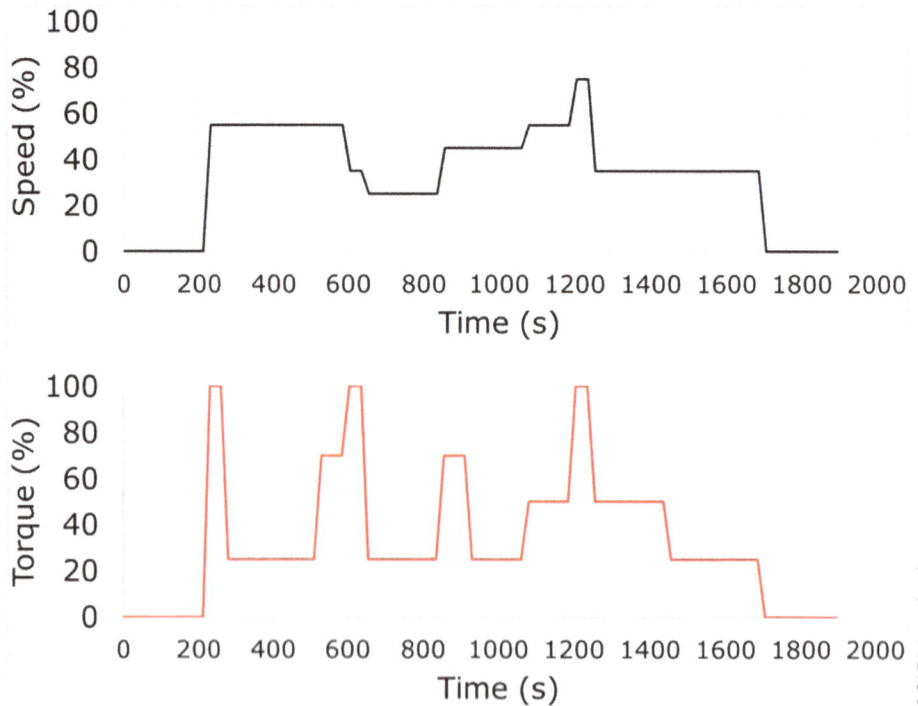

pollutants and GHG emissions from HDEs. There are actually two HD FTP cycles, one for HDDEs called the Heavy-Duty Diesel Transient Cycle (HDDTC), and one for HDSI engines called the Otto cycle FTP or the Heavy-Duty Gasoline Transient Cycle. Given the dominant market share of HDDE in the commercial vehicle space, most engineers assume that "HD FTP" refers to the HDDTC unless specified, and it is the usage in this section.

As defined in Appendix I to 40 CFR part 86 [4-7], the HD FTP lasts 1,200 s (20 min), and it is split into four segments, as shown in **Figure 4.22**: New York non-freeway, Los Angeles non-freeway, Los Angeles freeway, and New York non-freeway. During the cycle, 178 s are spent motoring the engine and another 448 s are spent at idle torque. Together these represent 34.8% of the total test cycle. Most of these idle events occur in the first and final segments of the test, which makes the cold-start HD FTP particularly challenging because the engine does not naturally operate in a way in the first segment that promotes warming up. The deceleration events in the cycle include operating on the motoring curve, although in Figure 4.22 these motoring points have been rounded up to 0% load.

It is also helpful to understand where the normalized test cycle operating points fall on the overall engine-operating map, shown in **Figure 4.23**. As with the traces of normalized load in Figure 4.22, the operating points on the motoring curve have been rounded up to 0% load. It is worth noting that the HD FTP includes operating points both below the warm idle speed and above the rated power speed; hence, the range of normalized

FIGURE 4.22   HD FTP traces with (a) normalized engine speed and (b) normalized engine torque with respect to time [4-7].

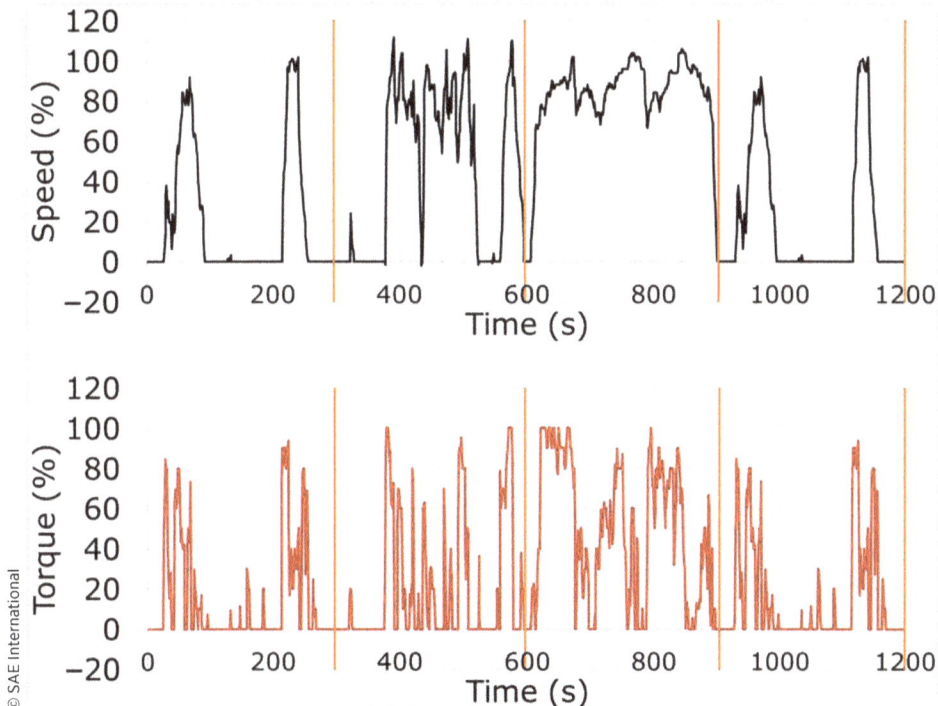

FIGURE 4.24   Example of WHTC test points on Volvo D13 engine operating map.

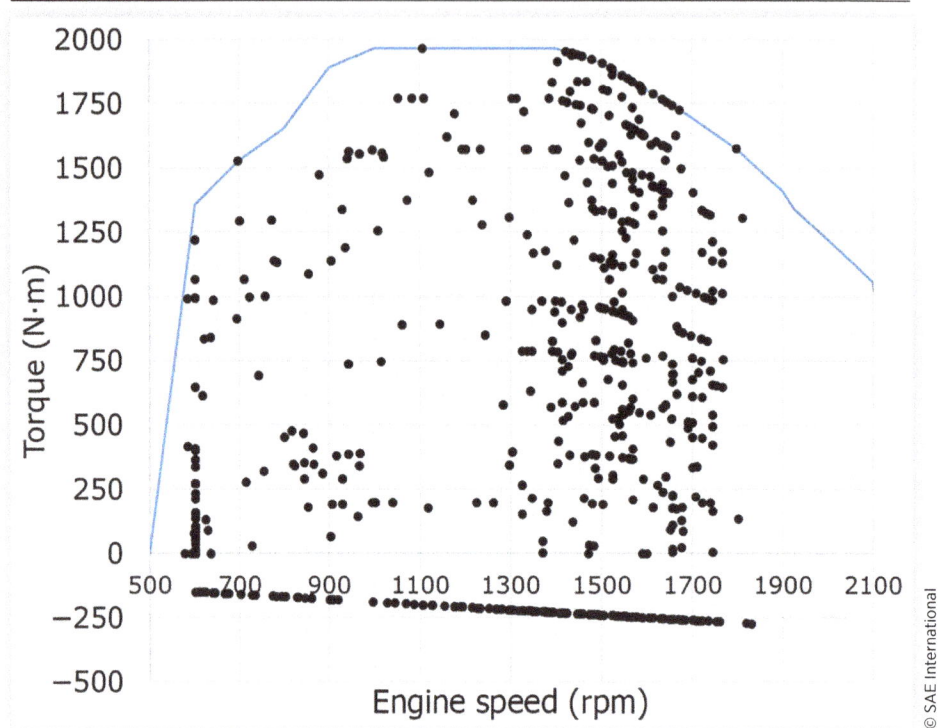

© SAE International

speeds extends beyond the 0%–100% range. Most of the operating points fall in the higher speed range, where the normalized engine speed is at least 60%. This distribution means that the operating points on the HD FTP will tend to have relatively hot exhaust gas temperatures, especially compared to the WHTC, where most of the operating points shown in Figure 4.18 are between 30% and 60% relative speed.

Note that for certification testing that complies with the requirements of 40 CFR part 86 [4-5], the cold-start HD FTP test is followed by at least three consecutive hot-start HD FTP tests, with a 1,200 s (20 min) soak period between each test run. The composite transient emission results E_i for each species i are calculated using a weighted average of the cold and hot cycle results, as follows:

$$E_i = \frac{1}{7} E_{i,\text{cold}} + \frac{6}{7} E_{i,\text{hot}}. \tag{4.2}$$

The HD FTP operating points are also shown on the Volvo D13 operating map to provide a more concrete example. The blue curve in **Figure 4.24** is the torque curve for the engine, which is also shown in Figure 4.15. As with the WHTC example shown in Figure 4.19, the motoring curve shown in Figure 4.24 is an approximation used to illustrate the test cycle. Note that by comparison to the WHTC, the operating points on the HD FTP cluster much closer to the rated speed of 1,625 rpm.

The engineering consultancy AVL has developed an approximation to the HD FTP cycle that uses eight steady-state operating modes to generate a weighted average of emissions results [4-26]. This cycle is called the AVL 8-mode cycle, and the operating points and associated weight factors are shown in **Table 4.5** and plotted in **Figure 4.25**.

FIGURE 4.23   HD FTP test points on normalized speed and load ranges [4-7].

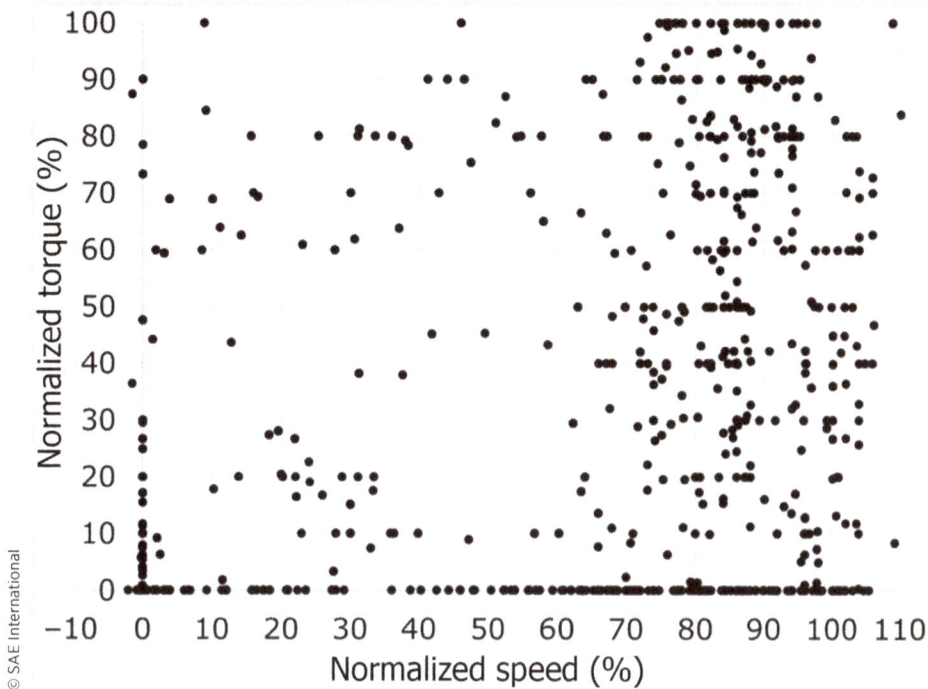

Note that the weight factors sum to 83.99, not to 100, and that the sum of weight factors in the higher speed range (modes 4–8) are about the same as that for idle (mode 1). The need for this approximating cycle comes from engine development programs, in which the engineers are able to implement an engine calibration that supports steady-state operation sooner than one that supports fully transient operation. The downside to the AVL 8-mode cycle is that it only gives insight into the hot-start HD FTP results, not the cold start. Nevertheless, it is an important development tool, especially when combined with the steady-state operating points that comprise the US Supplemental Emissions Test (SET), described in Section 4.2.4.

EPA has also developed a separate drive cycle for emissions testing of HDSI engines, also known as HD Otto cycle engines. As with the diesel HD FTP cycle, the Otto cycle FTP test points are defined in Appendix I to 40 CFR part 86 [4-7]. The Otto cycle FTP cycle only applies to some light and medium HDSI engines, as heavy HDSI engines and lighter HDSI engines derived from HDDEs are evaluated on the diesel HD FTP cycle. The normalized engine speed and load traces with respect to time are shown in **Figure 4.26**. The test cycle includes several transients in engine speed and load. For this cycle, the normalized speed covers the range from idle to the speed at rated power. Unlike the diesel HD FTP cycle, the Otto cycle FTP specifies the motoring torque inputs, since not all of the points with negative loads are at the motoring torque curve. The minimum normalized torque in the cycle is −10%. The Otto cycle FTP is slightly shorter

TABLE 4.5 AVL 8-mode operating points for HD FTP approximation; weight factors sum to 83.99 [4-26].

Mode number	Normalized speed (%)	Normalized load (%)	Weight factor
1	0	0	35.00
2	11	25	6.34
3	21	63	2.91
4	32	84	3.34
5	100	18	8.40
6	95	40	10.45
7	95	69	10.21
8	89	95	7.34

FIGURE 4.25 AVL 8-mode operating points and weighting factors; weight factors sum to 83.99 [4-26].

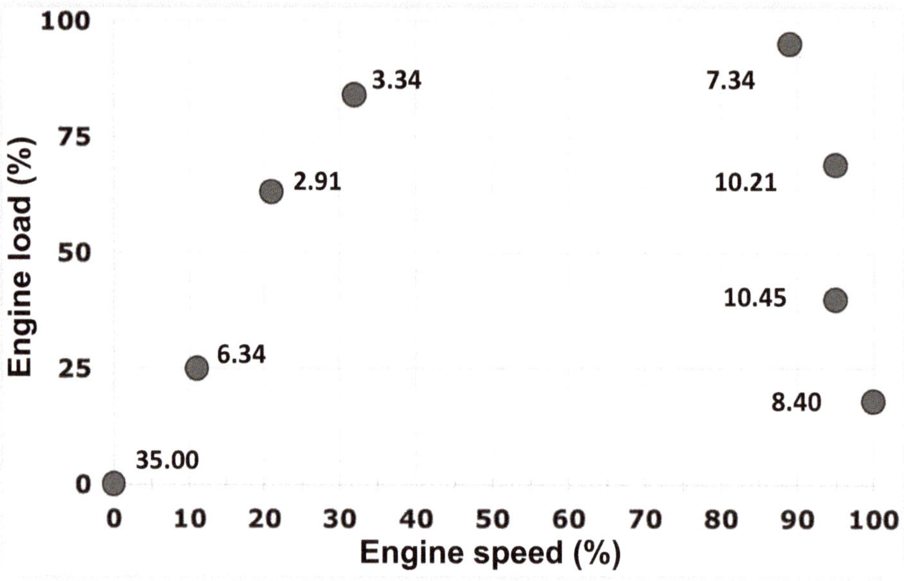

FIGURE 4.26 HDSI FTP traces with (a) normalized engine speed and (b) normalized engine torque with respect to time [4-7].

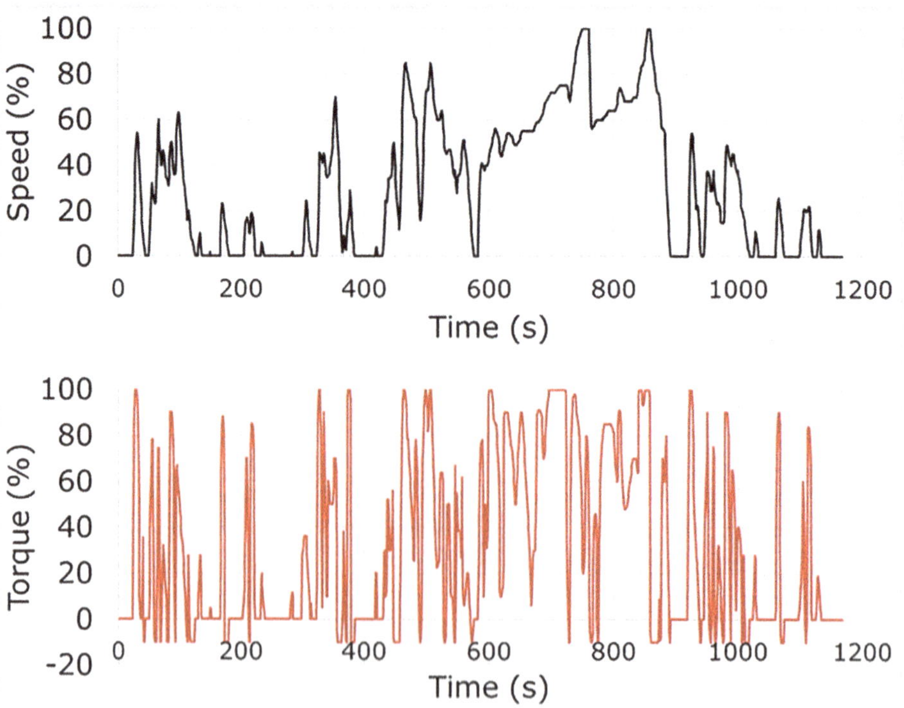

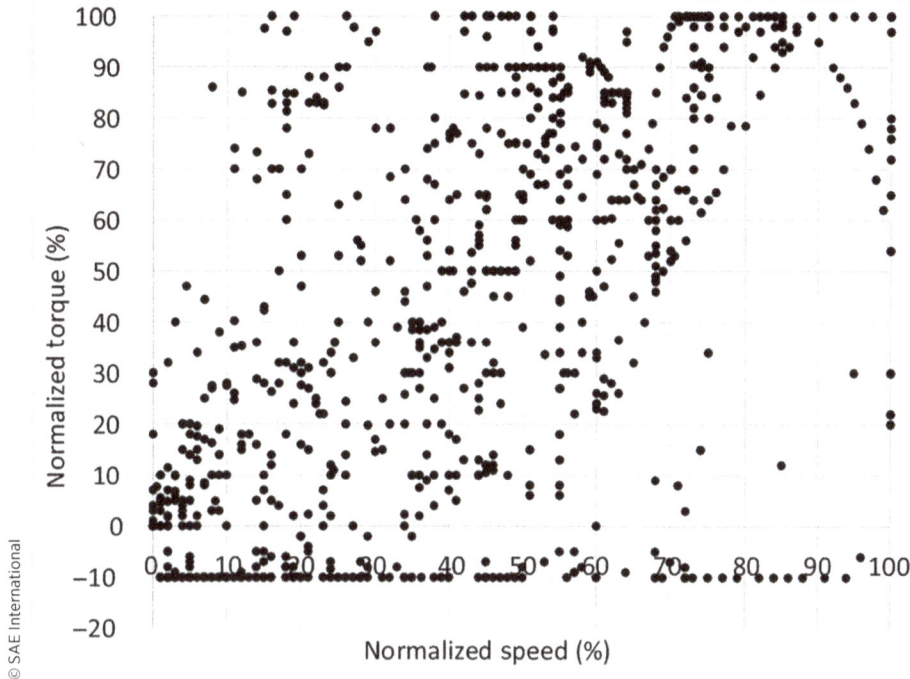

FIGURE 4.27   HDSI FTP test points on normalized speed and load ranges [4-7].

than the HD FTP, at 1,167 s duration, of which 142 s are spent motoring the engine and another 350 s are spent at idle load. Together, these represent 27.3% of the total test cycle, which is less than the 34.8% of HD FTP spent at negative or zero load.

Because of expected differences in engine behavior, the drive cycle covers a different set of engine operating points. The normalized engine loads are plotted against the normalized engine speeds in **Figure 4.27**. By comparison to the diesel HD FTP cycle, the HDSI operating points cover a considerably broader range of operation, with a cluster ranging from idle up to full load at around 80% normalized speed, and with more points at lower normalized loads.

4.2.5 US Supplemental Emissions Test and RMC

The SET is a 13-mode steady-state cycle, with engine loads defined in 40 CFR §86.1362 [4-27] and 40 CFR §1036.505 [4-28], and engine speeds defined in 40 CFR §1065.610 [4-29]. The SET was adapted from the ESC, which was defined in 1999 by the EU [4-30]. It includes operation at idle conditions and four engine loads at each of three speeds. The related RMC is an overall test schedule that is defined in the same regulations as the SET, as described later in this section. The RMC uses ramped transitions to move the engine operation between these 13 operating modes and includes these ramps in the overall measurements. The weighting for the RMC results are adjusted accordingly. The RMC may be used as an alternative to the SET, and most choose the RMC because it is a faster test to run. In the original ESC protocol, the witness to the certification test may ask for measurements at two operating points within the zone defined by the 12 non-idle

TABLE 4.6 US SET operating modes and weights for criteria pollutants [4-27] and greenhouse gas emissions [4-28]. The order shown is for criteria pollutants.

Mode number	Speed	Normalized load (%)	Weight (%) Pollutants (%)	GHG (%)
1	Idle	0	15.0	12.0
2	A	100	8.0	9.0
3	A	25	5.0	12.0
4	A	75	5.0	12.0
5	A	50	5.0	12.0
6	B	100	10.0	10.0
7	B	25	10.0	10.0
8	B	75	9.0	9.0
9	B	50	10.0	9.0
10	C	100	8.0	2.0
11	C	25	5.0	1.0
12	C	75	5.0	1.0
13	C	50	5.0	1.0

operating points. This part of the protocol serves as a check that the engine and EAS are operating correctly everywhere, not just on the 13 test modes.

The SET operating modes cover the engine idle point—idle speed and zero load—and a set of 12 points defined by three engine speeds, A, B, and C, and four relative engine loads, as shown in Table 4.6. The A, B, and C speeds are defined in 40 CFR §1065.610 [4-29] [4-19] and require calculating a minimum speed and a maximum speed based on the engine torque curve. The minimum speed, N_{min}, is the engine speed less than the rated speed where the power at full load is 50% of the rated engine power. Depending on the shape of the torque curve, the curve may need to be extended to the sub-idle zero torque speed. Similarly, the maximum speed, N_{max}, is the engine speed greater than the rated speed where the power at full load is 70% of the rated power. As with N_{min}, the torque curve may need to be extended out to the governed or cutoff speed to calculate N_{max}. Once these two speeds are known, the A, B, and C speeds may be calculated using the following equations:

$$A = N_{min} + 0.25\left(N_{max} - N_{min}\right) \tag{4.3}$$

$$B = N_{min} + 0.50\left(N_{max} - N_{min}\right) \tag{4.4}$$

$$C = N_{min} + 0.75\left(N_{max} - N_{min}\right) \tag{4.5}$$

These equations are Eqs. (1065.610-5), (1065.610-6), and (1065.610-7), respectively, from 40 CFR §1065.610 [4-29]. Note that with the introduction of the Phase 2 GHG Standard for HDVs in the US [4-31], there are now two sets of weightings for the SET, one for criteria pollutants defined by 40 CFR §86.1362 [4-27] and one for GHG defined by 40 CFR §1036.505 [4-28]. The SET modes tabulated in Table 4.6 are ordered per 40 CFR §86.1362 [4-27] for criteria pollutants, but both sets of weightings are included. Note that the idle condition indicated is the warm idle speed at zero net load, where the warm idle speed is the idle speed of a warmed-up engine. The operating points for the SET for criteria pollutants are plotted in Figure 4.28, which shows the relative weights for calculating criteria pollutants emissions from the SET. Similarly, the SET operating points and weights for calculating GHG emissions are plotted in Figure 4.29. As described earlier, the test witness may ask for two additional operating points to be run at any speed between the A and C speeds and any normalized load between 25% and 100% at the two chosen speeds.

For example, the D13 engine described in Section 4.2.1 has a rated power of 302 kW (405 hp) at 1,600 rpm. Using the torque curve in Figure 4.15, one can calculate that N_{min} is 831 rpm and that N_{max} is 2,109 rpm. Note that calculating N_{max} for this engine requires extrapolating out to an assumed cutoff speed of 2,200 rpm. From these values of N_{min} and N_{max}, Eqs. 4.3, 4.4, and 4.5 are used to calculate A = 1,151 rpm, B = 1,470 rpm, and C = 1,790 rpm. These speeds are 72%, 92%, and 112% of the speed at rated power, respectively.

FIGURE 4.28   SET test modes shown with weightings for criteria pollutants [4-27].

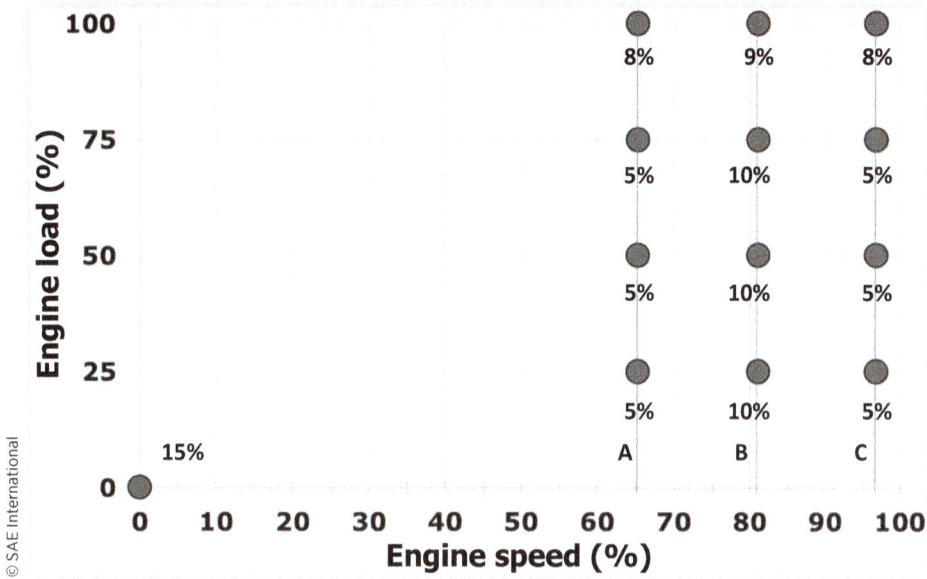

FIGURE 4.29   SET test modes shown with weightings for GHG emissions [4-28].

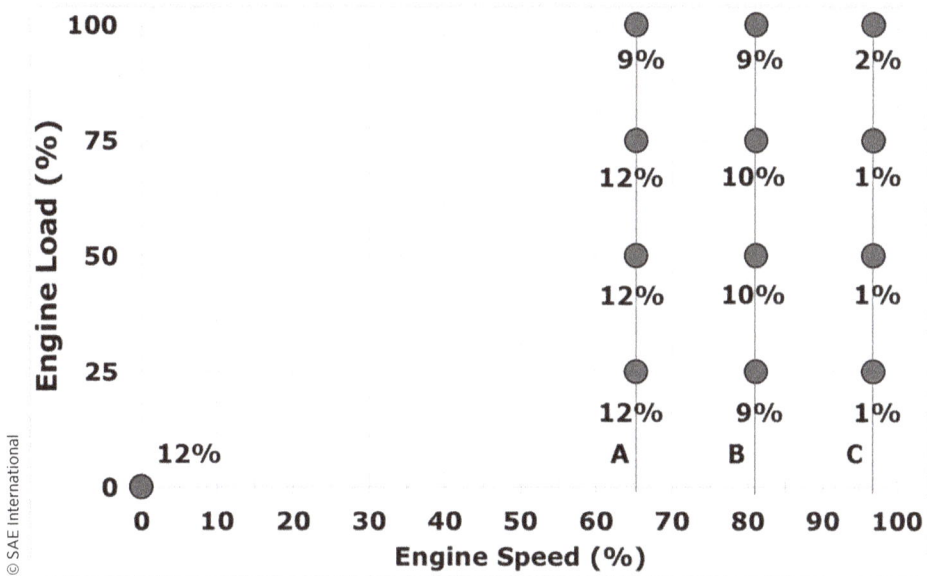

As with the SET, there are now two versions of the RMC: one for criteria pollutants defined in 40 CFR §86.1362 [4-27] that is tabulated in Table 4.7, and one for GHG emissions defined in 40 CFR §1036.505 [4-28] that is tabulated in Table 4.8. Both RMC cycles take a total of 2,400 s (40 min) to complete and step through a series of operating points that include 25%, 50%, 75%, and 100% of maximum torque at each of the A, B, and C speeds plus time at the warm idle mode at the start and end of the test. The dwell time at each mode is set so that the aggregated emission results from the complete RMC reflect

TABLE 4.7 Ramped-modal cycle (RMC) operating modes and dwell times for criteria pollutants [4-27].

Mode number	Speed	Normalized load (%)	Mode time (s)
1	Idle	0	170
2	A	100	170
3	A	25	102
4	A	75	100
5	A	50	103
6	B	100	194
7	B	25	219
8	B	75	220
9	B	50	219
10	C	100	171
11	C	25	102
12	C	75	100
13	C	50	102
14	Idle	0	168

After the initial idle period, there is a 20-s ramp to the next mode.

TABLE 4.8 Ramped-modal cycle (RMC) operating modes and dwell times for greenhouse gas emissions [4-28].

Mode number	Speed	Normalized load (%)	Mode time (s)
1	Idle	0	124
2	A	100	196
3	B	50	220
4	B	75	220
5	A	50	268
6	A	75	268
7	A	25	268
8	B	100	196
9	B	25	196
10	C	100	28
11	C	25	4
12	C	75	4
13	C	50	4
14	Idle	0	144

After the initial idle period, there is a 20-s ramp to the next mode.

the appropriate weightings for each of the operating modes, as shown in Figures 4.30 and 4.31. Note that because the GHG emissions SET or RMC tests put such a low weighting on the modes at the C speed, the end of the RMC for GHG spends more time in transition than dwelling at the operating modes, as shown in Figure 4.31.

4.2.6 US NTE Requirement

The NTE requirement is an in-use requirement for dynamometer-certified HDDEs in HDVs that ties into the heavy-duty onboard diagnostics requirements that are described in Chapter 6. The NTE performance is validated using PEMS testing of the HDV over the road. When operating within the NTE Control Area defined below, the engine must not exceed the NTE limits of 40 CFR §86.007-11(a)(4) [4-32] when averaged over a time period of at least 30 s. The minimum averaging period must include at least one regeneration event from those EAS devices that periodically regenerate, such as lean NOx trap (LNT) or DPF, which means that the minimum average period is likely to be hundreds of seconds.

The NTE Control Area is defined by 40 CFR §86.1370 [4-33] and includes the following range of the operating map:

- All operating speeds above the minimum engine speed $N_{\text{NTE,min}}$ as defined in the following equation:

$$N_{\text{NTE,min}} = N_{\text{min}} + 0.15\left(N_{\text{max}} - N_{\text{min}}\right) \tag{4.6}$$

- All normalized engine loads from 30% to the torque curve (100%) above $N_{\text{NTE,min}}$.
- In addition to the above zone, all operating points where the brake-specific fuel consumption (BSFC) is within 5% of the minimum BSFC for the engine, where BSFC is calculated per 40 CFR part 1065 [4-19].
- If any of the operating points within 5% of the best BSFC has a normalized load below 30%, those points may be excluded from the NTE Control Area.

An estimate of the NTE zone for the Volvo D13 operating map from Figure 4.15 is shown in Figure 4.32. The torque curve is shown in blue. $N_{\text{NTE,min}}$ is calculated as 1,023 rpm, and the torque at that speed ranges from 590 N·m to 1,966 N·m. The NTE Control Area between the 30% load curve and the torque curve is extended out to the rated speed of 1,600 rpm and is marked in gray. In addition, we have sketched a hypothetical boundary where the BSFC is within 5% of the minimum BSFC boundary in orange, the "Best BSFC boundary." The addition to the NTE Control Area where it extends below $N_{\text{NTE,min}}$ is shaded in orange.

40 CFR §86.1370 [4-33] indicates that some corrections to the measured results are allowed if the ambient humidity is outside of a defined range, or if the ambient temperature is below 55°F (12.8°C).

FIGURE 4.30 Ramped-modal cycle (RMC) for criteria pollutants [4-27] showing (a) normalized engine speed and (b) normalized engine torque.

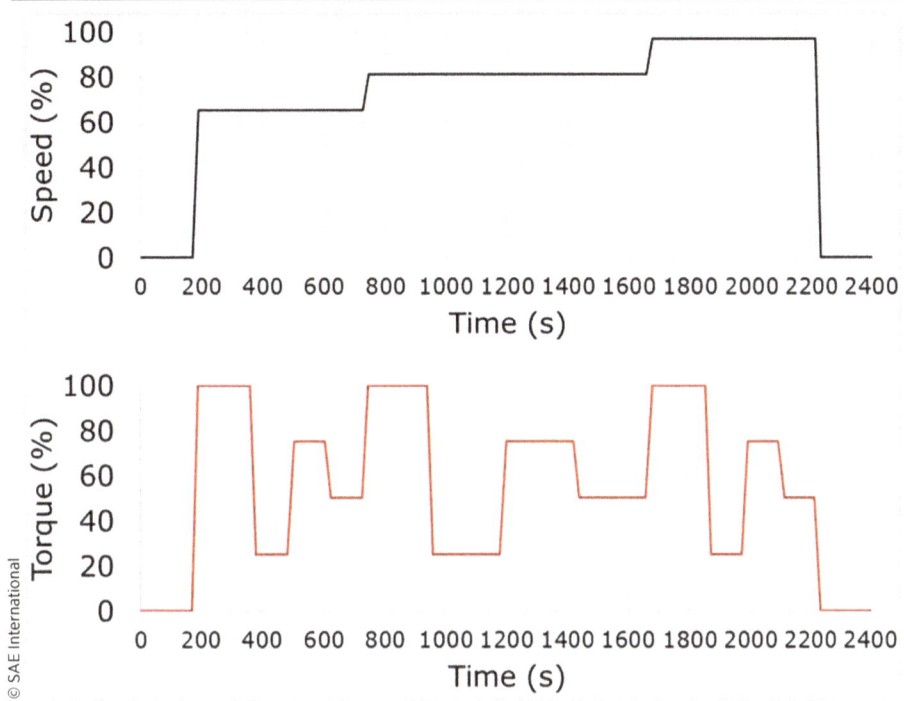

FIGURE 4.31 Ramped-modal cycle (RMC) for greenhouse gas emissions [4-28] showing (a) normalized engine speed and (b) normalized engine torque.

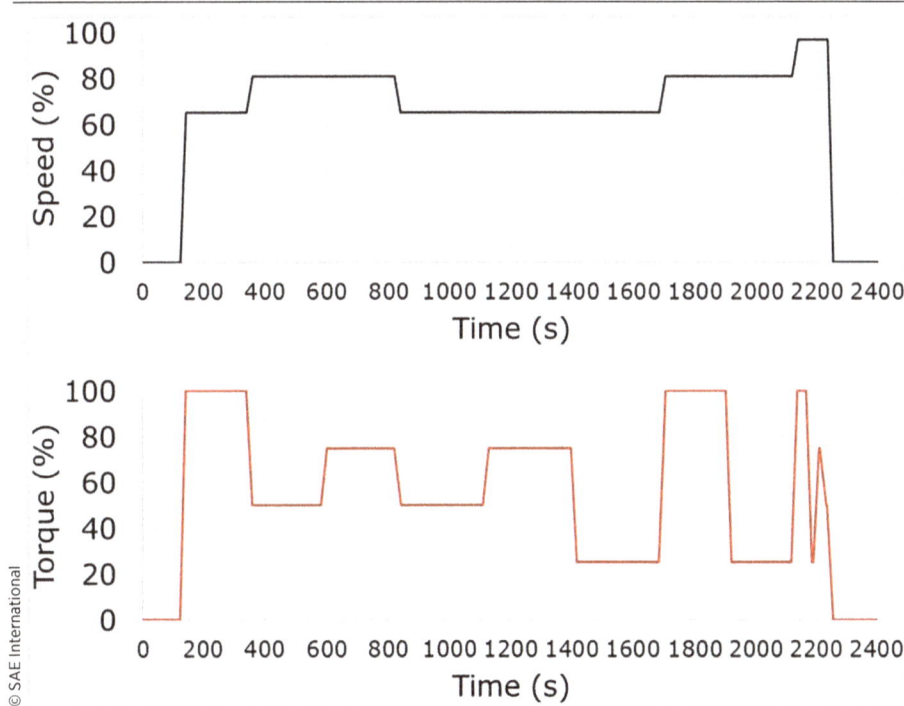

© SAE International

FIGURE 4.32 Not-to-Exceed (NTE) Control Area example using the Volvo D13 torque curve and a hypothetical best BSFC point.

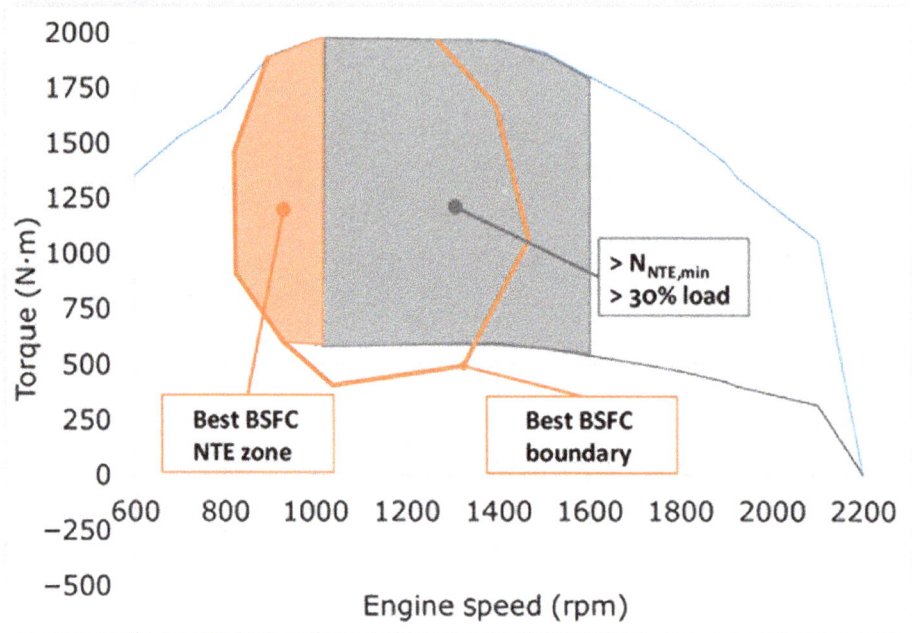

4.2.7 Non-Road Transient Cycle

The Non-Road Transient Cycle (NRTC) [4-34] is a transient test cycle used to evaluate non-road engines for EPA Tier 4 final, EU Stage V, EU Stage IV, Japan Stage 4, and other harmonized regulatory requirements that are described in Chapter 3, Emissions Control Regulations. The NRTC duration is 1,238 s, of which only 40 s (2.2%) are spent at idle conditions. The NRTC has no motoring torque conditions, which is consistent with expectations for off-road applications. As with the on-road HDE cycles, the inputs are normalized speed and load traces that are mapped to the actual engine's performance. The speed range goes from the warm idle speed to rated speed, and the load range goes from idle torque to the maximum torque at the demanded speed.

The full requirements for running the NRTC for certification in the US may be found in Appendix VI to 40 CFR part 1039 [4-34]. In general, though, the NRTC is run with both a cold start and a hot start with a soak period in between, and the results are averaged to generate the brake-specific certification values for criteria pollutants or fuel consumption. In the EU, the weighting is 10% from the cold test results and 90% from the hot test, whereas in the US, the weighting is 5% cold, 95% hot per 40 CFR 1039.510 [4-35].

As seen in **Figure 4.33**, the target engine speed has periods of highly transient behavior, and others where it is approximately steady. The torque has significant transient events throughout the drive cycle.

As with the on-road cycles, it is useful to understand how the operating points are clustered in the drive cycle. As shown in **Figure 4.34**, there is a cluster of operating points between 75% and 90% normalized speed, especially up to 75% normalized load.

FIGURE 4.33 NRTC speed and load traces versus time [4-34].

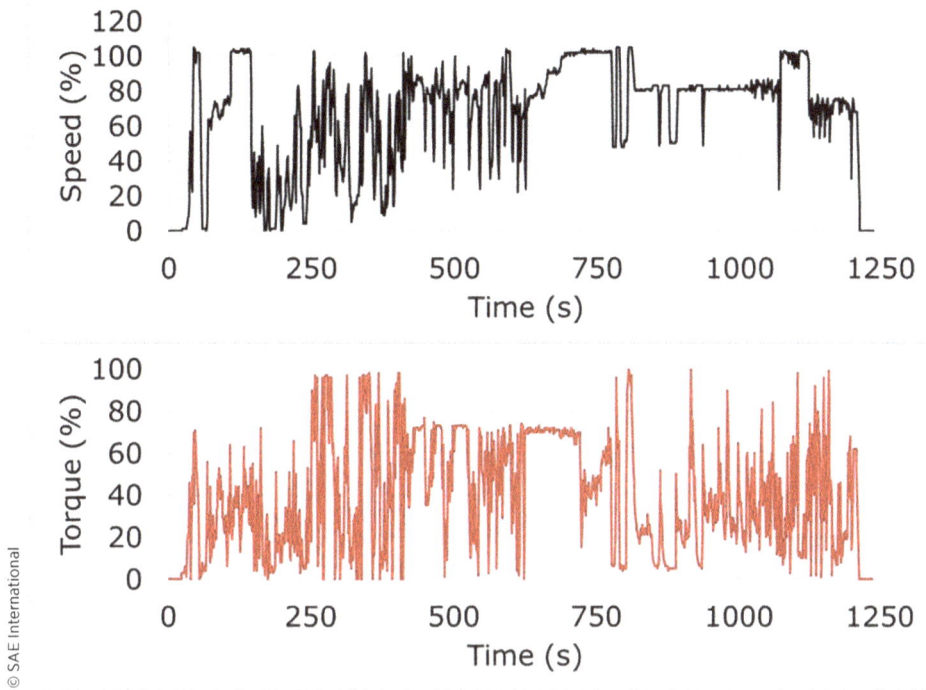

© SAE International

FIGURE 4.34 NRTC speed and load operating points plotted on normalized operating map [4-34].

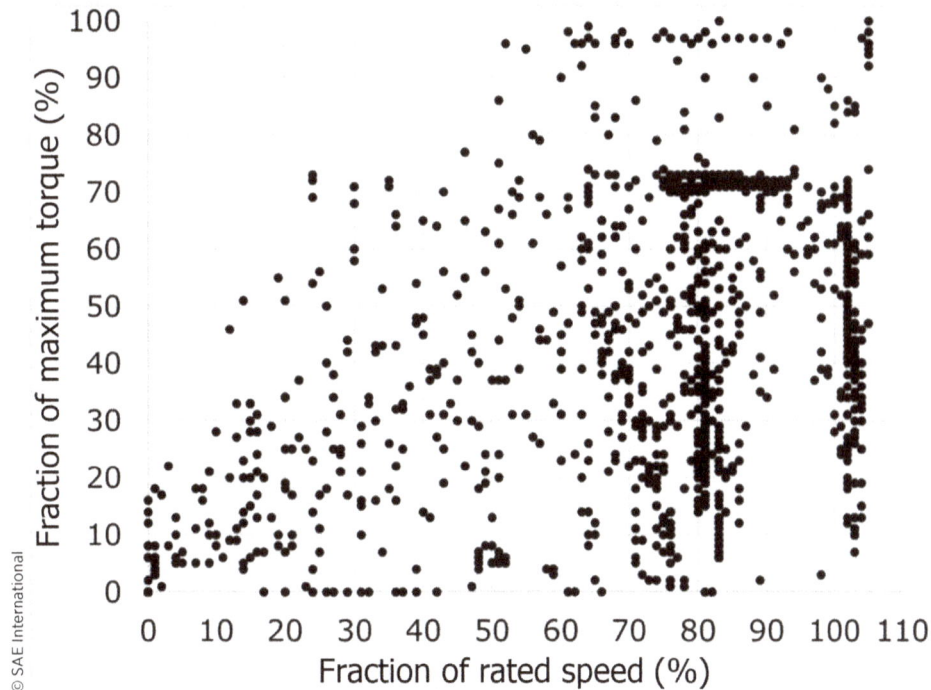

© SAE International

FIGURE 4.35 NRTC speed and load-operating points shown on the Scania DC13 085A operating map.

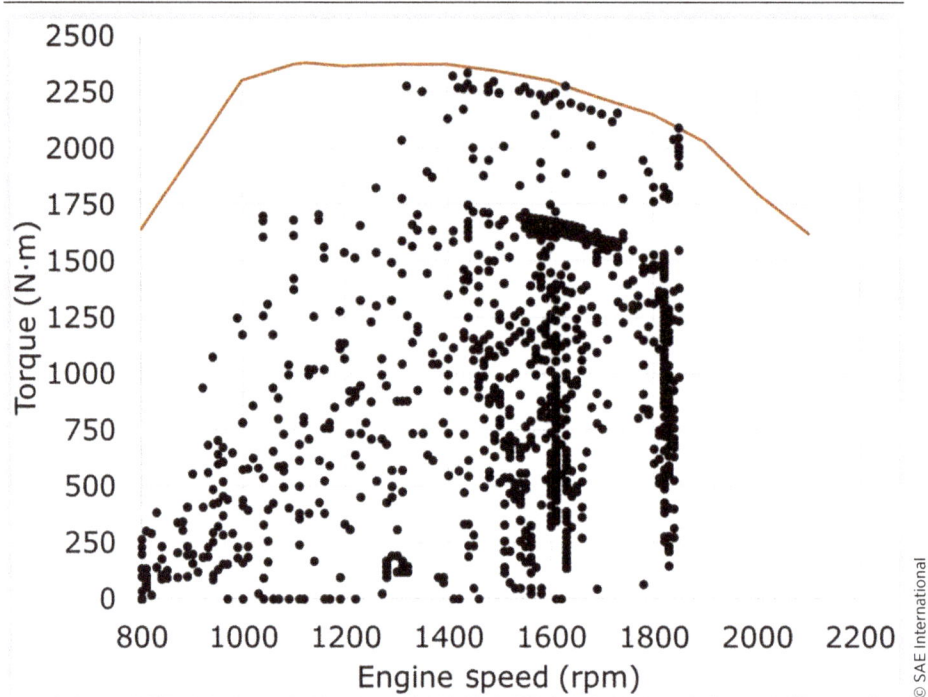

Interestingly, operating points between 100% and 110% of the normalized speed at several load conditions occur during 181 s of NRTC, which is about 14.6% of the total.

As with the on-road cycles, it is helpful to see how the NRTC operating points fall on a real engine-operating map. Using the torque curve from Figure 4.16, the NRTC normalized speed and load-operating points were mapped to real engine speeds and torques. These operating points are plotted in **Figure 4.35** and show how little of the NRTC runs at maximum torque.

4.2.8 Non-Road Steady-State Cycles

The Non-Road Steady-State Cycles (NRSC) are a series of modal cycles that have been globally harmonized in all markets, including the US. The NRSC is used in the following regulations for off-road engine criteria pollutants, which are described in the corresponding sections of Chapter 3, Emissions Control Regulations:

- US Tier 4 final, Section 3.3.1.2
- EU Stage V, Section 3.3.3.2
- EU Stage IV, Section 3.3.3.2
- Japan Stage 4, Section 3.3.4.2

See Chapter 3, Emissions Control Regulations, for the specific applications and power ranges that require the NRSC. The main NRSC is the eight-mode cycle described in ISO 8178 C1 and in Appendix II to 40 CFR part 1039 [4-36], which is tabulated below.

TABLE 4.9 NRSC C1 test modes for variable-speed engines [4-36].

Mode number	1	2	3	4	5	6	7	8
Speed	100%				Intermediate			Idle
Torque	100%	75%	50%	10%	100%	75%	50%	0%
Weighting factor	0.15	0.15	0.15	0.10	0.10	0.10	0.10	0.15

FIGURE 4.36 NRSC C1 steady-state operating points and weights on normalized operating map [4-36].

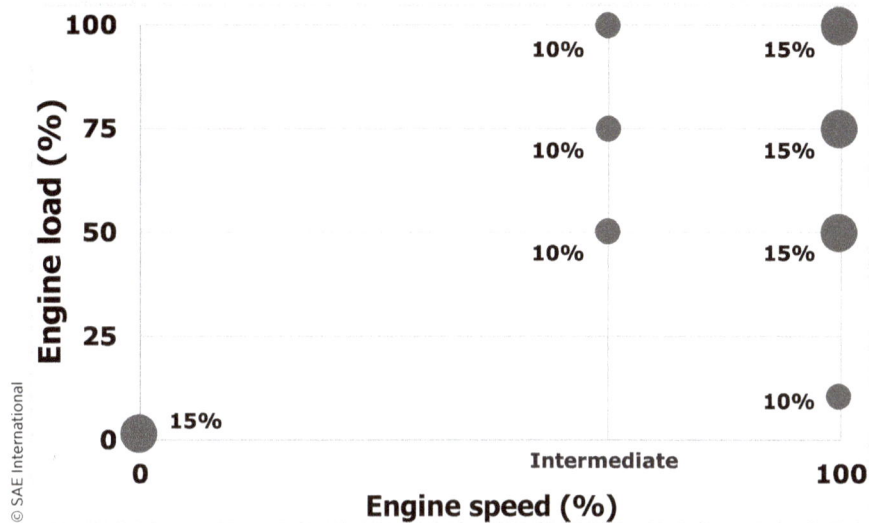

Variable-speed off-road engines covered by current rules should use the NRSC type C1 [4-36]. The test speeds and torques shown in Table 4.9 and plotted in Figure 4.36 are relative to the speed and torque at the rated power declared by the manufacturer. The intermediate speed is the speed of the declared peak torque, assuming this speed is between 60% and 75% of the rated speed. If the speed at peak torque is below this range, then the intermediate speed is 60% of rated speed; if the speed at peak torque is above this range, the intermediate speed is 75%. For example, with the Scania DC13 085A, the speed at peak torque is 1,200 rpm, which is 67% of the rated speed of 1,800 rpm. Therefore, the intermediate speed should be 1,200 rpm for this engine.

Constant-speed off-road engines covered by the latest off-road engine rules should use the NRSC type D2. The test speed and torques shown in Table 4.10 and plotted in Figure 4.37 are relative to the speed and torque at the rated power declared by the manufacturer.

TABLE 4.10 NRSC D2 test modes [4-36].

Mode number	1	2	3	4	5
Speed	100%	100%	100%	100%	100%
Torque	100%	75%	50%	25%	10%
Weighting factor	0.05	0.25	0.30	0.30	0.10

FIGURE 4.37 NRSC D2 steady-state operating points and weights on normalized operating map [4-36].

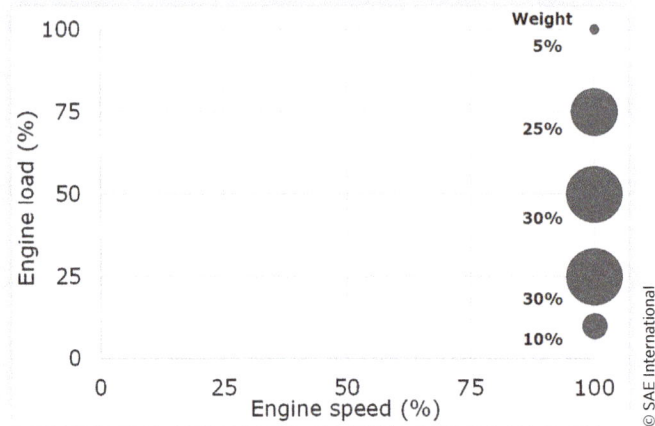

4.2.9 European Transient Cycle

The ETC is a transient test cycle used to certify HDE against the EU's Euro V regulatory standard and earlier that was defined in 1999 [4-30]. The test cycle lasts 1,800 s (30 min) and is divided as follows into three main parts, each of which lasts 600 s (10 min):

1. City driving

2. Rural driving

3. Motorway driving

The normalized engine speed and torque set points as a function of time are shown in **Figure 4.38**, and the transition from city to rural to motorway driving can be seen in the decreased amplitude of the transient events in speed, then load, over time. In addition, the average normalized torque in the city driving segment is about 46%, in the rural driving segment, about 62%, and in the motorway driving segment, about 29%. These averages are consistent with expectations, since the motorway driving segment represents a fairly steady cruise condition, whereas the rural driving segment has significant transients in both engine speed and load. Deceleration events during the test cycle include operation on the motoring curve of the engine, although in Figure 4.38, these motoring points have been rounded up to 0% normalized torque. During the cycle, 324 s are spent motoring the engine and another 168 s are spent at idle torque. Together, these represent 27.3% of the total test cycle.

The ETC engine-operating points are also shown together in **Figure 4.39**, with engine load plotted versus engine speed. As with Figure 4.38, the operating points on the motoring curve have been rounded up to 0% torque. Note that most of the operation on this cycle occurs between 40% and 60% of the speed range between idle and the rated power speed.

The Volvo D13 operating map has also been used to demonstrate the ETC operating points for a more concrete example. The blue curve in **Figure 4.40** is the torque curve for the engine, which is also shown in Figure 4.15. The operating points cluster between 1,000 and 1,300 rpm, which is a typical preferred speed range for this type of engine. The motoring curve shown is an approximation used to illustrate the drive cycle.

FIGURE 4.38 European Transient Cycle traces showing (a) normalized engine speed and (b) normalized engine torque with respect to time [4-30].

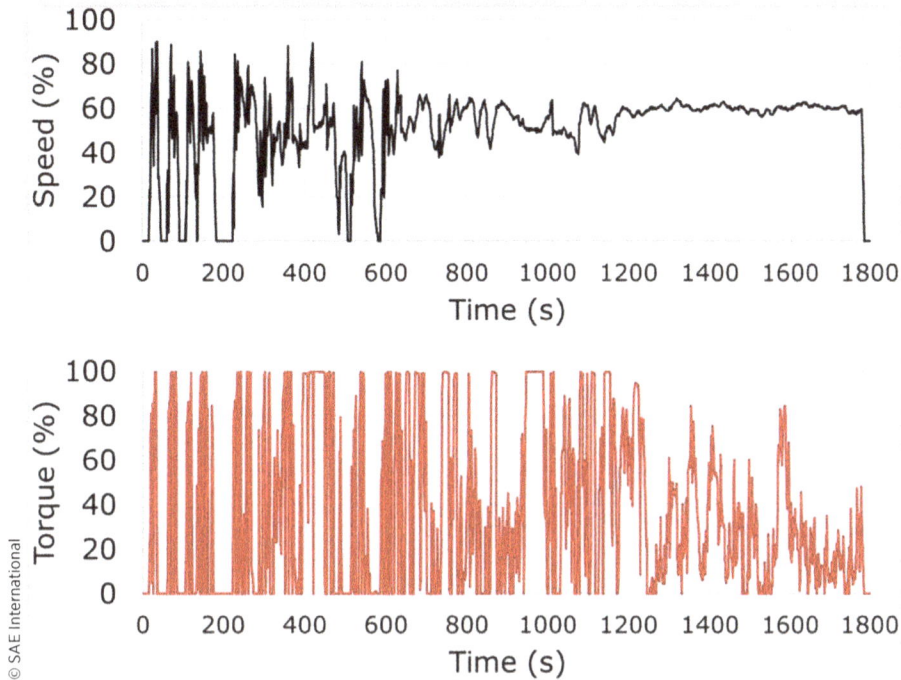

FIGURE 4.39 European Transient Cycle operating points and weights on normalized operating map [4-30].

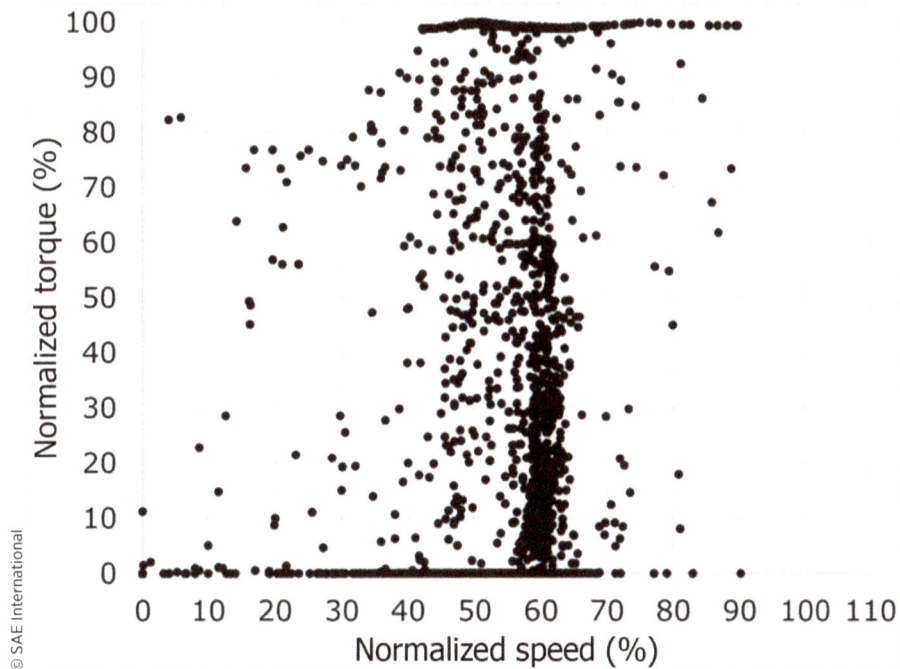

© SAE International

FIGURE 4.40 Example of ETC test points on Volvo D13 operating map.

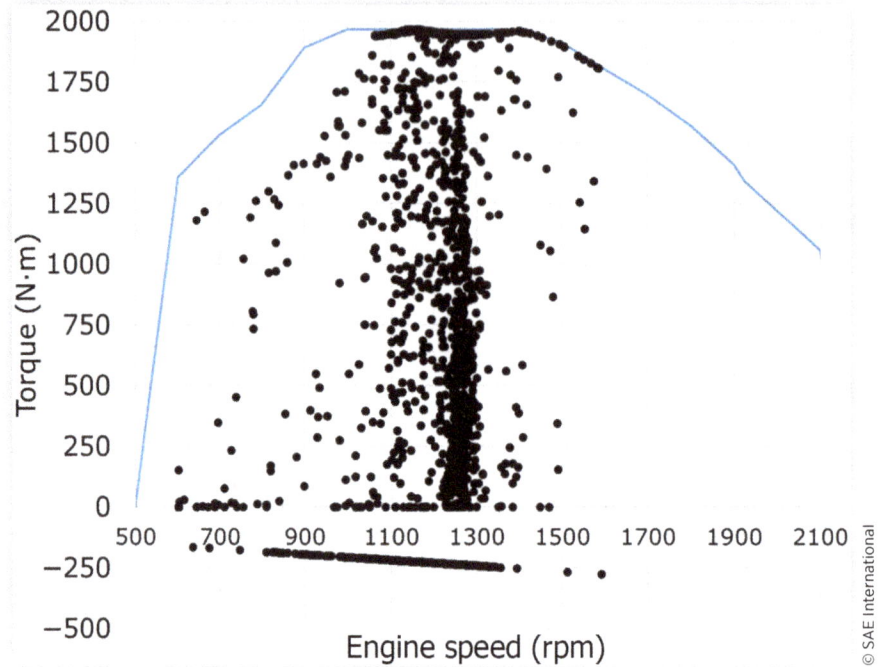

4.2.10 California Low-Load Cycle

CARB has been developing an ultra-low NOx standard for HDDE starting in MY2024 [4-37]. As part of the development work for this standard, CARB has found that there are significant NOx emissions at idle and at engine speeds below the current NTE threshold described in Section 4.2.6. Therefore, in early 2020 CARB plans to introduce a low load cycle (LLC) to evaluate NOx emissions. As of early 2019 [4-37], there were two leading candidates for the LLC, the high-level characteristics of which are shown in Table 4.11. Both LLC candidates incorporate the following three types of operating conditions to assess how well the engine and EAS maintain emissions control performance:

1. Transition from high load to low load

2. Sustained low load operation

3. Transition from low load to high load

In September 2019, CARB indicated that LLC candidate 7 would be the final cycle [4-38].

TABLE 4.11 CARB low load cycle candidate cycle characteristics [4-37].

Candidate	Duration (min)	Avg. speed (%)	Avg. torque (%)
7	90	27.0	10.6
8	81	16.2	3.5

4.3 HDV Test Cycles

To support the recent trend in regulating both criteria pollutant and GHG emissions from HDVs, new test cycles have been developed to support those regulations. The standards themselves are described in Chapter 3, Emissions Control Regulations. In general, the regulatory agencies involved have avoided requiring chassis dynamometer testing of HDVs, given the limited number of suitable test facilities and the lower precision of the test results. There are some exceptions, in that LHDVs are evaluated on the chassis dynamometer test cycles used for LDVs, and in the US, fully electric HDVs with batteries alone or fuel cells need to be tested on a chassis dynamometer.

For most HDVs, though, the vehicle compliance is evaluated by testing the powertrain components, either separately or in combination, in a test cell and then using those data in a simulation-based compliance tool. When tested separately, the engine is evaluated for its fuel consumption, and the transmission and driveline are evaluated for their efficiencies, but the complete powertrain can be evaluated for its transient fuel consumption over the test cycle. The EU is also implementing a HD version of its RDE standards, although to date it only has a reporting requirement, not a regulatory standard for manufacturers to work toward.

4.3.1 US Phase 1 GHG Standards

In developing the Phase 1 GHG standards for HDVs [4-39], EPA and NHTSA recognized the challenges of chassis dynamometer testing of HDVs, given the limited number of suitable test facilities and the low reproducibility of the testing. For the Phase 1 GHG standards, EPA and NHTSA provided engine maps, a fixed set of transmissions, and driveline efficiencies. Manufacturers were left to provide vehicle parameters for the GHG Emission Model (GEM), a software-based certification tool for HDVs. GEM is the official certification tool for HDVs under both phases of the GHG standards, although each phase has its own version of GEM. The HDV test cycles used in each phase are incorporated into the respective versions of GEM.

Class 2b and 3 LHDVs are certified using the LDV chassis dynamometer tests described in Section 4.1, although the fuel consumption and GHG emissions results are weighted by cargo mass. Nevertheless, the agencies allow a manufacturer to do component testing and have GEM evaluate the full vehicle results; therefore, heavier HDVs generally use powertrain component data on GHG emissions and fuel consumption, vehicle data, and GEM.

GEM for Phase 1 includes assumptions for the driveline and transmission, and so manufacturers only need to test the engine separately to generate a fuel consumption and GHG emissions value for certification. When testing the engine, the vehicle speed requirement is translated through the driveline to the torque required at the engine flywheel to create the appropriate load trace. For Phase 1 the agencies only require full HDV chassis dynamometer testing for hybridized HDVs.

The Phase 1 GHG standard was published as a final rule in September 2011 by EPA and NHTSA [4-39] and applies to HDVs from MY2014 through MY2020. Therefore, this standard will be at the end of its applicability period as of the publication of this book. The version of GEM for the Phase 1 GHG standard uses the following three drive cycles:

1. California heavy-duty transient truck cycle (HHDTTC).
2. 55 mph (89 km/h) constant speed on 0% grade.
3. 65 mph (105 km/h) constant speed on 0% grade.

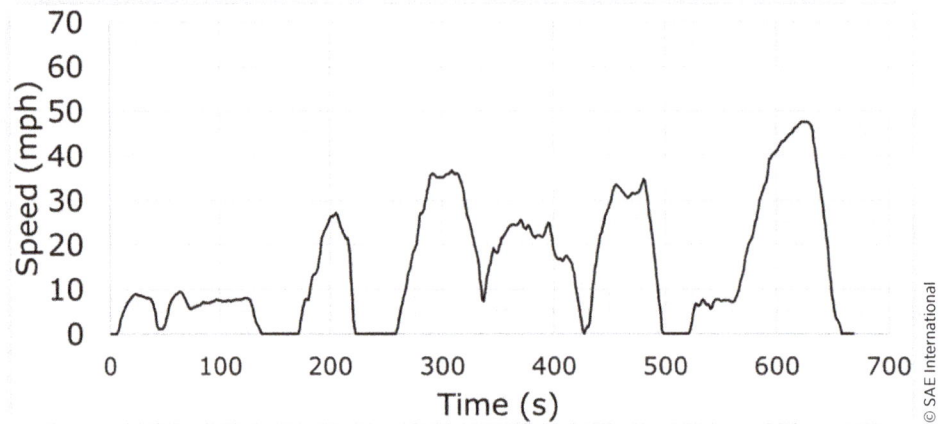

FIGURE 4.41 California HD transient truck drive cycle [4-39].

© SAE International

The speed trace for the HHDTTC is shown in **Figure 4.41**, where the data are available from Appendix I of 40 CFR part 1037 [4-40].

In addition to the testing and simulation results for the powertrain, different types of HDVs qualify for technology credits that reward expected on-cycle or off-cycle benefits. In the Phase 1 standard, for example, implementing a hybrid powertrain garnered credits for GHG emissions compliance.

4.3.2 US Phase 2 GHG Standards

Based on EPA's and NHTSA's learnings from the Phase 1 GHG standard, the two agencies developed an updated standard, the Phase 2 GHG standard, which was published into the CFR in October 2016 [4-31]. The Phase 2 GHG standards extend the HDV standards out to MY2027, with intermediate steps in MY2021 and MY2024, and are expected to apply through MY2030. As the full Phase 2 GHG standard and its implications for HDV design and testing are complicated, only an overview is presented in this book.

Starting with the Phase 2 GHG standards, a manufacturer may choose to evaluate the engine alone, the power pack (engine and transmission together), or the full power-train (engine, transmission, and axles together) separately from a vehicle. These test results are used with other vehicle characteristics in GEM. A manufacturer may choose to test components in combination if they expect a benefit from the component integration. For these tests, GEM is used to create the appropriate speed and load traces. The vehicle speed requirement is translated through the driveline to the speed and torque required at the engine flywheel, transmission output, or wheel hub, as appropriate. In addition to these drive cycles, GEM also evaluates fuel consumption at idle under two conditions: neutral idle and parked idle. Neutral idle has the transmission in gear and the torque converter active, whereas parked idle has the transmission in "park" with the torque converter disengaged.

GEM was updated to reflect the changes in the standard for Phase 2, and that included changes in the drive cycles used for certification. These changes were informed by the

FIGURE 4.42 Grade profile for 55 and 65 mph constant-speed drive cycles in the US Phase 2 GHG standard [4-41].

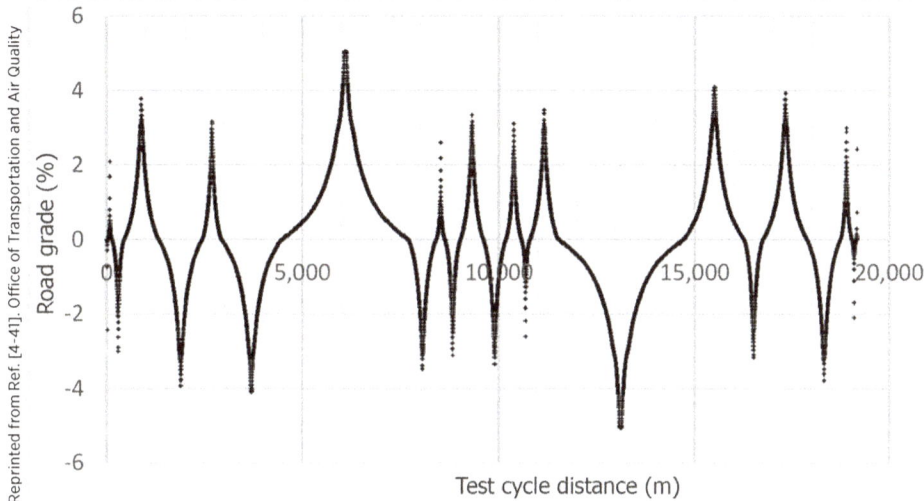

experience gained by the agencies and the regulated community, and so the version of GEM for the Phase 2 GHG standard uses the following three drive cycles [4-31]:

1. California heavy-duty transient truck cycle (HHDTTC).
2. 55 mph (89 km/h) constant speed on ≈22 km course with grade profile.
3. 65 mph (105 km/h) constant speed on ≈22 km course with grade profile.

The HHDTTC is unchanged from the Phase 1 standard and uses the vehicle speed schedule shown in Figure 4.41 [4-39, 4-40]. The grade profile used for the steady-speed cycles is new with the Phase 2 GHG standards, and it is shown in **Figure 4.42**, which is taken from p.3-63 of the Regulatory Impact Analysis [4-41].* In addition to these drive cycles, GEM also evaluates fuel consumption at idle under two conditions—neutral idle and parked idle for vocational vehicles only. To test the powertrain or its components, GEM is used to generate speed and torque traces appropriate for the type of HDV being evaluated for each of the HHDTTC and the steady-speed cycles. Depending on what level of testing is required or desired, these traces are calculated for the engine flywheel (engine output), for the transmission output shaft (power pack output), or for the wheel hubs (powertrain output). A manufacturer may submit the steady-state fueling map for the engine or provide cycle-averaged results for the highway cycles only. The HHDDTC requires the cycle average map. The generation of the cycle-averaged fuel consumption results is beyond the scope of this text.

---

* EPA acknowledges that there is a discrepancy between the grade profile shown in the RIA [4-41] and the data published in Appendix IV to 40 CFR part 1037 [4-42] and plans to update the published data in the CFR to match the curve from the RIA. The main difference is that the data do not include the initial and final grades shown in Figure 4.42.

4.3.3 EU HDV Cycles

As discussed in Section 3.4.3, the EU uses a simulation tool to evaluate HDV fuel consumption and GHG emissions for certification [4-43]. The Vehicle Energy Consumption Calculation Tool includes five HDV drive cycles, but these cycles are not published independently. For a given HDV, the drive cycle used, such as the Long Haul 2015, Regional Delivery 2016/2012, and Urban Delivery 2012 [4-43], depends on the type of HDV being evaluated and its expected duty cycle. Currently, HDV manufacturers only need to report their vehicles' fuel consumption and GHG emissions [4-44], though. European HDVs are also subject to in-use testing in a testing scheme similar to the RDE testing protocols for LDVs.

4.3.4 Japan JE05 and Intercity Cycles

Japan used the JE05 HDV drive cycle on vehicles of at least 3,500 kg GVW during the span covering MY2005 through MY2015 [4-45]. From MY2016, Japan has used a combination of the WHTC and WHSC engine tests to assess HDVs for certification (see Sections 4.2.2 and 4.2.3, respectively). Thus, the JE05 and related intercity drive cycles are now obsolete.

The JE05 cycle applied to HDVs with either diesel or gasoline engines. The cycle is a transient, full vehicle drive cycle that represents urban driving. The overall duration of the drive cycle is 1,830 s (30.5 min), and of that, the vehicle is at a stop for 461 s (25.6%). As shown in **Figure 4.43**, the drive cycle represents typical urban driving, including multiple starts and stops and operation below 60 km/h (37 mph).

The JE05 is combined with an intercity freeway driving cycle, where the vehicle is supposed to maintain a constant 80 km/h (50 mph) speed over the grade profile shown in **Figure 4.44**. The intercity cycle duration is 3,120 s (52 min), making it one of the longer regulatory drive cycles in use.

The overall fuel consumption rate for an HDV is calculated as a weighted average of the results from the two tests, as follows:

$$E_J = \left[\frac{1 - \alpha/100}{E_{JE05}} + \frac{\alpha/100}{E_{IC}} \right]^{-1}, \tag{4.7}$$

FIGURE 4.43   JE05 urban drive cycle on 0% grade [4-45].

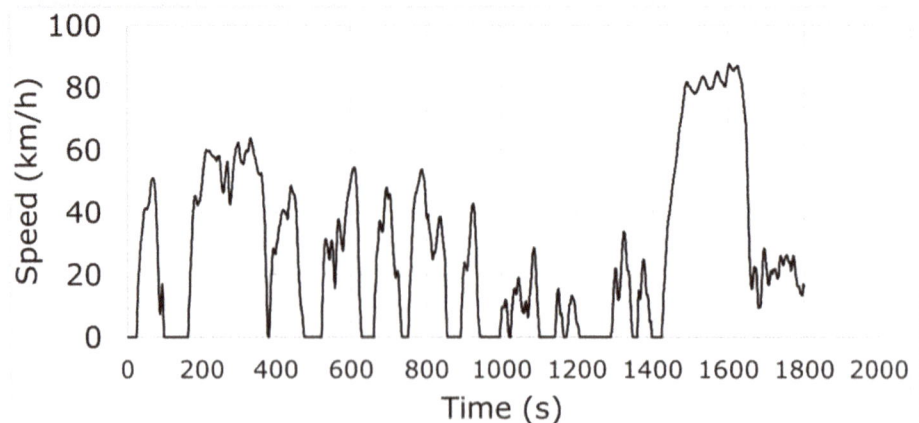

© SAE International

© SAE International

FIGURE 4.44 JE05 intercity drive cycle grade profile. The vehicle demand is a constant 80 km/h [4-45].

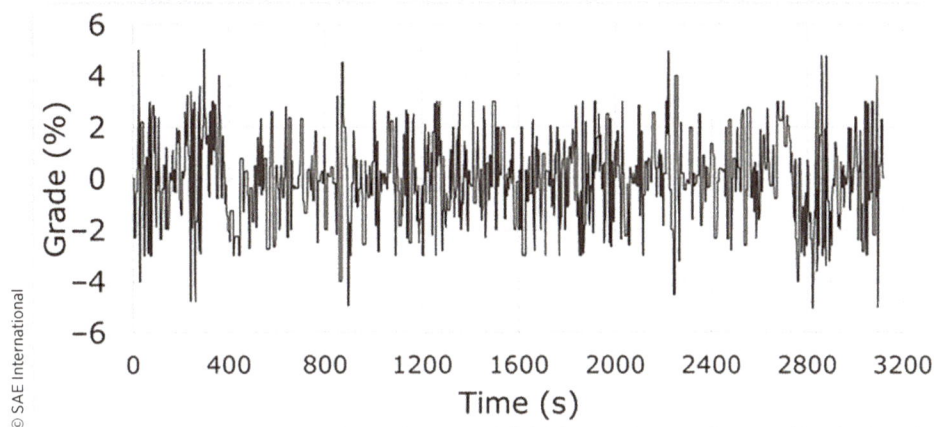

where E_J is the overall HDV fuel consumption rate (in km/m³), E_{JE05} is the JE05 fuel consumption rate (km/m³), E_{IC} is the intercity cycle fuel consumption rate (km/m³), and α is the intercity highway running ratio (%).

4.3.5 China Stage 3 Fuel Consumption

To support the Stage 3 rules for HDVs, the PRC is implementing the C-WTVC test procedures [4-46] to evaluate fuel consumption in HDVs. Although the main part of the procedure uses computer simulation, as described in Section 3.5.5, test data such as engine maps and other vehicle parameters are required to provide inputs to these simulations.

References

4-1.   Environmental Protection Agency, "Data on Cars Used for Testing Fuel Economy," https://www.epa.gov/compliance-and-fuel-economy-data/data-cars-used-testing-fuel-economy. Updated 14 Mar. 2019, Accessed 9 Jul. 2019.

4-2.   Environmental Protection Agency, "Fuel Economy Labeling of Motor Vehicles: Revisions to Improve Calculation of Fuel Economy Estimates; Final Rule," *Federal Register* **71**, no. 248 (Dec. 27, 2006): 77872–77969, https://www.gpo.gov/fdsys/pkg/FR-2006-12-27/pdf/06-9749.pdf.

4-3.   U.S. Department of Energy and Environmental Protection Agency, "FuelEconomy.gov – the official U.S. Government Source for Fuel Economy Information," https://fueleconomy.gov/. Updated 25 Mar. 2019, Accessed 25 Mar, 2019.

4-4.   Environmental Protection Agency, "Dynamometer Drive Schedules," https://www.epa.gov/vehicle-and-fuel-emissions-testing/dynamometer-drive-schedules. Updated 31 Jan. 2017, Accessed 14 Mar. 2019.

4-5.   Environmental Protection Agency, "Part 86—Control of Emissions from New and In-Use Highway Vehicles and Engines," *Code of Federal Regulations, Title 40*, Vol. **21** (2018): 429–1202, https://www.govinfo.gov/app/collection/cfr/2018/title40/chapterI/subchapterC/part86.

4-6.   Environmental Protection Agency, "§ 86.115-78 EPA Dynamometer Driving Schedule," *Code of Federal Regulations, Title 40*, Vol. **21** (2012): 514-515, https://www.govinfo.gov/app/details/CFR-2012-title40-vol19/CFR-2012-title40-vol19-sec86-115-00.

4-7.   Environmental Protection Agency, "Part 86—Control of Emissions from New and In-Use Highway Vehicles and Engines. Appendix I—Dynamometer Schedules," *Code of Federal Regulations, Title 40*, Vol. **21** (2018): 1139–1185, https://www.govinfo.gov/app/collection/cfr/2018/title40/chapterI/subchapterC/part86.

4-8.   Environmental Protection Agency, "§ 1066.831 Exhaust Emission Test Procedures for Aggressive Driving," *Code of Federal Regulations, Title 40*, Vol. **37** (2018): 368–369, https://www.govinfo.gov/app/collection/cfr/2018/title40/chapterI/subchapterU/part1066/subpartI.

4-9.   Environmental Protection Agency, "§ 1066.835 Exhaust Emission Test Procedures for SC03 Emissions," *Code of Federal Regulations, Title 40*, Vol. **37** (2018): 369–371, https://www.govinfo.gov/app/collection/cfr/2018/title40/chapterI/subchapterU/part1066/subpartI.

4-10.   Environmental Protection Agency, "§ 1066.710 Cold Temperature Testing Procedures for Measuring CO and NMHC Emissions and Determining Fuel Economy," *Code of Federal Regulations, Title 40*, Vol. **37** (2018): 357–360, https://www.govinfo.gov/app/collection/cfr/2018/title40/chapterI/subchapterU/part1066/subpartH.

4-11.   Holmes, J.R., "Research Note 96-11: Driving Patterns and Emissions: A New Testing Cycle," https://ww3.arb.ca.gov/research/resnotes/notes/96-11.htm. Accessed 18 Apr. 2019.

4-12.   United Nations, "Uniform Provisions Concerning the Approval of Vehicles with Regard to the Emission of Pollutants according to Engine Fuel Requirements," *Addendum 82: Regulation No. 83.07* (Jan. 22, 2015): 1–267.

4-13.   Ministry of Land, Infrastructure, Transport, and Tourism (MLIT), "Announcement that Prescribes Details of Safety Regulations for Road Vehicles, Announcement No. 619," *TRIAS 99-006-01*, 2002.

4-14.   United Nations, "Global Technical Regulation on Worldwide Harmonized Light Vehicles Test Procedures (WLTP)," *ECE/TRANS/180/Addendum 15/Amendment 3* (Feb. 1, 2018): 1–355.

4-15.   European Commission, "Regulation (EU) 2017/1151 of 1 June 2017 supplementing Regulation (EC) No 715/2007 of the European Parliament and of the Council on type-approval of motor vehicles with respect to emissions from light passenger and commercial vehicles (Euro 5 and Euro 6) and on access to vehicle repair and maintenance information, amending Directive 2007/46/EC of the European Parliament and of the Council, Commission Regulation (EC) No 692/2008 and Commission Regulation (EU) No 1230/2012 and repealing Commission Regulation (EC) No 692/2008," *Official Journal of the European Union*, Vol. **175** (2017): 1–643.

4-16.   Environmental Protection Agency, "§ 86.1803-01 Definitions," *Code of Federal Regulations, Title 40*, Vol. **21** (2018): 888-901, https://www.govinfo.gov/app/collection/cfr/2018/title40/chapterI/subchapterC/part86/subpartS.

4-17.   European Commission, "Regulation (EU) 2018/1832 of 5 November 2018 amending Directive 2007/46/EC of the European Parliament and of the Council, Commission Regulation (EC) No 692/2008 and Commission Regulation (EU) 2017/1151 for the purpose of improving the emission type approval tests and procedures for light passenger and commercial vehicles, including those for in-service conformity and real-driving emissions and introducing devices for monitoring the consumption of fuel and electric energy," *Official Journal of the European Union*, Vol. **301** (2018): 1–314.

4-18.   Ministry of Land, Infrastructure, Transport, and Tourism (MLIT), "Development of the Japan's RDE (Real Driving Emission) procedure," 76th GRPE GRPE-76-18e, 2018.

4-19.   Environmental Protection Agency, "Part 1065-Engine-Testing Procedures," *Code of Federal Regulations, Title 40*, Vol. **37** (2018): 45-282, https://www.govinfo.gov/app/collection/cfr/2018/title40/chapterI/subchapterU/part1065.

4-20.   United Nations, "Proposal for a draft global technical regulation (GTR): Test procedure for compression-ignition (C.I.) engines and positive-ignition (P.I.) engines fuelled with natural gas (NG) or liquefied petroleum gas (LPG) with regard to the emission of pollutants (World-wide harmonized heavy-duty certification (WHDC) procedure)," *ECE/TRANS/WP.29/2006/128* (Jul. 18, 2006): 1–124.

4-21.   United Nations, "Amendments to the proposal for a draft global technical regulation (GTR): Test procedure for compression-ignition (C.I.) engines and positive-ignition (P.I.) engines fuelled with natural gas (NG) or liquefied petroleum gas (LPG) with regard to the emission of pollutants (World-wide harmonized heavy-duty certification (WHDC) procedure)," *ECE/TRANS/WP.29/2006/128/Amendment 1* (Sep. 1, 2006): 1–12.

4-22.   International Organization for Standardization, "ISO 8178-1:2017. Reciprocating internal combustion engines – Exhaust emission measurement – Part 1: Test-bed measurement systems of gaseous and particulate emissions," International Organization for Standardization, 2017, 1–150.

4-23.   International Organization for Standardization, "ISO 8178-3:2019. Reciprocating internal combustion engines – Exhaust emission measurement – Part 3: Test procedures for measurement of exhaust gas smoke emissions from compression ignition engines using a filter type smoke meter," International Organization for Standardization, 2019, 1–11.

4-24.   Volvo Trucks, "Volvo D13 Engine Family," Volvo Trucks North America, 2016.

4-25.   Scania CV AB, "DC13 085A. 405 kW (550 hp)," Scania CV AB., 2018

4-26.   AVL, "AVL Emission Reports (Emrep)," https://www.avl.com/emission-reports. Accessed 15 Mar. 2019.

4-27.   Environmental Protection Agency, "§ 86.1362 Steady-State Testing with a Ramped-Modal Cycle," *Code of Federal Regulations, Title 40*, Vol. **21** (2018): 866–867, https://www.govinfo.gov/app/collection/cfr/2018/title40/chapterI/subchapterC/part86.

4-28.   Environmental Protection Agency, "§ 1036.505 Ramped-Modal Testing Procedures," *Code of Federal Regulations, Title 40*, Vol. **36** (2018): 113, https://www.govinfo.gov/app/collection/cfr/2018/title40/chapterI/subchapterU/part1036.

4-29.   Environmental Protection Agency, "§ 1065.610 Duty Cycle Generation," *Code of Federal Regulations, Title 40*, Vol. **37** (2018): 175–182, https://www.govinfo.gov/content/pkg/CFR-2018-title40-vol37/pdf/CFR-2018-title40-vol37-sec1065-610.pdf.

4-30.   European Commission, "Directive 1999/96/EC of the European Parliament and of the Council of 13 December 1999 on the approximation of the laws of the Member States relating to measures to be taken against the emission of gaseous and particulate pollutants from compression ignition engines for use in vehicles, and the emission of gaseous pollutants from positive ignition engines fuelled with natural gas or liquefied petroleum gas for use in vehicles and amending Council Directive 88/77/EEC," *Official Journal of the European Union*, Vol. **44** (2000): 1–155, https://eur-lex.europa.eu/legal-content/EN/TXT/?uri=CELEX%3A31999L0096.

4-31.   Environmental Protection Agency and Department of Transportation, "Greenhouse Gas Emissions and Fuel Efficiency Standards for Medium- and Heavy-Duty Engines and Vehicles—Phase 2," *Federal Register*, Vol. **81**, No. 206 (Oct. 25, 2016): 73478–74274, https://www.gpo.gov/fdsys/pkg/FR-2016-10-25/pdf/2016-21203.pdf.

4-32.   Environmental Protection Agency, "§ 86.007–11 Emission Standards and Supplemental Requirements for 2007 and later Model Year Diesel Heavy-Duty Engines and Vehicles," *Code of Federal Regulations, Title 40*, Vol. **81** (2018): 467–471, https://www.govinfo.gov/app/collection/cfr/2018/title40/chapterI/subchapterC/part86.

4-33.   Environmental Protection Agency, "§ 86.1370 Not-To-Exceed Test Procedures," *Code of Federal Regulations, Title 40*, Vol. **21** (2018): 867–873, https://www.govinfo.gov/app/collection/cfr/2018/title40/chapterI/subchapterC/part86.

4-34.   Environmental Protection Agency, "Part 1039—Control of Emissions from New and In-Use Nonroad Compression-Ignition Engines. Appendix VI—Nonroad Compression-Ignition Composite Transient Cycle," *Code of Federal Regulations, Title 40*, Vol. **36** (2018): 367–376, https://www.govinfo.gov/app/collection/cfr/2018/title40/chapterI/subchapterU/part1039.

4-35.   Environmental Protection Agency, "§ 1039.510 Which Duty Cycles do I Use for Transient Testing?," *Code of Federal Regulations, Title 40*, Vol. **36** (2018): 330–331, https://www.govinfo.gov/content/pkg/CFR-2018-title40-vol36/pdf/CFR-2018-title40-vol36-sec1039-510.pdf.

4-36.   Environmental Protection Agency, "Part 1039—Control of Emissions from New and In-Use Nonroad Compression-Ignition Engines. Appendix II—Steady-State Duty Cycles," *Code of Federal Regulations, Title 40*, Vol. **36** (2018): 365–367, https://www.govinfo.gov/app/collection/cfr/2018/title40/chapterI/subchapterU/part1039.

4-37.   California Air Resources Board, "California Air Resources Board Staff Current Assessment of the Technical Feasibility of Lower NOx Standards and Associated Test Procedures for 2022 and Subsequent Model Year Medium-Duty and Heavy-Duty Diesel Engines," California Air Resources Board, 2019.

4-38.   California Air Resources Board, "Heavy-Duty Low NOx Program: Low Load Cycle," Public Workshop, Diamond Bar, CA, Sep. 26, 2019.

4-39.   Environmental Protection Agency and Department of Transportation, "Greenhouse Gas Emissions Standards and Fuel Efficiency Standards for Medium- and Heavy-Duty Engines and Vehicles," *Federal Register*, Vol. **76**, No. 179 (15 Sep. 2011): 57106-57513, https://www.govinfo.gov/content/pkg/FR-2011-09-15/pdf/2011-20740.pdf.

4-40.   Environmental Protection Agency, "Appendix I to Part 1037—Heavy-Duty Transient Test Cycle," *Code of Federal Regulations, Title 40*, Vol. **36** (2018): 283-288, https://www.govinfo.gov/content/pkg/CFR-2018-title40-vol36/pdf/CFR-2018-title40-vol36-part1037-appI.pdf.

4-41.   Office of Transportation and Air Quality, "Greenhouse gas emissions and fuel efficiency standards for medium- and heavy-duty engines and vehicles – Phase 2. Regulatory Impact Analysis," U.S. Environmental Protection Agency Report No. EPA-420-R-16-900, 2016.

4-42.   Environmental Protection Agency, "Appendix IV to Part 1037—Heavy-Duty Grade Profile for Phase 2 Steady-State Test Cycles," *Code of Federal Regulations, Title 40*, Vol. **36** (2018): 289–290, https://www.govinfo.gov/content/pkg/CFR-2018-title40-vol36/pdf/CFR-2018-title40-vol36-part1037-appIV.pdf.

4-43.   Rexeis, M., Quaritsch, M., Hausberger, S., Silberholz, G., Kies, A., Steven, H., Goschütz, M., and Vermeulen, R., "VECTO Tool Development: Completion of Mthodology to Simulate Heavy Duty Vehicles' Fuel Consumption and CO_2 Emissions," European Commission Report No. I 15/17/Rex EM-I 2013/08 1670, 2017.

4-44.   European Commission, "Commission Regulation (EU) 2017/2400 implementing
        Regulation (EC) No 595/2009 of the European Parliament and of the Council as regards
        the determination of the CO_2 emissions and fuel consumption of heavy-duty vehicles
        and amending Directive 2007/46/EC of the European Parliament and of the Council and
        Commission Regulation (EU) No 582/2011," *Official Journal of the European Union*,
        Vol. **349** (Dec. 12, 2017): 1–247.

4-45.   Ministry of Land, Infrastructure, Transport, and Tourism (MLIT), "Test for Fuel
        Consumption Rate (Heavy Duty Motor Vehicles)," *TRIAS 99-007-01*, 2002.

4-46.   Ministry of Ecology and Environment of PRC, "Fuel Consumption Test Methods for
        Heavy-Duty Commercial Vehicles (C-WTVC)," *GB/T 27840-2011*, 2011.

Emissions Measurements

One accurate measurement is worth a thousand expert opinions.

—Unknown

In order for emissions to be regulated, controlled, or characterized they need to be accurately measured. Increasing environmental and health concerns have led to the development of a wide range of measurement techniques and equipment at various levels of sophistication, cost, and accuracy to be used for a variety of applications. Many of these techniques and instruments, especially those used for regulatory purposes in the test cycles described in Chapter 4, Emissions Testing Protocols, are highly standardized. This standardization helps produce reproducible emissions measurement results from the engine, powertrain, or vehicle of interest that can be compared across laboratories. This chapter discusses the function and capabilities of both stationary and portable emissions measurement systems (PEMS) that can be used to collect data on emissions species.

The two main types of emissions measurements can be classified as *laboratory testing*, which includes regulatory style testing, emissions research, combustion research, and development for engines, catalytic converters, or complete exhaust aftertreatment systems (EAS), and *field testing*, such as mobile emissions labs, on-vehicle measurements, vehicle inspections, vehicle maintenance, remote emissions testing [5-1], and occupational health measurements. The first method involves testing the engine, powertrain, or vehicle in a fixed test cell using a dynamometer and stationary emissions equipment, often referred to as "emissions benches." The second method is most often used to evaluate full vehicles at a field site [5-2], on the road or on a closed track using an onboard PEMS [5-3], where the PEMS measures and records selected key tailpipe exhaust components and other engine and vehicle data.

Laboratory emissions testing methods generally use highly sophisticated equipment for the best possible accuracy and repeatability. The test methods, especially those used for regulatory purposes such as engine or vehicle emissions certification or compliance, are precisely prescribed and highly standardized. They must include detailed descriptions of the measurement setup, type of equipment, and test procedures to be used. These standard methods include the measurement of regulated pollutants, including carbon monoxide (CO), hydrocarbons (HCs), nitrogen oxides (NOx), and particulate matter (PM), over the prescribed test procedures. Detailed descriptions of the test procedures can be found in Chapter 4, Emissions Testing Protocols. In addition to the description of the stationary equipment used in laboratory testing, the differences between a development test cell and a certification test cell are also described in this chapter.

Laboratory tests generally make use of a stationary test cell with either a fixed dynamometer or a chassis dynamometer. For fixed dynamometer testing, the test article may include the engine alone; the "power pack," comprising the engine and transmission together; or the full powertrain, comprising the engine, transmission, and axles. The chassis dynamometer allows one to test a complete vehicle. Testing with a fixed dynamometer in a test cell environment allows for the careful control of the test environment and engine inputs, allowing for better measurement precision. The test cells may measure raw emissions during development work, but for certification testing, the regulations specify the measurement of the tailpipe-style emissions characteristics, which requires a dilution scheme. For example, in the United States, Title 40, Part 1065 of the *Code of Federal Regulations* (40 CFR part 1065) [5-4] sets forth the requirements for measurements, which include the use of a diluted exhaust constant volume sampling (CVS) system.

Full-vehicle testing on a chassis dynamometer is also common, especially for light-duty vehicles (LDVs). Regulatory test cycles for LDVs are described in Section 4.1 for criteria pollutants and Section 4.3 for fuel economy or greenhouse gas emissions. Chassis dynamometer testing is more expensive and the vehicle test state is less reproducible than for engine or powertrain testing. Compared to the capacity available for LDVs, chassis dynamometers for heavy-duty vehicles (HDVs) are less common. Also, full-vehicle tests of HDVs are more expensive and less reproducible than those for LDVs, which is why there are few regulatory HDV chassis dynamometer tests. In either case, most full vehicle tests on a chassis dynamometer use stationary detectors. One exception is the FTP-75 cycle, where the exhaust is collected in a set of bags and then analyzed later as described in Section 4.1.1.

Field testing generally consists of on-road tests using PEMS. On-road testing is needed for the Real Driving Emissions (RDE) protocol for LDVs in the European Union (EU) that is described in Section 4.1.10. PEMS are also used to evaluate real-world driving emissions from in-use vehicles, both LD and HD. PEMS can also be a useful development tool since road tests can complement engine or vehicle calibration activities. In general, though, the recent industry trend is to perform more development testing work at the engine level and less at the vehicle level, since engine testing is considerably less expensive than vehicle testing is.

This chapter provides an overview of how both PEMS and stationary systems work by the type of emissions species that they analyze. Detailed descriptions of the test procedures can be found in Chapter 4, Emissions Testing Protocols. A description of the sensors used for on-board diagnostics is provided in Chapter 6, Sensors and On-Board Diagnostics.

5.1 Gas Analysis Methods and Techniques

Each regulatory body details the analytical methods to be used for measuring emissions for their standard. The most commonly used methods are nondispersive infrared (NDIR) analysis for CO and carbon dioxide (CO_2) flame ionization detectors (FID) for HCs, chemiluminescence detectors (CLD) for NOx, and paramagnetic analyzers for oxygen (O_2) [5-5, 5-6, 5-7, 5-8], though methods such as Fourier-transform infrared (FTIR) spectroscopy are also common. A comparison of the species analyzed, concentration ranges measured, and response times of these analyzers are shown in Table 5.1. There are many other methods commonly used for nonregulated emissions described in the literature.

All of the techniques listed in the table above rely on a particular property of the measured gas species. In some cases, the analyzer causes a chemical reaction, so that a property of a product species can be directly measured. In all cases, calibration must be done with a gas (or gases) of known concentration(s) for traceable and precise measurement of the target gas concentration in the exhaust. These techniques have been adapted or developed specifically for measuring internal combustion engine exhaust species concentrations with high accuracy and limited sensitivity to interference from other gas species occurring in the exhaust. In addition, a number of electrochemical cells and sensors, discussed in Chapter 6, Sensors and On-Board Diagnostics, have been developed, and many of these are used in PEMS in field testing. These instruments are generally less sensitive and, therefore, are not allowed to be used in many of the regulatory emissions testing procedures due to their tendencies toward cross-sensitivities and deterioration of the electrochemical cell [5-9]. For example, it has been shown that the cross-sensitivity for NO in a CO cell is as high as 30% in some cells. While the lifespan of a typical electrochemical cell can be as long as two to three years, they tend to experience a gradual decline in performance over that time.

In general, due to the increased scrutiny on regulated emissions species and the very low concentration targets defined by the regulatory bodies, it is very important that the analyzers used be accurate, sensitive, stable, and have a fast response time.

TABLE 5.1 Comparison of emissions analysis techniques (data from [5-8]).

Analysis technique	Gas species measured	Concentration range	Response time
Nondispersive infrared (NDIR)	CO	0–3,000 ppm	2 s
	CO_2	0%–20%	2 s
Chemiluminescence detection (CLD)	NOx	0%–1%	1–2 s
Flame ionization detection (FID)	THC	0%–1%	1 s
Fast-FID	THC	0%–1%	1 ms
Fourier-transform infrared (FTIR)	NOx, HC	Varies	5 s
Paramagnetic	O_2	0%–25%	1 s

5.1.1 Nondispersive Infrared (NDIR) Spectroscopy

NDIR analyzers are spectroscopic sensors that rely on the IR absorption property of a gas over a particular, narrow range of wavelengths to determine the concentration of the target molecules in the gas sample. NDIR analyzers are optically nondispersive because the IR energy can pass through the sampling chamber without deformation. The target molecules absorb IR radiation by converting the light energy into vibrational and rotational energy within the molecule, thereby changing its internal energy in a way that can be detected as temperature [5-10] and allowing for the measurement of the volumetric concentration of the gas in the sample. NDIR instruments use the total absorption over a specified range of wavelengths by employing an optical filter in front of the detector to eliminate all other wavelengths, rather than using a monochromatic prism to separate out a single wavelength to pass through the sample.

Briefly, the NDIR uses heated filaments as blackbody radiators to emit an IR light beam with a wide range of wavelengths through the gas-containing sample cell. A chopper wheel mounted in front of the detector spins to modulate the IR beam by continually correcting the offset and gain of the analyzer, allowing the single sample head to measure the concentration of different gases [5-10, 5-11, 5-12]. This also allows the detector to remain insensitive to the absorption of other chemical species as long as their absorption wavelengths do not overlap. Most instruments use two IR beams, where one passes through the sample cell and the other passes through a reference cell filled with an inert gas. An example of an NDIR detector is shown in **Figure 5.1**.

The NDIR is sensitive to pressure, which must remain constant during both calibration and gas sample analysis. Some of the materials commonly used in IR analysis,

FIGURE 5.1 NDIR schematic [5-10].

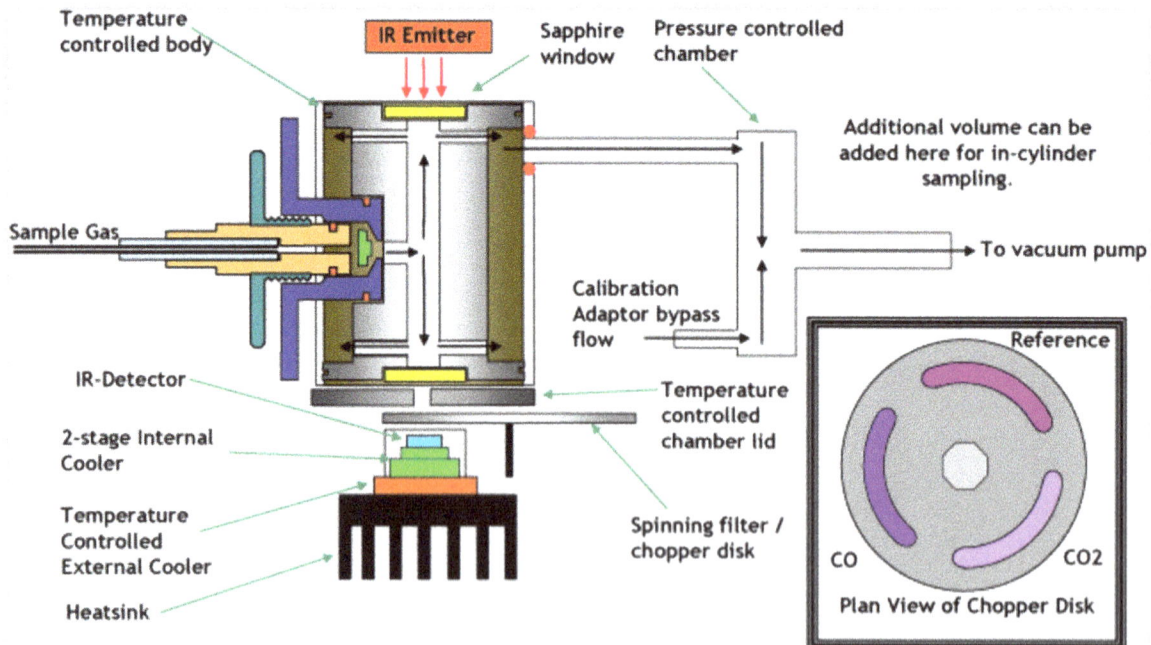

like the alkaline earth halogenide windows, are hydroscopic, thus, the measurement is sensitive to the presence of water. Water also absorbs IR with some interference with the bands for CO, carbon dioxide (CO_2), and NO. Therefore, gas samples for IR analysis should be dried. Heated NDIR (h-NDIR) analyzers that allow for measurement under wet gas conditions have been developed for on-board measurements [5-13]. NDIR analysis is most commonly used for CO and CO_2 and can also be used to measure methane (CH_4), NO, nitrous oxide (N_2O), sulfur dioxide (SO_2), hydrochloric acid (HCl), and some HC species.

5.1.2 Chemiluminescence Detector (CLD)

Chemiluminescence differs from phosphorescence or fluorescence because the excited state that is responsible for the emission of light is due to chemical reaction rather than the absorption of a photon [5-14]. CLDs are analyzers, such as the one shown in **Figure 5.2**, that rely on a chemical reaction that causes the emission of a light photon as an excited molecule decays to a lower energy state, as a result of the reaction. One photon of light is given off per molecule of reactant, allowing for the calculation of the species concentration in a known volume. The light-producing reaction between reactants A and B is very fast, quickly progressing through an excited transition state intermediate (I*), which decays into an electronic ground state and emits a photon, as shown in Eq. 5.1. The emitted light is filtered to eliminate interference from other gases and measured using a photomultiplier, where the signal is proportional to the gas species concentration in the sample.

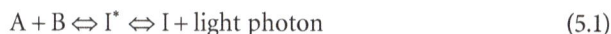

$$A + B \Leftrightarrow I^* \Leftrightarrow I + \text{light photon} \qquad (5.1)$$

CLDs are available in a wide range of measurement capabilities with fast response times. The primary limitation of using a CLD is that water vapor can cause interference, requiring either dry flow to the analyzer or additional, often expensive, components such as vacuum pumps and deionizers [5-15].

FIGURE 5.2 CLD diagram.

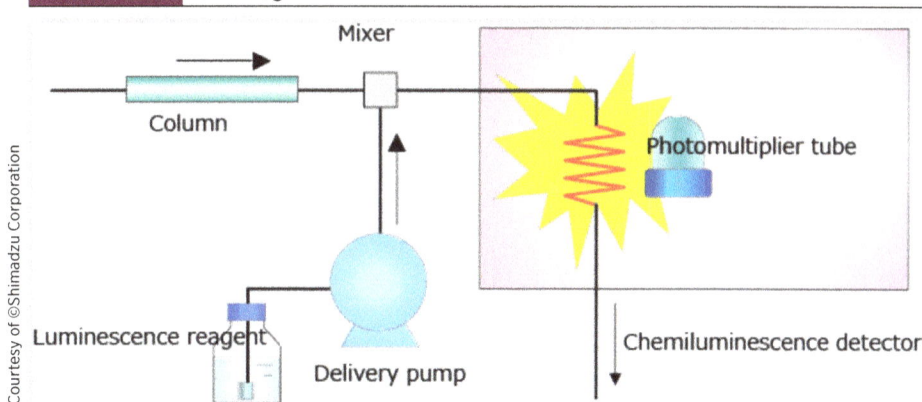

Courtesy of ©Shimadzu Corporation

5.1.3 Flame Ionization Detector (FID)

Flame ionization detectors, such as the one shown in **Figure 5.3**, relies on the free electrons and positive ions produced during combustion in a non-ionized hydrogen flame to determine the presence and quantity of HC molecules [5-16]. To detect these ions, two electrodes are used, in order to produce a potential difference, with the positive electrode also acting as the burner location for flame production. The second, negative electrode is located above the flame, and a current is developed between the positive electrode burner and the negative electrode and measured by a high impedance picoammeter [5-17]. The developed current is a function of the number of carbon atoms that pass through the flame [5-18], with the measurement given in ppm per volume of carbon. A paraffinic HC such as propane is used as a reference for the measurement. The FID has a weaker response to oxygenated HCs, such as alcohols, ketones, esters and aldehydes, and species such as CO and CO_2 are not detectable with the FID. One drawback of the FID is that the flame oxidizes all of the compounds that pass through it that can be oxidized. Therefore, it cannot be used as the first detector in sequential sample analysis.

The so-called fast FID is a version of the FID that has been miniaturized in order to reduce the response time to the order of milliseconds for use in transient testing.

5.1.4 Magnetic Methods

Paramagnetic detectors are used for gas species such as oxygen or NOx that have paramagnetic properties, that is, they respond to magnetic fields without retaining an inherent magnetism. These detectors work by exposing a gas sample to a heterogeneous

FIGURE 5.3   FID diagram.

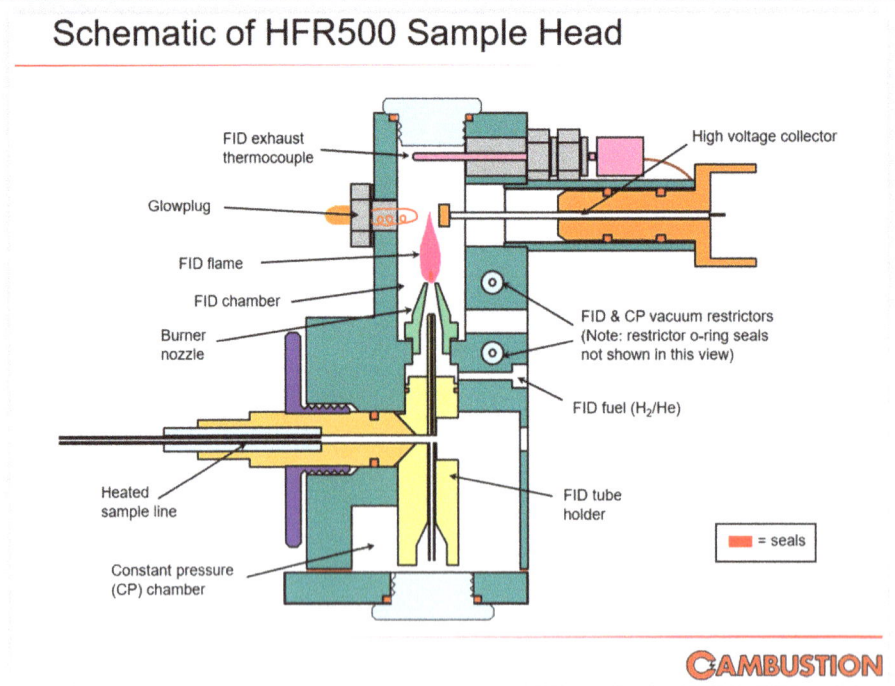

Schematic of HFR500 Sample Head

© Cambustion

magnetic field. Most gas species are diamagnetic and are repelled out of the field, whereas any paramagnetic gases present are attracted to the magnetic field. The resultant magnetic force can be measured in several ways, depending on the detector. Most commonly, it is measured by the current. Because paramagnetism is highly temperature-dependent, cold paramagnetic molecules are attracted by a strong magnetic force. When the molecules are heated, they leave the magnetic field and generate a measurable electrical current.

5.1.5 Fourier-Transform-IR (FTIR) Spectroscopy

FTIR is a technique used to measure the IR spectrum of absorption or emission from a gas. The FTIR spectrometer simultaneously generates high-spectral-resolution interferograms from a broadband light source with changing optical path length [5-19]. Rather than using a monochromatic light source, this broadband light source contains the full spectrum of wavelengths to be measured and shines all of the frequencies of light at the sample at once, measuring how much of the beam is absorbed by the sample with an interferometer, which comprises a particular configuration of mirrors—one fixed and one moved by a motor [5-20]. As the mirror moves, each of the wavelengths in the beam is temporarily blocked-transmitted-blocked-transmitted by wave interference. Different wavelengths have different modulation rates so that at each point of time, the beam has a different spectrum. This process is repeated many times, then the absorption at each wavelength can be calculated using the Fourier-transform algorithm, which gives the technique its name.

There are several advantages of an FTIR, as compared to a dispersive, or scanning, spectrometer. The first is that data from all wavelengths are collected simultaneously, which allows for a shorter scan time for a particular resolution. This also gives a higher signal-to-noise ratio for a given sampling time, though averaging over multiple scans increases it. A second advantage is in the interferometer throughput in an FTIR, which is determined by the size of the collimated IR beam, though an aperture is required to restrict its convergence. Another advantage is the wavelength accuracy, due to the wavelength scale/interferometer being calibrated by a laser beam of known wavelength, a practice that is more accurate and stable than the mechanical movement of diffraction gratings used for dispersive instruments. The final advantage is that the FTIR has less sensitivity to stray light from the radiation of one wavelength appearing at another in the spectrum, which can be a problem for dispersive instruments when there are imperfections in the diffraction gratings or unintended reflections.

FTIRs, such as the one shown in **Figure 5.4**, are useful for measuring exhaust gas species such as nitrogen monoxide (NO), nitrogen dioxide (NO_2), CO, CO_2, and small molecular alcohols, HCs, and aldehydes. It is not useful for the longer chain HCs, typical of fuels, and combustion products.

5.2 Exhaust Sampling Systems

In order to measure gas species concentrations and PM in engine exhaust, there are two necessary actions that must take place:

1. Sample from the exhaust
2. Analyze the sample

FIGURE 5.4 FTIR inner workings diagram.

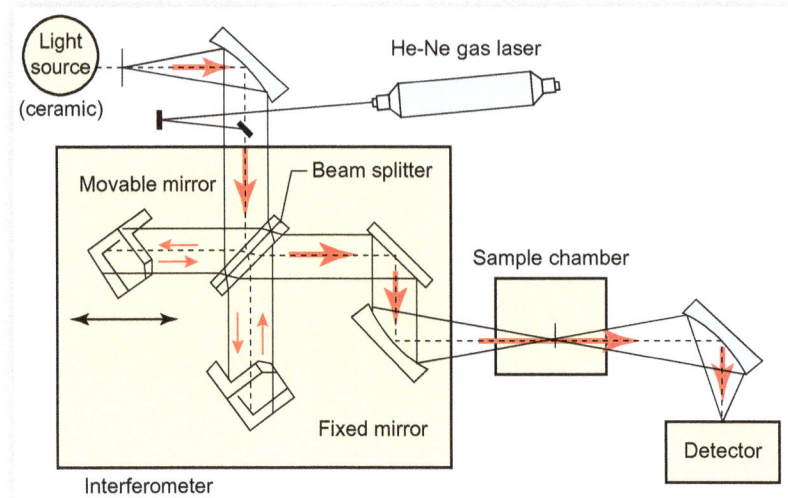

© SAE International

The purpose of the exhaust gas sampling system is to deliver a representative, uniform sample of gas from before, after, or within the EAS to the analyzers without changing the composition. The sample delivered to the analyzers must correspond to the actual emissions species and concentrations from the sampling site. While most sampling for gases or PM is now done in real-time with *in-situ* instruments, it was previously common for the gas samples to be collected using Teflon bags for later analysis. Collecting the sample in bags is still used for FTP-75, as described in Section 4.1.1. For real-time, continuous sampling, the gas is preconditioned by the sampling system for the temperature or humidity requirements of the analyzers. Particulate concentrations are measured by light transmission through the exhaust gas using opacity meters, by the light reflected from a smoke number filter paper, or by weighing Teflon or quartz sampling filters that have collected particulates from the exhaust. The sampling filters are weighed before being further analyzed for composition *ex-situ*. In the past decade, there have been a number of new instruments developed to allow for *in-situ* measurement of particulates, which will be further discussed in Section 5.3.5.

The primary role of the gas sampling system is to condition the exhaust gas sample, whether it is being continuously sampled from the engine *in situ* or was previously collected in bags or canisters, to the inlet sample specifications of the gas analyzer. Typically, this conditioning requires the removal of any PM from the gas. Thus, the gas sampling system includes removable, disposable particulate filters that require routine maintenance.

Another common requirement for analytical gas analyzers, such as FTIR systems, is a dry gas, which means the sampling system must dehumidify the exhaust. Dehumidification is typically done by cooling the gas and condensing out almost all of the water vapor in a manner that minimizes the loss of any of the other gas species, which may be soluble in the water condensate. More recently, many heated or "wet" versions of gas analyzers such as FID and CLD have been developed to prevent losses in the sampling system that result from condensing the water out. In these "wet" systems, all of the elements of the system that are in contact with the sample, including the sampling lines, pumps, filters, valves, and fittings, are kept heated to approximately 190°C–200°C to avoid condensation of water and some HCs in cool spots, which could occlude the flow. It is important to know which type of analyzer is being used—wet or dry—since measured concentrations in dry gas need to be adjusted back to the wet–gas value.

A common challenge with emissions measurement is that what is measured is not what needs to be reported. That is, the stationary gas analysis methods described in Section 5.1, and the portable ones in Section 5.4 measure concentrations, be it in volume fraction, mass fraction, mass concentration (in mass per unit volume), or molar concentration (in moles per unit volume). However, what needs to be reported is often a brake power-specific value (in g/kW·h or g/bhp·h) or a distance-averaged value (in g/km or g/mi.). Thus, in addition to the gas compositions, one must also know the overall exhaust mass flow, either raw or diluted. Raw exhaust mass flow can be calculated from the intake airflow and the fuel flow but diluted mass flows need extra calculation steps.

5.2.1 Dilution Tunnels

According to most regulations, PM sampling is done from diluted exhaust gas [5-21]. This is achieved using a dilution tunnel, where the exhaust gas is mixed with dilution air [5-22]. The dilution may be done in one [5-23] or more stages, with air at different temperatures in the successive stages [5-24, 5-25, 5-26, 5-27, 5-28]. As shown in **Figure 5.5**, air used for dilution enters the system through a filtered, and usually heated, stainless-steel sampling line and the engine exhaust enters the dilution tunnel upstream of this heated tube and turbulent mixing takes place between the two gas flows. An important parameter of a dilution tunnel is the dilution ratio or the factor by which the additional air has diluted the sample exhaust gas flow.

A full-flow dilution tunnel is a large, bulky, and expensive device; therefore, partial flow tunnels were developed [5-29, 5-30]. Several of these mini-dilution systems have been certified by the US Environmental Protection Agency (EPA) as giving equivalent measurements when compared to those from full dilution tunnels under steady-state conditions. There has been a significant amount of work in the past decade to develop partial flow dilution systems and validate them against full flow systems [5-31, 5-32, 5-33, 5-34, 5-35, 5-36, 5-37]. They are now a viable, commercial option.

FIGURE 5.5 Exhaust gas dilution system.

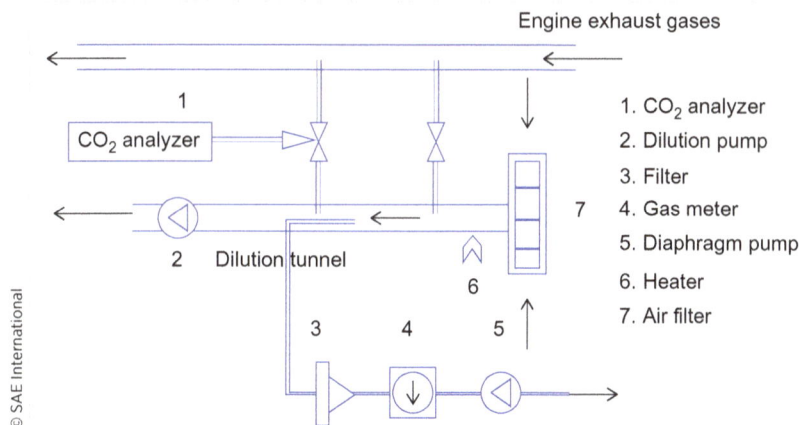

Steady-state exhaust gas analysis is most often done with the continuous gas flow to all of the instruments [5-22]. In this case, the exhaust gas sampling system needs to split and redistribute the sample flow to all of the analyzers at the individually required sampling flow rates. When analyzing transient cycles, the sampled exhaust gases can either be collected in Teflon bags or evacuated canisters for later analysis to provide cycle average concentration levels or continuously measured by fast response gas analyzers.

5.2.2 Constant Volume Sampling

The CVS method dilutes the exhaust gases with air to maintain a constant total flow rate of air and exhaust under all engine operating conditions and is commonly used for regulatory exhaust emissions testing globally [5-4, 5-31, 5-38]. Keeping a constant flow rate to the gas analyzers enables a straightforward calculation of the mass-based emissions required by regulatory agencies, although it complicates the calculation of the dilution ratio. Nevertheless, this methodology requires very sensitive analyzers capable of accurately measuring low concentrations of emissions species in the dilute exhaust. An example of a CVS system transient test cycle measurements is depicted in **Figure 5.6** [5-22].

The CVS system draws filtered air into the dilution tunnel, typically by using a positive displacement pump or a vacuum pump that pushes the gas through a critical flow venture. Use of the simple, inexpensive critical flow orifice assures control of the flow rate is maintained. The collected exhaust gases are mixed with this air in the full-flow primary dilution tunnel. Slipstream samples for measuring PM are drawn from the primary dilution tunnel into a secondary dilution tunnel, where the gas is mixed with yet more air before being collected on filters. The continuous gas flow measurement samples are drawn from the primary dilution tunnel. All of the sampling lines within the system must be maintained at temperatures of at least 190°C to avoid condensation of water or HCs in the lines that could lead to pressure-fluctuating flow occlusions or sample loss in the condensate. If Teflon bags are used for later analysis offline, the gas samples are drawn downstream of the heat exchanger. For gases that may condense or

FIGURE 5.6 Constant volume sampling (CVS) system. (Figure 9.4 in [5-22].)

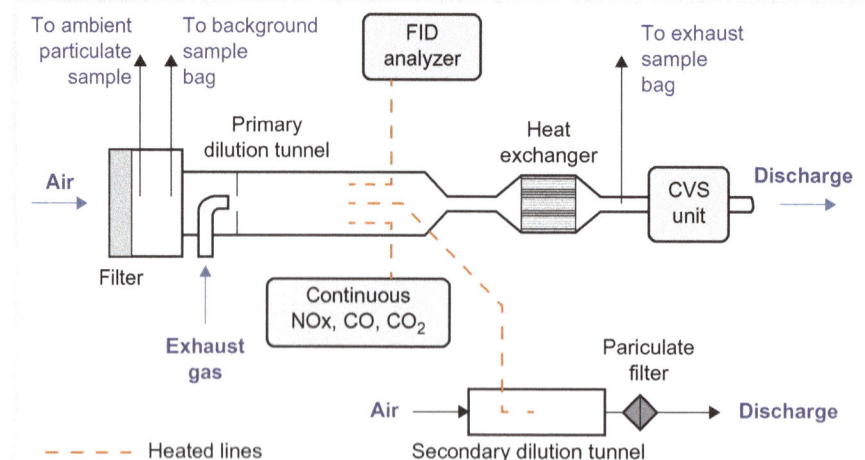

otherwise be lost in the heat exchanger, such as HCs or NOx, heated bag samples may be drawn from upstream of the heat exchanger for more reliable measurements. The risk of sample losses, especially from high boiling-point HCs, is non-negligible. To mitigate this sample loss risk, it is recommended that HCs measurements be performed by a continuous, heated analyzer, especially with diesel exhaust.

5.3 Stationary Analysis Strategies for Exhaust Pollutant Species

For most exhaust gas species, a preferred measurement methodology, technique, or instrument has become the most common one used across multiple manufacturers. However, in many cases, there are multiple techniques that could be used to measure the concentration of a particular gas species. Table 5.2 lists the instruments, methods, and techniques that can be used to measure each of the species, along with their range and limiting factors. Further descriptions can be found in the following sections, by exhaust component.

5.3.1 Carbon Monoxide and Carbon Dioxide Measurements

CO and CO_2 in the exhaust are most commonly measured by NDIR [5-39], described in Section 5.1.1. Since the NDIR technique is sensitive to the presence of water, the gas sample must be cooled and dried before it is measured. The accuracy of the measurement

TABLE 5.2 Measurement instruments, methods, and techniques by exhaust component

Species	Measurement approach	Measurement instrument
Carbon monoxide (CO)	Direct measurement	NDIR
		FTIR
Carbon dioxide (CO_2)	Direct measurement	NDIR
		FTIR
Hydrocarbons (HC or THC)	Direct measurement	FID or Fast FID
Methane (CH_4)	By difference between THC and NMHC	FID or Fast FID
Oxygen (O_2)	Direct measurement	Paramagnetic Sensor
Nitrous oxide (N_2O)	Direct measurement	FTIR
Nitrogen monoxide (NO)	Direct measurement	CLD
		FTIR
Nitrogen dioxide (NO_2)	By difference between NO and total NOx measurement	CLD
		FTIR
Particulate matter (PM)	Direct measurement	Weighing
		Smoke number
		Chemical composition
		Photoacoustic spectroscopy (PAS)
Particulate number (PN)	Direct measurement	Condensation particle counter (CPC)

is governed by the quality of the calibration done, so calibrating the equipment with two or three known span gas concentrations is recommended. FTIR is another technique capable of giving high-quality results for measuring CO and CO_2.

5.3.2 Hydrocarbons Measurement

Total hydrocarbons (THCs) are most commonly measured by FID, described in Section 5.1.3. Heated FID systems, where the sample lines are all held at temperatures greater than 190°C, should be used for diesel exhaust, to avoid HC losses from the condensation of heavy HC species, or when the water concentration in the exhaust is high. Non-methane hydrocarbons (NMHCs) are often measured after chromatographic separation and subtraction of methane from the THCs [5-17].

NDIR measurements of exhaust HCs were once common because they generate different responses for the different HC species. The technique has since been superseded by FIDs because of the improved quality of results, even though the measurement is now in terms of THCs rather than split out by species.

5.3.3 Oxygen Measurement

The most common analyzers used to measure oxygen in engine exhaust use paramagnetic techniques to determine the concentration of oxygen by exploiting its paramagnetic properties, first discovered by Faraday [5-40, 5-41]. The O_2 molecule has the largest magnetic moment of the gases in the exhaust or in air, and therefore, will dominate the magnetic susceptibility of the sample.

5.3.4 Nitrogen Oxides Measurement

In the test cell, the NOx is typically measured by a CLD, described in Section 5.1.2, making use of the reaction between NO and ozone (O_3), as

$$NO + O_3 \Leftrightarrow NO_2^* + O_2 \Leftrightarrow NO_2 + O_2 + \text{light photon} \tag{5.2}$$

The ozone for the reaction is produced within the detector by a high-voltage electrical discharge into a flow of oxygen. The excited nitrogen dioxide (NO_2^*) produced by the reaction then spontaneously returns to its ground state, emitting red light in the 0.6–3 μm band in the process. The light is filtered to eliminate interference from CO, SO_2, and HCs before being measured by the photomultiplier. The measured signal is proportional to the concentration of NO in the sample. The total NOx is measured from the same instrument by first passing the sample through a stainless-steel furnace to decompose any NO_2 present into NO and oxygen. The NO_2 concentration is calculated from the difference between the NO-only and total NOx measurements.

An FTIR spectrometer also measures the NOx species accurately, measuring both NO and NO_2 directly and simultaneously. Total NOx is calculated by summing the measurements of the two species.

Ultraviolet (UV) spectroscopy is another simple, accurate method for measuring NOx species in engine exhaust [5-15]. UV spectrometers are less expensive as compared to CLDs and do not have interference from common exhaust species such as water and CO_2. However, the drawback of UV analysis of NOx is that it cannot match the low-concentration range measurements that are likely to be encountered when testing dilute samples from modern, low emission engines or vehicles, and response time of the CLDs [5-42].

5.3.5 Particulate Matter Measurement

Although particulates are a regulated exhaust emission, the details of a particulate composition are complex. Engine-out particulates composition and structure are affected by fuel properties, engine technology, engine displacement, and operating condition. Furthermore, the particulates change with time as they travel in the exhaust gas into the EAS and out the tailpipe into the atmosphere. Particulates can undergo changes in composition and structure, including the following [5-43, 5-44, 5-45]:

- Volatile organic compounds can condense on or evaporate from the particles
- Particle nanostructure can change
- Particles can aggregate together into dense structures
- Particles can agglomerate together into loose structures

Therefore, it is extremely important to specify the particulate sampling method is done to avoid inducing artifacts before the particulates get to the measurement equipment. The diverse and changing composition of particles and the possibility of sampling artifacts make an absolute measurement of PM impossible. In fact, the measurement techniques are often used to name the mobile components of PM. For example, the soluble organic fraction (SOF) is the set of organic species than can be removed and collected by organic extraction techniques, and the volatile organic fraction (VOF) is the set of organic species that can be removed from the particulates through evaporation. The fixed elemental carbon (EC), also known as black carbon or dry PM, is the combustible material left behind after the removal of the SOF or VOF. PM* from SI engines, especially those with direct injection (DI) fuel injection systems, the measured particulates can also include nonsolid, condensed HC droplets, referred to as secondary organic aerosols [5-46]. Fuels with a higher sulfur content can also create droplets of aqueous sulfuric acid ($H_2SO_{4(aq)}$) that can be measured as PM. Therefore, all measurements of PM have a somewhat arbitrary character [5-22].

Many of the measurement techniques used for measuring and characterizing PM emissions from engine exhaust were adapted from other fields, such as EPA method 3546 for measuring SOF [5-47], which was adapted from geoscience, where it was used to extract HC contamination from soil. The choice of measurement methodology is often defined by its accuracy, repeatability, and suitability for its purpose, whether for research and development or for regulatory compliance testing. Measurement parameters of interest include particulate mass, particulate size, particulate number concentration, and particle composition [5-48].

Particulate mass is the most common parameter measurement used for research and for regulatory testing in the US. US regulations give limits for particulate emissions

---

* Note that dry PM can also include ash, which is noncombustible solids.

on a mass basis, as described in Chapter 3, Emissions Control Regulations. For example, EPA Tier 3 regulatory standards for LDV limit the emissions of vehicles to 30 mg/m. [5-49, 5-50]. The total particulate matter (TPM) emissions include both fixed, EC and the mobile organic carbon fractions, which together include a wide range of chemical species in the solid and liquid states. TPM emissions are determined by gravimetric methods.

Overall, the measurement techniques commonly used for PM can be split into two broad categories: collecting techniques, for later analysis and *in-situ* techniques, where the analysis is directly performed on the aerosol sample taken. Most of the PM measurement and characterization techniques are based on samples are collected for later analysis, where the particles are collected from the exhaust gas by deposition onto a sampling filter, then weighed before further analysis is undertaken. A schematic of the process is shown in **Figure 5.7**.

These collected particulate samples can be further analyzed by several analytical techniques, including solvent extraction, volatilization, microscopy, and thermogravimetric analysis. However, there is a strong concern that the sample collection process may change the particulate properties as compared to their state in the exhaust, including size, SOF, and VOF. In addition, the delay between the collection time and the analysis time can introduce additional artifacts that are not issues with continuous or quasi-continuous *in-situ* measurements. When the particle-laden exhaust sample is analyzed directly, there is less opportunity for artifacts to be introduced, and measurements for particle size, number, and organic speciation and concentration are considered to be more relevant to that in the exhaust. However, the measurement techniques associated with *in-situ* particulate sampling typically provide an indirect measurement of the particulate properties. The properties may be calculated based on aerodynamic, electrical, or optical properties, leading to complicated correlations or variability between instruments. There are a few instruments that have characteristics of both the *in-situ* and collection methods, such as cascade impactors that can continuously classify particles *in situ*, while collecting samples for later analysis. In recent years, there has been a move toward the development and use of *in-situ* techniques, which will be discussed in the following sections.

5.3.5.1 **Gravimetric Methods:** Gravimetric methods are those that measure the mass of the PM in a given volume of exhaust. Most regulatory and compliance testing requires gravimetric methods from CVS, using a dilution tunnel. The volume of mixing air used must be sufficient to cool the mixture of exhaust gas and air to the required temperature, of not more than 50°C (122°F). Also, the dilution ratio for the tunnel must be calculated.

The Teflon-coated quartz or fiberglass filter is conditioned and weighed before being installed in the sample collection system. Then, during the test, the particulate-laden exhaust sample volume is drawn from the dilution tunnel through the filter. After a known volume of exhaust gas has passed through the filter, it is removed from the system. The filter with the sample is equilibrated for temperature and humidity, typically in a dust-free conditioning chamber. The filter is next passed by charge neutralizer, and then the mass is measured using a high-precision balance. The TPM mass collected on the filter can then be calculated by subtracting the conditioned clean filter mass. Since the dilution ratio and the total volume of gas passing through the filter are known, the mass

FIGURE 5.7 Schematic of filter sampling process.

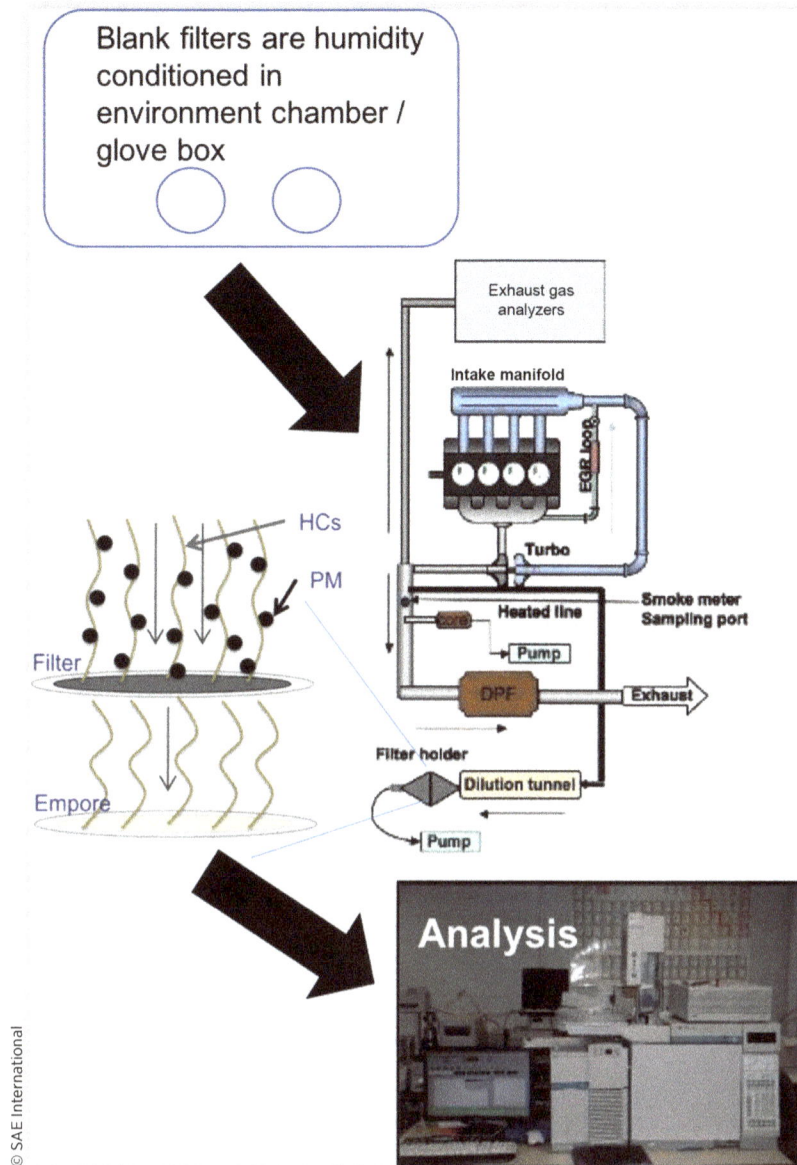

Blank filters are humidity conditioned in environment chamber / glove box

HCs
PM
Filter
Empore

Exhaust gas analyzers
Intake manifold
EGR loop
Turbo
Smoke meter Sampling port
Heated line
Pump
DPF
Exhaust
Filter holder
Dilution tunnel
Pump

Analysis

© SAE International

concentration of particulates in the exhaust can be calculated. For the sake of regulatory compliance, the results are often translated to a specific-power basis (in g/kW·h or g/bhp·h) or a distance-averaged basis (in g/km or g/mi). Following this measurement methodology, any material collected on the filters at or below 50°C is considered to be particulate, including the solids and any condensed liquid droplets. A drawback of this sampling method is that sulfuric acid or high-boiling-point HC species can be collected and measured, even though neither acts as a particulate in the environment.

Another measurement technique that can be used to measure PM mass is the tapered element oscillating microbalance (TEOM). The TEOM draws in the particle-laden gas sample through a filter on the top of an oscillating hollow glass rod [5-59, 5-60], as shown in **Figure 5.8**. The gas flow rate through the filter is kept constant, and the particles in the flow adhere to the filter. The added mass of the PM changes the oscillation frequency of the rod due to the change in inertia. This change in frequency is used to calculate the collected particulate mass, and from that, the concentration of particles per unit volume of air. The TEOM is used frequently in ambient air quality measurements, and for measuring the aerosolized coal dust in mines. Advantages of the TEOM include continuous operation and lower filter maintenance requirements than other sampling techniques, and a relatively low detection limit of 2 $\mu g/m^3$ for a 24-h integrated sample [5-61]. However, much like the filter method described above, the TEOM measures any mass that is collected on the filter, which may include condensed liquid droplets or ash.

FIGURE 5.8 TEOM inner workings.

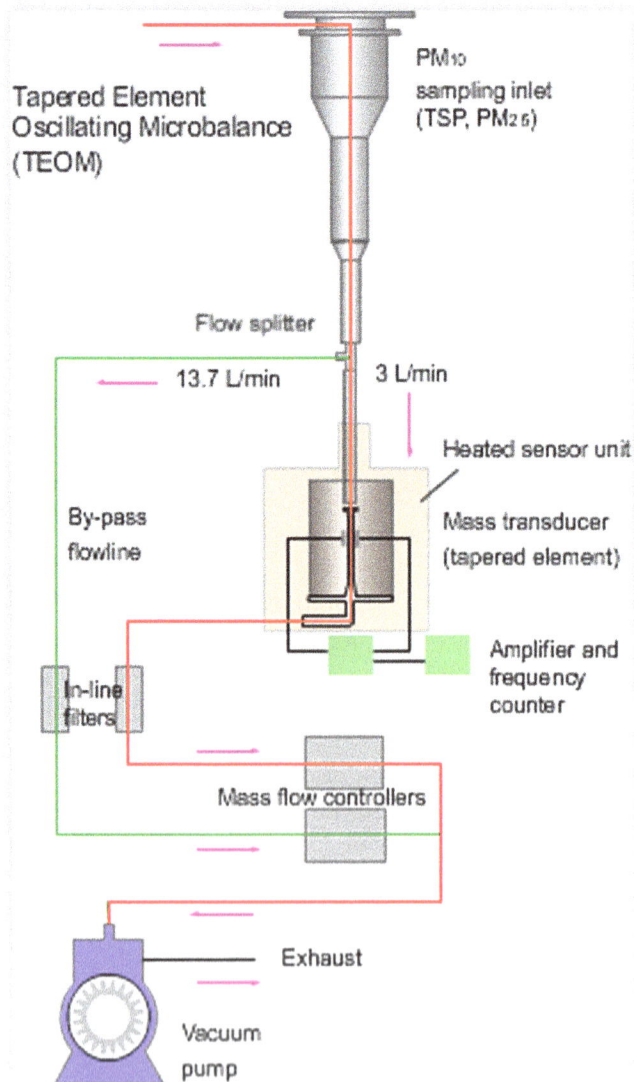

The Quartz crystal microbalance (QCM), which was very popular in the 1970s, uses the same working principle as the TEOM. In the QCM, the particles are deposited onto an oscillating quartz crystal by electrostatic precipitation [5-62].

5.3.5.2 Chemical Composition:
Quantifying the chemical composition of the PM is an extension of the gravimetric measurement method. The PM samples collected on the Teflon-coated filters can be subjected to further analysis by physical or chemical methods. Such methods may be used to determine what portion of the PM comes from the SOF and the composition of the SOF. SOF can be removed from the particulate by liquid solvent (Soxhlet) extraction, supercritical CO_2 extraction, or microwave-assisted solvent extraction before being further analyzed for characterization or speciation [5-63]. The solvent must first be removed from the organic fraction before the sample is analyzed for HC speciation by gas chromatography (GC), high-performance liquid chromatography (HPLC), or mass spectrometry (MS). Sometimes GC and MS are combined into one GC-MS device for a more thorough analysis of composition. Alternatively, the particulates' VOF of the particulates can be measured by vacuum evaporation, however, this method is prone to include any water or sulfates (SO_4^{2-}) in the measurement, as they will also evaporate, and no SOF is available for analysis. The sulfate content of the particulate can be determined by ion chromatography after the particle-coated filters have undergone a water extraction [5-64, 5-65].

5.3.5.3 Particle Number Measurements:
Particle number (PN) measurement is most commonly made using a condensation particle counter (CPC), the growth tube of which is detailed in **Figure 5.9**. As the particle-laden gas sample enters the CPC's conditioner, it is saturated with the CPC solvent vapor, the two most common of which are water and butanol [5-66], and equilibrated for temperature and humidity. The stream then passes to the growth tube where the walls are wetted walls to increase the vapor pressure. The high diffusivity of the solvent allows the vapor to permeate the sample stream faster than the air stream can be heated by the walls. The mixture is then cooled in the condenser tube, causing the vapor to become supersaturated and condense onto the particles, causing them to increase in diameter, to an effective total diameter of approximately 10 μm, so they can be detected by the optical sensor, where a pulse of light from each particle is converted to an electrical signal [5-67]. This particle size detection limit of the optical sensor requires an increasing saturation ratio for decreasing particle sizes. The current, commercially available CPCs have a detection limit of 5 nm [5-67], though detectors with limits below 2 nm are in development [5-68].

CPCs can be used in either the counting mode, as described above, or in the opacity mode. The counting mode provides the most accurate measurement, but can only be used at low particle concentrations, so the exhaust gas to be measured must often be diluted. In the opacity mode, the particulate concentration is determined from the total scattering intensity of the particles. This mode is typically prone to larger error than the counting mode because it typically requires that all of the particles, independent of size, be grown to the same effective diameter and that the optical sensor be frequently calibrated [5-69, 5-70, 5-71].

Additionally, the CPC is sensitive to position and vibration, due to the open solvent container which can spill solvent onto the electronics, and to temperature, which can

FIGURE 5.9   Condensation Particle Counter growth tube, TSI [5-67].

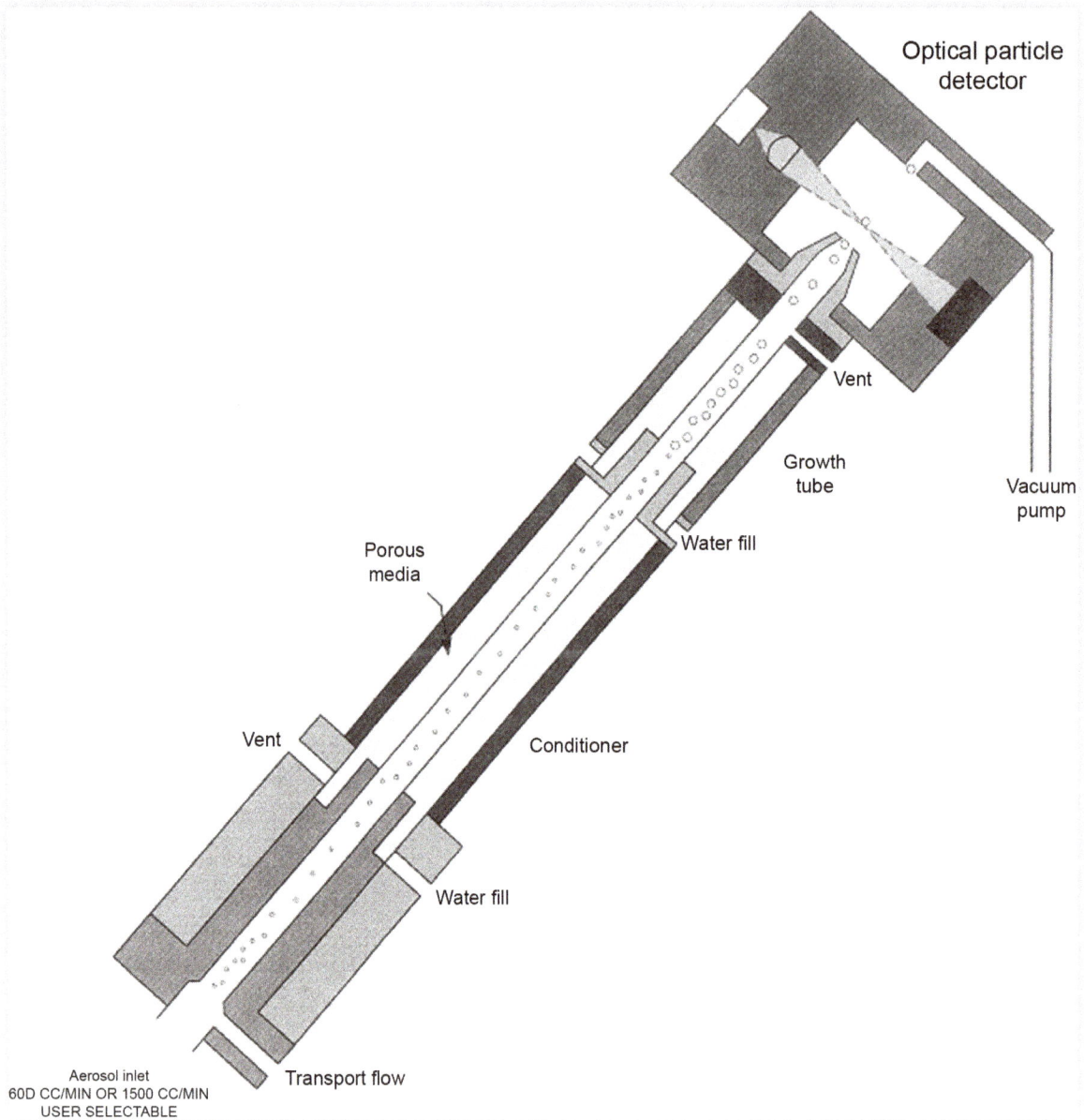

affect the degree to which the device can supersaturate the vapor. For these reasons, it is best suited for use in the lab, rather than in the field.

5.3.5.4 **Particle Size Measurements:** Particle size measurement is used to determine the size distribution of particulates, by diameter, in the exhaust, generally reported in graphical form with particle concentrations per volume broken into size bins. There are a number of instruments that are available to make particle size measurements, the most common of which are described below, and having the ability to resolve particles in different size ranges.

The PM/PN measurement equipment falls into a few methodological categories. First, there are *aerodynamic methods*, which make use of the inertia of accelerated particles, by measuring the relaxation time constant (a function of particle mass and mobility) of an accelerated particle. The *aerodynamic diameter* is equivalent particle diameter, defined as the diameter of a spherical particle with a density of 1 g/cm^3 (0.036 $lb/in.^3$) that has the same settling velocity as the particle being measured. The two major types of instruments that rely on the aerodynamic method are particle sizers and impactors.

Aerodynamic particle sizers use the linear acceleration of particles in a system of nozzles, along with a laser and optical detector, which limits the particle size detection to 0.5 μm [5-66]. Particles that enter the instrument are accelerated to their terminal velocity and smaller particles with greater aerodynamic mobility reach higher terminal velocities.

Aerodynamic methods are also used in impactors, where the aerosol stream flows through an orifice on an impaction plate, which forces a 90° change in the flow direction. The larger particles have greater inertia and cannot follow the flow streamlines, and instead, they impinge on the plate. Multistage impactors known as cascade impactors can be used to classify particulates by decreasing size. By using about 6 to 12 successively smaller stages the cascade impactors can detect particles down to about 10 nm in diameter. One notable issue is that since the gas sample is drawn through the system by a vacuum pump, evaporation of the VOF may occur in the low-pressure stages. An example of a cascade impactor is the electric low-pressure impactor (ELPI). The particle-laden sample enters the instrument. The particles are electrically charged by the corona effect and then neutralize their charge by depositing on the collection stage corresponding to their aerodynamic diameter [5-72, 5-73], as shown in **Figure 5.10**. The advantages of the ELPI include nearly real-time measurement resolution, and the ability to measure low concentrations of particles, such as those encountered in ambient air samples. However, because of the low-pressure conditions at which the ELPI operates, it may cause VOF species or condensed liquid particles to evaporate, changing the particle size distribution of the exhaust [5-74].

Another method for measuring particle size is by *electrical mobility analyzers*, which determine the mobility diameter of the particles in the exhaust sample by measuring the mobility of charged particles in a strong electrical field. A particle size distribution is measured based on the relationship between mobility (the measured parameter) and particle size (diameter). The size parameter determined by the electrical mobility measurement is the equivalent mobility diameter, a close approximation of the Stokes diameter, which is the diameter of a spherical particle with the same density and settling velocity as the particle being measured. Generally, the mobility diameter is different than the aerodynamic diameter described earlier as measured by aerodynamic particle sizers. Whereas the aerodynamic diameter is a representation of the particle's behavior during impaction and settling, the Stokes diameter is a representation of the particle's behavior in the diffusive process. Simply put, the aerodynamic diameter is more characteristic of large particles, with high inertial forces (high Stokes numbers), and the

FIGURE 5.10 ELPI diagram (Dekati).

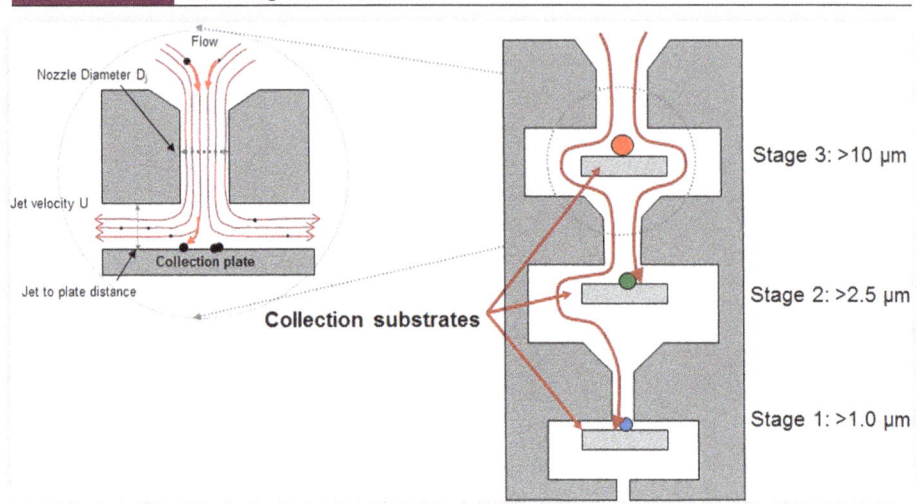

Stokes diameter is better for describing small particles. Since exhaust particulates typically have densities less than 1 g/cm³ (0.036 lb/in³), their aerodynamic diameter is generally smaller than their mobility diameter. The scanning mobility particle sizer (SMPS), shown in **Figure 5.11**, uses the electrical mobility analysis method.

The SMPS comprises an electrostatic classifier and a CPC, as described in Section 5.3.5.3. The classifier parses particles by diameter in the range from 10 to 1,000 nm for measurements of particle size distribution. The classifier consists of several subsystems including the impactor, the sheath-air flow controller, the aerosol neutralizer, the high-voltage controller, and the differential mobility analyzer (DMA). The impactor is mounted outside the inlet to the classifier and removes particles above a known size by inertial impaction. The flow passes through a nozzle and is then forced to make a sharp turn. Large particles with high inertia leave the stream and intercept a removable steel plate. The sheath flow controller maintains a constant flow through the sheath flow loop. This loop consists of a filter, a pump, a filter heat exchanger, and a flow meter. The flow meter has built-in temperature and pressure sensors. The flow is monitored by a microprocessor and used to control the pump based on the desired flow rate which is input from the front panel of the classifier. The high-voltage controller provides a charge on the center rod that acts as an electrode for the DMA. The voltage is monitored by a sensitive high-voltage divider and managed by a microprocessor. As the classifier scans through the desired range of particle sizes, the high-voltage controller changes the voltage of the electrode to select the correct electrical mobility, and thus the correct particle diameter [5-75]. The aerosol neutralizer uses a soft X-ray source (0.12 to 12 keV) to neutralize the electrostatic charges of the aerosol particles as they pass through the device [5-76] by creating both positive and negative ions. These ions are attracted to and will neutralize oppositely charged particles. The sample stream exits the neutralizer as polydisperse flow and forms one of the inlets to the DMA, which then manipulates the remaining charged aerosol particles. The DMA is mounted onto and controlled by the classifier. When the classifier scans through the particle diameter range of interest, it is the DMA that parses and selects each diameter bin in turn. It comprises two polished stainless

FIGURE 5.11 SMPS diagram (TSI).

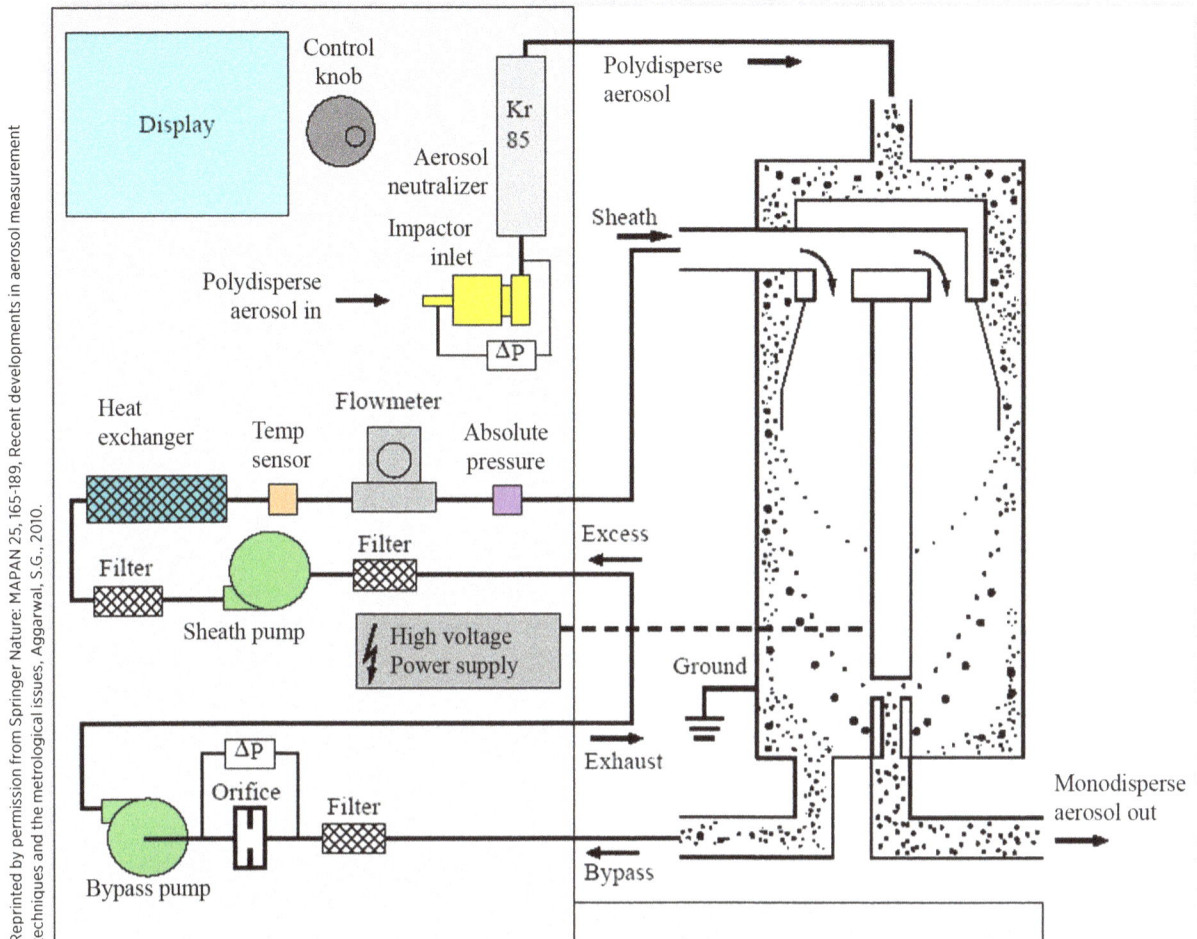

steel cylindrical electrodes which are insulated from each other by a Teflon annulus at the top and an acetyl-plastic annulus at the bottom. The lower insulator allows enough voltage leakage to prevent a static charge buildup near the exit. Filtered air enters the sheath flow inlet of the DMA and passes to an annular chamber at the top. The flow is made laminar by traveling through a double screen of nylon mesh. The air flows downward axially and through a region of the electric field for classification. Polydisperse flow enters the DMA through an inlet at the top and passes through a narrow annular gap to evenly distribute the concentration of particles. The thin annular flow travels to the classification region and is merged with the laminar sheath-air flow. Particles with a positive charge are attracted to and eventually intercept the outer electrode. Neutral particles are unaffected by the field and are removed with the excess flow. Negatively charged particles are attracted to the inner electrode as they move downward along the length of the DMA. Particles with a narrow range of electrical mobility may enter one of twelve apertures that are equal in size and location along the center electrode. These exit the DMA as monodisperse flow and are passed to the CPC for counting.

5.3.5.5 Mitigating Condensed Liquids and Volatiles: When making PN or particle size measurements, it is important to differentiate between droplets of condensed liquid and true carbonaceous particles. Most types of particle counting and sizing instruments are sensitive to both solid and liquid particles. Many of the particles present in engine exhaust, especially those with diameters in the nanoparticle range, are actually composed of condensed volatile species, such as HCs or sulfuric acid, or are water droplets [5-77, 5-78, 5-79]. These condensed species may lead to an inflated PN count. In addition, these species may be adsorbed to the solid particles, impacting the diameter of the particles and affecting the particle size measurement. Therefore, it is desirable to separate the impacts of liquid and solid materials. This separation can be achieved by preconditioning the exhaust gas sample to eliminate the condensed liquids and the volatile species that are precursors to liquid particles before they enter the measurement instruments. Common types of preconditioning instrumentation commercially available for removing the volatile species from the exhaust include the following:

- Thermodesorbers or thermodenuders [5-80, 5-81, 5-82, 5-83]
- Volatile particle removers [5-84]
- Catalytic strippers [5-85, 5-86, 5-87]

While all of the types of preconditioning will remove VOF species, differences in the resultant measurements have been demonstrated [5-77] that may lead to different conclusions regarding the presence, size, and concentration of particles present in the exhaust. Therefore, it is important to consider this effect when choosing which method to use.

Thermodesorbers also called thermodenuders, removes the semi-volatile material by adsorption. The inlet gas sample is heated to a constant temperature, usually between 250°C and 300°C, in order to desorb the volatiles from the particulates and evaporate any condensates that are present in the sample. The gas is then passed through an adsorption section that is cooled with 25°C water. The adsorber collects the vaporized species using activated carbon in a diffusion dryer. While the dried particle size measurements taken using thermodesorbers are very repeatable, they are not representative of atmospheric aerosols [5-61]. One disadvantage of using thermodesorbers is that it affects the solid particles while removing the volatile content, which results in sample losses of up to 30% [5-88]. These particle losses are attributed to thermophoretic deposition of the particles in the hot sample gas onto the cooler walls in the adsorption section and are especially an issue for the smallest particles. This particle deposition is also responsible for the need for frequent cleaning.

Volatile particle removers (VPR) are specified by the European Particle Measurement Programme (PMP) [5-88] to determine whether or not the exhaust is in compliance with the EU's PN standard. The VPR relies on the dilution of the semi-volatile components rather than removal by adsorption or oxidation. VPR consist of three components: a hot dilution section, an evaporation tube similar to the heated section in the thermodesorber, and a cool dilution section to cool the aerosol and reduce its concentration below the 10,000 particles/cm^3 threshold. The VPR wall temperature must be maintained between 300 and 400°C, and it has a residence time on the order of 10 times less than the thermodesorber.

Catalytic strippers remove the semi-volatile, organic carbon fraction of the particulates bypassing the dilute exhaust gas sample over an oxidation catalyst that is held at 300°C. The catalytic stripper typically includes a sulfur trap, which is a wash coated ceramic substrate upstream of the oxidation catalyst that traps sulfur from the gas, and the oxidation catalyst. Very small particles with high diffusion coefficients can be lost due to diffusion, while particles larger than 100 nm are unaffected. Thermophoretic losses can be as high as 25%, independent of the particle size [5-89]. One drawback of the catalytic stripper is that there is a notable trade-off between HC removal efficiency and particle loss—when the HC removal efficiency is high, so is the percentage of particles lost in the system.

5.3.5.6 Smoke Number and Opacity Measurements: The smoke number and opacity measurements take advantage of the optical properties of PM, specifically their ability to blacken paper. Smoke and opacity meters are generally quite simple and correspondingly less expensive than any of the other equipment described in the PM measurement section. These types of instruments quantify the amount of visible black smoke collected on media by light scattering and absorption [5-90]. Optical particulate measurements can be achieved in several ways. The smoke number is the fractional reduction in reflectance by a smoke filter due to surface blackening by the particulates. Transmittance is the fraction of incident light that passes through the collected particles, and reflectance is the fraction of incident light that is reflected from a surface. Opacity refers to the fraction of incident light that is absorbed or scattered by the collected PM. The two general types of smoke meters can be classified as opacity meters that measure the concentration of particulate in the exhaust, and those that optically evaluate the paper blackening of particulates collected on paper filters.

Despite their ubiquity in test cells, these types of instruments have several important limitations: low resolution, cross-sensitivity, insensitivity to small particles, condensed liquid particles and bio-fuel derived particles. Low-emission diesel and both DI and conventional gasoline engines are often below the resolution of opacity measurements. The cross-sensitivity to NO_2, which absorbs green light, is problematic especially when measuring the particle concentrations after an oxidation catalyst, which is likely to be oxidizing any NO in the exhaust to NO_2, to take advantage of passive oxidation in the particulate filter. The low sensitivity to small particle sizes is due to the low light absorption as a function of surface area, a property that scales with particle size. Finally, it has been shown that particulates from bio-fuel blends such as ethanol/gasoline or biodiesel/diesel have more organic content, and result in yellow-brown particles instead of the traditional black smoke [5-91]. These particles and the condensed liquid particles that are also common from bio-blend fuels do not blacken the paper or absorb light in the same manner as black soot. Finally, it has been shown that opacity measurements do not correlate well or consistently with other particulate measurements. Correlations that have been developed are often laboratory-specific, engine-specific, or fuel-specific, as the opacity measurements may be impaired by the presence of HCs, water vapor, particle SOF, and sulfates [5-92].

5.3.5.7 Photoacoustic Measurements: Photoacoustic spectroscopy (PAS) measures the concentration of a material using the acoustic waves it produces when it

is irradiated by modulated light of a specified wavelength. Though somewhat similar in nature to FTIR (described earlier), PAS is a unique technique in that it combines optical microscopy with thermal measurements, directly measuring the acoustic waves formed by the localized pressure waves from the thermal energy of the sample, rather than the effect relative to the background. The components of a PAS are shown in **Figure 5.12**. This leads to high sensitivity and measurement accuracy. Due to these properties, PAS has long been used for measuring gas concentrations but has increasingly been applied to liquid and solid samples as well.

The AVL 483 Micro Soot Sensor (MSS) is an example of using PAS for measuring solid samples. The MSS operates based on photoacoustic measurement principles, using a modulated laser beam that produces periodic heating of the absorbing particles while the laser is on. When the rotating chopper disc blocks the laser, the absorbed energy is dissipated and the resulting pressure fluctuations are detected by microphones as acoustic waves, as shown above in **Figure 5.13**. Filtered air does not produce a signal, but particle-laden air or exhaust produces a signal proportional to the concentration of PM. Use of the photoacoustic technique enables the detection limit of particulates by the MSS down to 1 $\mu g/m^3$, with high accuracy, which is assured by a signal processing algorithm to enhance the signal to noise ratio.

5.3.6 Aldehydes and Ketones Measurement

Carbonyls such as aldehydes and ketones, which are described in Section 2.8, can be measured in the laboratory [5-93, 5-94, 5-95, 5-96] or on-road [5-97, 5-98, 5-99, 5-100], but unlike the other emissions species described to this point, there is no commercial detector available. Instead, gaseous aldehydes and ketones must be collected from the exhaust gas using dinitrophenylhydrazine (DNPH) coated extraction cartridges. Similar to the collection of particulates on filters, the carbonyls are sampled from dilute exhaust over a period of time, in order to have enough exhaust volume pass through the cartridge to collect sufficient quantities for analysis. The DNPH derivatives are then solvent-extracted with acetonitrile and subsequently analyzed by HPLC with UV absorption detection to determine aldehyde and ketone concentrations [5-101]. The effluent from

FIGURE 5.12 PAS schematic.

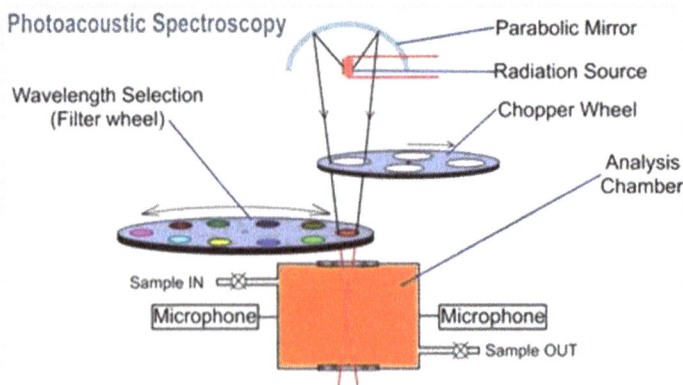

© 2012 IEEE. Reprinted, with permission, from D. Skelly, "Photo-acoustic spectroscopy for dissolved gas analysis: Benefits and Experience," 2012 IEEE International Conference on Condition Monitoring and Diagnosis, Bali, 2012, pp. 29–43.

the HPLC can be further analyzed by MS for hydrazine derivatives, using electrospray with negative ionization MS (ESI-MS) [5-102].

5.4 Portable Emissions Measurement Systems (PEMS)

The PEMS is a compact and lightweight device that can be carried inside a vehicle for measuring the emissions from a vehicle that is road-driven, rather than tested on a chassis dynamometer. PEMS are useful for calculating emissions inventories because they measure emissions over a wide range of operating conditions, including those that could not be replicated in the laboratory, on a chassis dynamometer, such as large gradations in the road, strong accelerations, or variations in altitude during the drive. Systems were first developed and commercialized in the early 1990s, in response to the UK Environment Research Program's on-road emissions measurement strategy and EPA's real-time on-road vehicle emissions testing. PEMS have been more widely used since the mid-2000s [5-103], and especially in response to the Volkswagen emissions scandal, as data from PEMS systems can identify discrepancies between regulatory testing and real-world driving fuel economy and emissions. To limit the use of defeat devices, the EU-RDE testing, as described in Section 4.1.9.

PEMS have many subsystems, each described in more detail in their subsection below, synchronized to a single platform. These typically include an exhaust flow meter (EFM), a particulate measurement system, and gas analyzers along with a global positioning system (GPS). One drawback of this configuration is that any uncertainty associated with the measurements will be propagated through the data. The main sources of uncertainty are the EFM, the gas analyzers and their drift. Therefore, PEMS has extra uncertainty in its measurements as compared to stationary gas analyzers or gas bag measurements, which must be accounted for by a *conformity factor*. The conformity factor defines the maximum allowable emission levels for road vehicles. PEMS is validated and calibrated against laboratory analyzers over a validation test cycle, and the allowable differences in the measurements are defined by the regulatory standards [5-104, 5-105, 5-106].

As of the writing of this book and particularly this section, the three most popular commercial PEMS units are the AVL M.O.V.E, the Horiba OBS-ONE, and the Sensors SEMTECH® DS+ systems, shown in **Figure 5.13(a-c)**. These systems appear in the published literature with the greatest frequency and have the most information available on the system components and analyzers used. They will be referenced and compared as examples in the following sections. A comparison of the specifications of the three models is done in **Table 5.3**.

5.4.1 Exhaust Flow Meter

Unlike the constant volume sampler (CVS) in the test cells, the EFM measures direct exhaust flow without dilution and over a wide dynamic range **Figure 5.14**. The EFM is often mounted to the base system and measures exhaust volume, temperature, and pressure to compute exhaust mass flow rate. EFMS may also be mounted remotely and use heated lines to transport

FIGURE 5.13 Commercial PEMS units. (a) AVL M.O.V.E, (b) Horiba OBS-ONE, and (c) Sensors SEMTECH DS+.

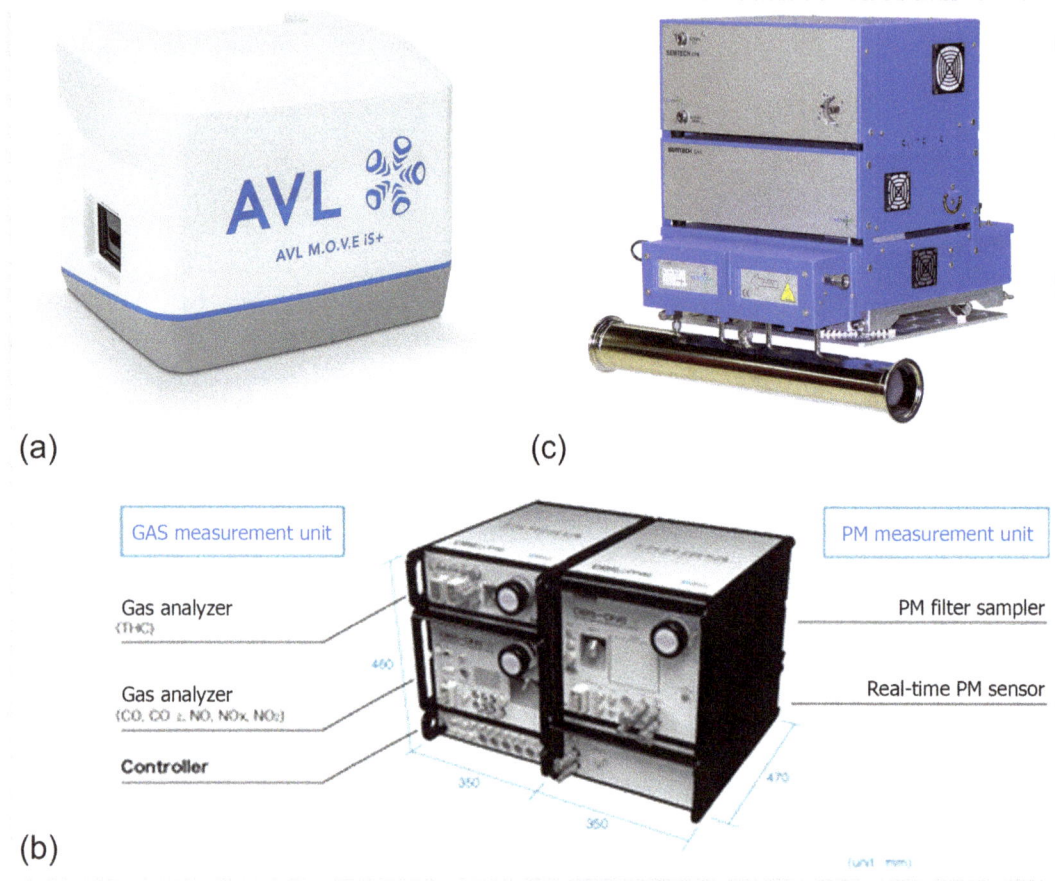

TABLE 5.3 Comparison of Commercial PEMS Units

Manufacturer	AVL	Horiba	Sensors
Model	M.O.V.E Gas PEMS iS+	OBS-ONE	SEMTECH DS+
EFM	Pitot 2″ & 2.5″	Pitot 2″	Pitot 1″, 1.5″, 2″, 2.5″, 3″, 4″, 5″, 6″
CO/CO_2 detector	NDIR	Heated NDIR	NDIR
CO/CO_2 range	10%|20%	10%|20%	8%|18%
NOx detector	NDUV	Heated CLD	NDUV
NOx range	5,000 ppm	3,000 ppm	3,000 ppm
THC detector	FID	FID	FID
THC range	0–30,000 ppm		0–30,000 ppm over 4 ranges
PM detector	photoacoustic (MSS) and gravimetric filters		ion mobility, gravimetric filters
PM range	1 ug/m³–1,000 mg/m³		1 ug/m³–250 mg/m³
PN detector	Advanced Diffusion Charger		CPC
PN range	3×10^3–2×10^7 #/cm³		0–3×10^7 #/cm³

FIGURE 5.14 The AVL M.O.V.E EFM is an example exhaust flow meter system.

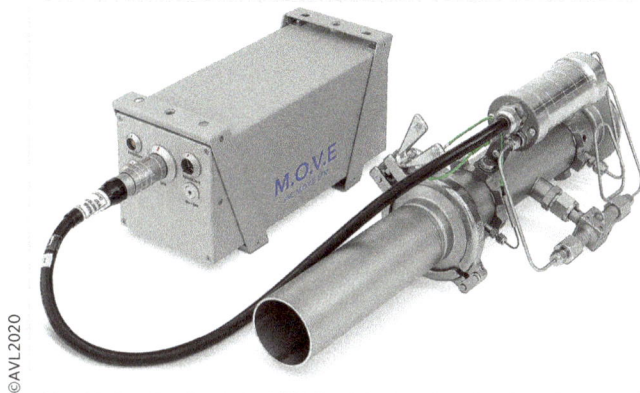

the sample from the EFM to the base system. PEMS are generally best operated in an ambient environment temperature window of −10°C to +45°C, where the sensors operate most reliably. Additional modules are often mounted directly to the base system via electrical or pneumatic connections in order to minimize gas line and cable lengths.

5.4.2 Carbon Monoxide and Carbon Dioxide Measurement

CO and CO_2 are measured by NDIR in the three most common commercial systems. Notably, Horiba's OBS-ONE used heated NDIR.

5.4.3 Hydrocarbons Measurement

THCs are measured by FIDs in the three most common commercial systems. Often, there is an option to use a dual FID system and also measure methane simultaneously.

5.4.4 Nitrogen Oxides Measurement

There is more variability among the systems for the measurement of NO and NO_2, the NOx. CLD is the most commonly used instrument in dyno test cells, with Horiba using a heated CLD in their PEMS. However, both AVL and Sensors use a nondispersive UV (NDUV) detector in their PEMS. The NDUV offers the advantage of simultaneously measuring NO and NO_2.

5.4.5 Particulate Measurement

For particulate measurement, there are options available for measuring PN or PM. PN is commonly measured using CPCs, while PM is typically measured gravimetrically.

5.4.6 Position Measurement and Power Requirements

Two aspects of PEMS that are not found in stationary gas benches are the need for position data and portable power. PEMS systems require an additional electronics box that contains the GPS, power source, power management system and switches. This unit is often coupled with a battery pack in the PEMS.

References

5-1.   Baum, M.M., Kiyomiya, E.S., Kumar, S., Lappas, A.M. et al., "Multi-Component Remote Sensing of Vehicle Exhaust by Dispersive Absorption Spectroscopy," SAE Technical Paper 2000-01-3103, 2000, https://doi.org/10.4271/2000-01-3103.

5-2.   Clark, N.N., Gautam, M., Bata, R.M., Wang, W.-G., Loth, J.L., Palmer, G.M., and Lyons, D.W., "Design and Operation of a New Transportable Laboratory for Emissions Testing of Heavy-Duty Trucks and Buses," *International Journal of Heavy Vehicle Systems* **2**, no. 3-4 (1995): 285–299, https://doi.org/10.1504/IJHVS.1995.054615.

5-3.   Jetter, J., Maeshiro, S., Hatcho, S., and Klebba, R., "Development of an On-board Analyzer for Use on Low Emissions Vehicles," SAE Technical Paper 2000-01-1140, 2000, https://doi.org/10.4271/2000-01-1140.

5-4.   Environmental Protection Agency, "Part 1065-Engine—Testing Procedures," *Code of Federal Regulations, Title 40,* Vol. **37** (2018): 45–282, https://www.govinfo.gov/app/collection/cfr/2018/title40/chapterI/subchapterU/part1065.

5-5.   Cohen, B.S. and Hering, S.V., *Air Sampling Instruments for Evaluation of Atmospheric Contaminants* 9th ed (American Conference of Governmental Industrial Hygenists, 2001).

5-6.   Ring, A.M., Canty, T.P., Anderson, D.C., Vinciguerra, T.P., He, H., Goldberg, D.L., Ehrman, S.H., Dickerson, R.R., and Salawitch, R.J., "Evaluating Commercial Marine Emissions and Their Role in Air Quality Policy Using Observations and the CMAQ Model," *Atmospheric Environment* **173** (2018): 96–107.

5-7.   Marć, M., Tobiszewski, M., Zabiegała, B.,de la Guardia, M., and Namieśnik, J., "Current Air Quality Analytics and Monitoring: A Review," *Analytica Chimica Acta* **853** (2015): 116–126.

5-8.   Plint, M. and Martyr, A. *Engine Testing: Theory and Practice* (Oxford, UK: Buttersworth-Heinemann, 1995).

5-9.   Borrego, C., Costa, A., Ginja, J., Amorim, M., Coutinho, M., Karatzas, K., Sioumis, T., Katsifarakis, N., Konstantinidis, K., and De Vito, S., "Assessment of Air Quality Microsensors versus Reference Methods: The EuNetAir Joint Exercise," *Atmospheric Environment* **147** (2016): 246–263.

5-10. Cambustion, "NDIR500 Operating Principle," https://www.cambustion.com/products/ndir500/operating-principle. Accessed 24 March 2019.

5-11. Lee, D.-D. and Lee, D.-S., "Environmental Gas Sensors," *IEEE Sensors Journal* **1**, no. 3 (2001): 214–224.

5-12. Gibson, D. and MacGregor, C., "A Novel Solid State Non-Dispersive Infrared CO2 Gas Sensor Compatible with Wireless and Portable Deployment," *Sensors* **13**, no. 6 (2013): 7079–7103.

5-13. Nakamura, H., Kihara, N., Adachi, M., and Ishida, K., "Development of a Wet-Based NDIR and Its Application to On-Board Emissions Measurement Systems," SAE Technical Paper 2002-01-0612, 2002, https://doi.org/10.4271/2002-01-0612.

5-14. Dodeigne, C., Thunus, L., and Lejeune, R., "Chemiluminescence as Diagnostic Tool: A Review," *Talanta* **51**, no. 3 (2000): 415–439.

5-15. Baronick, J.D., Heller, B., Lach, G., Bogner, F.et al., "Evaluation of a UV Analyzer for NOx Vehicle Emissions Measurement," SAE Technical Paper 2001-01-0213, 2001, https://doi.org/10.4271/2001-01-0213.

5-16. McWilliam, I.G. and Dewar, R.A., "Flame Ionization Detector for Gas Chromatography," *Nature* **181**, no. 4611 (1958): 760.

5-17. Holm, T., "Aspects of the Mechanism of the Flame Ionization Detector," *Journal of Chromatography A* **842**, no. 1-2 (1999): 221–227.

5-18. Scanlon, J.T. and Willis, D.E., "Calculation of Flame Ionization Detector Relative Response Factors Using the Effective Carbon Number Concept," *Journal of Chromatographic Science* **23**, no. 8 (1985):333–340.

5-19. Smith, B.C., *Fundamentals of Fourier Transform Infrared Spectroscopy* (Boca Raton, FL: CRC Press, 2011).

5-20. Stuart, B., "Infrared Spectroscopy," *Kirk-Othmer Encyclopedia of Chemical Technology* (2000): 1–18.

5-21. Kittelson, D.B., Johnson, J., Watts, W., Wei, Q. et al., "Diesel Aerosol Sampling in the Atmosphere," SAE Technical Paper 2000-01-2122, 2000, https://doi.org/10.4271/2000-01-2122.

5-22. Majewski, W.A. and Khair, M.K. *Diesel Emissions and Their Control* (Warrendale, PA: SAE International, 2006).

5-23. Wei, Q., Kittelson, D.B., and Watts, W.F., "Single-Stage Dilution Tunnel Performance," SAE Technical Paper 2001-01-0207, 2001, https://doi.org/10.4271/2001-01-0207.

5-24. Lyyränen, J., Jokiniemi, J., Kauppinen, E.I., Backman, U., and Vesala, H., "Comparison of Different Dilution Methods for Measuring Diesel Particle Emissions," *Aerosol Science and Technology* **38**, no. 1 (2004): 12–23, https://doi.org/10.1080/02786820490247579.

5-25. Abdul-Khalek, I., Kittelson, D., and Brear, F., "The Influence of Dilution Conditions on Diesel Exhaust Particle Size Distribution Easurements," SAE Technical Paper 1999-01-1142, 1999, https://doi.org/10.4271/1999-01-1142.

5-26. Graskow, B.R., Kittelson, D.B., Abdul-Khalek, I., Ahmadi, M. et al., "Characterization of Exhaust Particulate Emissions from a Spark Ignition Engine," SAE Technical Paper 980528, 1998, https://doi.org/10.4271/980528.

5-27. MacDonald, J.S., Plee, S.L., D'Arcy, J.B., and Schreck, R.M., "Experimental Measurements of the Independent Effects of Dilution Ratio and Filter Temperature on Diesel Exhaust Particulate Samples," SAE Technical Paper 800185, 1980, https://doi.org/10.4271/800185.

5-28. Maricq, M.M., Chase, R.E., Podsiadlik, D.H., and Vogt, R., "Vehicle Exhaust Particle Size Distributions: A Comparison of Tailpipe and Dilution Tunnel Measurements," SAE Technical Paper 1999-01-1461, 1999, https://doi.org/10.4271/1999-01-1461.

5-29. Silvis, W., Kreft, N., Marek, G., and Schindler, W., "Diesel Particulate Measurement with Partial Flow Sampling Systems: A New Probe and Tunnel Design that Correlates with Full Flow Tunnels," SAE Technical Paper 2002-01-0054, 2002, https://doi.org/10.4271/2002-01-0054.

5-30. Hirakouchi, N., Fukano, I., and Shoji, T., "Measurement of Diesel Exhaust Emissions with Mini-Dilution Tunnel," SAE Technical Paper 900643, 1989, https://doi.org/10.4271/900643.

5-31. Otsuki, Y., Haruta, K., and Rahman, M., "Dividing Flow-Weighted Sampling Approach in Partial Flow Dilution System for Particulate Emission Measurement in Internal Combustion Engine Exhaust," SAE Technical Paper 2018-01-0645, 2018, https://doi.org/10.4271/2018-01-0645.

5-32. Rahman, M., Rooney, R., Nevius, T., and Otsuki, Y., "Partial Flow Dilution System with Double Dilution for PM Sampling under Transient Test-Cycles," SAE Technical Paper 2018-01-0643, 2018, https://doi.org/10.4271/2018-01-0643.

5-33. Khalek, I.A., Ullman, T.L., Shimpi, S.A., Jackson, C.C.et al., "Performance of Partial Flow Sampling Systems Relative to Full Flow CVC for Determination of Particulate Emissions Under Steady State and Transient Diesel Engine Operation," SAE Technical Paper 2002-01-1718, 2002, https://doi.org/10.4271/2002-01-1718.

5-34. Badshah, H. and Khalek, I.A., "Solid Particle Emissions from Vehicle Exhaust during Engine Start-Up," *SAE Int. J. Engines* **8**(4):1492-1502, 2015, https://doi.org/10.4271/2015-01-1077.

5-35. Jaiprakash, H.G., and Kumar, S., "Evaluation of Portable Dilution System for Aerosol Measurement from Stationary and Mobile Combustion Sources," *Aerosol Science and Technology* **50**, no. 7 (2016): 717–731.

5-36. Keenan, M., "Exhaust Emissions Control: 60 years of Innovation and Development," SAE Technical Paper SAE Paper 2017-24-0120, 2017, https://doi.org/10.4271/2017-24-0120.

5-37.   Khan, M.Y., Sharma, S., Liew, C.M., Joshi, A., Barnes, D., Scott, N., Mensen, B., Cao, T., Li, Y., and Shimpi, S.A., "Comparison of Full Flow Dilution, Partial Flow Dilution, and Raw Exhaust Particle Number Measurements," *Emission Control Science and Technology* (2018): 1–10.

5-38.   Bainschab, M., Bergmann, A., Andersson, J., Karjalainen, P., Keskinen, J., Mamakos, A., Giechaskiel, B., Lähde, T., Haisch, C., and Thaler, K., "First Results of Vehicle Technology Effects on Sub-23 nm Exhaust Particle Number Emissions Using the DownToTen Sampling and Measurement System," in *2018 Cambridge Particle Meeting*, 2018.

5-39.   SAE International, *Measurement of Carbon Dioxide, Carbon Monoxide, and Oxides of Nitrogen in Diesel Exhaust* (Warrendale, PA: SAE International, 1995).

5-40.   Hill, R.W., "Determination of Oxygen Consumption by Use of the Paramagnetic Oxygen Analyzer," *Journal of Applied Physiology* **33**, no. 2 (1972): 261–263.

5-41.   Manning, A.C., Keeling, R.F., and Severinghaus, J.P., "Precise Atmospheric Oxygen Measurements with a Paramagnetic Oxygen Analyzer," *Global Biogeochemical Cycles* **13**, no. 4 (1999): 1107–1115.

5-42.   Harris, P., *Trace NOx Measurements by UV Spectroscopy: Recent Advances* (Newark, DE: AMETEK Process Instruments, 2003).

5-43.   Bukowiecki, N., Kittelson, D., Watts, W., Burtscher, H., Weingartner, E., and Baltensperger, U., "Real-Time Characterization of Ultrafine and Accumulation Mode Particles in Ambient Combustion Aerosols," *Journal of Aerosol Science* **33**, no. 8 (2002): 1139–1154.

5-44.   Elimelech, M., Gregory, J., and Jia, X., *Particle Deposition and Aggregation: Measurement, Modelling and Simulation* (Waltham, MA:Butterworth-Heinemann, 2013).

5-45.   Pei, Y.-Q., Qin, J., and Pan, S.-Z., "Experimental Study on the Particulate Matter Emission Characteristics for a Direct-Injection Gasoline Engine," *Proceedings of the Institution of Mechanical Engineers, Part D: Journal of Automobile Engineering* **228**, no. 6 (2014): 604–616.

5-46.   Joshi, A. and Johnson, T.V., "Gasoline Particulate Filters-A Review," *Emission Control Science and Technology* **4**, no. 4 (2018): 219–239.

5-47.   Environmental Protection Agency, "Method 3546 Microwave Extraction," 2007.

5-48.   Kittelson, D.B., Arnold, M., and Watts, W.F., "Review of Diesel Particulate Sampling Methods - Final Report," Center for Diesel Research, University of Minnesota, 1999.

5-49.   Piock, W., Hoffmann, G., Berndorfer, A., Salemi, P., and Fusshoeller, B., "Strategies towards Meeting Future Particulate Matter Emission Requirements in Homogeneous Gasoline Direct Injection Engines," *SAE Int. J. of Engines* **4**, no. 1 (2011):1455–1468, https://doi.org/10.4271/2011-01-1212.

5-50.   Gladstein, N., "Ultrafine Particulate Matter and the Benefits of Reducing Particle Numbers in the United States," A Report to the Manufacturers of Emissions Controls Association (MECA), 2013.

5-51.   Attfield, M.D., Schleiff, P.L., Lubin, J.H., Blair, A., Stewart, P.A., Vermeulen, R., Coble, J.B., and Silverman, D.T., "The Diesel Exhaust in Miners Study: A Cohort Mortality Study with Emphasis on Lung Cancer," *Journal of the National Cancer Institute* **104**, no. 11 (2012): 869–883, https://doi.org/10.1093/jnci/djs035.

5-52.   Anderson, J.O., Thundiyil, J.G., and Stolbach, A., "Clearing the Air: A Review of the Effects of Particulate Matter Air Pollution on Human Health," *Journal of Medical Toxicology* **8**, no. 2 (2012): 166–175, https://doi.org/10.1007/s13181-011-0203-1.

5-53.   Hoek, G., Krishnan, R.M., Beelen, R., Peters, A., Ostro, B., Brunekreef, B., and Kaufman, J.D., "Long-Term Air Pollution Exposure and Cardio-Respiratory Mortality: A Review," *Environmental Health* **12**, no. 1 (2013): 43, https://doi.org/10.1186/1476-069X-12-43.

5-54.   Weinmayr, G., Hennig, F., Fuks, K., Nonnemacher, M., Jakobs, H., Möhlenkamp, S., Erbel, R., Jöckel, K.-H., Hoffmann, B., and Moebus, S., "Long-Term Exposure to Fine

Particulate Matter and Incidence of Type 2 Diabetes Mellitus in a Cohort Study: Effects of Total and Traffic-Specific Air Pollution," *Environmental Health* **14**, no. 1 (2015): 53, https://doi.org/10.1186/s12940-015-0031-x.

5-55.   Abdel-Shafy, H.I. and Mansour, M.S., "A Review on Polycyclic Aromatic Hydrocarbons: Source, Environmental Impact, Effect on Human Health and Remediation," *Egyptian Journal of Petroleum* **25**, no. 1 (2016): 107-123, https://doi.org/10.1016/j.ejpe.2015.03.011.

5-56.   Kheirbek, I., Haney, J., Douglas, S., Ito, K., and Matte, T., "The Contribution of Motor Vehicle Emissions to Ambient Fine Particulate Matter Public Health Impacts in New York City: A Health Burden Assessment," *Environmental Health* **15**, no. 1 (2016): 89, https://doi.org/10.1186/s12940-016-0172-6.

5-57.   Dai, L., Zanobetti, A., Koutrakis, P., and Schwartz, J.D., "Associations of Fine Particulate Matter Species with Mortality in the United States: A Multicity Time-Series Analysis," *Environmental Health Perspectives* **122**, no. 8 (2014): 837–842, https://doi.org/10.1289/ehp.1307568.

5-58.   Mamakos, A., Martini, G., Marotta, A., and Manfredi, U., "Assessment of Different Technical Options in Reducing Particle Emissions from Gasoline Direct Injection Vehicles," *Journal of Aerosol Science* **63** (2013): 115–125, https://doi.org/10.1016/j.jaerosci.2013.05.004.

5-59.   Bae, M.-S., Schwab, J.J., Park, D.-J., Shon, Z.-H., and Kim, K.-H., "Carbonaceous Aerosol in Ambient Air: Parallel Measurements between Water Cyclone and Carbon Analyzer," *Particuology* (2018).

5-60.   Nosratabadi, A.R., Graff, P., Karlsson, H., Ljungman, A.G., and Leanderson, P., "Use of TEOM Monitors for Continuous Long-Term Sampling of Ambient Particles for Analysis of Constituents and Biological Effects," *Air Quality, Atmosphere & Health* **12**, no. 2 (2019): 161–171.

5-61.   Burtscher, H., *Literature Study on Tailpipe Particulate Emission Measurement for Diesel Engines* (Luxembourg:European Particle Measurement Programme, 2001).

5-62.   Booker, D., "Development of a Real-Time Transient Cycle Mass Monitor," in *3rd ETH Conference on Nanoparticle Measurement Paper #20*, 2000.

5-63.   Strzelec, A., Storey, J.M., Lewis, S.A., Daw, C.S.et al., "Effect of Biodiesel Blending on the Speciation of Soluble Organic Fraction from a Light Duty Diesel Engine," SAE Technical Paper 2010-01-1273, 2010, https://doi.org/10.4271/2010-01-1273.

5-64.   Perez, J.M., Jass, R.E., and Leddy, D.G., *Chemical Methods for the Measurement of Unregulated Diesel Emissions* Alpharetta, GA: (Coordinating Research Council, 1987).

5-65.   Society of Automotive Engineers, "*Chemical Methods for the Measurement of Nonregulated Diesel Emissions* (Warrendale, PA: SAE, 1995).

5-66.   Baron, P.A. and Willeke, K., *Aerosol Measurement: Principles, Techniques & Applications* 2nd ed. (John Wiley: New York, 2001).

5-67.   TSI Inc., "Model 3787 General Purpose Water-based Condensation Particle Counter - Operation and Service Manual," 2011.

5-68.   Barmpounis, K., Ranjithkumar, A., Schmidt-Ott, A., Attoui, M., and Biskos, G., "Enhancing the Detection Efficiency of Condensation Particle Counters for sub-2 nm Particles," *Journal of Aerosol Science* **117** (2018): 44–53.

5-69.   Wiedensohlet, A., Orsini, D., Covert, D., Coffmann, D., Cantrell, W., Havlicek, M., Brechtel, F., Russell, L., Weber, R., and Gras, J., "Intercomparison Study of the Size-Dependent Counting Efficiency of 26 Condensation Particle Counters," *Aerosol Science and Technology* **27**, no. 2 (1997): 224–242.

5-70.   Tauber, C., Steiner, G., and Winkler, P., "Counting Efficiency Determination from Quantitative Intercomparison between Expansion and Laminar Flow Type Condensation Particle Counter," *Aerosol Science and Technology* (2019): 1–11.

5-71. Wiedensohler, A., Wiesner, A., Weinhold, K., Birmili, W., Hermann, M., Merkel, M., Müller, T., Pfeifer, S., Schmidt, A., and Tuch, T., "Mobility Particle Size Spectrometers: Calibration Procedures and Measurement Uncertainties," *Aerosol Science and Technology* **52**, no. 2 (2018): 146–164.

5-72. Keskinen, J., Pietarinen, K., and Lehtimäki, M., "Electrical Low Pressure Impactor," *Journal of Aerosol Science* **23**, no. 4 (1992): 353–360.

5-73. Ahlvik, P., Ntziachristos, L., Keskinen, J., and Virtanen, A., "Real Time Measurements of Diesel Particle Size Distribution with an Electrical Low Pressure Impactor," SAE Technical Paper 980410, 1998, https;//doi.org/10.4271/980410.

5-74. Järvinen, A., Heikkilä, P., Keskinen, J., and Yli-Ojanperä, J., "Particle Charge-Size Distribution Measurement Using a Differential Mobility Analyzer and an Electrical Low Pressure Impactor," *Aerosol Science and Technology* **51**, no. 1 (2017): 20–29.

5-75. TSI Inc., "Series 3080 Electrostatic Classifiers - Operation and Service Manual," 2011.

5-76. TSI Inc., Model 3087 Aerosol Neutralizer - Operation and Service Manual, 2011.

5-77. Swanson, J. and Kittelson, D., "Evaluation of Thermal Denuder and Catalytic Stripper Methods for Solid Particle Measurements," *Journal of Aerosol Science* **41**, no. 12 (2010): 1113–1122.

5-78. Bertola, A., Schubiger, R., Kasper, A., Matter, U., Forss, A.M., Mohr, M., Boulouchos, K., and Lutz, T., "Characterization of Diesel Particulate Emissions in Heavy-Duty DI-Diesel Engines with Common Rail Fuel Injection Influence of Injection Parameters and Fuel Composition," SAE Technical Paper 2001-01-3573, 2001, https://doi.org/10.4271/2001-01-3573.

5-79. Burtscher, H., Künzel, S., and Hüglin, C., "Characterization of Particles in Combustion Engine Exhaust," *Journal of Aerosol Science* **29**, no. 4 (1998): 389–396.

5-80. An, W.J., Pathak, R.K., Lee, B.-H., and Pandis, S.N., "Aerosol Volatility Measurement Using an Improved Thermodenuder: Application to Secondary Organic Aerosol," *Journal of Aerosol Science* **38**, no. 3 (2007): 305–314.

5-81. Fierz, M., Vernooij, M.G., and Burtscher, H., "An Improved Low-Flow Thermodenuder," *Journal of Aerosol Science* **38**, no. 11 (2007): 1163–1168.

5-82. Huffman, J.A., Ziemann, P.J., Jayne, J.T., Worsnop, D.R., and Jimenez, J.L., "Development and Characterization of a Fast-Stepping/Scanning Thermodenuder for Chemically-Resolved Aerosol Volatility Measurements," *Aerosol Science and Technology* **42**, no. 5 (2008): 395–407.

5-83. Cappa, C., "A Model of Aerosol Evaporation Kinetics in a Thermodenuder," *Atmospheric Measurement Techniques* **3**, no. 3 (2010): 579–592.

5-84. Sandbach, E., "Volatile Particle Remover Calibration Procedure," 2007.

5-85. Amanatidis, S., Ntziachristos, L., Giechaskiel, B., Katsaounis, D., Samaras, Z., and Bergmann, A., "Evaluation of an Oxidation Catalyst ("catalytic stripper") in Eliminating Volatile Material from Combustion Aerosol," *Journal of Aerosol Science* **57** (2013): 144–155.

5-86. Khalek, I.A. and Bougher, T., "Development of a Solid Exhaust Particle Number Measurement System Using a Catalytic Stripper Technology," *SAE Int. J. Engines* **4**(1): 610–618, 2011, https://doi.org/10.4271/2011-01-0635.

5-87. Kittelson, D., Watts, W., Savstrom, J., and Johnson, J., "Influence of a Catalytic Stripper on the Response of Real Time Aerosol Instruments to Diesel Exhaust Aerosol," *Journal of Aerosol Science* **36**, no. 9 (2005): 1089–1107.

5-88. Andersson, J., "UK Particle Measurement Programme: Heavy-Duty Methodology Development," 2002.

5-89. Hinds, C.W., *Aerosol Technology, Properties, Behavior, and Measurement of Airborne Particles* (New York: J. Wiley & Sons, Inc., 1999).

5-90. Lapuerta, M., Martos, F.J., and Cárdenas, M.D., "Determination of Light Extinction Efficiency of Diesel Soot from Smoke Opacity Measurements," *Measurement Science and Technology* **16**, no. 10 (2005): 2048.

5-91. Armas, O., Hernández, J.J., and Cárdenas, M.D., "Reduction of Diesel Smoke Opacity from Vegetable Oil Methyl Esters during Transient Operation," *Fuel* **85**, no. 17-18 (2006): 2427–2438.

5-92. Clark, N.N., Jarrett, R.P., and Atkinson, C.M., "Field Measurements of Particulate Matter Emissions, Carbon Monoxide, and Exhaust Opacity from Heavy-Duty Diesel Vehicles," *Journal of the Air & Waste Management Association* **49**, no. 9 (1999): 76–84.

5-93. Reter, R.M., Gorse, R.A. Jr., Painter, L.J., Benson, J.D.et al., "Effects of Oxygenated Fuels and RVP on Automotive Emissions - Auto/Oil Air Quality Improvement Program," SAE Technical Paper 920326 , 1992, https://doi.org/10.4271/920326.

5-94. Stump, F.D., Knapp, K.T., and Ray, W.D., "Seasonal Impact of Blending Oxygenated Organics with Gasoline on Motor Vehicle Tailpipe and Evaporative Emissions," *Journal of the Air & Waste Management Association* **40**, no. 6 (1990): 872–880.

5-95. Hoekman, S.K., "Speciated Measurements and Calculated Reactivities of Vehicle Exhaust Emissions from Conventional and Reformulated Gasolines," *Environmental Science & Technology* **26**, no. 6 (1992): 1206–1216.

5-96. Sigsby, J.E., Tejada, S., Ray, W., Lang, J.M., and Duncan, J.W., "Volatile Organic Compound Emissions from 46 In-Use Passenger Cars," *Environmental Science & Technology* **21**, no. 5 (1987): 466–475.

5-97. Schmid, H., Pucher, E., Ellinger, R., Biebl, P., and Puxbaum, H., "Decadal Reductions of Traffic Emissions on a Transit Route in Austria-Results of the Tauerntunnel Experiment 1997," *Atmospheric Environment* **35**, no. 21 (2001): 3585–3593.

5-98. Fraser, M.P., Cass, G.R., and Simoneit, B.R., "Gas-Phase and Particle-Phase Organic Compounds Emitted from Motor Vehicle Traffic in a Los Angeles Roadway Tunnel," *Environmental Science & Technology* **32**, no. 14 (1998): 2051–2060.

5-99. Staehelin, J., Keller, C., Stahel, W., Schläpfer, K., and Wunderli, S., "Emission Factors from Road Traffic from a Tunnel Study (Gubrist Tunnel, Switzerland). Part III: Results of Organic Compounds, SO2 and Speciation of Organic Exhaust Emission," *Atmospheric Environment* **32**, no. 6 (1998): 999–1009.

5-100. Zweidinger, R.B., Sigsby, J.E., Tejada, S.B., Stump, F.D., Dropkin, D.L., Ray, W.D., and Duncan, J.W., "Detailed Hydrocarbon and Aldehyde Mobile Source Emissions from Roadway Studies," *Environmental Science & Technology* **22**, no. 8 (1988): 956–962.

5-101. Grosjean, E., Green, P.G., and Grosjean, D., "Liquid Chromatography Analysis of Carbonyl (2, 4-dinitrophenyl) Hydrazones with Detection by Diode Array Ultraviolet Spectroscopy and by Atmospheric Pressure Negative Chemical Ionization Mass Spectrometry," *Analytical Chemistry* **71**, no. 9 (1999): 1851–1861.

5-102. Prikhodko, V.Y., Curran, S.J., Barone, T.L., Lewis, S.A., Storey, J.M., Cho, K., Wagner, R.M., and Parks, J.E., "Emission Characteristics of Diesel Engine Operating with In-cylinder Gasoline and Diesel Fuel Blending," SAE Technical Paper 2010-01-2266, 2010, https://doi.org/10.4271/2010-01-2266.

5-103. Hart, C., Koupal, J., and Gianelli, R., *EPA's Onboard Analysis Shootout: Overview and Results* (U.S. Enivronmental Protection Agency, 2002).

5-104. Andersson, J., May, J., Favre, C., Bosteels, D.et al., "On-Road and Chassis Dynamometer Evaluations of Emissions from Two Euro 6 Diesel Vehicles," *SAE Int. J. Fuels Lubr.* **7**: 919-934, 2014, https://doi.org/10.4271/2014-01-2826.

5-105. Kwon, S., Park, Y., Park, J., Kim, J., Choi, K., Cha, J., "Characteristics of On-Road NOx Emissions from Euro 6 Light-Duty Diesel Vehicles Using a Portable Emissions Measurement System," *Science Total Environment* **576** (2017): 70–77.

5-106. Czerwinski, J.Z.Y., Comte, P., and Bütler, T., "Experiences and Results with Different PEMS," *Journal of Earth Sciences and Geotechnical Engineering* **6** (2016): 91–106.

6

Sensors and On-Board Diagnostics

The senses, being explorers of the world, open the way to knowledge.

—Maria Montessori [6-1]

Sensors are required in modern exhaust aftertreatment systems (EAS) to provide feedback to the controllers used to manage engine and EAS performance in both light-duty vehicles (LDVs) and heavy-duty applications. In addition, regulatory requirements for on-board diagnostics (OBD) require the engine and EAS to monitor their emissions-related performance and to detect faults, both in the engine and EAS systems, and in the sensors and actuators used for control.

Unlike the analytical measurement systems described in Chapter 5, Emissions Measurements, the exhaust component sensors described here are designed to last through the full useful life (FUL) of the application, as described in Chapter 3, Emissions Control Regulations. This means that they are more robust than the bench or portable emission measurement systems, but more limited in capability and typically less precise.

This chapter provides an overview of the types of sensors commonly used in an EAS and an overview of OBD requirements. The sensors discussed here include both those that are used in almost every EAS, such as temperature and pressure sensors, and those that are only needed for a specific EAS unit, such as the urea-water solution quality sensors and tank level sensors. The OBD overview discusses requirements for both light-duty and heavy-duty applications in several global jurisdictions. A full description of the test procedures needed to validate the OBD system, however, is beyond the scope of this book.

6.1 **Sensors**

An engine or aftertreatment control unit needs to monitor the performance of the EAS to ensure that it is functioning properly, especially where OBD is required. Sensors play an important role in providing feedback and other inputs to the controller to ensure that the EAS efficiently removes pollutants and maintains good exhaust gas flow through the system. There are several sensor types common to all EAS, such as those for temperature, pressure, and air-fuel ratio (AFR) or λ. The most common exhaust gas component sensors are those for oxygen gas (O_2) and nitrogen oxides (NOx).

Common sensors found in almost every EAS include those for temperature and AFR or λ, and increasingly for NOx concentration. More specialized ones, such as those for ammonia (NH_3) or particulates, are used when those species need to be monitored by the EAS controller. As the sensors are an integral part of the EAS function, they are subject to the same FUL requirements, as the rest of the EAS is. For more details on these FUL requirements, please see Chapter 3, Emissions Control Regulations.

6.1.1 **Temperature Sensors**

The span of temperatures encountered in the EAS is very broad, covering the range from approximately −7°C (20°F) to approximately 1,000°C (≈1,800°F), with ambient temperatures that may be as cold as −40°C (−40°F). This range and the durability requirements for EAS components lead engineers to typically choose thermistors or thermocouples.

Thermistors measure temperature by taking advantage of the change in electrical resistance in a material, which can increase or decrease with the temperature depending on the material. The change in resistance usually spans several orders of magnitude. The resistance can be measured and then translated to the measured temperature.

Thermocouples measure temperature because of the change in voltage potential generated at the junction of two dissimilar metals. This voltage signal from the thermocouple can be mapped to the temperature at the junction of the two metals. For the temperature range needed for an EAS, K-type thermocouples are preferred, as they can cover the temperature range from −270°C (−454°F) to 1,260°C (2,300°F). In the K-type thermocouple, the two metal alloys used are nickel-chromium and nickel-alumel.

Both thermistors and thermocouples are not placed directly into the exhaust gas stream, but are instead placed in thermowells. For measurements of the bulk gas temperature, the thermowell can act like a fin extending from the wall, so the measured temperature at the end of the thermowell needs to be corrected to the actual gas temperature [6-2]. The thermowell can also slow the response time of the temperature measurement, as it adds thermal mass between the fluid and the sensor.

Exhaust gas temperature measurements are typically made at the EAS inlet position, and, if present, upstream of the particulate filter (PF) and of the selective catalytic reduction (SCR) system. The urea–water solution temperature also needs to be monitored. An example of engine with K-type thermocouples installed for test measurements is shown in **Figure 6.1** [6-3]. The yellow plugs on the thermocouples will be connected to a data measurement system in the test cell. The thermocouple wires are sheathed in metal, and are routed to measurement points on the engine. Temperature sensor placement is also illustrated in **Figure 6.2**.

6.1.2 Pressure Sensors

Statically acting pressure sensors are preferred for automotive applications [6-4], as the response is suitable for most exhaust gas pressure measurements. Of these, the most common type uses a strain gauge attached to a diaphragm to measure pressure. Pressure measurements in the EAS are typically gauge pressure based, which is measured relative to the current barometric pressure, but sometimes sensors that report absolute pressure are used. Pressure sensors need a fitting on the exhaust pipe wall so that they can be attached, and will also have operating temperature restrictions. The pressure sensors are generally placed on the top half of the exhaust pipe to prevent condensation pooling on or in the sensor.

Differential pressure sensors may also be installed to measure the pressure drop across a PF more accurately than measuring the difference between an upstream and downstream pressure sensor that are not linked to each other, otherwise. An example of differential pressure sensor placement around a PF is shown using the green-colored sensors in Figure 6.2. The red sensors in Figure 6.2 are thermocouples for EAS temperature measurements, and the blue sensor is for NOx.

In addition to the differential pressure across a PF, if present, there is usually one pressure sensor at the EAS inlet position, which for a boosted engine is where the turbine discharges into the downstream exhaust pipe. In general, pressures in the EAS do not exceed two bars, which suggest that a lower-range sensor can be used, and will improve accuracy and absolute precision in the pressure range of interest.

FIGURE 6.1 K-type thermocouples (with yellow plugs) on engine awaiting installation in test cell [6-3].

Reprinted from Ref. [6-3]. Oak Ridge National Laboratory

FIGURE 6.2 Differential pressure sensor placement around particulate filter is shown in green. Thermocouples are shown in red, and a NOx sensor in blue. (Figure 34(c) in [6-5].)

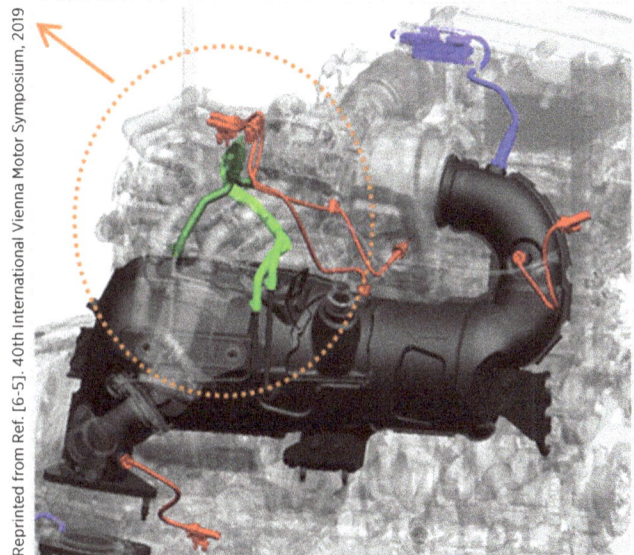

Reprinted from Ref. [6-5]. 40th International Vienna Motor Symposium, 2019

6.1.3 Oxygen (λ) Sensors

Controlling the mass-based AFR in the exhaust is an important part of managing the engine and the EAS performance. AFR is commonly converted to the non-dimensional parameter λ, which is the ratio of the actual AFR to the stoichiometric AFR for the given fuel. The inverse of λ is known as the equivalence ratio, ϕ. AFR or λ control is especially important for gasoline spark-ignition (SI) engines that use a stoichiometric AFR, where

$\lambda = 1$, where the sensor is used for feedback control. As discussed in Chapter 9, the three-way catalyst (TWC) system works best when the AFR of the feed gas to the TWC is held close to but oscillates slightly around the stoichiometric state.

There are two main types of oxygen or λ sensors:

1. Switching or Two-step λ sensors
2. Wide-band λ or universal exhaust gas oxygen (UEGO) sensors

Two-step λ sensors were first developed in late 1960s by Robert Bosch GmbH [6-6], and work by switching their voltage signal from low to high as λ goes from low to high, and vice versa [6-4]. They are particularly useful when controlling to a stoichiometric ($\lambda = 1$) condition, as the engine control unit (ECU) can make adjustments to keep the engine AFR oscillating at a low amplitude around that set point. Also, near $\lambda = 1$, the rate of change of the voltage signal with respect to λ is very rapid, which allows for a fine measurement of actual stoichiometry in that regime as the fuel composition changes or the fuel system components age. Two-step λ sensors are typically placed downstream of both the close-coupled TWC and the under floor TWC in the EAS.

Wide-band λ sensors, also known as UEGO sensors, provide a signal that is proportional to λ or the AFR, and can provide the feedback necessary for finer control [6-4] of lean burn engines. Wide-band λ sensors are needed when the engine operates in a lean condition or switches between lean and stoichiometric operation regularly. The sensor for such an EAS is typically placed at the EAS inlet. In an EAS for a stoichiometric SI engine, the engine-out condition upstream of the first TWC may be measured with a wide-band λ sensor. Compared with the two-step λ sensor, the change in voltage signal is much smoother across a wide range of λ conditions in the exhaust. An example response curve for a UEGO is shown in **Figure 6.3** [6-7].

FIGURE 6.3 UEGO current signal as a function of λ [6-7].

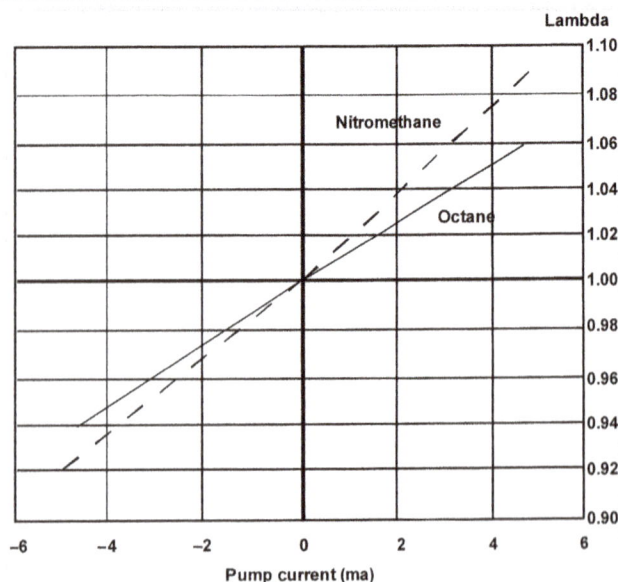

Reprinted from Ref [6-7], US Patent Application 20110218726A1, 2011

6.1.4 Nitrogen Species Sensors

The key nitrogen-containing species of interest for on-board measurement are the regulated ones. These species include the criteria pollutant NOx, a mixture of nitrogen monoxide (NO) and nitrogen dioxide (NO_2), but also include ammonia (NH_3) and the potent greenhouse gas (GHG) nitrous oxide (N_2O). Sensors for nitrogen species can confirm the emissions control performance of the engine and EAS. The sensors also provide feedback to the ECU for the active management of the NOx control system in the EAS.

N_2O is a potent GHG, and is regulated under the GHG rules described in Chapter 3, Emissions Control Regulations. There is currently not a sensor for on-board use that can measure N_2O directly.

6.1.4.1 **Nitrogen Oxides Sensors:** NOx sensors are a critical part of a modern EAS because NOx species are regulated to improve air quality, as discussed in Chapter 3, Emissions Control Regulations. They are necessary to monitor the tailpipe-out concentration of NOx in the exhaust to ensure the proper performance of the NOx control devices in the EAS. Thus, a NOx sensor is typically placed downstream of each NOx control unit, such as the TWC, the lean NOx trap (LNT), or the SCR system and ammonia slip catalyst (ASC). The final NOx sensor in the EAS provides a direct measurement of the NOx concentration leaving the EAS, which can then be compared with acceptable in-use limits for the engine or the vehicle. NOx sensors may also be placed upstream of the NOx control devices, especially LNT or SCR systems, to provide better understanding of the NOx conversion performance requirements and feed-forward control signals for control of these devices [6-4].

The NOx sensor includes a heated ceramic element, and the applied voltage is just enough to reliably dissociate NO_2 into NO and O_2, but not enough to dissociate NO into nitrogen (N_2) and O_2. Thus, the sensor measures total NOx as NO, and it cannot distinguish between NO and NO_2. The sensor signal for NOx is linear between 0 and 1,200 ppm of NOx in exhaust. Unfortunately, NOx sensors are also cross-sensitive to ammonia (NH_3).

6.1.4.2 **Ammonia Sensors:** Ammonia (NH_3) sensors are also of interest, especially since NH_3 is commonly used as the reducing agent for NOx in SCR systems and since NH_3 is increasingly regulated on and off cycle. NH_3 generates a cross-signal on the NOx sensor, so an early approach was to try to minimize NOx and NH_3 emissions by minimizing the sensor signal. But since the early 2000s when SCR systems became a more popular NOx control option for internal combustion engines, Delphi [6-8, 6-9] and others have been developing ways to measure the NH_3 in the exhaust directly. Lambert [6-10] describes work done to evaluate NH_3 sensors at Ford, including assessing the cross-sensitivity to NOx. One sensor evaluated had a desirably low cross-sensitivity to water (H_2O) as seen in the response curve in **Figure 6.4**.

The Delphi NH_3 sensor, for example, generates a signal that is proportional to the logarithm of the NH_3 concentration as follows:

$$\mathcal{E} \approx \frac{kT}{3e} \ln p_{NH_3} - \frac{kT}{4e} \ln p_{O_2} - \frac{kT}{2e} \ln p_{H_2O}, \tag{6.1}$$

where $\mathcal{E}$ is the electromotive force (EMF) from the sensor, k is the Boltzmann constant, T is the absolute temperature, and p_i is the partial pressure of species i [6-8]. Although the presence of O_2 and H_2O affect the EMF signal, their respective concentrations change with respect to combustion conditions in a way that the total confounding signal stays approximately constant.

With almost all SCR catalysts, NH_3 is stored on the catalyst surface and then consumed by NOx, as described in Chapter 12, Selective Catalytic Reduction Systems. Thus, there is an interest in knowing what the stored quantity of NH_3 in the SCR system is, especially to improve the control of the SCR system, optimize NH_3 use, and minimize NH_3 slip. With current technology, including the gaseous NH_3 sensor described above, it is possible to use a model to estimate the quantity stored. An alternate approach that is currently under development uses radio-frequency (RF) sensors to detect the stored

FIGURE 6.4 Voltage signal response for a given ammonia concentration. (Figure 23 in [6-10].)

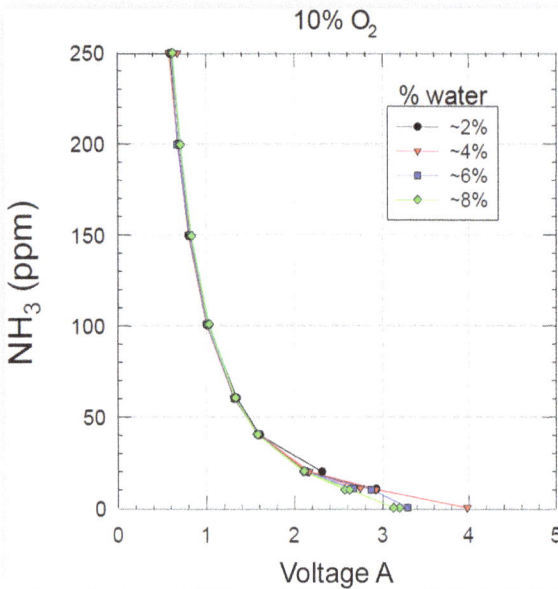

FIGURE 6.5 Corrected signal from RF ammonia sensor linearly maps to stored ammonia quantity [6-11].

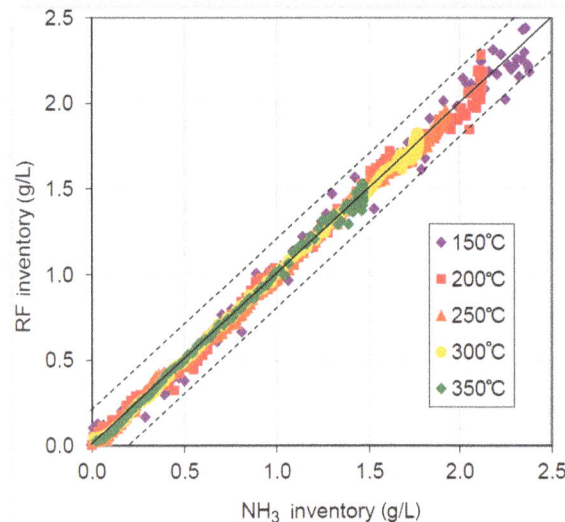

quantity of NH_3 directly. Such a system is currently under development by CTS Corp. [6-11, 6-12] with support from the US Department of Energy (DOE). CTS has managed to develop the RF sensor to the point where it provides a linear response that is proportional to the stored NH_3, as shown in **Figure 6.5**.

6.1.5 Hydrocarbon Sensors

Currently, there are no on-vehicle sensors available to measure hydrocarbons (HC), including total HC or non-methane HC (NMHC). There has been on-going research since 1990s toward this end, including work by Mukundan and colleagues at the Los Alamos National Laboratory [6-13, 6-14, 6-15], and others [6-16, 6-17]. These sensors tend to use ceria or zirconia stabilized metal oxides, but the response times are on the order of seconds, which is too slow to be useful as an on-vehicle sensor.

While a direct on-board HC sensor is still in development, there are alternative approaches from which one can infer the HC composition of the exhaust gas. The common approach uses the signal from the UEGO upstream of the TWC or other EAS components, and use that signal and a downstream measurement to estimate the HC species concentrations in the exhaust gas [6-18]. The UEGO is described further in Section 6.1.3.

6.1.6 Particulates Sensors

As discussed in Chapter 3, Emissions Control Regulations, there is an increased regulatory interest in controlling both particulate mass (PM) and particulate number (PN) emissions from engines, both compression ignition (CI) and SI. Currently, the only on-board sensors available for particulates measure mass, although there is a correlation between PM and PN for engines [6-19, 6-20, 6-21], especially from diesel-fueled CI engines, as shown in **Figure 6.6**.

PN measurements can be made on a vehicle using a portable emissions measurement system (PEMS). PEMS are described further in Section 5.4. There are currently no on-board sensors that directly measure PN.

For PF, the control and OBD requirements usually include a differential pressure sensor, described in Section 6.1.2, to estimate the PM loading in the PF through a model-based calculation based on pressure drop and flow through the device. A direct measurement of PM is also possible. Particulates concentration in the gas phase can be measured

FIGURE 6.6   Correlation between particulate mass and particle number [6-21].

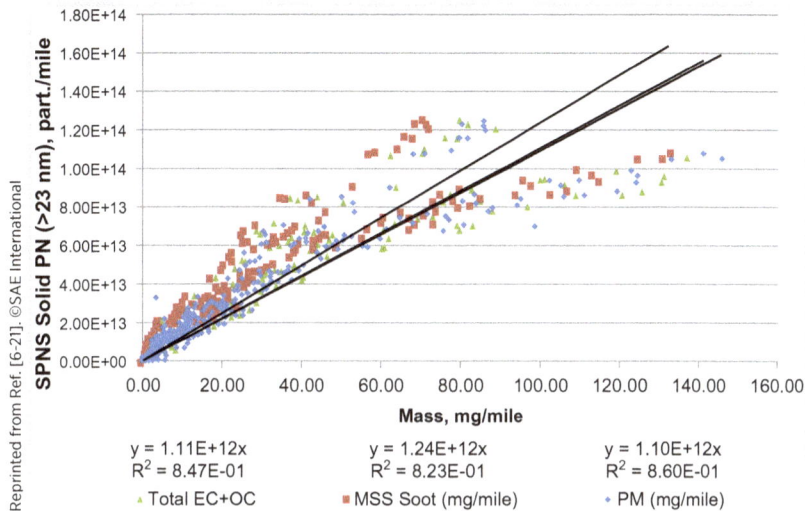

$y = 1.11E{+}12x$
$R^2 = 8.47E{-}01$
▲ Total EC+OC

$y = 1.24E{+}12x$
$R^2 = 8.23E{-}01$
■ MSS Soot (mg/mile)

$y = 1.10E{+}12x$
$R^2 = 8.60E{-}01$
◆ PM (mg/mile)

with a sensor that measures the decrease in electrical resistance as particulates land on the surface between two electrodes and create conductive paths between them [6-22, 6-23]. The electrodes are held at a constant voltage, so as the resistance drops the current through the sensor increases. The sensor also includes an integrated heating element to manage the temperature at which the measurements are made. An example of such a sensor from Delphi Technologies is shown in **Figure 6.7**.

When the sensor current reaches a certain electrical current threshold, it triggers a particulate regeneration event, both in the PF and on the sensor [6-4]. The sensor heating element raises the sensor temperature to about 600°C to burn the accumulated soot off and reset the sensor.

The particulate sensor is usually placed immediately downstream of the DPF so that the particulates concentration leaving the PF can be measured directly. Careful consideration is given to the shroud around the sensor element, since particulates are supposed to land on the sensor during its normal operation.

In addition to the resistive sensor placed in the exhaust gas, Sappok et al. [6-24], developed a method to use radio frequency (RF) measurements of a PF to determine the soot loading within the filter. Subsequent work reported by Sappok and Bromberg [6-25] also showed how the RF technique could measure ash in the PF. As of writing, this technology is still being developed for commercial use.

6.1.7 Urea–Water Solution Sensors

As described in Chapter 12, Selective Catalytic Reduction Systems, the urea–water solution used to convert NOx to N_2 is an important part of a properly functioning SCR system. The urea–water solution is known as Diesel Exhaust Fluid (DEF) in the United States (US) or AdBlue in the European Union (EU). For proper function of the SCR system, the urea–water solution tank must always have an appropriate amount of urea–water solution in it, the temperature of the DEF must be maintained above

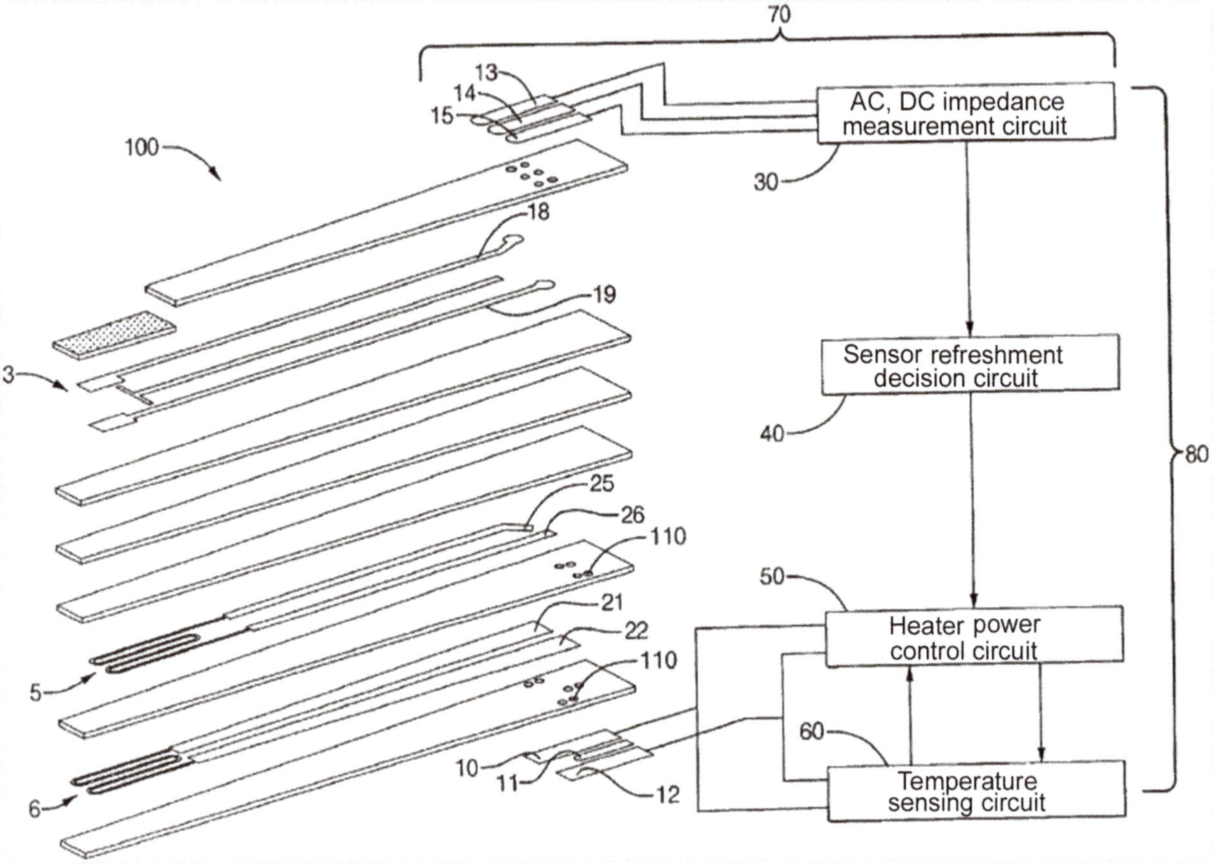

Reprinted from Ref [6-22]. US Patent 6,634,210 B1, 2003

FIGURE 6.7 Example particulates sensor using resistive signal [6-22].

solidification and that urea–water solution must have the correct urea $(CO(NH_2)_2)$ concentration of 32.5 wt%.

6.1.7.1 Urea–Water Solution Temperature Sensor:
The urea–water solution temperature in the tank needs to be monitored, since the urea–water solution freezes at –11°C (12°F). Also, knowing the temperature allows the ECU to calculate what the urea–water solution density should be. The temperature sensor is usually integrated with the other sensor functions into one unit [6-26].

6.1.7.2 Urea–Water Solution Level Sensor:
For proper function of the SCR system in the EAS, the urea–water solution tank must not be allowed to run empty. Also, as the level decreases through use, the ECU must start reminding the operator to refill the tank or start de-rating the engine to maintain emissions compliance. Thus, the urea–water solution tank needs a level sensor. The simplest type is a float sensor, where the float is attached to an arm that pivots as the UWS level changes. However, because the urea–water solution can freeze, other level sensor options, such as a pressure sensor at the base of the tank, may also be attractive. TE Connectivity offers a reed switch level sensor for DEF fluid level measurement that is integrated with the temperature measurement [6-26].

6.1.7.3 **Urea Quality Sensor:** Urea quality sensors ensure that the urea–water solution on the vehicle is at the correct urea concentration for use in the EAS. Common methods for measuring urea concentration in urea–water solution include measuring near infrared (Near IR) absorption spectra [6-26] or measuring the response to an ultrasonic signal [6-27]. For ultrasonic measurements, the rate of wave propagation in the solution increases with increasing urea concentration, which supports a reasonably rapid response [6-27]. The quality sensor is usually integrated with a temperature sensor, level sensor, or both [6-26, 6-27].

Continental AG, for example, has developed a line of urea concentration sensors using ultrasonic measurements for on-road heavy-duty vehicles (HDVs) [6-28], agricultural and other non-road applications [6-29], and LDVs [6-30]. These sensor modules integrate the urea quality sensor, level sensor, and temperature sensor into one package.

In contrast, the urea quality sensor from SUN-A Corporation measures the subtle changes of heat capacity as a function of urea concentration [6-31]. This quality sensor is integrated with a platinum (Pt) thermistor.

6.2 Model-Based Control and Feedback

The traditional approach to calibrating an engine and EAS for performance and emissions was to create a series of multi-dimensional maps for the ECU to use and interpolate within. However, as engines and EAS have increased in complexity, the effort to calibrate one engine and one actively managed EAS has grown to the point where it is barely manageable. When one factors in the additional effort to calibrate several variants of an engine and EAS, it becomes better use of time to develop a model-based controller.

In model-based control, the engineers develop a model of the engine and EAS that is implemented within the ECU. This model is designed to run in real time, or slightly faster than real time, and use measured inputs to take control action. The full operating space still needs to be sampled to develop and fit the model, but with model-based control, a more efficient test plan can be used for the engine calibration. This control approach is especially useful when calibrating multiple variants of the engine. Model-based control methods for EAS systems have been an active area of research for some time. Catalytic converter systems that have been the focus of study include, but are not limited to, the diesel oxidation catalyst (DOC) [6-32, 6-33] and the SCR system [6-34, 6-35, 6-36, 6-37]. Other examples of model-based control include predicting engine-out emissions such as NOx or PM. Models are also used to estimate NH_3 loading in a SCR system and soot loading in a DPF.

6.3 On-Board Diagnostics Requirements

The purpose of OBD requirements is to implement a system on a vehicle that monitors and reports on the vehicle's performance. Regulators such as the US Environmental Protection Agency (EPA), California Air Resources Board (CARB), EU, China, Japan, and India have been defining requirements to ensure that the engine and EAS remove pollutants from the exhaust during real use throughout the vehicle's FUL, as defined in the applicable regulation.

This section presents an overview of OBD requirements, including testing, but only at an introductory level of detail. OBD certification, including initial and follow-up testing requirements, are complicated and are treated in detail by Palocz–Andresen [6-38] and in several SAE Technical Papers from the SAE OBD Symposia and other meetings.

6.3.1 On-Board Diagnostics Performance and Testing

In general, the OBD systems need to function throughout the FUL, as defined by the governing regulations for the given vehicle or application. Thus, the OBD system monitors the tailpipe emissions levels to ensure that the emissions monitor threshold limits are not being exceeded. In addition, OBD system monitors the sensor and actuator performance, as a fault in their performance can also lead to a failure to meet the emissions monitor threshold limits. These limits are described in detail in Section 6.3.2.

The testing needed to demonstrate the OBD monitor performance and appropriate fault detections is extensive, especially for the US and California. There are tests before the engine or vehicle is approved for sale in order to demonstrate compliant OBD function. In this US, these are the diagnostic demonstration engine (DDE) tests, which must be completed once the engine and EAS calibrations are fixed for production. Then, once sales is started, the manufacturer needs to run regular tests on field-aged systems to confirm that the aging OBD monitors still function as required.

The details of what monitors and tests are required may be found in the appropriate regulations [6-39, 6-40, 6-41, 6-42, 6-43, 6-44, 6-45, 6-46, 6-47, 6-48, 6-49, 6-50, 6-51, 6-52].

6.3.2 On-Board Diagnostics Rules

Several jurisdictions have been implementing or strengthening their OBD regulatory requirements in recent years, especially following the Volkswagen emissions cheating revelations in September 2015.* There is a growing recognition by regulatory bodies that the in-use performance more strongly influences ambient air quality than the certification based test results do, especially given the gaps between the reproducible test conditions and the ways in which LDVs and HDVs are actually used. The following sections contain overviews of the regulatory OBD requirements, with an emphasis on criteria pollutant emissions including carbon monoxide (CO), NMHC, NOx, and PM.

6.3.2.1 **United States and California Light-Duty Vehicle On-Board Diagnostics:** In the US, OBD certification is generally led by CARB under California's OBD II standards for LDVs in Title 13, §1968.2 of the *California Code of Regulations* (13 CCR §1968.2) [6-39]. The multipliers for the LEV III OBD emissions monitor thresholds, or in-use not-to-exceed (NTE) limits, are shown in **Table 6.1**. In this case, the factors are multipliers, meaning that the on-cycle regulatory limit is multiplied by the corresponding factor. These monitors are intended to apply throughout the FUL of the vehicle. In addition,

---

* See the explanation in Chapter 3, Emissions Control Regulations.

TABLE 6.1 LEV III OBD emissions thresholds by bin [6-39].

Vehicle type	Vehicle emission category	Monitor thresholds		
		CO Mult.	NMHC + NOx Mult.	PM Mult.
All LDVs with GVWR ≤ 8,500 lb (3,856 kg)	LEV160	1.50	1.50*	2.00
	ULEV125			
	ULEV70		2.00	
	ULEV50			
	SULEV30	2.50	2.50	
	SULEV20			
MDVs with GVWR ≥ 8,501 lb (3,856 kg) and GVWR ≤ 10,000 lb (4,536 kg)	All categories	1.50	1.50*	—
MDVs with GVWR ≥ 10,001 lb (4,536 kg) and GVWR ≤ 14,000 lb (6,350 kg)	All categories	1.50	1.50*	2.00

* The diesel aftertreatment monitor threshold multipliers for NMHC+NOx in MDVs are 1.75.

several systems need to have their performance monitored by the OBD system, including the EAS components and other subsystems that may affect criteria pollutant emissions.

6.3.2.2 United States and California Heavy-Duty Vehicle On-Board Diagnostics: The US requirements for OBD are defined in Title 40, Part 86, §86.010-18 of the *Code of Federal Regulations* (40 CFR §86.010-18) [6-40]. For heavy-duty diesel engines (HDDE) from model year (MY) 2013 and on, the OBD emissions thresholds in Table 6.2 are applicable. A multiplicative threshold means that the threshold is a multiplier of the applicable emissions standard, *e.g.*, 2× for NMHC. Likewise, an additive threshold means that the threshold value is added to the applicable emissions standard, *e.g.*, +0.3 (g/hp·h) for NOx. For PM, the threshold is either the absolute level, *e.g.*, 0.05 (g/hp·h), or an additive threshold, *e.g.*, +0.04 (g/hp·h).

For heavy-duty engines in HDVs, the certification requirements in California are specified in 13 CCR §1956.8 [6-41], where HDDEs are covered by 13 CCR 1956.8(a). California OBD requirements for certification and later use are specified in 13 CCR §1971.1 [6-43].

For HDSI engines, the OBD emissions threshold is 1.75× for NMHC and NOx for the catalyst system, and 1.5× for CO, NMHC, and NOx for other monitors. The applicable

TABLE 6.2 EPA OBD emissions thresholds for HDDE [6-40].

Component	OBD thresholds			
	CO	NMHC	NOx	PM
NOx aftertreatment	—	—	+0.3	—
DPF	—	2×		0.05 or +0.04
AFR sensors upstream of EAS	2×	2×	+0.3	0.03 or +0.02
AFR sensors downstream of EAS	—	2×	+0.3	0.05 or +0.04
NOx sensors	—	—	+0.3	0.05 or +0.04
Other monitors	2×	2×	+0.3	0.03 or +0.02

TABLE 6.3 Euro 6 on-board diagnostics threshold limits (OTL) [6-44].

Vehicle category	Level	Reference mass (RM) (kg)	Limits									
			CO (mg/km)		NMHC (mg/km)		NOx (mg/km)		PM (mg/km)	
			SI	CI	SI	CI	SI	CI	SI*	CI
M	—	All	1,900	1,750	170	290	90	140	12	
N1	I	RM ≤ 1,305	1,900	1,750	170	290	90	140	12	
	II	1,305 < RM ≤ 1,760	3,400	2,200	225	320	110	180	12	
	III	1,760 < RM	4,300	2,500	270	350	120	220	12	
N2	—	All	4,300	2,500	270	350	120	220	12	

* The PM OTL for SI engines only applies when there is a direct injection fuel system.

California regulations for certification are in 13 CCR 1956.8(c) [6-42], but the OBD requirements are the same as for HDDE.

6.3.2.3 European Union Light-Duty Vehicle On-Board Diagnostics: The Euro 6 standards for LDVs includes OBD requirements for several required components, as described in Article 4 of Regulation (EU) 2017/1151 [6-44] as amended by Regulation (EU) 2018/1832 [6-45]. Table 6.3 shows the current OBD threshold limits (OTL) for LDVs under the Euro 6 scheme. A particulate number (PN) standard for OBD may be introduced at some point in the future.

6.3.2.4 European Union Heavy-Duty Vehicle On-Board Diagnostics: The Euro VI standards for on-road HD engines includes a set of OBD requirements for several required components, described in Annex 9A of ECE R49-06 [6-46]. This regulation also includes the associated OTL for NOx and either PM for HDDE or CO for HDSI engines, which are shown in Table 6.4. Additional in-use performance requirements (IUPR) are also defined in ECE R49-06 Annex 9A.

6.3.2.5 China Light-Duty Vehicle On-Board Diagnostics: In the People's Republic of China (PRC), Annex J of the China Phase 6 rules ("China 6") [6-47] requires an OBD system and associated IUPR for LDVs. For LDVs, the OBD system is supposed to monitor several engine and vehicle functions. The most relevant for this work are the OBD emission thresholds for CO, NMHC+NOx, and PM, as shown in Table 6.5. As with other OBD systems, fault codes are to be retained for monitoring purposes.

6.3.2.6 China Heavy-Duty Vehicle On-Board Diagnostics: In the PRC, Annex F of the China Phase VI rules ("China VI") [6-48] requires an OBD system and associated IUPR for HDVs. The OTL for HDVs are the same, as shown in Table 6.4. A key difference under China VI is that HDV OBD systems will need to store OBD data and periodically report back over the air to a government database for continuous in-use oversight by the government.

TABLE 6.4 European Union on-board diagnostics threshold limits for heavy-duty engines [6-46].

	OBD thresholds (mg/kW·h)		
	NOx	CO	PM
General requirement, CI	1,200	—	25
General requirement, SI	1,200	7,500	—

6.3.2.7 Japan Light-Duty Vehicle J-OBD: In Japan, the Post New Long Term Regulation of 2009 [6-49] and the

TABLE 6.5 China 6 OBD emission thresholds [6-47].

Vehicle category	Level	Test mass (TM) (kg)	Limits CO (mg/km)	NMHC+NOx (mg/km)	PM (mg/km)
I	—	All	1,900	260	12
II	I	TM ≤ 1,305	1,900	260	12
	II	1,305 <TM ≤ 1,760	3,400	335	12
	III	1,760 < TM	4,300	390	12

subsequent post Post-New Long Term Regulation of 2018–2019 [6-51] both require OBD, called J-OBD, on all new LDVs sold. Note that the post Post-New Long Term Regulation is still phasing in, with full implementation on all new vehicles sold applying as of 1 October 2021.

LDV J-OBD are defined by Attachment 48 to "Safety Regulations for Road Vehicles, Article 31" [6-50]. There are two applicable technical standards from the regulations. The first, J-OBD I, applies to CI engines in LDVs. The second, J-OBD II is for SI engines in LDVs that are fueled by gasoline or liquefied petroleum gas (LPG). The emission threshold values for J-OBD II are shown in Table 6.6; J-OBD I does not have emission threshold values.

6.3.2.8 India Light-Duty Vehicle On-Board Diagnostics: LDVs in India

that meet the Bharat Standard VI (BS VI) [6-52] must be equipped with an OBD system. BS VI defines OBD threshold values for LDVs with SI or CI engines that are manufactured after 1 April 2020, BS VI-1 OBD, or 1 April 2023, BS VI-2 OBD. The BS VI-1 OBD

TABLE 6.6 Japan J-OBD II emission threshold values [6-50].

Motor vehicle category	Limits CO (g/km) JC08	WLTC	NMHC (g/km) JC08	WLTC	NOx (g/km) JC08	WLTC
1. Passenger vehicles or LDVs	4.06	4.06	0.28	0.32	0.30	0.30
2. Mini-sized trucks	12.46	14.12	0.28	0.32	0.30	0.30
3. Medium-duty motor vehicles	14.28	8.96	0.28	0.46	0.30	0.41

TABLE 6.7 Bharat VI-1 OBD requirements [6-52].

Vehicle category	Level	Reference mass (RM) (kg)	OBD thresholds CO (mg/km) SI	CI	NMHC (mg/km) SI	CI	NOx (mg/km) SI	CI	PM (mg/km) SI*	CI
M	—	All	1,900	1,750	170	290	150	180	25	
N1	I	RM ≤ 1,305	1,900	1,750	170	290	150	180	25	
	II	1,305 < RM ≤ 1,760	3,400	2,200	225	320	190	220	25	
	III	1,760 < RM	4,300	2,500	270	350	210	280	30	
N2	—	All	4,300	2,500	270	350	210	280	30	

* The PM and PN limits for SI engines only apply when there is a direct injection fuel system.

TABLE 6.8 Bharat VI-2 OBD requirements [6-52].

Vehicle category	Level	Reference mass (RM) (kg)	CO (mg/km) SI	CO (mg/km) CI	NMHC (mg/km) SI	NMHC (mg/km) CI	NOx (mg/km) SI	NOx (mg/km) CI	PM (mg/km) SI*	PM (mg/km) CI
M	—	All	1,900	1,750	170	290	90	140	12	
N1	I	RM ≤ 1,305	1,900	1,750	170	290	90	140	12	
	II	1,305 < RM ≤ 1,760	3,400	2,200	225	320	110	180	12	
	III	1,760 < RM	4,300	2,500	270	350	120	220	12	
N2	—	All	4,300	2,500	270	350	120	220	12	

* The PM and PN limits for SI engines only apply when there is a direct injection fuel system.

threshold values are shown in Table 6.7, and the BS VI-2 OBD threshold values are shown in Table 6.8. The main difference between the two is that the NOx and PM thresholds decrease for both SI and CI engines with the BS VI-2 OBD.

References

6-1.   Montessori, M., *The Absorbent Mind* (Radford, VA: Wilder Publications, LLC, 1959).

6-2.   Welty, J.R., Wicks, C.E., and Wilson, R.E., *Fundamentals of Momentum, Heat, and Mass Transfer* 3rd ed. (New York: John Wiley & Sons, 1984).

6-3.   Oak Ridge National Laboratory, "User Facilities – National Transportation Research Center," https://www.ornl.gov/facility/ntrc/research-areas/fuels-engines-emissions. Accessed 15 Jul. 2019.

6-4.   Dietsche, K.-H. and Reif, K., *Automotive Handbook* 10th ed. (Karlsruhe, Germany: Robert Bosch, GmbH, 2018).

6-5.   Verdino, V., Lavazza, P., Rampone, D., Garzarella, L., Mogavero, A., Vassallo, A., Barta, J., Catalogna, J., Franco, S., Marlett, C., and Umierski, M., "The New General Motors 3.0 Liter Duramax Diesel Inline 6-Cylinder Engine for the 2019 Chevrolet Silverado and GMC Sierra," *40th International Vienna Motor Symposium*, Austria, 2019.

6-6.   Kuhlgatz, D., "Sensing the Adequate Mixture – The Bosch Lambda Sensor," https://blog.bosch.com/history/en/2016/07/20/939/. Updated 20 Jul. 2016, Accessed 20 May 2019.

6-7.   Bowling, B. and Grippo, A., "Calorimetric Hydrocarbon Gas Sensor," US20110218726A1, Sep. 8, 2011.

6-8.   Wang, D.Y., Yao, S., Cabush, D., and Racine, D., "Ammonia Sensor for SCR NOx Reduction," *DEER 2007*, Detroit, 2007.

6-9.   Wang, D.Y., Yao, S., Shost, M., Yoo, J.-H., Cabush, D., Racine, D., Cloudt, R., and Willems, F., "Ammonia Sensor for Closed-Loop SCR Control," *SAE Int. J. Passeng. Cars – Electron. Electr. Systems* 1, no. (1) 2009: 323–333, https://doi.org/10.4271/2008-01-0919.

6-10.   Lambert, C., "Advanced CIDI Emission Control System Development," Report No. 898797, 2006, https://doi.org/10.2172/898797.

6-11.   Ragaller, P., Sappok, A., Bromberg, L., and Pihl, J., "Real-Time Catalyst Monitoring and Diagnostics Using Radio Frequency Sensors," *CLEERS 2017*, Ann Arbor, MI, 2017.

6-12.   Ragaller, P., "On-Board Measurements of Ammonia Storage on SCR and Combined SCR Filter Systems Using Radio Frequency Sensors," *CLEERS 2018*, Ann Arbor, MI, 2018.

6-13.   Mukundan, R., Brosha, E.L., Brown, D.R., and Garzon, F.H., "Ceria-Electrolyte-Based Mixed Potential Sensors for the Detection of Hydrocarbons and Carbon Monoxide," *Electrochemical and Solid-State Letters* **2**, no. 8 (1999): 412–414, https://doi.org/10.1149/1.1390855.

6-14.   Mukundan, R., Brosha, E.L., and Garzon, F.H., "Mixed Potential Hydrocarbon Sensors Based on a YSZ Electrolyte and Oxide Electrodes," *Journal of The Electrochemical Society* **150**, no. 12 (2003): H279–H284, https://doi.org/10.1149/1.1621880.

6-15.   Kreller, C., Mukundan, R., and Brosha, E.L., "Evaluation of HC, NO_x and NH_3 Mixed-Potential Sensors in Diesel and Gasoline Engine Exhaust," *2016 CLEERS*, 2016.

6-16.   Hibino, T., Tanimoto, S., Kakimoto, S., and Sano, M., "High-Temperature Hydrocarbon Sensors Based on a Stabilized Zirconia Electrolyte and Metal Oxide Electrodes," *Electrochemical and Solid-State Letters* **2**, no. 12 (1999): 651–653, https://doi.org/10.1149/1.1390937.

6-17.   Young, D.A., Moya, A., Koripella, C.R., Naber, J., Adams, N., Markyvech, C., and Miller, A., "Calorimetric Hydrocarbon Gas Sensor," US5989398A, Nov. 23, 1999.

6-18.   Fiengo, G., Grizzle, J.W., Cook, J.A., and Karnik, A.Y., "Dual-UEGO Active Catalyst Control for Emissions Reduction: Design and Experimental Validation," *IEEE Transactions on Control Systems Technology* **13**, no. 5 (2005): 722–736, https://doi.org/10.1109/TCST.2005.854326.

6-19.   Giechaskiel, B., Cresnoverh, M., Jörgl, H., and Bergmann, A., "Calibration and Accuracy of a Particle Number Measurement System," *Measurement Science and Technology* **21**, no. 4 (2010): 045102, https://doi.org/10.1088/0957-0233/21/4/045102.

6-20.   Giechaskiel, B., Mamakos, A., Andersson, J., Dilara, P., Martini, G., Schindler, W., and Bergmann, A., "Measurement of Automotive Nonvolatile Particle Number Emissions within the European Legislative Framework: A Review," *Aerosol Science and Technology* **46**, no. 7 (2012): 719–749, https://doi.org/10.1080/02786826.2012.661103.

6-21.   Premnath, V., Khalek, I.A., and Morgan, P., "Relationship among Various Particle Characterization Metrics Using GDI Engine Based Light-Duty Vehicles," SAE Technical Paper 2018-01-0353, 2018, https://doi.org/10.4271/2018-01-0353.

6-22.   Bosch, R.H. and Wang, D.Y., "Particulate Sensor System," US Patent 6,634,210 B1, Oct. 21, 2003.

6-23.   Zhang, X., "Method and System for Exhaust Particulate Matter Sensing," US Patent 10,078,043 B2, Sep. 18, 2018.

6-24.   Sappok, A., Bromberg, L., Parks, J.E., and Prikhodko, V., "Loading and Regeneration Analysis of a Diesel Particulate Filter with a Radio Frequency-Based Sensor," SAE Technical Paper 2010-01-2126, 2010, https://doi.org/10.4271/2010-01-2126.

6-25.   Sappok, A. and Bromberg, L., "Radio Frequency Diesel Particulate Filter Soot and Ash Level Sensors: Enabling Adaptive Controls for Heavy-Duty Diesel Applications," *SAE Int. J. Commer. Veh.* **7**(2): 468–477, 2014, https://doi.org/10.4271/2014-01-2349.

6-26.   TE Connectivity, "Fluid Property Sensors," TE Connectivity, 2016.

6-27.   Continental AG, "Continental Urea Sensor Helps to Make Diesel Engines Cleaner," https://www.continental-corporation.com/en/press/press-releases/2016-02-04-urea-sensor-99102. Accessed 24 May 2019.

6-28.   Continental AG, "Urea Ultrasonic Level & Concentration Sensor," https://www.continental-automotive.com/en-gl/Trucks-Buses/Powertrain/Diesel-Vehicles/Exhaust-

Management-Aftertreatment/Sensors/Urea-Ultrasonic-Level-Concentration-Sensor. Accessed 24 May 2019.

6-29.   Continental AG, "Fluid Sensor – Urea Concentration & Level," https://www.continental-automotive.com/en-gl/Agriculture/Powertrain/Exhaust-Management-After-treatment/Sensors/Fluid-Sensor-Urea-Concentration-Level. Accessed 24 May 2019.

6-30.   Continental AG, "Fluid Sensor – Urea Concentration & Level," https://www.continental-automotive.com/en-gl/Passenger-Cars/Powertrain/Diesel-Technology/Exhaust-Aftertreatment/Fluid-Sensor-%E2%80%93-Urea-Concentration-Level. Accessed 24 May 2019.

6-31.   SUN-A Corporation, "Urea Quality Sensor," http://www.sun-awks.co.jp/products_en/uqs/. Accessed 24 May 2019.

6-32.   Lepreux, O., Creff, Y., and Petit, N., "Model-Based Temperature Control of a Diesel Oxidation Catalyst," *Journal of Process Control* **22**, no. 1 (2012): 41–50, https://doi.org/10.1016/j.jprocont.2011.10.012.

6-33.   Kim, Y.-W., Van Nieuwstadt, M., Stewart, G., and Pekar, J., "Model Predictive Control of DOC Temperature during DPF Regeneration," SAE Technical Paper 2014-01-1165, 2014, https://doi.org/10.4271/2014-01-1165.

6-34.   Song, Q. and Zhu, G., "Model-Based Closed-Loop Control of Urea SCR Exhaust Aftertreatment System for Diesel Engine," SAE Technical Paper 2002-01-0287, 2002, https://doi.org/10.4271/2002-01-0287.

6-35.   Ericson, C., Westerberg, B., Andersson, M., and Egnell, R., "Modelling Diesel Engine Combustion and NOx Formation for Model Based Control and Simulation of Engine and Exhaust Aftertreatment Systems," SAE Technical Paper 2006-01-0687, 2006, https://doi.org/10.4271/2006-01-0687.

6-36.   Devarakonda, M., Parker, G., Johnson, J.H., Strots, V., and Santhanam, S., "Model-Based Estimation and Control System Development in a Urea-SCR Aftertreatment System," *SAE Int. J. Fuels Lubr.* **1**(1): 646–661, 2008, https://doi.org/10.4271/2008-01-1324.

6-37.   Hsieh, M.-F. and Wang, J., "Development and Experimental Studies of a Control-Oriented SCR Model for a Two-Catalyst Urea-SCR System," *Control Engineering Practice* **19**, no. 4 (2011): 409–422, https://doi.org/10.1016/j.conengprac.2011.01.004.

6-38.   Palocz-Andresen, M., *Onboard Diagnostics and Measurement in the Automotive Industry, Shipbuilding, and Aircraft Construction* 1st ed (Warrendale, PA: SAE International, 2012).

6-39.   State of California, "Malfunction and Diagnostic System Requirements for 2004 and Subsequent Model-Year Passenger Cars, Light-Duty Trucks, and Medium-Duty Vehicles and Engines (OBD II)," *California Code of Regulations, Title 13* (2016): 1–126, https://www.arb.ca.gov/msprog/obdprog/obdregs.htm.

6-40.   Environmental Protection Agency, "§ 86.010-18 On-Board Diagnostics for Engines Used in Applications Greater than 14,000 pounds GVWR," *Code of Federal Regulations, Title 40*, Vol. **21** (2018): 483–551, https://www.govinfo.gov/app/collection/cfr/2018/title40/chapterI/subchapterC/part86.

6-41.   State of California, "California Exhaust Emission Standards and Test Procedures for 2004 and Subsequent Model Heavy-Duty Diesel Engines and Vehicles," *California Code of Regulations, Title 13* (2018): 1–85, https://ww2.arb.ca.gov/resources/documents/road-heavy-duty-certification-program-california-test-procedures-diesel-engines.

6-42.   State of California, "California Exhaust Emission Standards and Test Procedures for 2004 and Subsequent Model Heavy-Duty Otto-Cycle Engines and Vehicles," *California Code of Regulations, Title 13* (2018): 1–44, https://ww2.arb.ca.gov/resources/documents/road-heavy-duty-certification-program-california-test-procedures-otto-engines.

6-43.   State of California, "On-Board Diagnostic System Requirements—2010 and Subsequent Model-Year Heavy-Duty Engines," *California Code of Regulations, Title 13* (2016): 1–126, https://govt.westlaw.com/calregs/Document/I448CD58FC0F74FB685D306A21545FC30? viewType=FullText&originationContext=documenttoc&transitionType= CategoryPageItem&contextData=(sc.Default).

6-44.   European Commission, "Regulation (EU) 2017/1151 of 1 June 2017 supplementing Regulation (EC) No 715/2007 of the European Parliament and of the Council on type-approval of motor vehicles with respect to emissions from light passenger and commercial vehicles (Euro 5 and Euro 6) and on access to vehicle repair and maintenance information, amending Directive 2007/46/EC of the European Parliament and of the Council, Commission Regulation (EC) No 692/2008 and Commission Regulation (EU) No 1230/2012 and repealing Commission Regulation (EC) No 692/2008," *Official Journal of the European Union*, Vol. **175** (2017): 1–643, https://eur-lex. europa.eu/legal-content/EN/TXT/?uri=CELEX%3A32017R1151.

6-45.   European Commission, "Regulation (EU) 2018/1832 of 5 November 2018 amending Directive 2007/46/EC of the European Parliament and of the Council, Commission Regulation (EC) No 692/2008 and Commission Regulation (EU) 2017/1151 for the purpose of improving the emission type approval tests and procedures for light passenger and commercial vehicles, including those for in-service conformity and real-driving emissions and introducing devices for monitoring the consumption of fuel and electric energy," *Official Journal of the European Union*, Vol. **301** (2018): 1–314.

6-46.   United Nations, "Agreement concerning the adoption of uniform technical prescriptions for wheeled vehicles, equipment, and parts which can be fitted and/or be used on wheeled vehicles and the conditions for reciprocal recognition of approvals granted on the basis of these prescriptions," *Addendum 48: Regulation No. 49, Revision 6* (Mar. 6, 2013): 1–434.

6-47.   Ministry of Ecology and Environment of China, "Limits and Measurement Methods for Emissions from Light-Duty Vehicles (China 6)," *GB 18352.6-2016*, 2016.

6-48.   Ministry of Ecology and Environment of China, "Limits and Measurement Methods for Emissions from Diesel-Fueled Heavy-Duty Vehicles (CHINA VI)," *GB 17691-2018*, 2018.

6-49.   Ministry of Environment, "Post New Long-Term Standards," 2009.

6-50.   Ministry of Land, Infrastructure, Transport, and Tourism (MLIT), "Safety Regulations for Road Vehicles, Article 31 (Emission Control Device), Attachment 48 Technical Standard for On-Board Diagnostic (OBD) System for Exhaust Emission Control Devices for Motor Vehicles," 2016.

6-51.   Road Transport Bureau, Ministry of Land, Infrastructure, Transport, and Tourism (MLIT), "Partial amendments of the 'Announcement that prescribes Details of Safety Regulations for Road Transport Vehicles', etc.," 2016.

6-52.   Ministry of Road Transport and Highways, "Notification No. G.S.R. 889(E), New Delhi, the 16th September, 2016, regarding mass emission standards for BS-VI," *G.S.R. 889(E)*, 2016, http://egazette.nic.in/WriteReadData/2016/171776.pdf.

Exhaust Aftertreatment Systems Overview

Cleanliness and order are not matters of instinct; they are matters of education, and like most great things, you must cultivate a taste for them.

—Benjamin Disraeli

Exhaust aftertreatment systems (EAS)* for internal combustion engines (ICEs) have been in development for over 60 years, with the first public papers being presented in 1957 by Cannon et al. [7-1] of Ford and in 1959 by Nebel and Bishop [7-2] of General Motors. As noted by Keenan [7-3], EAS have made great strides over this time. The earliest catalytic converters that were brought into production in the 1970s used base metal catalysts loaded onto pellet supports to oxidize carbon monoxide (CO) and hydrocarbons (HCs) into carbon dioxide (CO_2) and water (H_2O). Modern EAS that control CO, HC, nitrogen oxides (NOx), particulate matter (PM), and particulate number (PN) can include a half-dozen separate catalytic converters as well as sensors, injectors, and computer controls. Over this period of years, engineers have found ways to improve the efficiency and efficacy of catalytic converters, which translates into lower component and operating costs for the EAS overall.

In this chapter, we introduce an overview of the function of EAS used in both on-road and off-road applications and several related concepts on aftertreatment chemistry and catalytic converters. Background information on the catalytic converter chemistry is discussed in Section 7.2, and the transport phenomena encountered in the EAS are discussed

---

* Note that some manufacturers and suppliers abbreviate "Exhaust aftertreatment system" as EATS, or omit "exhaust" and refer to the aftertreatment system, or ATS. We have chosen to use "EAS" throughout this work.

in Section 7.3. Common issues in catalytic converter design are discussed in Section 7.4, including considerations of EAS

- Layout and packaging
- Exhaust gas flow and pressure drop
- Thermal management

Lastly, Section 7.5 presents several examples of EAS that integrate combinations of the catalytic converters described in Chapters 8 through 15. Stoichiometric spark-ignition (SI) engines generally use three-way catalysts (TWCs), but treating the exhaust from engines that run in a fuel-lean mode, such as diesel-fueled compression ignition (CI) engines, is more complicated. Several devices may be required in the EAS to treat all the criteria pollutants as required, as illustrated in the examples in Section 7.5. The specific EAS catalytic converters discussed in this book include the following:

- Oxidation catalysts (oxycats), described in Chapter 8
- Three-way catalysts (TWCs), in Chapter 9
- Lean NOx traps (LNTs), in Chapter 10
- Particulate filters (PFs), in Chapter 11
- Selective catalytic reduction (SCR) systems, in Chapter 12
- Ammonia slip catalysts (ASCs), in Chapter 13
- Passive aftertreatment systems, in Chapter 14
- Combined devices and systems, in Chapter 15.

The passive aftertreatment systems in Chapter 14 include both passive NOx adsorbers (PNAs) and HC traps. The combined devices and systems in Chapter 15 include systems for use with either SI or CI engines.

7.1 Overview of EAS Function

The fundamental purpose of the overall EAS is to take the criteria pollutants emitted by the engine, including CO, HC, NOx, and PM, and convert them into less noxious species in the exhaust gas, such as CO_2, water, and nitrogen (N_2), before the gas leaves the tailpipe. The EAS function must meet the regulatory requirements described in Chapter 3 for the target market and jurisdiction for the engine or vehicle or both. The EAS must work both on the regulatory cycles described in Chapter 4, Emissions Testing Protocols, and in typical real-world use. Additional requirements for the EAS include the following:

- Work over the regulated full useful life (FUL) for the target application and market
- Survive the operating environment imposed by the engine and application
- Maximize selectivity for preferred product species (CO_2, N_2, water)
- Minimize selectivity for by-product species, especially pollutants such as nitrous oxide (N_2O)

Section 7.2 discusses the concepts related to chemistry in a catalytic converter. For example, as described in Section 7.2.1, most catalytic converters are chemical reactors that use heterogeneous catalysts. Therefore, their pollutant-removal function is influenced by the chemistry on the catalyst and by the transport phenomena (fluid dynamics and heat and mass transfer) within the reactor system. The chemical reaction rate on the catalyst is a function of the catalyst activity, the catalyst temperature, and the concentrations of the reactants and products at the catalyst.

In addition to the chemistry, the catalytic converter performance can be affected by the transport phenomena, which are described in Section 7.3. These include the gas flow, heat transfer, and mass transfer within the catalytic converter and connecting piping. The space velocity (SV) of the gas through the catalytic converter, gas temperature, and physical arrangement of the reactor are all factors that influence the transport phenomena. Depending on the state of the exhaust gas at a given operating point, the chemistry can be limited by the chemical thermodynamics, the chemical reaction rate, or the transport phenomena within the reactor.

7.2 Catalytic Chemistry Concepts

The main part of the EAS function is carried out by the catalytic converters within it, which facilitate certain types of chemistry to convert the pollutants into the preferred products. This section provides background information on catalytic converters, chemistry, and catalysts that will be referenced throughout Chapters 8 through 15.

7.2.1 Catalytic Converters

A catalytic converter is a general term for any emissions control device in the EAS that uses a catalyst to facilitate the chemical conversion of criteria pollutants, described in Section 1.2.2, and some greenhouse gas (GHG), such as methane or N_2O, into less harmful species. The colloquial use of the term "catalytic converter" has typically referred to the gasoline TWC, which was the first deployed automotive catalytic converter and is the most common type. Nevertheless, the other catalytic devices used in an EAS, such as an oxycat or SCR system, will also be described as catalytic converters in this book.

Catalytic converters are usually designed to support reactions for a limited set of chemical compounds present in the exhaust being treated but have the common components described below. These components need to be sufficiently durable in the operating environment of an exhaust system, which is hot, humid, and subject to vibration from the engine and the vehicle. The various constituents of a catalytic converter, shown in **Figure 7.1**, are described in the following subsections, from the canning into the washcoat and catalyst.

The various parts of a catalytic converter system are described in the following subsections.

7.2.1.1 **Catalyst:** A catalyst is a chemical compound that facilitates or speeds up the rate of chemical reaction at a given temperature, without being changed or consumed by the process. The catalyst has several active sites where other chemical species interact with the catalyst and with each other. As illustrated in **Figure 7.2**, a catalyst actively participates in the chemical reaction, and some ways of describing the chemistry can consider the active sites both reactants and products of the reactions.

FIGURE 7.1   Schematic of a catalytic converter, showing substrate, washcoat, insulation, and canning.

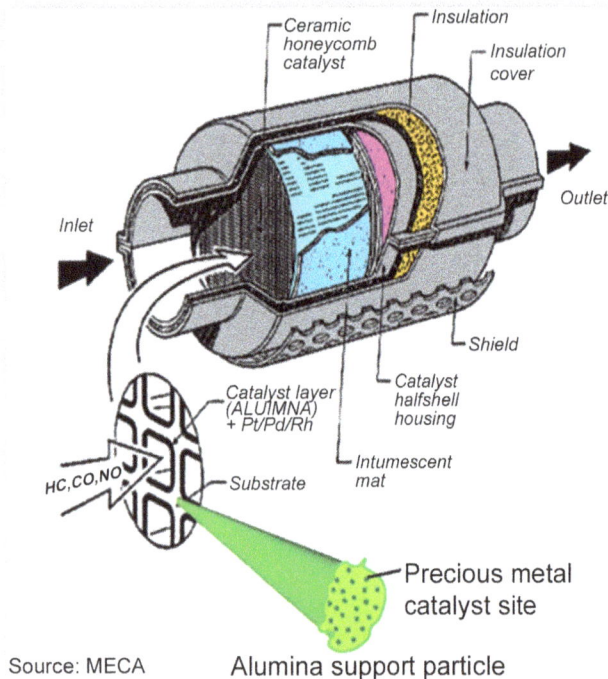

Source: MECA

FIGURE 7.2   Chemical reaction on catalyst.

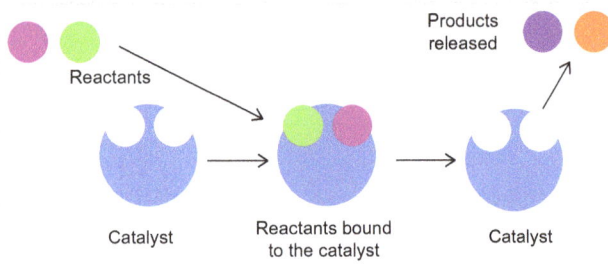

The catalysts used in the EAS for emissions control are heterogeneous, as they are solid compounds being used to facilitate the reactions of gaseous species. Most catalysts used in catalytic converters are platinum group metals (PGMs), which include platinum (Pt), palladium (Pd), and rhodium (Rh). The mass of catalyst in a given catalytic converter is called the loading, and is usually on a specific volume basis with units of catalyst mass per unit substrate volume (g/L or g/ft^3). The catalyst composition and loading used in a particular EAS device depend on several factors, including emission standards, engine-out emission concentration, catalyst cost, and catalyst effectiveness. PGM catalysts are expensive commodities with volatile prices. For example, as of writing (May 2020), the 12-month average prices for the three main PGM catalysts are approximately \$28.19/g (\$877/tr.oz.) for Pt, \$60.19/g (\$1,872/tr.oz.) for Pd, and \$210.90/g (\$6,559/tr.oz.) for Rh, where one troy ounce is 31.10 g. Thus, engineers have a strong incentive to get the desired device performance with the least amount of PGM catalysts. Zeolite catalysts are also commonly used in catalytic converters, and although they are engineered ceramic materials, they are considerably less expensive than PGM catalysts. Zeolites are discussed in more detail in Section 7.2.5.

As is suggested in the illustration in Figure 7.2, it is the surface of the heterogeneous catalyst that provides the catalyst function. Thus, a large active surface area is desired. This active area is related to the overall catalyst loading but is also a function of how finely dispersed the catalyst can be since many small catalyst particles have more surface area for a given loading than a lesser number of larger particles. Part of the challenge of developing an effective catalyst for a catalytic converter is maintaining that active area over the device's design life. The washcoat, described in Section 7.2.1.2, can stabilize the catalyst and slow the loss of active surface area.

Thus, the preference is for finely dispersed catalyst particles in the washcoat, as shown in **Figure 7.3(a)**. **Figure 7.3(b)**, by contrast, shows how the catalysts migrate and form fewer and larger particles, so even if the mass stays the same, the surface area decreases.

7.2.1.2 **Washcoat:** The washcoat is a mixture of compounds that support the catalyst and usually enhance its performance. It helps deliver and bind the catalyst to the substrate, and also serves as a carrier for the catalyst that helps in maintaining a high specific surface area and provides thermal stability for the catalyst [7-5, 7-6]. The washcoat support is a bulk material, such as γ-alumina (γ-Al_2O_3), which is the matrix that supports the other compounds. These other compounds include promoters such as ceria (cerium oxide, CeO_2), zirconia (zirconium oxide, ZrO_2), or lanthanum oxide (La_2O_3). Each of these promoters provides an oxygen storage capacity (OSC) that can improve the function and durability of the catalyst. Nickel oxide (NiO) is often included in the washcoat to suppress hydrogen sulfide (H_2S) formation.

There are two commonly used methods to prepare supported catalysts in a washcoat material: impregnation and co-precipitation. Catalyst impregnation occurs when a suspension of the solid support is treated with a pre-catalyst solution, often a metal salt then activated to convert the pre-catalyst to its active, usually metal, state. Co-precipitation involves using a homogeneous solution, such as an acidic solution of aluminum salt and pre-catalyst along with a base to precipitate the mixed-metal hydroxide, which is subsequently heated to a high temperature in air or oxygen, a process called calcination. There are three common mechanisms for co-precipitation: inclusion, occlusion, and adsorption, which is used for coating substrates. Impregnation is generally the preferred method for catalyzing pellets and commonly applied via slurry for substrates. The PGM catalysts are impregnated into the highly porous alumina washcoat that is applied to the channels of the substrate in a layer between 20 and 40 μm thick [7-7]. The washcoat is primarily Al_2O_3 with a specific surface area of 150–200 m^2/g, and represents 5%–15% of the total monolith mass.

The simplest configuration has one washcoat layer that covers (or glazes) the substrate wall throughout the length of the channel. As manufacturers have become more sophisticated, it is now possible to have so-called zone coatings of washcoats, where, for example, the front section has a different thickness or catalyst or formulation than the back section does to provide multiple functions within one unit. It is also possible to place washcoat layers over each other to combine functions throughout the unit. These different ways of combining washcoats within one monolith are discussed further in Section 15.1.

FIGURE 7.3 Images of (a) fresh, well-dispersed catalyst and (b) aged, lower-area catalyst [7-4].

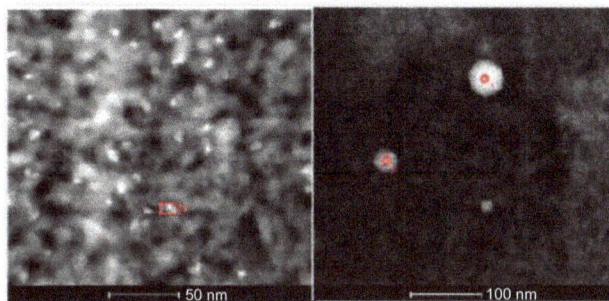

50 nm   100 nm

Reprinted from Ref. [7-4] © SAE International

7.2.1.3 **Substrate:** The substrate provides the structural support for the washcoat and catalyst in the catalytic converter. Most modern catalytic converters have a monolith substrate, like the ones shown in **Figure 7.4**, which have a series of small channels running in parallel in the direction of flow. The washcoat forms a layer on the channel walls within the substrate, and thus the surface of the washcoat layer is the interface between the exhaust gas flowing in the channel and the washcoat and the catalyst it contains.

A metric for monoliths is the cell density, or the number of channels per unit frontal area, which is measured in cells per square centimeter (cell/cm^2) or cells per square inch (cpsi). The earliest monoliths for TWC had cell densities of approximately 41.9 cell/cm^2 (270 cpsi), whereas modern monoliths have cell densities of 93.0 cell/cm^2 (600 cpsi) to 186 cell/cm^2 (1,200 cpsi) [7-6, 7-7]. Another area of refinement for monoliths is wall thickness. Thinner walls mean more area for flow and less obstruction of the flow through

FIGURE 7.4 Substrates for catalytic converters: (a) cordierite (single brick extrusion) [7-8] and (b) silicon carbide (SiC, segmented brick) [7-9].

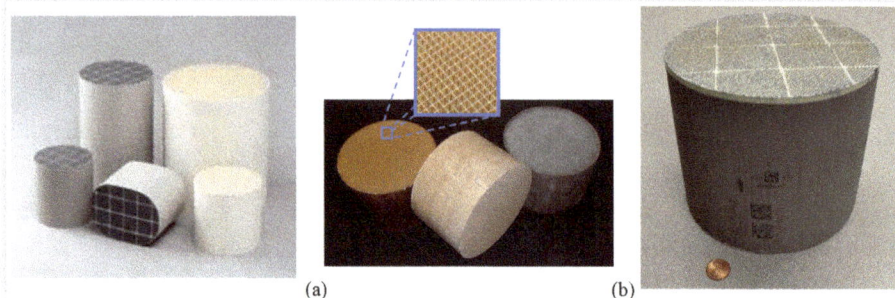
(a)　　　　(b)

Reprinted from Ref. [7-8]. Oak Ridge National Laboratory; Reprinted from Ref. [7-9]. Stewart, M.L., Kamp, C.J., Gao, F., Wang, Y., and Engelhard, M.H., "Coating Distribution in a Commercial SCR Filter." Emission Control Science and Technology 4(4):260–270, 2018. Licensed under Creative Commons Attribution 4.0 International License (http://creativecommons.org/licenses/by/4.0/)

the catalytic converter, which translates to less pressure drop over the monolith. Modern monolith supports have wall thicknesses of approximately 0.10–0.15 mm and square cross-sectional channels with a hydraulic diameter of approximately 1 mm [7-7].

Monoliths have undergone continuous improvement since they were first developed for catalytic converters. Design targets for the substrate include the following [7-6, 7-10]:

- High surface area
- Low thermal mass and heat capacity
- High thermal stability in the hot exhaust
- Low coefficient of thermal expansion
- Compatibility with the washcoat
- Ability to help washcoat adhere
- Mechanical strength
- Oxidative resistance

Common substrate materials for catalytic converters include metal foils or ceramics. The two most common ceramics for substrates are silicon carbide (SiC) and cordierite, a magnesium-alumino-silicate compound ($[MgO]_2[Al_2O_3]_2[SiO_2]_5$). Cordierite substrates used for automotive applications are extruded in one piece, known as a monolith, of up to 30 cm (12 in.) in diameter. The monolith has channels that extend the length of the piece. The monolith structure resists mechanical degradation from the vibrational loads that arise in catalytic converters. Cordierite porosity is reaction formed, as ground rice hulls are mixed into the ceramic matrix as a negative template.

SiC has a higher melting point than cordierite but is not as thermally stable. SiC monoliths are made by extruding segments that have the desired length, but are approximately 7.5 cm by 7.5 cm (3 in. by 3 in.) in cross-section. These segments are then cemented together, as shown in Figure 7.4(b). The substrate structure and porosity are dependent on the material of construction. The SiC substrate is made by sintering granules, which creates porosity in the material similar to a packed bed.

7.2.1.4 **Canning:** The canning or can serves as the packaging for the catalytic converter. The canning immobilizes the catalyst-coated ceramic substrate and protects it from vibration, insulates the monolith, connects the catalytic converter to the EAS or the vehicle, and delivers uniform flow to the inlet face of the catalyst-coated substrate. The canning includes the alloy steel housing for application to the vehicle and an insulating layer, called matting, placed between the metal housing and the catalyst substrate or brick. The matting is compressed during the canning process to keep the exhaust flow from bypassing the monolith and to firmly secure the monolith within the housing. The insulating layer is usually a ceramic fiber mat that is wrapped around the substrate and has a dual function of providing thermal insulation and some protection against vibrational loads.

The canning also includes the metal ductwork or piping needed to connect the catalytic converter to the rest of the EAS. The substrate is often wider in diameter than the exhaust pipe, and so the canning includes divergent inlet and convergent exit cone sections. To ensure uniform flow delivery, baffles or similar features can be included in the canning to improve the uniformity of gas flow distribution to the catalytic converter.

Several styles of catalytic converter cans are used, depending on the application. The most common is a one-piece stuffed can that can be integrated into a wide variety of applications. A clamshell-style is used in research applications when the substrate needs to be removed or exchanged. For the one-piece stuffed can, the canning process must account for tolerances in the can, mat, and substrate materials. To achieve the highest durability, the measure-stuff-calibration (MSC) process is used, as shown in **Figure 7.5**. This process includes measuring the diameter of the substrate and the mass of the mat material, soft-stuffing the mat-wrapped substrate into a slightly oversized shell and then calibrating by compressing the shell to reduce the diameter to compress the mat to the desired level.

7.2.2 Chemical Reactions

Chemical reactions are at the heart of EAS technologies. Chemical species called reactants are combined or exchanged as they are turned into products. These products are formed because they are thermodynamically favorable, but also because the kinetics—the rates of a chemical reaction—are fast enough for the reaction to occur in a useful amount of time. A common way to calculate reaction rate constants is the Arrhenius rate expression,

$$k = k_0 \exp\left(\frac{-E_a}{RT}\right) \tag{7.1}$$

where k is the rate constant, k_0 is the pre-exponential coefficient or frequency factor, E_a is the activation energy, R is the ideal gas law constant, and T is the absolute temperature [7-11]. The activation energy E_a is particularly important, since it indicates how rapidly the reaction will proceed.

Since catalytic converters have heterogeneous catalysts, overall rate expressions that account for chemical species on the catalyst surface, such as Eley–Rideal or Langmuir–Hinshelwood* mechanisms, are common.

In catalytic converters, the reactants include the criteria pollutants to be removed from the exhaust

FIGURE 7.5 Illustration of the measure-stuff-calibration (MSC) process.

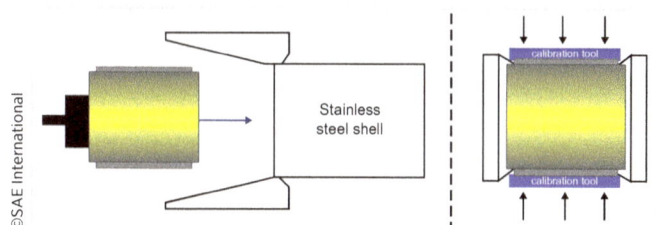

© SAE International

---

* As graduates of the University of Wisconsin, the authors are obligated to note that Langmuir–Hinshelwood mechanisms are also called Hougen–Watson mechanisms.

gas, and the desired products are typically CO_2, water, and nitrogen gas (N_2). Side products or byproducts are also possible, such as N_2O being formed in the catalytic converters being used to remove NOx.

In addition to the chemical kinetics, chemical thermodynamics also govern chemical reactions. The thermodynamic favorability of a reaction is indicated by the change in Gibbs free energy,

$$\Delta G = \Delta H - T\,\Delta S, \tag{7.2}$$

where ΔG is the change in Gibbs free energy from the reaction, ΔH is the change in enthalpy, T is the absolute temperature at which the reaction occurs, and ΔS is the change in entropy. The more negative ΔG is at a given temperature T, the more favorable the forward direction is for the chemical reaction.

7.2.2.1 Stoichiometry: The stoichiometry of a chemical reaction describes the molar proportion between reactants and products while maintaining an atomic-level mass balance. For example, the conversion of NO to N_2 in the standard SCR reaction is represented by

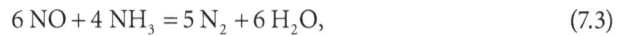

$$6\,NO + 4\,NH_3 = 5\,N_2 + 6\,H_2O, \tag{7.3}$$

where the stoichiometric coefficients are chosen to be integers. The stoichiometry supports an atomic mass balance, in that the reaction in Eq. 7.3 has 10 nitrogen, 6 oxygen, and 12 hydrogen atoms on both the reactant and product sides of the equation.

7.2.2.2 Conversion: Conversion is a measure of how much of a reactant species is consumed by the chemistry in the catalytic converter. The conversion of a species j is defined on a molar basis as follows:

$$x_j = \frac{c_{j,i} - c_{j,o}}{c_{j,i}}, \tag{7.4}$$

where $c_{j,i}$ is the molar concentration of species j at the converter inlet, and $c_{j,o}$ is the concentration at the outlet [7-11]. A value of x_j equal to one means the reactant is completely consumed in the catalytic converter, and a value of zero means there is no net removal of the reactant. Conversion may also be expressed on a percentage basis, from 0% to 100%, in which case it is commonly called conversion efficiency.

The conversion of a species can also be calculated over the complete EAS, and not just the individual catalytic converters within it. In this case, the target concentration of species j at the tailpipe, $c_{j,TP}$, is set by the applicable regulation. The target concentration at the engine outlet, $c_{j,EO}$, is generally set by the engineering team developing the engine and EAS together. The target conversion $x_{j,EAS}$ of species j over the EAS is then defined by

$$x_{j,EAS} = \frac{c_{j,EO} - c_{j,TP}}{c_{j,EO}}. \tag{7.5}$$

As an example, if the tailpipe concentration of NOx needs to be 0.20 (g/bhp·h), and the engine-out concentration is 4.0 (g/bhp·h), then the net conversion of NOx in the EAS needs to be 0.95 or 95%. The system-level conversion is a useful metric for sizing the catalyst. If the engine-out concentration is high and the tailpipe concentration must be low to meet a stringent standard, then a larger catalyst will be required to reach the target conversion. If the catalytic converter volume is constrained by packaging or cost, then the engine-out pollutant concentration must be reduced to avoid an unrealistically high system-level conversion target.

7.2.2.3 Yield and Selectivity:

Product yield from a catalytic converter is not a typical metric for exhaust emissions control, but it is defined as the quantity of product made per unit of reactant consumed, as scaled by the stoichiometry. A yield of 100% means that all of the reactant went to making the product. For example, in the standard SCR reaction in Eq. 7.3, the yield of N_2 is

$$Y_{N_2} = \frac{6\left(c_{N_2,o} - c_{N_2,i}\right)}{5\left(c_{NO,i} - c_{NO,o}\right)}, \tag{7.6}$$

where the factor of 6/5 comes from the stoichiometry of Eq. 7.3 and c_j is the molar concentration of species j.

Selectivity is a relative measure of the yield of one product compared to another. For example, if a side reaction in the catalytic converter made N_2O, then the selectivity of N_2O compared to N_2 is

$$S_{N_2O} = \frac{\left(c_{N_2O,o} - c_{N_2O,i}\right)}{\left(c_{N_2,o} - c_{N_2,i}\right)}. \tag{7.7}$$

In this case, a low selectivity is preferred, since N_2O is a potent GHG. In the literature on catalytic converters, "selectivity" is sometimes used when "yield" is meant.

7.2.3 Catalyst Terminology

In this section, several aspects of catalysts are defined and described. The following terms and concepts are referenced frequently throughout this book, especially in the description of the various EAS components. The catalysts used in EAS devices are heterogeneous, as they are solid compounds that facilitate the reactions of gaseous species.

7.2.3.1 Catalytic Activity and Light-Off:

The activity of a catalyst is its ability to expedite a chemical reaction. Similarly, an active catalyst means one that can improve the rate of the chemical reaction. For the heterogeneous catalysts used in catalytic converters, the activity is usually governed by the surface structure of the catalyst and its immediate environment on the catalyst support. Catalytic activity generally increases with temperature, following Arrhenius behavior, as shown in Eq. 7.1. The activation energy, which is always positive, indicates how readily the reaction proceeds. Smaller values lead to faster reaction rates; larger ones, slower reaction rates.

We define a light-off temperature for the catalyst to indicate its activity within the catalytic converter. A common practice is to measure the temperature corresponding to 50% conversion, T_{50}. Synthetic gas bench testing is used to characterize catalyst light-off behavior, as seen in **Figure 7.6** below. An "S" shaped curve is typical, though different species exhibit different light-off characteristics.

FIGURE 7.6 Light-off behavior of propene on oxidation catalyst [7-12].

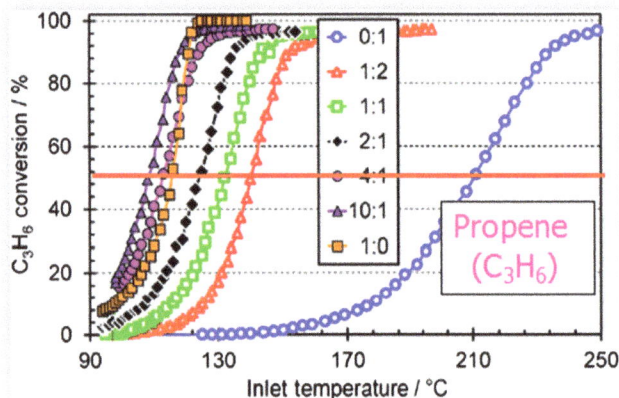

Reprinted from Ref. [7-12] © SAE International

Similarly, catalysts will lose activity as the temperature decreases. Another important temperature is the light-out temperature, where the catalytic activity drops off.

7.2.3.2 Catalyst Aging and Degradation:

Freshly made catalysts are at their most active because the catalyst is well-dispersed in the washcoat as small particles with a high ratio of surface area per the mass of catalyst deposited. Catalysts do not stay in this "green" state for long, and within a few hundred hours of use they become "de-greened." The de-greened state represents a loss of catalytic activity from the green state but to a level of activity that is relatively stable over the useful life of the catalyst. The regulatory testing described in Chapter 4, Emissions Testing Protocols, requires the catalysts in the EAS to have been de-greened before testing. Industrially accepted de-greening protocols for exhaust aftertreatment catalysts can be found on the website of the Crosscut Lean Exhaust Emissions Reductions Simulations (CLEERS) group, which is sponsored by the US Department of Energy (DOE) [7-13].

After the de-greening period, the catalyst continues to age and lose activity gradually. The effects of catalyst aging include the following:

- Increase in light-off temperature for pollutant species
- Decrease in pollutant conversion at normal operating temperatures
- Change in responses to specific exhaust compounds

Catalyst aging mechanisms and their effects are often analyzed together as a way to assess the durability of the catalyst. Catalysts can be deliberately aged by exposing them to air with 10% water at high temperatures, *e.g.*, 650°C–850°C, in what is called hydrothermal aging (HTA). Again, industrially accepted practices for hydrothermally aging a catalyst can be found at the CLEERS website [7-13].

Excessive temperature exposure can cause the catalyst or its support to decompose or otherwise be destroyed. This high-temperature threshold depends on the catalyst and its support, but is generally around 900°C–1,200°C, depending on the materials and the application. Even repeated or prolonged exposure to temperatures above 800°C can lead to significant losses in precious metal surface area and deterioration of the catalyst support. For example, the exposure of TWC to oxidizing conditions at high temperatures can result in substantial changes to the CO oxidation and NO reduction reactions due to Rh deactivation [7-14]. This damage can be reversed somewhat when the catalyst is exposed to reducing conditions, but repeated cycling may cause permanent damage.

The transformation of Rh into a hard to reduce form under oxidizing conditions at high temperatures has an important consequence for the use of Rh in catalytic converters. This transformation and the resultant loss of catalytic activity limits the operating range of Rh catalyst. The temperature at which this transformation occurs, approximately 625°C, can easily be reached during the normal operation of a catalytic converter. Another operating range temperature issue for Rh is that it can diffuse into the subsurface and bulk regions of the alumina when it is heat treated above 540°C [7-15].

Another thermal effect is the surface separation of Rh in Pt-Rh catalysts and supported Pt-Rh alloys at high temperatures under oxidizing conditions [7-16]. Growth of the crystal phase of Pt-Rh and surface enrichment of the Rh could be seen after just 50 h of operation in oxidizing conditions [7-17].

Catalysts can lose activity and functionality from a reduction in surface area, which results from sintering. Catalyst sintering occurs when catalyst particles, especially PGM catalysts, migrate within the washcoat at high temperatures, *e.g.*, over 650°C, and agglomerate to form fewer but larger catalyst particles, as illustrated in **Figure 7.7**. The mass of catalyst material does not change, but the total surface area of the active sites is reduced. Pt and Pd are both susceptible to sintering, with Pt more susceptible than Pd. Unfortunately, Pd can form alloys with Rh at higher temperatures, and in these alloys, the Pd is enriched on the surface of the catalyst particles and the Rh is buried in the core where it is not catalytically active [7-18].

Catalysts can also lose their activity by poisoning, in which a chemical compound interacts with the catalyst and prevents it from facilitating the chemistry it is designed for in the EAS. This poisoning can be temporary or reversible or can result in permanent changes in the catalyst. Often, these poisons come from compounds in the fuel or lubricant that are released into the exhaust gas through combustion in the engine.

An example of reversible poisoning is the adsorption of sulfur oxides (SOx) onto the catalysts used for oxidation reactions, usually in the form of sulfate (SO_4^{2-}) deposits. The SOx species adsorb onto the catalyst surface and keep them from being used for pollutant control. The catalysts may oxidize sulfur dioxide (SO_2) into sulfur trioxide (SO_3) or sulfates during fuel-lean or oxidizing conditions. Nevertheless, SOx can be driven off by heating the catalyst and causing them to desorb. An example of permanent poisoning is that phosphorous (P) species from lubricant additives such as ZDDP will permanently block catalyst active sites, generally with phosphate (PO_4^{3-}) deposits [7-19, 7-20]. A direct relationship between catalyst poisoning by P and lowered HC conversion has been well-established [7-21].

It is noteworthy that the sulfur (S) poisoning resistance of Pd is significantly less than that of either Pt or Rh. Specifically, it has been shown that sulfur, in particular, decreases the ability of Pd to effectively convert CO and HC in reducing exhaust conditions, narrowing the air-fuel ratio (AFR) window for which the catalyst can achieve high conversion of all three emission species (CO, HC, and NOx) simultaneously [7-22].

Although substrates do not age in the same way as the catalysts do, they are susceptible to failures. One mechanism is a thermal shock from a short duration thermal excursion of the exhaust gas. This thermal shock can cause cracks in the substrate. While small cracks are generally manageable for flow-through monoliths, in particulate filters (PFs) they can lead to an effective loss of function as particulates flow through the cracks. Cracks can also propagate to the point where the substrate collapses, leading to a potential exhaust flow blockage.

FIGURE 7.7   Illustration of catalyst sintering.

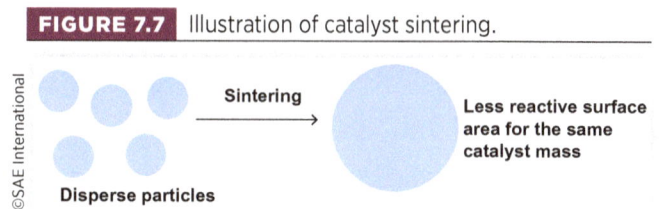

It is also possible for sections of the ceramic to melt if they are locally overheated. This failure mode is most likely in PFs when the soot is burning. Melting can lead to flow restrictions or, in PFs, a loss of function.

7.2.3.3 Space Velocity: SV is used to characterize how much time a chemical species has to react within the catalytic converter, and is defined by

$$SV = \frac{F_g}{V_{cc}},\qquad(7.8)$$

where F_g is the volumetric flowrate of the exhaust gas and V_{cc} is the overall volume of the catalyst substrate. SV has units of inverse time, usually given in reciprocal hours. This means that SV is the inverse of the residence time parameter that is commonly used in chemical reactor design [7-11]. The typical SV range for automotive catalysts is 15,000–150,000 hr^{-1}, and this range is independent of engine or vehicle size because larger volume substrates are used in heavy-duty applications with larger displacement engines. SV is, therefore, an important sizing factor for matching experimental parameters between research reactors and vehicle operation and influences the emissions conversion in the catalytic converter.

Catalytic converter design guidelines dictate that one should size the converter for the maximum flowrate the engine will produce. This design point is usually at the engine's rated power point.

7.2.3.4 Pressure Drop: The substrate properties, catalytic converter size, and the general aftertreatment system design all affect the pressure drop over the EAS, which in turn, affects the engine performance. A honeycomb structure (stacked, parallel, co-axial channels in the substrate) provides the best performance trade-off between surface area and pressure drop. In general, the pressure drop across a catalytic converter increases with decreasing open inlet area (related to substrate diameter), and with increasing length. Therefore, a short, large-diameter substrate would have a lower pressure drop as compared to a long, small-diameter substrate of the same total volume. Increasing the cell density (measured in cells per square inch, or CPSI) not only increases the surface area of the converter but also increases the pressure drop, as the cell density is related to the hydraulic diameter of the gas channels and total area available for exhaust flow through the substrate. Reducing the channel wall thickness not only reduces the pressure drop and thermal mass but also reduces the robustness of the substrate – in terms of its strength and resistance to thermal shock. In addition, the catalytic converter canning and other EAS piping contribute to the pressure drop. The packaging space in the vehicle often dictates the overall can design. An advantage of integrated systems, such as those shown in Section 7.5, is that they have fewer cones and shorter pipe runs, providing an overall lower pressure drop for the EAS.

7.2.4 Catalyst Screening

Engine manufacturers and catalyst suppliers have developed both laboratory bench based and dynamometer-controlled engine facilities where catalysts are screened before certification for use in vehicles. Engine dynamometer-based catalyst aging allows the durability of the catalyst to be measured (often more quickly than in-use testing) under precisely controlled conditions as compared to vehicle-aged samples. These types of studies, where catalysts are exposed to high temperatures and poisons, are often referred to as accelerated aging. Many aging protocols have been developed over the years to replicate what the catalyst would be likely to experience on the vehicle. The US DOE-sponsored CLEERS group maintains a repository of these protocols on their website at www.cleers.org.

Generally, a dyno-based aging experiment includes a dynamometer-controlled engine, gas mass flow controllers, emissions gas analyzers, and a data acquisition system. The engine is operated to simulate a driving schedule that produces the prescribed exhaust composition and temperatures [7-23]. The temperature of the catalyst itself depends on both the exhaust temperature and the heat generated by the exothermic oxidation reactions. Emissions performance tests are done regularly (*e.g.*, every 100 h) and the conversions of CO, HCs, and NOx are measured throughout the durability test and compared against the fresh catalyst performance. Durability tests can be used to compare catalyst formulations, investigate fuel effects on performance, determine the impact of prolonged exposure to poisons (species such as lead (Pb), phosphorus (P), or sulfur (S)), and to test sensitivity to exhaust conditions (composition, temperature, etc.), or some combination of these.

An important catalyst property that can be determined from dyno-controlled engine studies is the transient response of the catalyst. This is the time required for the catalyst to reach light-off temperature.

7.2.5 Zeolite Catalysts

Zeolites are aluminosilicate crystals, a class of compounds that are a blend of alumina (Al_2O_3) and silica (SiO_2), typically of formula $x\ SiO_2 * y\ Al_2O_3$. They may be naturally occurring or synthetic and can accommodate a large variety of cationic (positively charged) ions such as calcium (Ca^{2+}), magnesium (Mg^{2+}), and potassium (K^+), which can be exchanged out or substituted for the ions originally within the synthesized zeolite in solution chemistry. Some zeolites occur in nature, although most zeolites used as industrial catalysts are synthesized to better control their properties and avoid structural defects. Most are metastable, meaning that the structures can collapse when exposed to excessive heat or pressure. There are several types of zeolites, which are characterized by their overall structure, including pore sizes and inner cage volumes. Most zeolites are thermally stable up to around 500°C, but ones used for automotive catalysts are stable to about 800°C.

Zeolites are microporous solids, also known as "molecular sieves," allowing them to selectively trap molecules of a particular size by exclusion in their well-defined crystalline structures. The zeolite pore sizes are on the same order as molecular diameters, typically ranging from 3–8 Å. Molecules of this size or smaller can enter the zeolite pores, while larger molecules are excluded based on size—it is this size selectivity that gives rise to the notion of zeolites being "molecular sieves." A commonly used approach for structural classification of zeolites, with respect to their pore size, is based on the number of tetrahedrons, or *T-units* that comprise the pore framework [7-24]. The

FIGURE 7.8 Structure of ZSM-5.

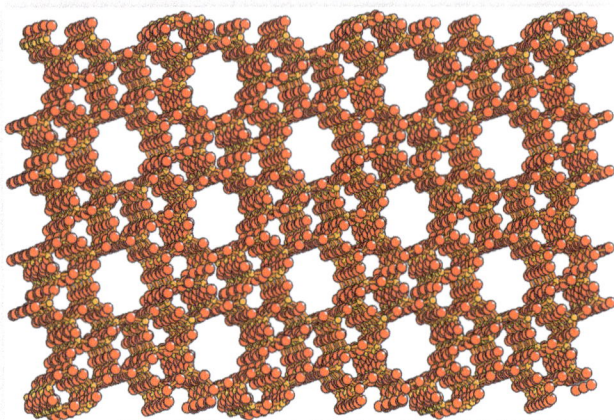

© Shutterstock

classification bins are (a) *small pore zeolites*, with 8 T-units forming the pore opening, (b) *medium pore zeolites*, with 10 T-units forming the pore opening, and (c) *large pore zeolites*, with openings formed by 12 T-units. Because the pore sizes within the zeolite structure are about the same size as a molecule, the combination of zeolite and exchange ion (or substitution ion) can be designed together to improve selectivity for the desired compound or to inhibit a side reaction.

What makes zeolites particularly interesting is that they can experience a charge imbalance resulting from aluminum (Al) displacing silicon (Si) in the structure that is beneficial in molecular selectivity. For example, the displacement of the Si results in the Al sites having a negative charge that can coordinate with hydrogen ions (H^+). Other positive ions can be substituted in for the H^+, which creates catalytic activity within the zeolite. The quantity of Al incorporated into the zeolite structure affects the catalytic activity since it governs how many binding sites there are, and represents a measure of the electron charge and polarity of the zeolite. Therefore, the Si:Al ratio (SAR) is an important characterization of a zeolite.

Zeolites are also characterized by high specific surface areas. For example, the pentasil zeolite (ZSM-5) has a specific surface area of approximately 400 m^2/g. This high surface area makes zeolites excellent adsorption materials, and they are used in industrial drying, purification, and separation processes. ZSM-5 is used in automotive applications, and its structure is shown in **Figure 7.8**.

Zeolites have increasingly been used in EAS devices both as catalysts and as adsorbers. Zeolites commonly used in catalytic converters include pentasil (ZSM-5), chabazites (SSZ-13), SAPO-34, and beta zeolites (BEA), although these are only a small subset of the hundreds of zeolites that have been characterized. The structures of SSZ-13 and SAPO-34 zeolites are shown in **Figures 7.9** and **7.10**, respectively.

For catalytic applications, zeolites are often ion-exchanged with metal cations. This is done by treating the acid form of the zeolite (HZ) with an aqueous solution of ammonium nitrate (NH_4NO_3) to form the ammonium exchanged zeolite ($NH_4^+Z^-$), which is then treated with a salt solution that contains the catalytic cation to form the metal exchanged zeolite (MZ).

Zeolites are crystalline hydrates of aluminosilicates with a porous structure. They are solids, with an open three-dimensional crystal structure built from interlinked tetrahedra of alumina (AlO_4) and silica (SiO_4), often with alkali or alkaline-earth metals or water molecules trapped in the cavities. Zeolites are stable solids, resistant to high pressures, with high

FIGURE 7.9 Structure of SSZ-13.

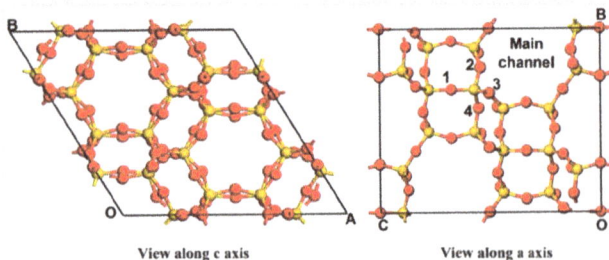

View along c axis   View along a axis

melting points (over 1,000°C). They are insoluble in water and organic solvents and do not oxidize in air, making them ideal for use in the automotive exhaust environment.

Zeolites can be formed with a number of different crystalline structures with open pores on the same size order as small molecules. There are approximately 40 naturally occurring zeolite formations, which can be mined. Chabazite (CHA) is an example of a naturally occurring zeolite form. There are even more synthetic zeolite formations which have been designed to have specific properties, such as uniform pore sizes, or uses. ZSM-5 is an example of a synthetic zeolite designed to be a petroleum catalyst and the SAPO-34 zeolite is a synthetic micropore zeolite, with a similar crystal structure to that of chabazite. Their open, cage-like framework structures can trap other molecules inside them. This is how water or other metal cations can become part of the zeolite crystal, through reversible hydration or cation exchange.

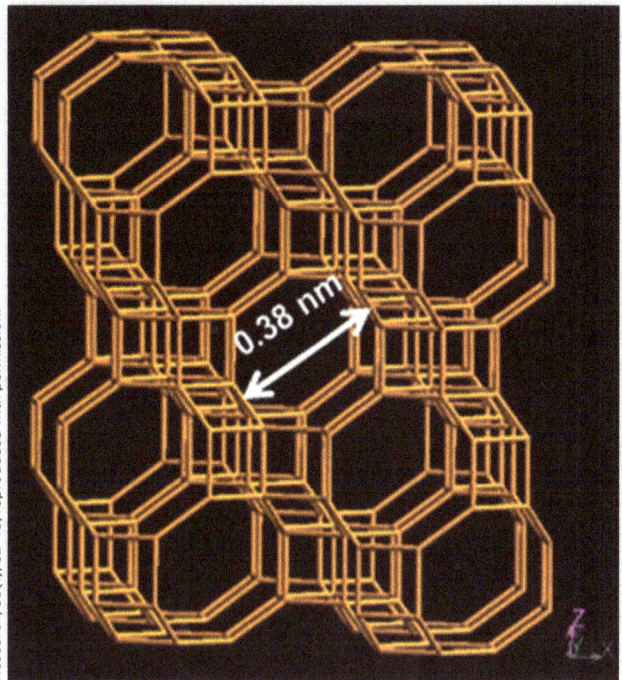

FIGURE 7.10   Structure of SAPO-34.

Carreon, M. (2018). Molecular sieve membranes for N2/CH4 separation. Journal of Materials Research, 33(0), 32-43, reproduced with permission.

7.3 Transport Phenomena within the Exhaust Aftertreatment System

A common issue in designing EAS is managing the transport phenomena within it, both within the monolith and washcoat for each catalytic converter and within the piping and other flow elements in the EAS. This section discusses all three elements of transport phenomena—fluid dynamics, heat transfer, and mass transfer—although the main focus is on the fluid dynamics and heat transfer since the mass transfer issues are generally only relevant within the catalytic converter monoliths and washcoats.

7.3.1 Fluid Dynamics and Pressure Drop

One of the key criteria for the overall EAS is the pressure drop of the exhaust gas from the engine outlet—the exhaust manifold or turbine outlet—through to the tailpipe outlet. Since the tailpipe outlet is always at ambient pressure, the pressure drop determines the backpressure, or the exhaust gas pressure in the exhaust manifold. The engine must push the gas out of the cylinder against this pressure, so higher backpressure means that the engine must do more work to achieve the same net (brake) power output. An important part of the EAS design, then, is sizing the components and routing the piping so that they minimize the pressure loss, especially at the design point, which is usually rated power. Rated power is usually the operating point with the maximum volumetric flow of exhaust gas because the exhaust gas is usually at its hottest temperature and its highest mass flow. The design target for backpressure, therefore, provides a constraint on how the EAS is sized and packaged.

The most significant flow restrictions come from the EAS subsystem monoliths since they are essentially bundles of small square channels in which the flow is laminar. The length of the channel also influences the pressure loss, as do the entrance effects as the flow enters the monolith and the exit effects as the flow leaves. Because the monolith channels are approximately square, usually with rounded corners, but with a small diameter, the gas flow in the channel is laminar, and thus the friction factor f [7-25] for the flow is

$$f = \frac{16}{\text{Re}}, \tag{7.9}$$

where Re is the Reynolds number for the flow in the channel, which is defined as

$$\text{Re} = \frac{D\bar{v}\rho}{\mu}, \tag{7.10}$$

where D is the channel width or diameter, $\bar{v}$ is the average gas velocity in the channel, ρ is the bulk gas density, and μ is the bulk gas viscosity. Note that the gas properties are functions of temperature, so the operating state of the catalytic converter will influence the pressure drop through it. This estimate neglects the entrance and exit effects in each of the channels. In general, the channel length L is significantly longer than the axial distance needed for the flow to develop, also called the entrance length L_e, which is defined [7-25] as

$$L_e = 0.035\, D\, \text{Re}. \tag{7.11}$$

The pressure drop over the complete monolith is the factor of interest, not just the portion where the flow is developed. Thus, an estimate of the entrance length is necessary to confirm that it is negligible by comparison to the frictional losses in the channel. The pressure losses come from a rearranged version of the Hagen–Poiseuille law [7-25], and yield

$$f = \frac{1}{4}\left(\frac{L}{D}\right)\left(\frac{p_0 - p_L}{\frac{1}{2}\rho\bar{v}^2}\right), \tag{7.12}$$

where p_i is pressure at the inlet (0) or outlet (L) of the channel, and the other variables are as before. Knowing the friction factor f from Eq. 7.9, one can solve for the pressure drop using Eq. 7.12.

The rest of the piping in the exhaust system also influences the overall pressure drop. For example, the EAS for many heavy-duty diesel engines (HDDEs) include switch-backs to keep the overall EAS packaging compact and to accommodate the length of piping needed to decompose urea ($CO(NH_2)_2$) into ammonia (NH_3) for the SCR system. The 180° bends in the flow can contribute significantly to the overall pressure drop. Likewise, there are pressure losses associated with the expansion of the flow from a pipe segment into a catalytic converter's can and with the contraction of the flow back into the exhaust pipe. It is possible to use 1-D and 3-D CFD tools and flow experiments to evaluate the pressure losses in the EAS and improve the design within the other constraints.

The same approach using a friction factor may be used, although in the main exhaust piping the exhaust gas flow is more likely to be turbulent under most engine operating conditions, and a Moody diagram [7-25, 7-26, 7-27] should be used to estimate the friction factor once the Reynolds number is known. For the changes in cross-sectional area and similar flow restrictions, one can estimate the additional pressure head losses [7-27] using

$$\Delta P = \left(\frac{\rho \bar{v}^2}{2g} \right) \left(\frac{L}{R_h} \right) f, \tag{7.13}$$

where g is the acceleration of gravity and R_h is the hydraulic diameter for the exhaust pipe, which is the ratio of the cross-sectional area to the perimeter of the flow channel. A scaled friction factor e_v,

$$e_v = \left(\frac{L}{R_h} \right) f, \tag{7.14}$$

can then be estimated for various flow restrictions under turbulent flow, and from those values of e_v the equivalent friction factor f can be calculated. Some common scaled friction factors e_v are presented in **Table 7.1**, where β is the ratio of the smaller cross-sectional area to the larger cross-sectional area.

Given the complex geometry of most EAS, including the piping, the best estimates of pressure losses and velocity profiles in the catalytic converter monoliths and exhaust piping will come from computational fluid dynamics (CFDs).

7.3.2 Heat Transfer and Thermal Management

Thermal management of the EAS components through well-controlled heat transfer is critical in achieving a high conversion of pollutants over the FUL of the EAS. Two key thermal management functions for the EAS are the ability to quickly warm up the catalytic converters and to keep the EAS from overheating during normal operation.

7.3.2.1 Warm-Up and Keep-Warm Strategies:
In normal use and on many of the regulatory test cycles described in Chapter 4, Emission Testing Protocols, the engine and EAS start at ambient temperature and warm up to their operating temperature as the engine runs. Fortunately, PFs trap particulates even when cold, so PM and PN control starts when the engine does. However, for high conversion of gaseous pollutant species such as NOx and non-methane hydrocarbons (NMHCs), most catalysts need to reach their light-off temperature of roughly 200°C to 350°C. During the warm-up period between engine start and catalyst light-off, the catalytic converters in the EAS let much of the emissions through mostly untreated, and up to 90% of on-cycle emissions can be attributed to the cold start period [7-28]. The engine and EAS control strategy must warm up the catalysts in the EAS in the shortest feasible time needed to reach their light-off temperature. At the light-off temperature, the catalyst activity supports a high level of pollutant conversion. A related thermal management function is keeping the catalysts warm during low-load operation, when the exhaust gas temperature may drop enough

TABLE 7.1 Scaled friction factors for turbulent flow in pipes [7-25, 7-27].

Obstacle	e_v
Sudden contraction	$0.45\,(1-\beta)$
Sudden expansion	$\left(\frac{1}{\beta} - 1 \right)^2$
Orifice, sharp-edged	$2.7(1-\beta)(1-\beta^2)\frac{1}{\beta^2}$
45° elbow	0.3–0.4
90° elbow, rounded	0.4–0.9
180° bend	1.6

to inhibit the catalytic function of the EAS. As described in Chapter 3, California is considering adding a low-load cycle as part of its NOx mitigation regulations for heavy-duty engines, and several jurisdictions have not-to-exceed (NTE) requirements for pollutant emissions that may be extended to lower engine loads during real-world driving.

Fortunately, there are several strategies in use and in development to support both accelerated warm-up and keep-warm functionality. These strategies usually increase component costs, and may also impose a fuel consumption penalty at a given level of engine performance. The strategies described in this section include a mix of those that heat EAS elements directly and those that change the engine management and hardware to help convey enthalpy from the engine cylinder into the EAS.

There are two passive EAS thermal management strategies that are already in production. One strategy is to minimize heat losses from the exhaust pipe into and through the EAS, and the other is to close-couple EAS subsystems by placing them as close to the exhaust manifold or turbine outlet as possible.

Several manufacturers already use double walls with an air gap or other types of insulation on the exhaust pipes, especially at the hotter end of the EAS, to ensure that the enthalpy in the exhaust gas makes it to the EAS subsystems where it is needed. Adding the insulation does not incur a fuel consumption penalty, although it does increase the cost of the exhaust pipe and can complicate packaging requirements. Some non-road applications need exhaust pipe insulation to ensure that the surface temperatures stay within a safe range for the operating environment. For example, hot surfaces in environments with grain dust or vapors increase the risk of fire.

Close-coupled EAS subsystems are placed close to the exhaust manifold or turbine outlet, which exposes them to hot exhaust gas and accelerates their warm-up. Usually, these systems are relatively small so that they do not penalize the overall EAS performance by inhibiting the warm-up of the downstream or underfloor EAS components. For example, the use of close-coupled TWC is commonplace in gasoline-fueled SI engines, where a smaller TWC is placed close to the exhaust manifold. There has also been research into installing a close-coupled diesel oxidation catalyst (DOC) or a close-coupled SCR [7-29] unit into the EAS for diesel engines to improve the time needed to reach the light-off temperature in the catalytic converters. The challenges with close-coupled EAS subsystems include their exposure to the hottest exhaust gas temperatures in the EAS, which can accelerate the thermal aging of the catalysts, and the limited packaging space available for them, as described in Section 7.3.

The EAS can also be directly heated through one of a few strategies, including the use of an electrical heater or a diesel mini-burner in the EAS. Both of these strategies place an additional system into the EAS, where they need to function through the FUL of the application. A related approach is to inject diesel fuel into the exhaust line and rely on catalytic combustion in the leading EAS subsystem, such as a DOC, to heat the downstream EAS components. This approach, however, requires the leading EAS subsystem to already be at its light-off temperature to be effective.

With an electrically heated catalyst, a resistive heating element is placed in the exhaust system just upstream of the leading element of the EAS, such as the DOC. Electrical power is then applied to heat the EAS subsystem; the input power can also be cycled to help keep the EAS warm. The main challenge with an electrical heater is the power required for effective heating, and the effect of that electrical power on the engine's fuel consumption.

A typical 12 V electrical system can only support a power draw of about 2 or 3 kW because of current and cable size restrictions for safety, and with a typical alternator efficiency of 50% this electrical power demand can require up to an additional 6 kW of mechanical power from the engine. A power draw of 2–3 kW may be enough for light-duty vehicle (LDV) EAS, but for larger EAS, more power is required. A 48 V electrical system, such as is used with a mild hybrid, can support 10–12 kW of electrical load for a similar restriction on current and cable size. Electrical heating is more efficient at the higher voltage level, but it also makes sense to use it when a hybrid system has been implemented on the vehicle. With the hybridization, recovered energy can be used to power the electrical heater and thereby minimize the fuel consumption penalty.

If more heating is needed than can be provided by an electrical heater, one can find a suitable diesel mini-burner and place it in the exhaust gas line. Mini-burners are capable of providing 10–30 kW of heating to EAS components. The heat from the burner is usually applied to the DOC or similar upstream system. The challenge with a burner is maintaining stable and clean combustion, especially as the system gets larger; the burner also consumes fuel as it operates.

As mentioned earlier, one can inject fuel into the exhaust line to manage the exhaust gas temperature. This fuel is catalytically combusted in a warmed-up leading EAS subsystem, typically a DOC, and the resulting catalytic combustion heats the subsystem and the exhaust gas, thus warming up the downstream EAS components. One approach is to add in a late post-injection fueling event in the sequence of fuel injection events. This is most common with CI engines, for which direct injection (DI) fueling systems are standard, but it can also be used with DI SI engines. Challenges with this approach include the added complexity in the fuel injection strategy and the risk of spraying fuel onto the cylinder wall and thereby contaminating the lubricant with fuel. Another approach is to install an additional fuel injector in the exhaust line. This so-called seventh fuel injector* delivers fuel to the first catalytic converter in the EAS, and as with the late post-injection approach, the resulting catalytic combustion heats the EAS. The late post-injection and the seventh injector approaches have both been used to heat the diesel PF (DPF) when an active regeneration event was needed. DPF load management is covered in more detail in Chapter 11, Particulate Filters.

A few strategies rely on modifying the intake or exhaust flow restriction to make the engine work harder to achieve a target load. One strategy is to install an exhaust flap or throttle to increase the flow restriction in the exhaust, and thereby reduce the efficiency of the engine. The effect of closing the restriction is to increase the exhaust gas temperature. For CI engines, one can also implement an air intake flap or throttle to push down the AFR and the engine efficiency at a given load. These results also raise the exhaust gas temperature. These two approaches can also be combined, although that further increases the complexity of the engine management strategy. In addition to the added complexity, each of these approaches involves finding package space to install the flap or throttle and its actuator in the gas flow path near the engine. Another potential advantage of the exhaust flap or intake throttle is that they can be used, singly or in combination, to help drive exhaust gas recirculation (EGR) flow with a more favorable pressure drop.

---

* So called because the approach is frequently used in the EAS of six-cylinder HDDE. It is not limited to six-cylinder engines.

The last broad category of strategies for warm-up acceleration most directly move enthalpy from the combustion cylinder into the exhaust gas to warm up the EAS. These strategies include the following:

- Late intake valve closure (LIVC) or Atkinson cycle operation
- Early exhaust valve opening (EEVO)
- Secondary exhaust valve opening event
- Turbine wastegate to have exhaust gas bypass the turbine

For a CI engine, LIVC reduces both the AFR [7-30] and the effective compression ratio [7-31]. The lower AFR tends to increase the gas temperature in the cylinder since there is less excess air to absorb the enthalpy released by combustion; however, the lower compression ratio tends to keep the in-cylinder temperature down. Which effect dominates will depend on the demanded engine load, as that will determine the fueling needed in the engine. If calibrated correctly, although, LIVC can increase the turbine outlet temperature by 80°C–120°C by delaying the intake valve close event by 60 to 70 crank angle degrees (CA°) [7-32]. Using LIVC for increased exhaust gas temperatures would only be used during operating periods when the EAS needs to be warmed up or kept warm, so a variable valve actuation (VVA) system is needed to switch between the normal and LIVC timing.

A similar approach for CI engines is to use EEVO during warm-up and keep-warm operating periods. Two main approaches that have been investigated for EEVO: use of cam profile switching (CPS) and use of a variable cam phaser (VCP) [7-33]. The CPS system switches between cam lobes, where the main one is for the normal exhaust valve profile, and the secondary lobe is for the EEVO profile. With a CPS system, the exhaust valve can be opened up to 100 CA° earlier than usual and can be held open for those extra crank angles. It provides a high level of heating—up to 100°C hotter exhaust gas in some conditions—but it comes with a significant fuel consumption penalty since the enthalpy going into the exhaust to warm up the EAS is not available to do work. The VCP system is cheaper and simpler to implement, but the limitation is that the opening event can only be pulled forward in time; the opening profile does not change. Also, a wide-range VCP with the authority of up to 65 CA° or so is needed for this application. Daimler Trucks have implemented such a system on their production DD5 and DD8 engines to facilitate EAS warm-up and support best in class NOx certification levels as of writing [7-33].

Another approach that requires VVA is to add a second exhaust valve opening event during the next intake stroke. The secondary event has a much lower lift than the primary exhaust valve opening event does. This approach pushes uncooled EGR through the cylinder and into the exhaust manifold, thereby warming the exhaust gas. It supports a moderate temperature increase for a moderate fuel consumption penalty.

Lastly, the turbine can be fitted with a wastegate so that the exhaust gas bypasses the turbine and retains more of its enthalpy to heat the EAS. This approach has a marginal cost from adding in the wastegate and will penalize the engine performance during the warm-up period since the turbine will be less effective at recovering energy from the exhaust gas.

A related strategy, one that is more effective for EAS keep-warm functionality, is to implement cylinder deactivation (CDA). CDA generally requires a VVA system, so it is a natural complement to some of the approaches described earlier. A typical engine for CDA has an even number of cylinders, such as six or eight, and then the CDA system causes the engine to only use half of those cylinders to meet the power demand. The load in each cylinder is, therefore, higher, and so the gas temperature in the exhaust manifold and at the turbine outlet increases in the low-load regime, as shown in **Figure 7.11**. There is a transition torque at which the CDA system activates or deactivates, usually with some hysteresis built in to help minimize switching the system on and off. When the CDA is active, the exhaust gas flow is roughly halved, although the temperature increases. Thus, the total enthalpy delivered to the EAS may drop even though the exhaust gas temperature is higher. This feature is why CDA is more useful as a keep-warm feature.

Ding *et al.* [7-30] showed that the Eaton CDA system can implement an increase in exhaust gas temperature with a minimal fuel consumption penalty. The typical trade-off between turbine outlet temperature and brake-specific fuel consumption (BFSC) penalty is shown in **Figure 7.12** along with the curve that results from implementing the CDA system. For this CDA system, a given target increase in exhaust gas temperature results in a modest BSFC penalty.

FIGURE 7.11 Effect of cylinder deactivation on turbine outlet temperature (TOT) map [7-34].

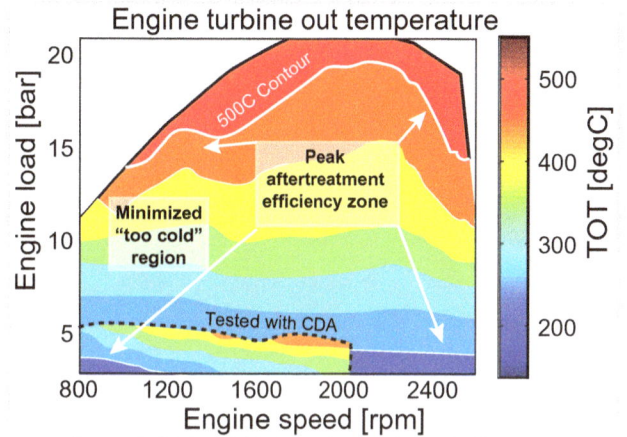

Reprinted by permission from Springer Nature: VDI Commercial Vehicle Engine Technology Conference, 2017. McCarthy Jr, J., Theissl, H., and Walter, L., "Improving Commercial Vehicle Emissions and Fuel Economy with Engine Temperature Management Using Variable Valve Actuation.", Copyright 2017.

7.3.2.2 High Temperature Mitigation: A very common degradation mechanism for the catalysts used in the EAS is HTA, or exposure to high temperatures in the presence of water. These high temperatures can reduce the activity of the catalysts, and in the most extreme cases, cause irreversible damage to the washcoat or monolith. HTA is described further in Section 7.2.3.2, and is also discussed for the various catalytic converters in Chapters 8 through 14.

To mitigate HTA, the EAS thermal management strategy must protect the catalysts and monoliths from high-temperature excursions that could cause acute damage or accelerate the EAS aging. Managing the engine operation is the best approach, especially

FIGURE 7.12 Trade-offs in turbine outlet temperature (TOT) and fuel consumption (BSFC) increase with and without cylinder deactivation [7-30].

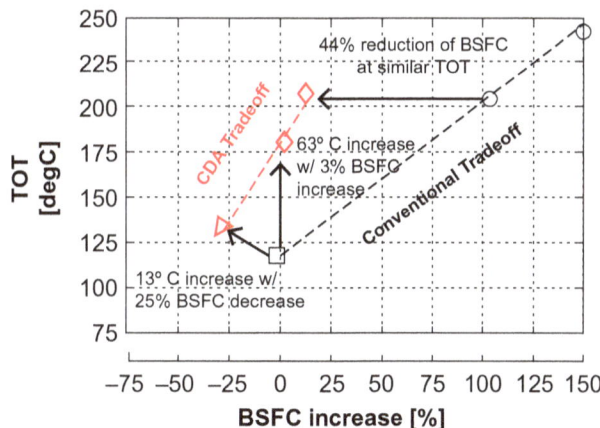

Ding, C., Roberts, L., Fain, D.J., Ramesh, A.K., Shaver, G.M., McCarthy Jr, J., Ruth, M., Koeberlein, E., Holloway, E.A., and Nielsen, D., "Fuel efficient exhaust thermal management for compression ignition engines during idle via cylinder deactivation and flexible valve actuation." International Journal of Engine Research 17(6):619–630, 2016, copyright © 2016 by SAGE Publications, Ltd., Reprinted by Permission of SAGE Publications, Ltd

at or near the rated powerpoint where the engine runs the hottest. For example, in stoichiometric SI engines, a certain amount of fuel enrichment is generally allowed as an Auxiliary Emissions Control Device (AECD) because it cools the exhaust gas temperature and thereby protects the turbine and the close-coupled TWC. AECDs are described further in Sections 3.2.1, 3.2.2, and 3.3.1.

Another potential cause for overheating the EAS is a poorly managed DPF regeneration event, but nowadays this only results from a fault in the engine or in the soot monitoring model that leads to the DPF holding more combustible material than the controller expected. The result can lead to high temperatures within the DPF and downstream. The mitigation is to have multiple layers of control that manage the soot accumulation in the first place.

7.3.3 Mass Transfer

The convection of species by the bulk exhaust gas is the main way in which they move through the EAS. Within the monoliths, diffusion processes also become important. In general, there are two steps to the mass transfer within the monolith: transport from the bulk gas to the washcoat surface and transport into and through the washcoat. Both of these processes are shown in **Figure 7.13**, which includes species diffusing to and from the gas–washcoat interface and within the washcoat.

The rate of diffusion is a function of the concentration gradient, which drives the flow of molecules through the gas; the bulk motion of the gas in the channel; the width of the channel; and the temperature. As described by Welty et al. [7-27], the mass transfer across a laminar boundary layer to a flat surface at some distance z downstream of the leading edge of the monolith channel is characterized by

$$\text{Nu}_{AB,z} = \frac{k_c z}{D_{AB}} = 0.332\,\text{Re}_z^{1/2}\,\text{Sc}^{1/3}, \tag{7.15}$$

where $\text{Nu}_{AB,x}$ is the mass transfer Nusselt number, k_c is the mass transfer coefficient, z is some axial distance at which the laminar flow is established, D_{AB} is the diffusion coefficient for species A in B, Re_z is the Reynolds number for the through the channel,

FIGURE 7.13   Illustration of mass transfer within the monolith channel, including the bulk gas in the channel and the catalytic washcoat.

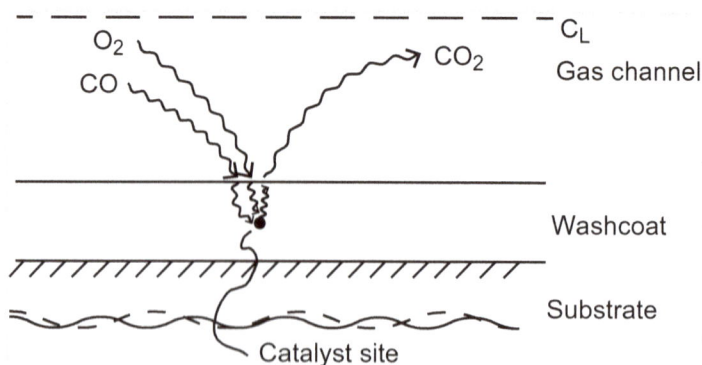

and Sc is the Schmidt number. Here, the Reynolds number Re_z is the Reynolds number in the axial flow direction along the channel wall, which is defined as

$$\text{Re}_z = \frac{z v_\infty \rho}{\mu},$$ (7.16)

where z is the length along the channel wall, and the other properties are as for Eq. 7.10. Note that the gas properties are functions of temperature, so the operating condition of the catalytic converter will influence the mass transfer. Similarly, the Schmidt number, Sc, is defined as

$$\text{Sc} = \frac{\rho D_{AB}}{\mu}.$$ (7.17)

Within the washcoat, it is assumed that there is no bulk gas flow, but the width (depth) of the washcoat and the temperature are still factors. One of the more complicated aspects of heterogeneous catalyses, such as that found in the washcoats of catalytic converters, is that the mass transfer and the chemical reactions both influence each other. Known as the diffusion-reaction problem, the governing equation for mass transfer becomes the following:

$$\frac{\partial c_A}{\partial t} = D_{AB} \nabla^2 c_A + R_A,$$ (7.18)

where c_A is the molar concentration of species A. R_A is the rate of generation of species A, and is usually a function of c_A and the local temperature per the Arrhenius rate expression in Eq. 7.1. Note that if A is a reactant, $R_A < 0$.

In some operating conditions the limiting factor for the rate of change of species A is the rate of a chemical reaction; these conditions are called kinetically controlled. In other conditions, the limiting factor is the rate at which species A can diffuse across the boundary layer per Eq. 7.15, and through the washcoat to the catalyst per Eq. 7.18; these conditions are called mass-transfer controlled. In addition, the thermodynamics governing the various chemical reactions may also constrain the extent of reaction, and these situations are described where they occur in the following chapters.

In wall-flow PFs, described in Chapter 11, species are also carried through the wall by the exhaust gas flow. Even within the filter cake and wall, though, there are opportunities for chemical species to diffuse through the filter cake or monolith wall in perpendicular to or against the main gas flow across the monolith wall.

7.4 Exhaust Aftertreatment System Sizing, Packaging, and Configuration

One of the engineering challenges with designing and implementing an EAS for production is balancing various attributes, including the following: catalytic converter sizes, overall flow restriction or pressure drop, available package space, configuration in the exhaust pipe, and cost.

Each catalytic converter in the EAS contributes to the pollution control function, and each needs to be large enough to convert criteria pollutants at a limiting case, usually an operating point with the maximum exhaust gas flow through the system. The EAS subsystem monolith size and the volumetric gas flow determine the SV, as shown in Eq. 7.8, and the SV helps determine the pollutant conversions. Larger EAS subsystems tend to be more expensive, although for most EAS subsystems it is the total PGM loading that has the strongest influence on the system cost. Pd, Pt, and Rh are all expensive commodities, as discussed in Section 7.2.1.1, even before they are processed into catalysts.

As discussed in Section 7.3.1, the pressure drop is driven by the catalytic converters' sizes and the geometry of the flow path through the EAS.

Other constraints on the packaging include the space available on the application for the EAS. EAS subsystems that are attached to the exhaust manifold or turbine outlet are known as close-coupled systems. These close-coupled systems are attractive because they warm up quickly because of their proximity to the engine, as discussed in Section 7.3.2. On the other hand, there needs to be room to fit the close-coupled EAS subsystem and its associated hardware next to the engine on the given application. Similarly, it is common to see bends and kinks in a commercial EAS that reflect the need to route the exhaust pipes around other features on the application, such as the frame or suspension.

FIGURE 7.14   Honda I-4 1.5 L VTEC TURBO engine [7-35]. (Image courtesy of Honda Motor Co.)

© Honda

TABLE 7.2   Honda 1.5 L I-4 VTEC TURBO engine as implemented in the Honda Accord [7-36].

Attribute	Value
Displacement	1.498 L
Bore	73.0 mm
Stroke	89.4 mm
Compression ratio	10.3:1
Rated power	143 kW (192 hp) at 5,500 rpm
Peak torque	260 N·m (192 lb·ft) at 1,600–5,000 rpm
Fuel injection system	Direct injection (DI)
Air handling system	Turbocharger with electronic wastegate

7.5 Example Systems

Several example engines and EAS are presented in this section to provide some concrete examples for later use in the text. These examples are intended to provide context for the descriptions of the catalytic converters provided in Chapters 8 through 15, and will be referenced throughout the text as appropriate. The example engines and EAS are drawn from typical light-duty and heavy-duty applications, including passenger cars; commercial vehicles; and off-road applications.

Each example includes an overview of the engine, including a high-level description of performance attributes such as the rated power and peak torque. The overview of the EAS includes the system configuration, such as what EAS subsystems and sensors are used and in what order in the exhaust system.

7.5.1 Light-Duty Spark Ignition

Light-duty SI engines are currently the predominant choice in the US and Japan for LDV, and they have significant market share in other markets as well. A typical light-duty SI engine is the inline four cylinder 1.5 L VTEC TURBO engine from Honda Motor Company, also known as the L15B7 engine [7-35]. This engine is shown in **Figure 7.14**, and the version of the engine implemented in the Honda Accord sedan has the attributes shown in **Table 7.2**. This engine has variable valve timing control (VTC) applied to both the intake and exhaust valves, which supports a supercharging effect at low engine speeds [7-35].

FIGURE 7.15 Three-way catalytic converters for Honda engine. (a) Upstream unit with boss for inlet oxygen sensor on left. (b) Downstream unit with boss for outlet oxygen sensor and affixed heat shield. (Images by John J. Kasab)

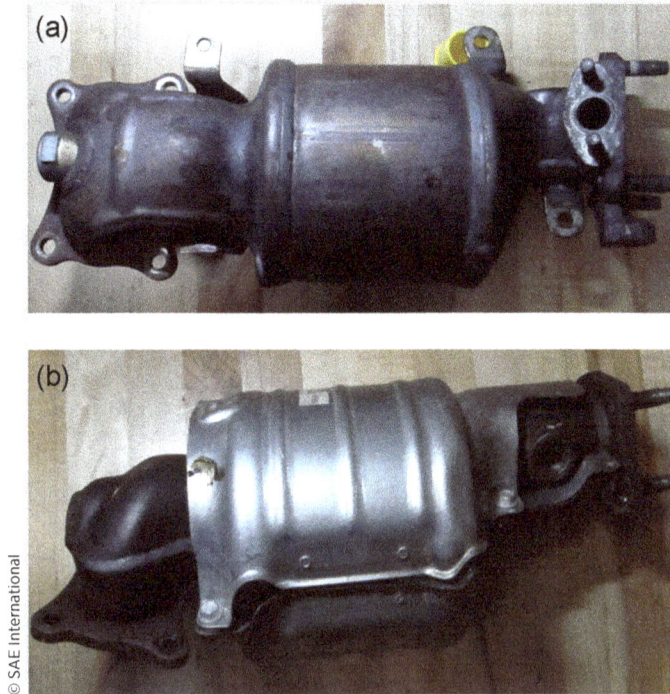

© SAE International

The EAS for this engine features one close-coupled TWC and one underfloor TWC, and together with the engine control strategy, the vehicle is certified to the EPA Tier 3 bin 30 level [7-37] described in Section 3.2.1. **Figure 7.15** shows two examples of canned TWCs from Honda. The first, in Figure 7.15(a), is further upstream and includes a boss, or fitting, for the inlet oxygen sensor on the left side of the unit as shown. The other, in Figure 7.15(b), has a boss on the right for the outlet oxygen sensor, and still has its heat shield attached. In addition to the TWC units, the exhaust system also contains resonators to manage the noise, vibration, and harshness (NVH) of the exhaust gas as it flows through the EAS.

Another EAS containing TWCs for a V-type SI engine is shown in **Figure 7.16**. There are two exhaust pipes from the engine, one for each bank of cylinders. After they merge there is a smaller, close-coupled TWC which is followed by the main, underfloor TWC. Note that the EAS piping has several gentle bends to route it around features under the vehicle. The image in Figure 7.16 just shows what is called the hot end of the exhaust, and omits the subsequent mufflers and remaining exhaust piping.

FIGURE 7.16 Example of close-coupled and underfloor three-way catalysts for light-duty SI engine. (Image courtesy of Walker.)

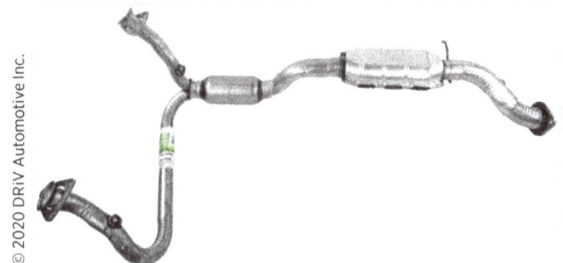

© 2020 DRiV Automotive Inc.

FIGURE 7.17 FCA 3.0 L V-6 EcoDiesel engine [7-38]. (Image courtesy of FCA.)

© FCA US LLC

TABLE 7.3 FCA 3.0 L V-6 SI EcoDiesel [7-38].

Attribute	Value
Displacement	2.988 L
Bore	83.0 mm
Stroke	92.0 mm
Compression ratio	16.5:1
Rated power	179 kW (240 hp) at 3,600 rpm
Peak torque	570 N·m (420 lb·ft.) at 2,000 rpm
Fuel injection system	Common rail DI
Air handling system	Turbocharged

7.5.2 Light-Duty Diesel

Two typical light-duty diesel engines (LDDEs) are described in this section. The first is the 3.0 L V-6 EcoDiesel engine from Fiat Chrysler Automobiles (FCA). The second is the 3.0 L I-6 Duramax engine from General Motors.

The 3.0 L V-6 EcoDiesel is used in the RAM 1500 and Jeep Grand Cherokee LDVs. The statistics for the engine are shown in **Table 7.3**, and the engine is pictured in **Figure 7.17**.

The EAS for the FCA EcoDiesel engine includes a DOC, DPF, and SCR units, as shown in **Figure 7.18**. The EAS configuration has a DOC and DPF placed in one can near the engine, followed by an SCR in the underfloor position. After leaving the SCR, the exhaust then flows through a muffler and out the tailpipe. It can be seen in Figure 7.18 how the exhaust pipe is routed around other features under the vehicle, including the path from the engine to the underfloor position, and the bend in the exhaust pipe needed to accommodate the urea injection system upstream of the SCR. The relatively short pipe length between the urea injector and the SCR inlet suggests that there is a static mixer in the pipe. Chapter 12, Selective Catalytic Reduction Systems, has more detail on urea injection and ammonia generation and distribution in the feed to the SCR system. The FCA light trucks with this engine and EAS are certified to the EPA Tier 3 bin 160 [7-37] levels described in Section 3.2.1.

The second LDDE is the 3.0 L I-6 Duramax engine from General Motors. This LDDE is used in the Chevrolet

FIGURE 7.18 Exhaust aftertreatment system schematic for the FCA 3.0 L V-6 EcoDiesel engine in the RAM 1500 light truck [7-39]. (Image courtesy of FCA.)

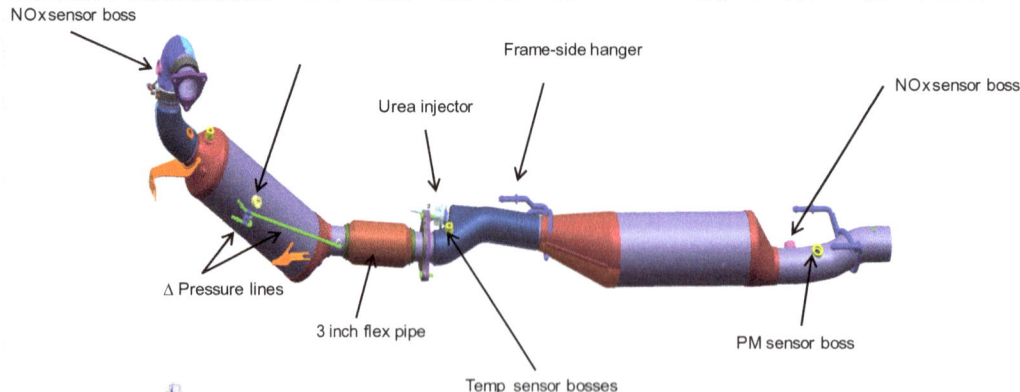

© FCA US LLC

Silverado light truck and is new as of 2019 [7-40]. The statistics for the engine are shown in **Table 7.4**, and the engine is pictured in **Figure 7.19**.

The EAS for the 3.0 L I-6 Duramax includes a close-coupled DOC, combined SCR–DPF*, underfloor SCR, and ASC units, as shown in **Figure 7.20**, where the SCR–DPF is labeled "SCRF." High-pressure and low-pressure EGR flow paths are included for criteria pollutant management. Verdino et al. [7-40] also illustrated where the EAS sensors are placed. As shown in **Figure 7.21**, the Duramax EAS includes four temperature sensors, one differential pressure sensor, three NOx sensors, and one PM sensor. As of writing the EPA certification levels were not available from FuelEconomy.gov [7-37], but are estimated to be Tier 3 bin 125 or Tier 3 bin 160.

7.5.3 Light Heavy-Duty Diesel

Light HDDEs (LHDDEs) are typically in the displacement range of about 5 L to about 7 L. They are used in lighter HDVs, especially those Class 2b and 3 HDVs that are certified on a chassis dynamometer (chassis-certified) as a homologated vehicle.

A typical LHDDE is the Ford Power Stroke® diesel engine, pictured in **Figure 7.22**. The engine is a 6.7 L V-8, and has one version used in chassis-certified LHDVs such as the Ford F-250 or F-350 SuperDuty pickup trucks and a related version used in dyno-certified HDVs spanning Class 2b through Class 8 ratings. The attributes of the MY2015 version of the engine are shown in **Table 7.5**.

As described by Lambert [7-42], the EAS for the MY2011 chassis-certified Power Stroke® engine uses the following subsystems: DOC, urea–SCR, and DPF. One of the interesting features of this EAS is that the urea-SCR system is placed upstream of the DPF, as shown in the schematic of the EAS in **Figure 7.23**. The urea–water solution, also known as Diesel Exhaust Fluid (DEF) or AdBlue, is discussed in more detail in Chapter 12, Selective Catalytic Reduction Systems. The DEF is injected after the DOC and upstream of the SCR unit, and the schematic in Figure 7.23 suggests that there is a static mixing element in the exhaust pipe upstream of the SCR unit. It is assumed that the SCR has been placed upstream of the DPF to facilitate its thermal management since the SCR has a minimum operating temperature during warm-up of about 150°C and a preferred minimum operating temperature of about 200°C. By contrast, the DPF will filter particulates even when cold, although triggering an active regeneration is more challenging when it is the last element of the EAS.

TABLE 7.4 General Motors 3.0 L I-6 Duramax engine [7-40].

Attribute	Value
Displacement	2.993 L
Bore	84.0 mm
Stroke	90.0 mm
Compression ratio	15.0:1
Rated power	205 kW (278 hp) at 3,500 rpm
Peak torque	610 N·m (450 lb·ft.) at 1,500 rpm
Fuel injection system	Common rail DI
Air handling system	Variable geometry turbocharger

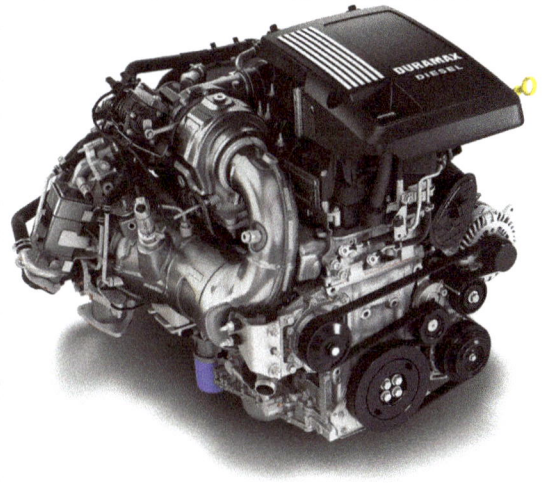

FIGURE 7.19 General Motors 3.0 L I-6 Duramax engine. (Image courtesy of General Motors.)

© General Motors

---

* The combined SCR-DPF unit is also known as an SDPF and is described further in Section 15.2.

FIGURE 7.20 Exhaust aftertreatment system for the General Motors 3.0 L I-6 Duramax engine. (Figure 32 in [7-40]; image courtesy of General Motors.)

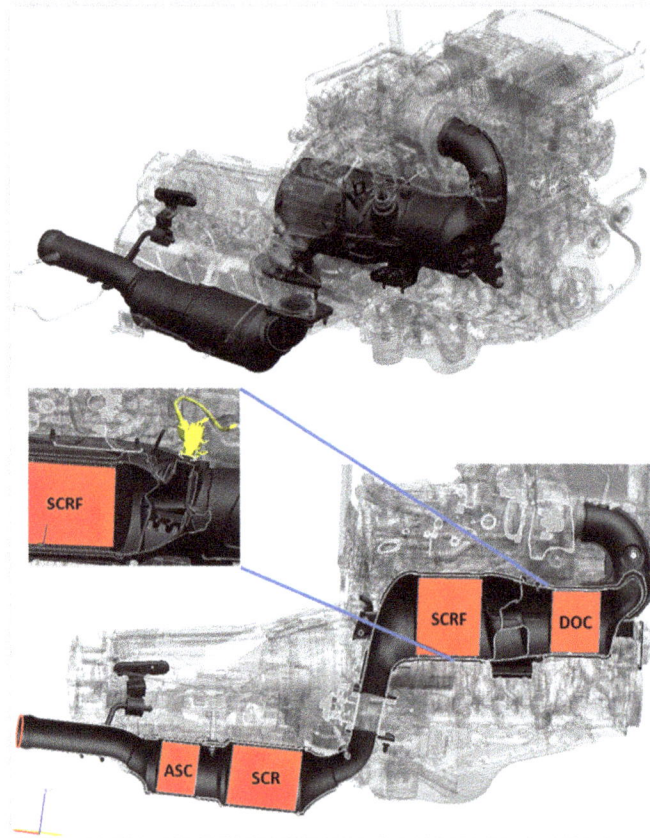

Reprinted from Ref. [7-40]. 40th International Vienna Motor Symposium, 2019

FIGURE 7.21 Exhaust aftertreatment system sensors for the General Motors 3.0 L I-6 Duramax engine. Red: temperature sensors; green: differential pressure sensor; blue: NOx sensors; purple: PM sensor. (Figure 34 in [7-40]; image courtesy of General Motors.)

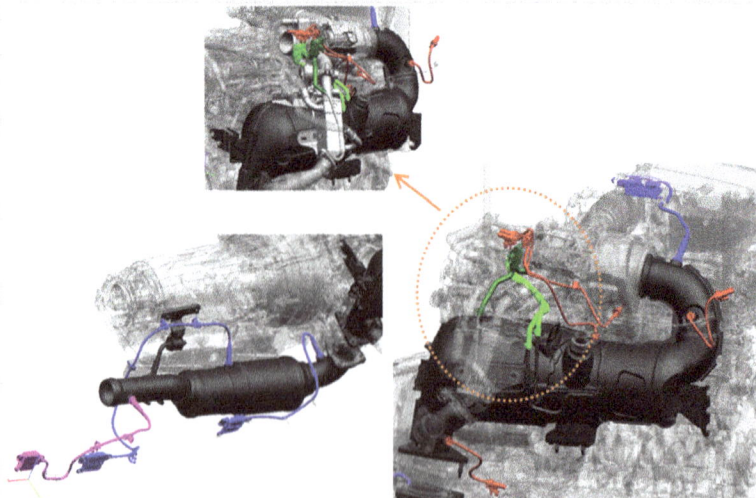

Reprinted from Ref. [7-40]. 40th International Vienna Motor Symposium, 2019

FIGURE 7.22 Model year 2011 Ford Power Stroke® engine [7-41].

TABLE 7.5 The Ford Power Stroke® light heavy-duty diesel engine for MY2015.

Attribute	Value
Displacement	6.702 L
Bore	99 mm
Stroke	108 mm
Compression ratio	16.2:1
Rated power	336 kW (450 hp) at 2,800 rpm
Peak torque	1,268 N·m (935 lb·ft) at 1,600 rpm
Fuel injection system	Common rail DI
Air handling system	Variable geometry turbocharger

FIGURE 7.23 Exhaust aftertreatment system for the MY2011 chassis-certified Ford Power Stroke® engine [7-42] (image courtesy of Ford Motor Company).

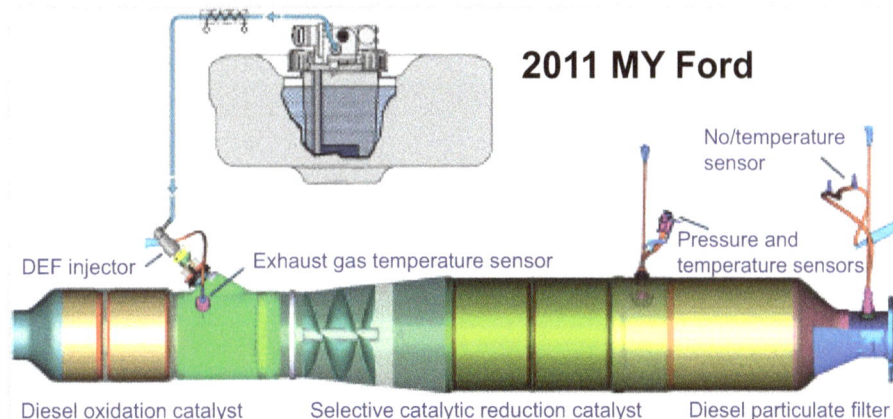

2011 MY Ford

No/temperature sensor

Pressure and temperature sensors

DEF injector

Exhaust gas temperature sensor

Diesel oxidation catalyst

Selective catalytic reduction catalyst

Diesel particulate filter

7.5.4 **Medium Heavy-Duty Diesel**

Medium HDDEs (MHDDEs) span a displacement range from about 5 L to about 11 L. They are typically used in light and medium HDV that have the engine certified separately from the vehicle (dyno certified), and can be found in Class 2b through Class 8 vocational vehicles, especially Class 6 through 8 applications.

A typical MHDDE is the Daimler Detroit™ DD8™, pictured in **Figure 7.24**. The engine is a 7.7 L I-6 and is sold as a dynamometer-certified engine for HDV applications. The attributes of the MY2017 version of the engine are shown in **Table 7.6**. As is typical for engines in this size range, there is a corresponding 5.1 L I-4 engine in the same family, the DD5 [7-33]. The DD5 and DD8 engines are the North American versions of the Daimler OM934 and OM936 engines, respectively.

As described by Daimler [7-44], the EAS for the Detroit™ DD8™ engine has a "1-Box" package that contains the following subsystems: DOC, DPF, and urea-SCR. The DEF is

FIGURE 7.24 Daimler Detroit™ DD8™ medium heavy-duty diesel engine [7-43] (image courtesy of Daimler Trucks North America).

© Daimler

TABLE 7.6 The Daimler Detroit™ DD8™ medium heavy-duty diesel engine [7-43].

Attribute	Value
Displacement	7.7 L
Bore	110 mm
Stroke	135 mm
Compression ratio	17.6:1
Rated power	261 kW (350 hp) at 2,200 rpm
Peak torque	1,424 N·m (1,050 lb·ft.) at 1,400 rpm
Fuel injection system	High-pressure common rail DI
Air handling system	Dual-stage turbocharger

FIGURE 7.25 Exhaust aftertreatment system for the Daimler Detroit™ DD8™ medium heavy-duty engine [7-44] (image courtesy of Daimler Truck North America).

injected after the DPF and upstream of the SCR unit, into the decomposition tube, which is the length of exhaust pipe upstream of the SCR unit. As shown in **Figure 7.25**, the exhaust gas enters through the port on the top left of the 1-Box unit and flows to the top of the back face of the unit. The exhaust gas then flows from back to front through the DOC and DPF. The DEF is then injected into the exhaust gas as the flow turns to go from front to back. The gas flow then makes another 180° bend and flows from back to front through the SCR unit and then out the exhaust pipe to the tailpipe end. Herrmann et al. [7-45] described the function of the EAS for the related OM936 engine also.

7.5.5 Heavy Heavy-Duty Diesel

An example of a common on-road heavy heavy-duty diesel engine (HHDDE) is the D13 engine from Volvo Trucks. The 405 hp variant of this on-road, inline-6 HHDDE was used as an example in Chapter 4, Emissions Testing Protocols, and the torque and power curves for the engine are found in Figure 4.15 in Section 4.2.1. The engine itself is pictured in **Figure 7.26**, and its attributes, including the rated power and peak torque for the example variant, are listed in **Table 7.7**.

The EAS used with the D13 is called the "One Box" EAS or EATS [7-46, 7-47] and is used to meet the EPA 2010 emissions standards described in Chapter 3. This compact approach is common for on-road HHDDE because of the packaging challenges on an HDV, and the manufacturer's interest in keeping vehicle mass down. For example, the Volvo One Box EATS configuration saves 17 lb. (7.7 kg) compared to a more distributed configuration [7-47]. The One Box EATS includes the DOC, DPF, urea injection system, SCR unit, and ASC all within one enclosure. The D13 One Box EATS specifically uses a copper-zeolite catalyst in the SCR unit. Other manufacturers sometimes place the DOC in a close-coupled position so that it warms up faster. As shown in **Figure 7.27**

TABLE 7.7 Attributes of the Volvo Trucks D13 engine [7-46].

Attribute	Value
Displacement	12.8 L
Bore	131 mm
Stroke	158 mm
Compression ratio	17:1
Rated power	302 kW (405 bhp) at 1,600 rpm
Peak torque	1,966 N·m (1,450 lb·ft) at 1,000 rpm
Fuel injection system	Common rail DI
Air handling system	Sliding nozzle variable geometry turbocharger

FIGURE 7.26   Volvo Trucks D13 engine [7-46] (image courtesy of Volvo Group).

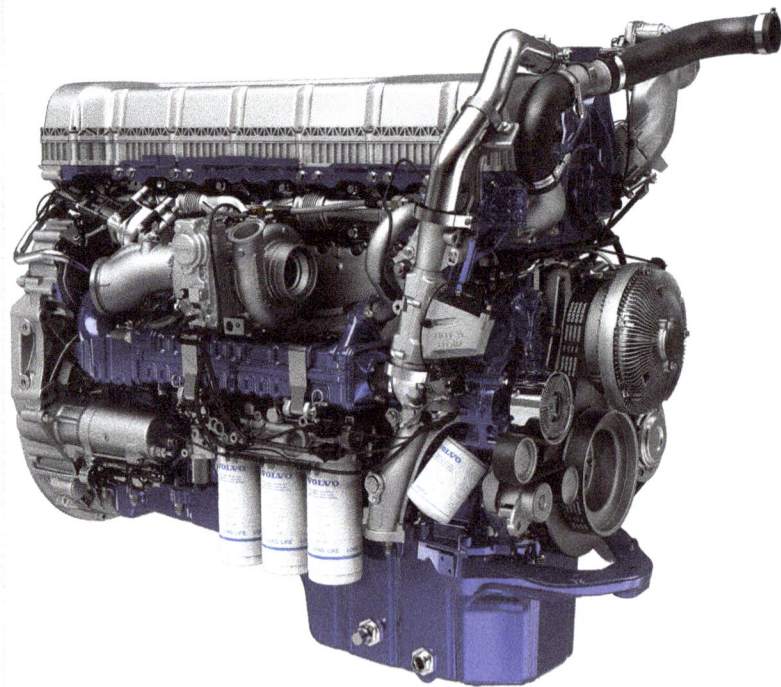

© Volvo

[7-47], the intake is on the top of the enclosure, and the final exhaust is at the back left. The DEF injector is at the bottom center of the right side face, near where the orange wires connect to the One Box EATS. The flow path in the Volvo One Box EATS is typical, where the exhaust gas flow switches directions several times through the system to go through the various components and then exit to the last part of the tailpipe. Referencing Figure 7.27, the flow enters the top of the system, flows from left to right through the DOC and DPF, turns to flow right to left through the urea decomposition tube, turns to flow from left to right through the SCR and ASC, and finally turns and flows from right to left to exit the One Box EATS.

FIGURE 7.27 One Box Exhaust aftertreatment system (EATS) for D13 engine [7-47] (image courtesy of Volvo Group).

© Volvo

The other HDDE used as an example in Chapter 4 is the Scania DC13 085A [7-48], which is an off-road inline 6 engine shown in **Figure 7.28**. As with the on-road Volvo D13 HDDE, the torque and power curves for this engine are in Section 4.2.1, in Figure 4.16. The engine's attributes are shown in **Table 7.8**.

FIGURE 7.28 Scania DC13 085A Stage IV or Tier 4f industrial engine [7-48] (image courtesy of Scania CV AB).

© Scania

TABLE 7.8 Scania DC13 085A Stage IV or Tier 4f industrial engine [7-48].

Attribute	Value
Displacement	12.7 L
Bore	130 mm
Stroke	160 mm
Compression ratio	17.5:1
Rated power	405 kW (550 bhp) at 1,800 rpm
Peak torque	2,363 N·m (1,755 lb·ft) at 1,200 rpm
Fuel injection system	Scania XPI Common rail DI
Air handling system	Variable geometry turbocharger

FIGURE 7.29 Exhaust aftertreatment system for Scania DC13 engine [7-48] (image courtesy of Scania CV AB).

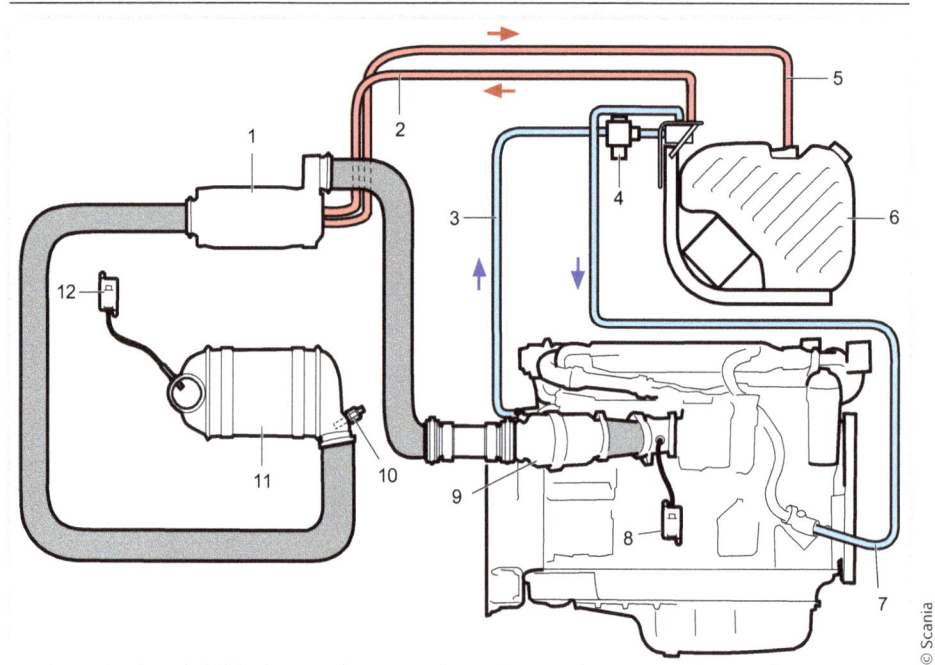

© Scania

The EAS for the DC13 uses a combination of EGR and SCR to meet EU Stage IV and EPA Tier 4 final non-road emissions requirements, which are described in Chapter 3. A schematic of the EAS for the DC13 is shown in **Figure 7.29**. The unit marked "9" is where the DOC is placed in the related 9 L DC9 engine; the DC13 EAS does not include a DOC. The unit marked "1" is where the DEF is injected, evaporated, and decomposed to ammonia (NH_3). The unit marked "11" is the SCR catalyst. There are upstream and downstream nitrogen oxide (NOx) sensors at "8" and "12," respectively, to help manage the DEF dosing control. In addition, note that the urea storage tank "6" has engine coolant routed to it to melt the DEF if it starts freezing. As discussed in detail in Chapter 12, DEF freezes when below −11°C (12.2°F).

References

7-1.   Cannon, W., Hill, E., and Welling, C., "Single Cylinder Engine Tests of Oxidation Catalysts," SAE Technical Paper 570280, 1957, https://doi.org/10.4271/570280.

7-2.   Nebel, G.J. and Bishop, R.W., "Catalytic Oxidation of Automobile Exhaust Gases – An Evaluation of the Houdry Catalyst," SAE Technical Paper 590224, 1959, https://doi.org/10.4271/590224.

7-3.   Keenan, M., "Exhaust Emissions Control: 60 Years of Innovation and Development," SAE Technical Paper 2017-24-0120, 2017, https://doi.org/10.4271/2017-24-0120.

7-4.   Ruetten, O., Pischinger, S., Küpper, C., Weinowski, R.et al., "Catalyst Aging Method for Future Emissions Standard Requirements," SAE Technical Paper 2010-01-1272, 2010, https://doi.org/10.4271/2010-01-1272.

7-5.   McLaren, M.G. and Ott, W.R., "Ceramic Monolithic Substrates—State of the Art," SAE Technical Paper 741050, 1974, https://doi.org/10.4271/741050.

7-6.   Howitt, J.S., "Thin Wall Ceramics as Monolithic Catalyst Supports," SAE Technical Paper 800082, 1980, https://doi.org/10.4271/800082.

7-7.   Koltsakis, G.C. and Stamatelos, A.M., "Catalytic Automotive Exhaust Aftertreatment," *Progress in Energy and Combustion Science* **23**, no. 1 (1997): 1–39.

7-8.   Parks II, J.E., Storey, J.M.E., Williams, A.M., Ferguson III, H.D., Tassitano, J., Ponnusamy, S., and Theiss, T.J., "Lean NOx Trap Catalysis for Lean Natural Gas Engine Applications," Report ORNL/TM-2007/150, https://info.ornl.gov/sites/publications/files/Pub7755.pdf, 2007.

7-9.   Stewart, M.L., Kamp, C.J., Gao, F., Wang, Y., and Engelhard, M.H., "Coating Distribution in a Commercial SCR Filter," *Emission Control Science and Technology* **4**, no. 4 (2018): 260–270, https://doi.org/10.1007/s40825-018-0097-3.

7-10.   Oh, S.H. and Cavendish, J.C., "Transients of Monolithic Catalytic Converters: Response to Step Changes in Feedstream Temperature as Related to Controlling Automobile Emissions," *Industrial & Engineering Chemistry Product Research and Development* **21**, no. 1 (1982): 29–37.

7-11.   Levenspiel, O., *Chemical Reaction Engineering*, 2nd ed. (New York: John Wiley & Sons, 1972).

7-12.   Etheridge, J.E., Watling, T.C., Izzard, A.J., and Paterson, M.A., "The Effect of Pt: Pd Ratio on Light-Duty Diesel Oxidation Catalyst Performance: An Experimental and Modelling Study," *SAE Int. J. Engines* **8** (2015): 1283–1299, https://doi.org/10.4271/2015-01-1053.

7-13.   Crosscut Lean Exhaust Emissions Reductions Simulations (CLEERS), "CLEERS," https://cleers.org/. Accessed 30 Jul. 2019.

7-14.   Summers, J.C. and Hegedus, L.L., "Modes of Catalyst Deactivation in Stoichiometric Automobile Exhaust," *Industrial & Engineering Chemistry Product Research and Development* **18**, no. 4 (1979): 318–324, https://doi.org/10.1021/i360072a018.

7-15.   Yao, H.C., Japar, S., and Shelef, M., "Surface Interactions in the System Rh/Al$_2$O$_3$," *Journal of Catalysis* **50**, no. 3 (1977): 407–418, https://doi.org/10.1016/0021-9517(77)90053-7.

7-16.   Wang, T. and Schmidt, L.D., "Intraparticle Redispersion of Rh and Pt-Rh Particles on SiO2 and Al2O3 by Oxidation-Reduction Cycling," *Journal of Catalysis* **70**, no. 1 (1981): 187–197, https://doi.org/10.1016/0021-9517(81)90328-6.

7-17.   Wang, T. and Schmidt, L.D., "Surface Enrichment of Pt-Rh Alloy Particles by Oxidation-Reduction Cycling," *Journal of Catalysis* **71**, no. 2 (1981): 411–422, https://doi.org/10.1016/0021-9517(81)90245-1.

7-18.   Nunan, J.G., Williamson, W.B., Robota, H.J., and Henk, M.G., "Impact of Pt-Rh and Pd-Rh Interactions on Performance of Bimetal Catalysts," SAE Technical Paper 950258, 1995, https://doi.org/10.4271/950258.

7-19.   Williamson, W.B., Gandhi, H.S., Heyde, M.E., and Zawacki, G.A., "Deactivation of Three-Way Catalysts by Fuel Contaminants—Lead, Phosphorus and Sulfur," SAE Technical Paper 790942, 1979, https://doi.org/10.4271/790942.

7-20.   Williamson, W.B., Stepien, H.K., Watkins, W.L., and Gandhi, H.S., "Poisoning of Platinum-Rhodium Automotive Three-Way Catalysts by Lead and Phosphorus," *Environmental Science & Technology* **13**, no. 9 (1979): 1109–1113, https://doi.org/10.1021/es60157a016.

7-21.   Falk, C.D. and Mooney, J.J., "Three-Way Conversion Catalysts: Effect of Closed-Loop Feed-Back Control and Other Parameters on Catalyst Efficiency," SAE Technical Paper 800462, 1980, https://doi.org/10.4271/800462.

7-22.   Engler, B.H., Lox, E.S., Ostgathe, K., Ohata, T., Tsuchitani, K., Ichihara, S., Onoda, H., Garr, G.T., and Psaras, D., "Recent Trends in the Application of Tri-Metal Emission Control Catalysts," SAE Technical Paper 940928, 1994, https://doi.org/10.4271/940928.

7-23.   Casassa, J. and Beyerlein, D., "Engine Dynamometers for the Testing of Catalytic Converter Durability," SAE Technical Paper 730558, 1973, https://doi.org/10.4271/730558.

7-24.   Smedler, G., Ahlström, G., Fredholm, S., Frost, J., Lööf, P., Marsh, P., Walker, A., and Winterborn, D., "High Performance Diesel Catalysts for Europe Beyond 1996," SAE Technical Paper 950750, 1995, https://doi.org/10.4271/950750.

7-25.   Bird, R.B., Stewart, W.E., and Lightfoot, E.N., *Transport Phenomena*, 1st ed. (New York: John Wiley & Sons, 1960).

7-26.   Moody, L.F., "Friction Factor for Pipe Flow," *Transactions of the ASME* **66** (1944): 671–684.

7-27.   Welty, J.R., Wicks, C.E., and Wilson, R.E., *Fundamentals of Momentum, Heat, and Mass Transfer*, 3rd ed. (New York: John Wiley & Sons, 1984).

7-28.   Lupescu, J.A., Chanko, T.B., Richert, J.F., and DeVries, J.E., "Treatment of Vehicle Emissions from the Combustion of E85 and Gasoline with Catalyzed Hydrocarbon Traps," *SAE Int. J. Fuels Lubr.* **2**, no. 1 (2009): 485–496, https://doi.org/10.4271/2009-01-1080.

7-29.   de Monte, M., Hadl, K., Noll, H., and Mannsberger, S., "SCR Control Strategies with Multiple Reduction Devices for Lowest NOx Emissions," *SAE Heavy-Duty Diesel Emissions Control Symposium*, Gothenburg, 2018.

7-30.   Ding, C., Roberts, L., Fain, D.J., Ramesh, A.K., Shaver, G.M., McCarthy Jr, J., Ruth, M., Koeberlein, E., Holloway, E.A., and Nielsen, D., "Fuel Efficient Exhaust Thermal Management for Compression Ignition Engines during Idle via Cylinder Deactivation and Flexible Valve Actuation," *International Journal of Engine Research* **17**, no. 6 (2016): 619–630, https://doi.org/10.1177/1468087415597413.

7-31.   Honardar, S., Busch, H., Schnorbus, T., Severin, C.et al., "Exhaust Temperature Management for Diesel Engines Assessment of Engine Concepts and Calibration Strategies with Regard to Fuel Penalty," SAE Technical Paper 2011-24-0176, 2011, https://doi.org/10.4271/2011-24-0176.

7-32.   Gehrke, S., Kovács, D., Eilts, P., Rempel, A., and Eckert, P., "Investigation of VVA-Based Exhaust Management Strategies by Means of a HD Single Cylinder Research Engine and Rapid Prototyping Systems," *SAE Int. J. Commer. Veh.* **6**: 47–61, 2013, https://doi.org/10.4271/2013-01-0587.

7-33.   Gruden, I., Cozzolini, A., Aleksandrov, M., and Robertson, J., "The Detroit DD5 Engine –Daimler's Commercial Medium Duty Engine Generation Enters NAFTA Market," *25th Aachen Colloquium*, Aachen, 2016.

7-34.   McCarthy Jr, J., Theissl, H., and Walter, L., "Improving Commercial Vehicle Emissions and Fuel Economy with Engine Temperature Management Using Variable Valve Actuation," *VDI Commercial Vehicle Engine Technology Conference*, Baden-Baden, Germany, 2017.

7-35.   Honda, "VTEC TURBO – The New Era Turbo Engine," https://global.honda/innovation/technology/automobile/Vtec-turbo-picturebook.html. Accessed 22 Apr. 2019.

7-36.   Honda Motor Company, "2019 Accord Sedan Specifications and Features," https://automobiles.honda.com/accord-sedan#. Accessed 22 Apr. 2019.

7-37.   U.S. Department of Energy and Environmental Protection Agency, www.FuelEconomy.gov – The official government source for fuel economy information. https://www.fueleconomy.gov/. Updated 25 Mar. 2019, Accessed 25 Mar. 2019.

7-38.   FCA, "RAM 1500 EcoDiesel," https://www.ramtrucks.com/ecodiesel.html. Updated 2019, Accessed 20 Jun. 2019.

7-39.   FCA, RAM 1500 Big Horn Diesel Aftertreatment System," https://www.moparfactoryparts.com/v-2018-ram-1500--big-horn--3-0l-v6-diesel/emission-system--diesel-aftertreatment-system. Updated 2019, Accessed 20 Jun. 2019.

7-40.   Verdino, V., Lavazza, P., Rampone, D., Garzarella, L., Mogavero, A., Vassallo, A., Barta, J., Catalogna, J., Franco, S., Marlett, C., and Umierski, M., "The New General Motors 3.0 Liter Duramax Diesel Inline 6-Cylinder Engine for the 2019 Chevrolet Silverado and GMC Sierra," *40th International Vienna Motor Symposium*, Vienna 2019.

7-41.   Deraad, S., Fulton, B., Gryglak, A., Hallgren, B.et al., "The New Ford 6.7L V-8 Turbocharged Diesel Engine," SAE Technical Paper 2010-01-1101, 2010, https://doi.org/10.4271/2010-01-1101.

7-42.   Lambert, C., "Practical Aspects of Urea SCR for Automotive NOx Control," *2018 CLEERS Workshop*, Ann Arbor, MI, 2018.

7-43.   Daimler, "Detroit DD8 Technical Specifications," Daimler Trucks North America, 2017.

7-44.   Daimler, "Aftertreatment Support," https://demanddetroit.com/parts-service/detroit-genuine-parts-program/aftreament-support/. Accessed 15 Feb. 2019.

7-45.   Herrmann, H.-O., Nielsen, B., Gropp, C., and Lehmann, J., "Mittelschwerer Nfz-Motor von Mercedes-Benz," *MTZ-Motortechnische Zeitschrift* **73**, no. 10 (2012): 730-738, https://doi.org/10.1007/s35146-012-0473-4.

7-46.   Volvo Trucks, "Volvo D13 Engine Family," Volvo Trucks North America, 2016.

7-47.   Volvo Trucks, "Powertrain Emissions Technology," https://www.volvotrucks.us/trucks/emissions/. Accessed 2 Apr. 2019.

7-48.   Scania, "DC13 085A. 405 kW (550 hp)," Scania CV AB, 2018.

Oxidation Catalysts

Oxidation, I never tire of reminding myself, is what happens when oxygen attacks.
—Alan Bradley, A Red Herring without Mustard

In this chapter on oxidation catalysts or oxycats, we begin the discussion of the individual catalytic converters that comprise the exhaust aftertreatment system (EAS) that continues through Chapter 15. It is important to note that all of the systems described in Chapters 8 through 15 are generally classified as "catalytic converters," meaning they contain a catalyst which participates in the conversion of the harmful emissions species into more benign products. Catalytic converters have now been in use on vehicles for decades, but the very first commercialized aftertreatment system, commonly referred to as *the* catalytic converter, is the oxycat.

Oxycats were developed to lower the emissions of carbon monoxide (CO) and hydrocarbons (HC) in the exhaust from gasoline spark ignition (SI) engines. The first production catalytic converter was developed by Engelhard Corporation in 1973 [8-1]. This first, commercial exhaust aftertreatment device was an oxidation catalyst deployed on the 1975 model year (MY) light-duty vehicles (LDVs) with gasoline-fueled SI engines [8-2]. These vehicles were required to meet the emissions standards resulting from the 1970 Clean Air Act. This federal act is more fully discussed in Chapter 3, Emissions Control Regulations, but briefly, it curtailed the allowable level of HC and CO in vehicular emissions. Due to the successful use of these oxycats, exhaust emissions from LDVs with SI engines have been drastically reduced in the decades since, with a net improvement of over 95% [8-3]. Starting in 1977, several states, including emissions control leader California, required the use of catalysts for nitrogen oxide (NOx) control. Thus, the oxycat was developed into the three-way catalyst (TWC), which is used with modern stoichiometric SI engines and is discussed further in Chapter 9, Three-Way Catalysts.

The oxycat began a resurgence in the 1990s with a change in its application to compression ignition (CI) engines, where it became known as the diesel oxidation catalyst (DOC). There are two primary reasons for the rebirth of the oxycat. First, as described in Chapter 3, Emissions Control Regulations, changing regulations in the 1990s required the removal of HC and CO emissions from CI engine exhaust. Though diesel-fueled CI engines are generally thought of as low emitters of HCs and CO in the raw exhaust, at the time they were relatively high compared to the tailpipe emissions from SI engines, which had TWCs to remove pollutants. The second reason for the reappearance of the oxycat was related to particulate matter (PM) emissions. PM formation and characteristics are described in detail in Section 2.7.1. Briefly, PM comprises a carbonaceous solid and a complex mixture of higher molecular weight organic species, which is known as the soluble organic fraction (SOF). The oxycat can remove some of the SOF but the exhaust gas temperatures are too low to oxidize the solid carbon fraction, so it passes through largely unchanged. The DOC can also have a significant role in changing the ratio of the nitrogen oxide NOx species in the exhaust gas, as nitric oxide (NO) can be oxidized to nitrogen dioxide (NO_2) or vice versa in the DOC.

Oxycats were the first type of catalytic converter to be commercialized and widely deployed in the 1970s. The first version to be implemented was a two-way oxidation catalyst that simultaneously oxidizes HCs and CO. In addition, the SOF of diesel particulates and nonregulated emissions, such as aldehydes or polycyclic aromatic hydrocarbons (PAHs), can be oxidized to harmless products over the oxycat, and thus, can be controlled. In modern EAS for fuel-lean exhaust gas, an important function of the oxycat is to oxidize NO to NO_2, which supports soot removal from diesel particulate filters (DPF) and NOx reduction in selective catalytic reduction (SCR) systems.

The DOC is an essential part of the EAS for fuel-lean exhaust gas and its performance depends on the precious metal catalyst used and the washcoat that supports it. The lean nature of CI engine exhaust means that there is sufficient oxygen (O_2) for the desired oxidation reactions. DOC performance is also affected by the operating conditions, especially temperature. It has a "sweet-spot" operating temperature range, outside of which its activity can fall off precipitously. A comprehensive discussion of reactions, reaction kinetics, and other aspects of the technology is presented in the following sections.

8.1 Oxidation Reaction Chemistry

The chemical reactions over the oxidation catalyst are promoted by active catalytic sites that can adsorb oxygen. In general, the catalytic oxidation reactions include the following four stages:

1. Oxygen in the bulk gas bonds to a surface catalyst site, effectively storing it on the catalyst

2. Emissions species, such as CO and HCs, diffuse from the bulk gas to the surface of the catalyst

3. Emissions species react with the stored oxygen on the surface of the catalyst, following Eley–Rideal kinetics

4. Reaction products, such as carbon dioxide (CO_2) and water (H_2O) vapor, desorb from the catalyst surface and diffuse into the bulk exhaust gas

The HC species in the exhaust are typically a mixture of paraffin (alkanes), olefins (alkenes), and aromatics ranging in size from one to eight carbons. Over the oxidation catalyst, these HCs are oxidized to form CO_2 and water as shown in 8.1 using a generic HC compound (C_xH_y),

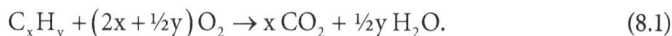

$$C_xH_y + \left(2x + \tfrac{1}{2}y\right)O_2 \rightarrow x\,CO_2 + \tfrac{1}{2}y\,H_2O. \tag{8.1}$$

The oxycat also facilitates the oxidation of CO to CO_2,

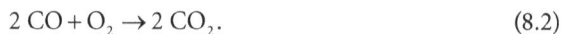

$$2\,CO + O_2 \rightarrow 2\,CO_2. \tag{8.2}$$

The reactions in Eqs. 8.1 and 8.2 complete the combustion of the fuel-derived emissions species into CO_2 and water vapor.*

As described in Chapter 2, Emissions Formation, the organic fraction of PM comprises adsorbed fuel and partially combusted HCs. Therefore, the reactions represented by Eqs. 8.1 and 8.2 apply to two processes: the oxidation of gas-phase HC emissions species and of the PM SOF. These reactions are both exothermic and the resulting heat release can be desirable. The normal operation of the oxycat is unlikely to create enough heat to influence the temperature of catalytic converters downstream of the oxycat in the EAS. Late fuel injection in the cylinder can be used to generate heat in the oxycat via catalytic combustion and thereby warm up the following catalytic converters in the EAS. This approach is typically used to help trigger an active regeneration event in a DPF, as discussed further in Chapter 11, Particulate Filters. In addition, the oxidation of NO to NO_2 is essential for both passive regeneration in a DPF [8-5], as discussed in Chapter 11, and to enhance the NOx reduction chemistry in the SCR system, as discussed in Chapter 12. The oxidation of NO to NO_2, can be kinetically or thermodynamically limited depending on the oxycat operating conditions.

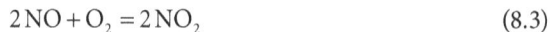

$$2\,NO + O_2 = 2\,NO_2 \tag{8.3}$$

To heat the DPF to the temperatures at which the soot ignites and active regeneration starts—450°C–600°C—active thermal strategies to raise the exhaust gas temperature are required. These thermal management strategies incur a fuel penalty and expose the oxycat and DPF to high temperatures that can accelerate catalyst aging. Since soot is oxidized by NO_2 at lower temperatures, typically 300°C–400°C, and the reaction does not usually require active thermal management, and is therefore considered a passive process.

Likewise, changing the ratio of NO to NO_2 in the exhaust can enhance the performance of the SCR system. As discussed in Section 12.1.1.2, the SCR reaction with equimolar amounts of NO and NO_2, shown in Eq. 12.6, is faster than the reactions consuming only NO or NO_2.

---

* The review paper on DOCs by Russell and Epling [8-4] has more detail on oxidation reaction mechanisms and kinetics, including curated tables of CO and HC oxidation reaction kinetic parameters collected from the literature.

Therefore, the oxycat also supports the performance of other catalytic converters in the EAS. That said, increasing the NO_2 concentration in the oxycat raises the risk that more NO_2 escapes into the environment, which is an issue because NO_2 is significantly more reactive and hazardous than NO. Thus, in situations where not all of the promoted NO_2 is consumed by the downstream catalysts, the increased tailpipe NO_2 emissions will contribute to air quality problems [8-6].

8.1.1 Diesel Oxidation Catalysts

For SI engines, the oxycat was superseded by the TWC due to the added requirement for NOx abatement. However, starting in 1988, when oxycats appeared on LDVs with CI engines, they regained popularity in EAS. In the 1990s, their use expanded to heavy-duty diesel engines (HDDEs) in the United States. While the raw, engine-out emissions of diesel-fueled CI engines have relatively low concentrations of CO and HCs as compared to their SI counterparts, catalytic treatment of the exhaust is necessary for regulatory compliance [8-7].

The DOC promotes the oxidation of exhaust gas components by oxygen, which is present in ample quantities in diesel exhaust. The function of the DOC in the EAS has evolved significantly since the DOC was first introduced as a stand-alone device to control gaseous emissions of CO and HCs. Another contributing factor to the increased use of DOCs was the ability to provide some degree of PM emission control. PM, including soot, is described in detail in Chapter 2, Emissions Formation, and Chapter 11, Particulate Filters. Briefly, PM consists of a solid, carbonaceous skeleton with a mixture of adsorbed HCs, the SOF. An oxycat can remove some of the SOF, through the reactions shown in Eqs. 8.1 and 8.2, although the temperatures are too low to oxidize the elemental carbon portion of the PM. DOCs designed for PM, HCs, and CO control can also reduce several unregulated emissions species, such as aldehydes and PAH, as well as reducing the compounds responsible for exhaust odor. However, unlike stoichiometric SI engine exhaust, the lean air-fuel ratio (AFR) used in CI combustion results in the exhaust gas with oxygen. Thus, a DOC cannot also reduce NOx using redox chemistry, as is done in a TWC. The DOC, therefore, needs additional catalytic converters in the EAS for the application to meet its emissions targets.

The location of the oxycat within the EAS affects the performance of the overall system. The preferred location is close enough to the engine or turbine outlet that the oxycat warms up quickly [8-8], but not so close that high exhaust gas temperatures damage or prematurely age the catalyst. Typically, then, the oxycat is the first catalytic converter in the EAS, and it is placed upstream of both the DPF and SCR system since the oxycat can enhance the performance of both of these catalytic converters as discussed in Section 8.1.1.

In addition, the DOC can support EAS warm-up by oxidizing HCs that are either introduced via direct diesel fuel injection into the exhaust system just ahead of the DOC or by a late-fuel injection into the cylinder after the majority of the work extraction has occurred. If both the NO oxidation and warm-up functions are required, they are typically combined in one DOC by zone-coating two different DOC formulations optimized for the respective activities. Zone coating is discussed further in Section 15.1.

8.1.2 **Warm-up Strategies**

As discussed in Section 7.3.2, the highest levels of pollutant emissions from a vehicle occur during cold start when both the exhaust gas and catalytic converters are colder than the light-off temperatures for the pollutant removal reactions. The duration of this cold start period is influenced by several factors including the distance between the exhaust manifold and the oxycat, the exhaust gas flow rate, the substrate type, the ambient conditions, and the emissions profile. The best performance of the DOC must balance the oxidation reactions for CO, HCs, SOF, and even sulfur dioxide (SO_2). These competing reactions result in conflicting temperature requirements for the catalyst, with the trade-off between the oxidation of SOF and SO_2 being the most important. In addition, because the DOC is often the first catalytic converter in the EAS, it can be used to accelerate the warm-up of the rest of the EAS. For all these reasons, thermal management of the DOC is desirable, especially heating the catalyst to its light-off temperature quickly [8-9].

Exhaust thermal management strategies can be achieved using either an air or fuel control strategy, as discussed in Section 7.3.2.1. In addition to airflow strategies to heat up the exhaust, fuel control strategies can be used to heat the exhaust gas thermal management at the expense of reduced engine efficiency [8-10]. Work at Oak Ridge National Laboratory (ORNL) [8-9] demonstrated the influence of fuel injection timing on exhaust gas temperature entering the DOC. Fuel dosers can also inject fuel into the exhaust upstream of the DOC to accelerate light-off [8-11, 8-12]. The air control strategies generally result in lower fuel consumption and better DOC warm-up via lower air-fuel ratios [8-10, 8-13, 8-14].

8.2 **Catalyst Formulations**

The engine exhaust environment is challenging for catalyst operation. A catalyst must be able to function at both low and high temperatures, over a wide range of exhaust flows, and tolerate poisons, mechanical vibrations, and thermal excursions. They must also remain metallic under the transient operating conditions and not form volatile oxide species that would lead to their disappearance during use. The ideal catalyst oxidizes several HC species and the SOF along with CO and NO over a wide range of temperatures while oxidizing no sulfur species.

From the beginning, most oxidation catalysts have comprised precious metals known as platinum group metals (PGM), including platinum (Pt) and palladium (Pd), which provide the active sites for the oxidation reactions. The choice of PGM as the active catalytic species is due to the following three key characteristics of Pt and Pd [8-15]:

1. They have enough catalytic activity to be effective given typical space velocities in the oxycat

2. They resist sulfur poisoning

3. They are not deactivated by the typical metal oxides used in the washcoat, such as alumina (Al_2O_3), ceria (CeO_2), titania (TiO_2), or zirconia (ZrO_2)

In addition, PGM are well-characterized for oxidation reactions and has strong oxidation performance with sufficient thermal durability [8-1]. Specifically, these materials are known to have superior oxidation activity for CO and HC species [8-16]. The PGM species have vastly different activities for HC oxidation depending on their activation states.

8.2.1 Platinum

The catalytic activity of Pt has been studied for more than 200 years. Its catalytic combustion activity was first discovered in 1817 by Humphry Davy [8-1, 8-17]. A few years later in 1823, Dobereiner was the first to prepare a supported heterogeneous catalyst, to study the Pt-catalyzed hydrogen (H_2) oxidation reaction used in lighters [8-18]. Faraday also studied the Pt-catalyzed combustion of H_2 and was the first to describe the reaction mechanism that includes the simultaneous adsorption of reactants onto the Pt surface [8-1, 8-19]. Another important use for Pt catalysts was in the production of ammonia (NH_3), which was very important when the Haber-Bosch process was scaled for industrial production before the First World War [8-20]. As work on Pt-based catalysts in chemical plants and research laboratories continued, elucidating their oxidation properties, they became attractive in controlling exhaust emissions.

Pt is the noblest of the PGM and has been shown to have the highest oxidation activity for HCs, especially higher weight HCs [8-21, 8-22]. However, under prolonged oxidizing conditions at high temperatures, it is susceptible to sintering through the migration of the oxide species [8-23, 8-24, 8-25]. Pt is susceptible to poisons, such as sulfur, though it is yet to be understood if the deactivation that is seen on Pt/Al_2O_3 catalysts is due to the sulfur interacting with the Pt metal itself [8-26, 8-27, 8-28], or with its Al_2O_3 support [8-29, 8-30, 8-31].

8.2.2 Palladium

Palladium catalysts can form Pd oxides (PdO) [8-32], which have been found to be more active than metallic Pd for some oxidation reactions [8-33, 8-34, 8-35]. Pd oxidation follows the Cabrera-Mott mechanism [8-36], in which dissociative adsorption of oxygen first forms an oxygen monolayer on the catalyst, before forming an amorphous oxide layer and finally PdO crystals [8-37]. Pd has been shown to be more resistant to sintering at high temperatures than Pt [8-32, 8-38], as well as useful in the oxidation of methane (CH_4), a strong greenhouse gas (GHG) that is usually excluded from criteria pollutant regulations [8-31, 8-39].

Though Pd experiences lower levels of SO_2 adsorption than Pt [8-40, 8-41], it is very susceptible to poisoning and activity inhibition by sulfur, as well as water, lead (Pb), and phosphorous (P) [8-42, 8-43, 8-44]. Reduction of PdO under mild conditions (175°C–275°C) is possible with a variety of reductants, each following a different mechanism [8-4].

8.2.3 Comparison of Materials

While both Pt and Pd can provide adsorption sites for both oxygen and HCs, Pt has a lower relative surface oxygen coverage [8-22] and is considered to have the highest oxidation activity [8-21]. It has also been shown that Pt has better performance for higher molecular weight HCs as compared to Pd [8-45]. In situations where thermal stability at high temperatures is not required, Pt can provide good oxidation activity, even in the presence of SO_2. However, in cases where the DOC will need to operate at the high temperatures required for DPF regeneration, thermal degradation will lead to the deactivation of Pt oxidation activity [8-46]. Though Pd forms a more stable oxide than Pt, Pd-based catalysts are also prone to deactivation at high temperatures, where they can sinter or decompose from PdO to Pd, which decreases the activity [8-47]. The desirable, highly dispersed state for the metal catalysts is due to activity considerations. As described in Section 7.2.1.1, catalytic activity is directly related to the active surface area of the metal particles.

For both the Pt and Pd catalysts, poison species can easily move from the metal active site to the support material, forming sulfates or sulfites [8-21, 8-43, 8-48, 8-49]. The effect of sulfation is more significant for Pd-based catalysts because the PdO active site can interact with both SO_2 and SO_3 [8-50, 8-51].

8.2.4 Mixed Metal Compositions

Frequently, two or more metals are used in combination as a catalyst. Pt and Pd can be combined into a mixture with properties desirable for oxidation catalysts. Mixtures of Pt and Pd have been shown to have slower sintering behavior than monometallic Pt catalysts [8-32, 8-38, 8-52]. Radially zone-coated bimetallic Pt–Pd particles, where the interior Pd was coated with an exterior shell of Pt, exhibited better light-off and steady-state performance for oxidizing CO and model HCs than monometallic catalysts did [8-53]. One significant advantage, though one which is highly dependent on the volatile PGM market pricing, is that adjusting the Pt and Pd content can help manage oxycat cost.

8.2.5 Alternatives to PGM

Many alternative materials, including alkaline-earth metals, base metals, lanthanides, and transition metals such as copper (Cu) and nickel (Ni) have been investigated as replacements for PGM catalysts [8-54], as they are all able to oxidize the PM SOF. Nevertheless, all of these alternatives are susceptible to poisoning by fuel contaminants or have insufficient thermal durability to last in automotive applications [8-55].

In addition, there has been a significant body of work regarding the use of zeolites in oxidation catalysts [8-41, 8-56, 8-57, 8-58, 8-59]. Zeolites are a family of catalysts that are described in depth in Section 7.2.5. In the oxycat application, zeolites can store HCs in their pores at low temperatures and release them at higher temperatures, when the oxycat is lit-off and active [8-54]. The pore size of the zeolites is a critical property for selectivity favoring water and CO_2 and avoiding byproducts. Zeolites are also commonly used as catalyst support for the oxycat PGM, as described in Section 8.3.

8.3 Washcoat Function and Substrates

A washcoat support with a high specific surface area is desirable to keep the catalyst highly dispersed and providing good oxidation performance. High PGM dispersion increases the exposed catalytic active sites for reaction. Using very small PGM particle sizes maximizes the surface area to volume ratio of the catalyst used and minimizes the mass of catalyst that must be used, thereby decreasing the cost of the device. However, the smaller the PGM particles, the more likely they are to migrate and coalesce, and therefore, the greater the dependence on the washcoat support to keep them dispersed. The washcoat support also influences the performance of the catalyst by improving its thermal stability, surface area, pore-volume, and surface reactivity [8-21, 8-60].

Alumina (Al_2O_3) is the most common washcoat material, though silica (SiO_2), titania (TiO_2), and zeolites, or any or all of these in combination are also used in order to tailor the material for its target application [8-61, 8-62]. Alumina can exist in several different phases, with γ-Al_2O_3 having the highest surface area, making it the most desirable form for the catalyst support. High temperatures can cause the crystal structure to change from γ-Al_2O_3 to the less desirable α-, δ-, or θ-Al_2O_3 structures. Zirconia (ZrO_2) has a significantly lower surface area than Al_2O_3 due to its highly crystalline structure, but Pt/ZrO_2 has been shown to light-off at lower temperatures compared to Pt/Al_2O_3 [8-60, 8-63].

Zeolites, described in detail in Section 7.2.5, are another common support material. Zeolites can be used to adsorb HCs in the exhaust before the catalyst has achieved light-off, which can lead to drastic improvements in the cold-start behavior of the catalyst. Once the catalyst is warm enough to support good HC oxidation activity, the zeolite releases the stored HCs and they are also oxidized over the catalyst. Zeolites have been shown to improve both CO and HC oxidation during cold-start [8-64].

To improve thermal stability, metal oxide support additives such as barium oxide (BaO), CeO_2, lanthanum oxide (La_2O_3), SiO_2, and ZrO_2 have been demonstrated [8-64, 8-65, 8-66]. Unfortunately, some of these metal oxides can block catalyst active sites, and therefore require further consideration.

As discussed in Section 7.2.1.3, the substrate is the solid, underlying material to which the washcoat with the catalyst is affixed and through which the exhaust gas flows. Factors that determine the substrate choice include size, cost, coatability, thermal properties and durability, and supplier availability.

Thin-walled ceramic honeycomb monolith supports are currently the most common substrates in use for oxycats. Depending on the application and temperature regime, this ceramic may be cordierite or silicon carbide (SiC). Some oxycat applications may also use a metal foil substrate.

8.4 Failure Modes

Oxycats have a few known failure modes, including catalyst aging and byproduct formation. Overall, the PGM catalysts used in oxycats age following the mechanisms discussed in Section 7.2.3.2, including sintering and poisoning.

Repeated exposure to high temperatures causes the PGM particles to migrate within the washcoat and agglomerate. These agglomerated catalyst particles have less surface area for the same mass, which leads directly to a loss of activity. Oxycat catalysts can also be poisoned by sulfur species, phosphorus, and other compounds deriving from the fuel or lubricant. Palladium (Pd) is particularly susceptible to poisoning and activity inhibition by sulfur, water, Pb, and phosphorus [8-42, 8-43, 8-44]. For example, SO_2 may be stored on the washcoat and block the active sites needed for the oxidation reactions.

Reduction of PdO at lower temperatures (175°C–275°C) is possible with a variety of reductants, each following a different mechanism [8-4], and this leads to a loss of catalytic activity.

One disadvantage of the oxycat is that it can oxide other species to form undesirable byproducts. For example, the oxycat can contribute to sulfate emissions by oxidizing SO_2 to sulfur trioxide (SO_3),

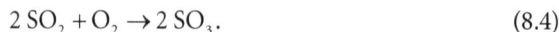

$$2\,SO_2 + O_2 \rightarrow 2\,SO_3. \tag{8.4}$$

The SO_3 can subsequently form sulfate ash compounds by reacting with metals such as calcium (Ca) or magnesium (Mg) that come from engine wear or lubricant additives. Alternately, it can form sulfuric acid (H_2SO_4) by reacting with water, per

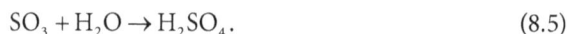

$$SO_3 + H_2O \rightarrow H_2SO_4. \tag{8.5}$$

When the exhaust gas leaves the EAS and mixes with environmental air, its temperature decreases. Under these conditions, gaseous sulfuric acid can combine with water vapor and condense to form liquid droplets of hydrated sulfuric acid or sulfate particulates. These liquid particles are measured as particulates by the systems used to measure the total PM emissions from the engine. Any catalytic formation of sulfate particles, which is especially a problem with high sulfur content in the fuel, can therefore significantly increase the total particulate emissions. Likewise, if these particles are trapped by the DPF, they will form non-combustible sulfate compounds, or ash, which degrades the DPF performance over time.

References

8-1.   Twigg, M.V., "Roles of Catalytic Oxidation in Control of Vehicle Exhaust Emissions," *Catalysis Today* **117**, no. 4 (2006): 407–418.

8-2.   Taylor, K.C., "Automobile Catalytic Converters," Editors: John R. Anderson, Michel Boudart, *Catalysis* (Berlin: Springer, 1984), 119–170.

8-3.   Collins, N.R. and Twigg, M. V., "Three-Way Catalyst Emissions Control Technologies for Spark-Ignition Engines—Recent Trends and Future Developments," *Topics in Catalysis* **42**, no. 1-4 (2007): 323–332.

8-4.   Russell, A. and Epling, W.S., "Diesel Oxidation Catalysts," *Catalysis Reviews* **53**, no. 4 (2011): 337-423.

8-5.   Majewski, W.A., Ambs, J.L., and Bickel, K., "Nitrogen Oxides Reactions in Diesel Oxidation Catalyst," SAE Technical Paper 950374, 1995, https://doi.org/10.4271/950374.

8-6.   Ambs, J.L. and McClure, B.T., "The Influence of Oxidation Catalysts on NO_2 in Diesel Exhaust," SAE Technical Paper 932494, 1993, https://doi.org/10.4271/932494.

8-7.   Eastwood, P., *Critical Topics in Exhaust Gas Aftertreatment* (Hertfordshire, England: Research Studies Press, LTD, 2000).

8-8.   Konstantinidis, P.A., Koltsakis, G.C., and Stamatelos, A.M., "Computer Aided Assessment and Optimization of Catalyst Fast Light-Off Techniques," *Proceedings of the*

Institution of Mechanical Engineers, Part D: Journal of Automobile Engineering **211**, no. 1 (1997): 21–37, https://doi.org/10.1243/0954407971526191.

8-9. Parks, J., Huff, S., Kass, M., and Storey, J., "Characterization of In-Cylinder Techniques for Thermal Management of Diesel Aftertreatment," SAE Technical Paper 2007-01-3997, 2007, https://doi.org/10.4271/2007-01-3997.

8-10. Gosala, D.B., Ramesh, A.K., Allen, C.M., Joshi, M.C., Taylor, A.H., Van Voorhis, M., Shaver, G.M., Farrell, L., Koeberlein, E., McCarthy, J., and Stretch, D., "Diesel Engine Aftertreatment Warm-Up through Early Exhaust Valve Opening and Internal Exhaust Gas Recirculation during Idle Operation," *International Journal of Engine Research* **19**, no. 7 (2018): 758–773, https://doi.org/10.1177/1468087417730240.

8-11. Singh, P., Thalagavara, A.M., Naber, J., Johnson, J.H., and Bagley, S.T., "An Experimental Study of Active Regeneration of an Advanced Catalyzed Particulate Filter by Diesel Fuel Injection Upstream of an Oxidation Catalyst," SAE Technical Paper 2006-01-0879, 2006, https://doi.org/10.4271/2006-01-0879.

8-12. Joshi, A., Chatterjee, S., Sawant, A., Akerlund, C., Andersson, S., Blomquist, M., Brooks, J., and Kattan, S., "Development of an Actively Regenerating dpf System for Retrofit Applications," SAE Technical Paper 2006-01-3553, 2006, https://doi.org/10.4271/2006-01-3553.

8-13. Garg, A., Magee, M., Ding, C., Roberts, L., Shaver, G., Koeberlein, E., Shute, R., Koeberlein, D., McCarthy Jr, J., and Nielsen, D., "Fuel-Efficient Exhaust Thermal Management Using Cylinder Throttling via Intake Valve Closing Timing Modulation," *Proceedings of the Institution of Mechanical Engineers, Part D: Journal of Automobile Engineering* **230**, no. 4 (2016): 470–478.

8-14. Bouchez, M. and Dementhon, J.B., "Strategies for the Control of Particulate Trap Regeneration," SAE Technical Paper 2000-01-0472, 2000, https://doi.org/10.4271/2000-01-0472.

8-15. Shelef, M. and McCabe, R.W., "Twenty-Five Years after Introduction of Automotive Catalysts: What Next?," *Catalysis Today* **62**, no. 1 (2000): 35–50.

8-16. Haass, F. and Fuess, H., "Structural Characterization of Automotive Catalysts," *Advanced Engineering Materials* **7**, no. 10 (2005): 899–913.

8-17. Davy, H., *The Collected Works of Sir Humphry Davy (etc.)* (London, UK: Smith, Elder and Company, 1840), **5**.

8-18. Dobereiner, W., "Ober neu Entdeckte Höchst Merk-wiirdige Eigenschaften des Platins,", Jena Germany, 1823.

8-19. Burwell Jr, R.L., "Heterogeneous Catalysis before 1934," *Catalysis* **2** (1894), Chapter 1, 3-12.

8-20. Furter, W., *A Century of Chemical Engineering* 1st ed. (New York: Springer, 1982).

8-21. Neyestanaki, A.K., Klingstedt, F., Salmi, T., and Murzin, D.Y., "Deactivation of Postcombustion Catalysts: A Review," *Fuel* **83**, no. 4-5 (2004): 395–408.

8-22. Yao, Y.-F.Y., "Oxidation of Alkanes over Noble Metal Catalysts," *Industrial & Engineering Chemistry Product Research and Development* **19**, no. 3 (1980): 293–298.

8-23. Twigg, M.V., "Twenty-Five Years of Autocatalysts," *Platinum Metals Review* **43**, no. 4 (1999): 168–171.

8-24. Twigg, M.V., "Automotive Exhaust Emissions Control," *Platinum Metals Review* **47**, no. 4 (2003): 157–162.

8-25. Olsson, L. and Fridell, E., "The Influence of Pt Oxide Formation and Pt Dispersion on the Reactions $NO_2 \Leftrightarrow NO + 1/2\ O_2$ over Pt/Al_2O_3 and $Pt/BaO/Al_2O_3$," *Journal of Catalysis* **210**, no. 2 (2002): 340–353.

8-26. Lee, A.F., Wilson, K., Goldoni, A., Larciprete, R., and Lizzit, S., "A Fast XPS Study of Sulphate Promoted Propene Decomposition over Pt {111}," *Surface Science* **513**, no. 1 (2002): 140–148.

8-27. Corro, G., Fierro, J., Montiel, R., Castillo, S., and Moran, M., "A Highly Sulfur Resistant Pt-Sn/γ-Al$_2$O$_3$ Catalyst for C$_3$H$_8$-NO-O$_2$ Reaction under Lean Conditions," *Applied Catalysis B: Environmental* **46**, no. 2 (2003): 307–317.

8-28. Chang, H., Chen, X., Li, J., Ma, L., Wang, C., Liu, C., Schwank, J.W., and Hao, J., "Improvement of Activity and SO$_2$ Tolerance of Sn-Modified MnO$_x$-CeO$_2$ Catalysts for NH$_3$-SCR at Low Temperatures," *Environmental Science & Technology* **47**, no. 10 (2013): 5294–5301.

8-29. Wu, H.-C., Liu, L.-C., and Yang, S.-M., "Effects of Additives on Supported Noble Metal Catalysts for Oxidation of Hydrocarbons and Carbon Monoxide," *Applied Catalysis A: General* **211**, no. 2 (2001): 159–165.

8-30. Ono, Y. and Hattori, H., *Solid Base Catalysis* (Berlin, Germany: Springer Science & Business Media, 2012), **101**.

8-31. Gélin, P. and Primet, M., "Complete Oxidation of Methane at Low Temperature over Noble Metal Based Catalysts: A Review," *Applied Catalysis B: Environmental* **39**, no. 1 (2002): 1–37.

8-32. Chen, M. and Schmidt, L., "Morphology and Composition of Pt-Pd Alloy Crystallites on SiO$_2$ in Reactive Atmospheres," *Journal of Catalysis* **56**, no. 2 (1979): 198–218.

8-33. Henry, C.R., "Surface Studies of Supported Model Catalysts," *Surface Science Reports* **31**, no. 7-8 (1998): 231–325.

8-34. Ciuparu, D., Lyubovsky, M.R., Altman, E., Pfefferle, L.D., and Datye, A., "Catalytic Combustion of Methane over Palladium-Based Catalysts," *Catalysis Reviews* **44**, no. 4 (2002): 593–649.

8-35. Herz, R.K. and Marin, S.P., "Surface Chemistry Models of Carbon Monoxide Oxidation on Supported Platinum Catalysts," *Journal of Catalysis* **65**, no. 2 (1980): 281–296.

8-36. Cabrera, N. and Mott, N.F., "Theory of the Oxidation of Metals," *Reports on Progress in Physics* **12**, no. 1 (1949): 163.

8-37. Su, S.C., Carstens, J.N., and Bell, A.T., "A Study of the Dynamics of Pd Oxidation and PdO Reduction by H2 and CH4," *Journal of Catalysis* **176**, no. 1 (1998): 125–135.

8-38. Kaneeda, M., Iizuka, H., Hiratsuka, T., Shinotsuka, N., and Arai, M., "Improvement of Thermal Stability of NO Oxidation Pt/Al$_2$O$_3$ Catalyst by Addition of Pd," *Applied Catalysis B: Environmental* **90**, no. 3-4 (2009): 564–569.

8-39. Ribeiro, F., Chow, M., and Dallabetta, R., "Kinetics of the Complete Oxidation of Methane over Supported Palladium Catalysts," *Journal of Catalysis* **146**, no. 2 (1994): 537–544.

8-40. Heck, R.M. and Farrauto, R.J., "Automobile Exhaust Catalysts," *Applied Catalysis A: General* **221**, no. 1-2 (2001): 443–457.

8-41. Farrauto, R.J. and Voss, K.E., "Monolithic Diesel Oxidation Catalysts," *Applied Catalysis B: Environmental* **10**, no. 1-3 (1996): 29–51.

8-42. Webster, D., "25 Years of Catalytic Automotive Pollution Control: A Collaborative Effort," *Topics in Catalysis* **16**, no. 1-4 (2001): 33–38.

8-43. Mowery, D.L. and McCormick, R.L., "Deactivation of Alumina Supported and Unsupported PdO Methane Oxidation Catalyst: The Effect of Water on Sulfate Poisoning," *Applied Catalysis B: Environmental* **34**, no. 4 (2001): 287–297.

8-44. Yu, T.-C. and Shaw, H., "The Effect of Sulfur Poisoning on Methane Oxidation over Palladium Supported on γ-Alumina Catalysts," *Applied Catalysis B: Environmental* **18**, no. 1-2 (1998): 105–114.

8-45. Schlangen, J.W., Neuhaus, G.W., Madani, M., and Maier, W.F., "Unterschiede in der Totaloxidation organischer Verbindungen an heterogenen Platin-und Palladiumkatalysatoren," *Journal für Praktische Chemie/Chemiker-Zeitung* **334**, no. 6 (1992): 465–473.

8-46. Watanabe, T., Kawashima, K., Tagawa, Y., Tashiro, K., Anoda, H., Ichioka, K., Sumiya, S., and Zhang, G., "New DOC for Light Duty Diesel DPF systemS," *SAE Paper Number 2007-01-1920*, 2007, https://doi.org/10.4271/2007-01-1920.

8-47. Farrauto, R.J., Hobson, M., Kennelly, T., and Waterman, E., "Catalytic Chemistry of Supported Palladium for Combustion of Methane," *Applied Catalysis A: General* **81**, no. 2 (1992): 227–237.

8-48. Kröcher, O., Widmer, M., Elsener, M., and Rothe, D., "Adsorption and Desorption of SOx on Diesel Oxidation Catalysts," *Industrial & Engineering Chemistry Research* **48**, no. 22 (2009): 9847–9857.

8-49. Lampert, J.K., Kazi, M.S., and Farrauto, R.J., "Palladium Catalyst Performance for Methane Emissions Abatement from Lean Burn Natural Gas Vehicles," *Applied Catalysis B: Environmental* **14**, no. 3-4 (1997): 211–223.

8-50. Galisteo, F.C., Larese, C., Mariscal, R., Granados, M.L., Fierro, J., Fernández-Ruiz, R., and Furio, M., "Deactivation on Vehicle-Aged Diesel Oxidation Catalysts," *Topics in Catalysis* **30**, no. 1-4 (2004): 451–456.

8-51. Forzatti, P. and Lietti, L., "Catalyst Deactivation," *Catalysis Today* **52**, no. 2-3 (1999): 165–181.

8-52. Morlang, A., Neuhausen, U., Klementiev, K., Schütze, F.-W., Miehe, G., Fuess, H., and Lox, E., "Bimetallic Pt/Pd Diesel Oxidation Catalysts: Structural Characterisation and Catalytic Behaviour," *Applied Catalysis B: Environmental* **60**, no. 3-4 (2005): 191–199.

8-53. Summers, J.C. and Hegedus, L.L., "Effects of Platinum and Palladium Impregnation on the Performance and Durability of Automobile Exhaust Oxidizing Catalysts," *Journal of Catalysis* **51**, no. 2 (1978): 185–192.

8-54. Phillips, P.R., Chandler, G.R., Jollie, D.M., Wilkins, A.J.J., and Twigg, M.V., "Development of Advanced Diesel Oxidation Catalysts," SAE Technical Paper 1999-01-3075, 1999, https://doi.org/10.4271/1999-01-3075.

8-55. Yao, Y.Y. and Kummer, J., "A Study of High Temperature Treated Supported Metal Oxide Catalysts," *Journal of Catalysis* **46**, no. 3 (1977): 388–401.

8-56. Burch, R., Breen, J., and Meunier, F., "A Review of the Selective Reduction of NOx with Hydrocarbons under Lean-Burn Conditions with Non-Zeolitic Oxide and Platinum Group Metal Catalysts," *Applied Catalysis B: Environmental* **39**, no. 4 (2002): 283–303.

8-57. Corma, A., "State of the Art and Future Challenges of Zeolites as Catalysts," *Journal of Catalysis* **216**, no. 1-2 (2003): 298–312.

8-58. Sampara, C.S., Bissett, E.J., and Assanis, D., "Hydrocarbon Storage Modeling for Diesel Oxidation Catalysts," *Chemical Engineering Science* **63**, no. 21 (2008): 5179–5192.

8-59. Koltsakis, G.C. and Stamatelos, A.M., "Catalytic Automotive Exhaust Aftertreatment," *Progress in Energy and Combustion Science* **23**, no. 1 (1997): 1–39.

8-60. Hubbard, C., Otto, K., Gandhi, H., and Ng, K., "Effects of Support Material and Sulfation on Propane Oxidation Activity over Platinum," *Journal of Catalysis* **144**, no. 2 (1993): 484–494.

8-61. Maunula, T., Suopanki, A., Torkkell, K., and Härkönen, M., "The Optimization of Light-Duty Diesel Oxidation Catalysts for Preturbo, Closed-Coupled and Underfloor Positions," SAE Technical Paper 2004-01-3021, 2004, https://doi.org/10.4271/2004-01-3021.

8-62.   Winkler, A., Ferri, D., and Aguirre, M., "The Influence of Chemical and Thermal Aging on the Catalytic Activity of a Monolithic Diesel Oxidation Catalyst," *Applied Catalysis B: Environmental* **93**, no. 1-2 (2009): 177–184.

8-63.   Hubbard, C., Otto, K., Gandhi, H., and Ng, K., "The Influence of Sulfur Dioxide on Propane Oxidation Activity over Supported Platinum and Palladium," *Catalysis Letters* **30**, no. 1-4 (1994): 41–51.

8-64.   Heck, R.M., Farrauto, R.J., and Gulati, S.T., *Catalytic Air Pollution Control: Commercial Technology* (New York, NY: John Wiley & Sons, 2016).

8-65.   Beguin, B., Garbowski, E., and Primet, M., "Stabilization of Alumina toward Thermal Sintering by Silicon Addition," *Journal of Catalysis* **127**, no. 2 (1991): 595–604.

8-66.   Béguin, B., Garbowski, E., and Primet, M., "Stabilization of Alumina by Addition of Lanthanum," *Applied Catalysis* **75**, no. 1 (1991): 119–132.

Three-Way Catalysts

Three is a magic number
Yes it is, it's a magic number …
No more, no less, you don't have to guess
When it's three, you can see
It's a magic number

—Bob Dorough, "3 Is the Magic Number"

The purpose of the three-way catalyst (TWC) is to convert carbon monoxide (CO), hydro-carbons (HCs), and nitrogen oxides (NOx) into less harmful species, including carbon dioxide (CO_2), water (H_2O), and nitrogen gas (N_2). TWCs are typically used to treat the exhaust gas from gasoline-fueled, spark-ignition (SI) engines.

Catalytic converters for exhaust gas have been in use on vehicles for decades, but their use first began with the oxidation catalyst (oxycat), described in Chapter 8. The oxycat was introduced in the mid-1970s to remove emissions in the exhaust from SI engines. Oxycats were the direct precursors to the TWC, however, they only oxidized CO and HCs, making them two-way catalysts. Oxyats do not reduce NOx. Until 1978, these oxycats were sufficient, as the emissions of NOx were sufficiently treated non-catalytically through exhaust gas recirculation (EGR) or fuel-lean combustion. Beginning with some model year (MY) 1977 vehicles sold in California, NOx emissions required additional, catalytic control. The two-way oxycats were then superseded by the two-catalyst, three-way system, which then included the reduction of NOx as the "third way" [9-1]. These two-catalyst systems were then combined into the single TWC, now a reduction-oxidation (redox) catalyst, and became the predominant formulation in use in the United States and Canada [9-2].

These new catalysts had an immediate and dramatic impact. The 1989 EPA National Air Quality and Emissions Trends Report noted that between 1980 and 1989 the emissions of NOx had decreased significantly, in large part due to highway vehicles which experienced a 25% reduction in NOx [9-3]. Due to the successful use of these catalysts in the decades since, exhaust emissions from light-duty vehicles with SI engines have been drastically reduced, to under 5% of what they were in the 1970s [9-4]. The use of catalysts on vehicles has been one of the greatest successes of heterogeneous catalysis [9-5]. TWC aftertreatment of SI engines achieves regulatory compliance and is of great benefit to the quality of atmospheric air quality.

An additional challenge for early oxycats turned out to be useful for developing control strategies for the TWC. These engines were fueled by carburetors, which have imprecise control of the air-fuel ratio (AFR) supplied to the engine. The AFR strongly affects the exhaust conditions and emissions. Fortuitously, with the AFR moving randomly about the stoichiometric point, it was found that a platinum-rhodium (Pt-Rh) catalyst could, under the appropriate conditions, simultaneously oxidize CO and HCs and reduce NOx with high efficiency [9-6, 9-7]. This concept along with the development of electronic fuel injection and oxygen (O_2) sensors led to the AFR control strategy for the TWC [9-6].

By 1981, the majority of catalytic converters used on vehicles were either a TWC or a dual-bed converter. A few manufacturers also deployed dual converters. The difference between a dual-bed converter and dual converter is that the dual-bed converter had the two catalysts within a single container, and the dual converter had two separately contained catalysts in series. Supplementary air was introduced into the exhaust ahead of the oxycat, to provide sufficient oxygen for the oxidation reactions. The pioneering GM catalytic converter was a pellet dual-bed that enclosed the TWC and oxycat in a single vessel, separated by an air plenum [9-8], as shown in **Figure 9.1**.

FIGURE 9.1   Cross-sectional view of the GM dual-bed Three-Way Catalyst [9-8].

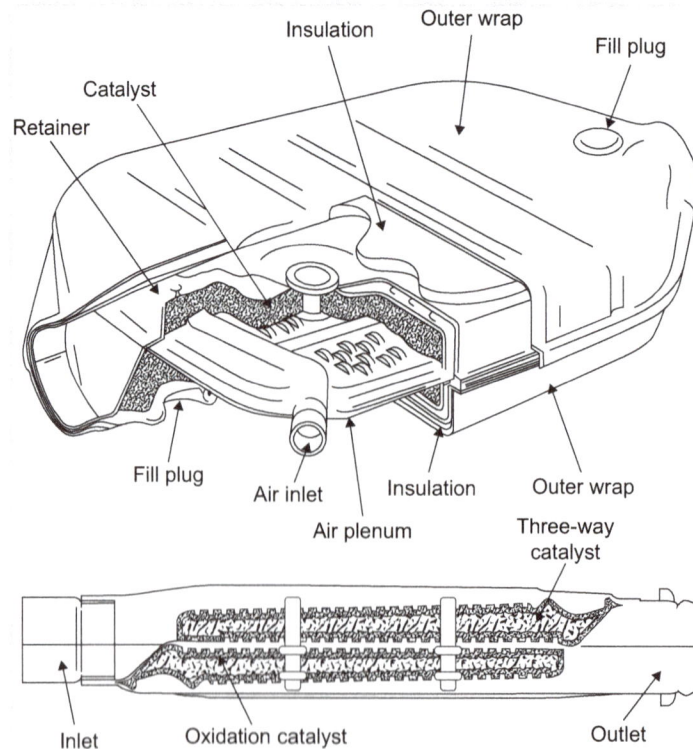

In this chapter, TWC function, catalysts, and washcoats are discussed, along with some discussion of TWC controls, substrates, and durability.

9.1 **Three-Way Catalyst Functions**

The engine-out exhaust emissions concentrations of CO, HCs, and NOx vary depending on many factors, including AFR, ignition timing, and EGR level. Of these, the AFR is one of the most significant factors, as described in Chapter 2, Emissions Formation. A related parameter is λ, which is the ratio of the current AFR to the stoichiometric AFR for the fuel. When the engine operates under lean AFR, or $\lambda > 1$, the exhaust is in a net oxidizing condition which results in lower HC and CO concentrations, and more residual oxygen in the exhaust. These conditions are well-suited for the catalytic oxidation reactions to occur in the subsequent aftertreatment device. These same net oxidizing conditions, however, are necessarily insufficient for reducing the NOx present in the exhaust. NOx are formed during combustion due to the high temperature and pressure conditions in the cylinder, as described in Section 2.4. As the AFR becomes closer to stoichiometric ($\lambda \approx 1$), there is less oxygen present in the exhaust but correspondingly more CO which must be oxidized. At this condition, the concentrations of the oxidizing gases and reducing gases are balanced and the equilibration of this mixture would yield just CO_2, water, and N_2 [9-9].

SI engines are operated near the stoichiometric AFR ($\lambda = 1$) to ensure that the TWC facilitates the three key redox reactions for criteria pollutant abatement. These are CO oxidation

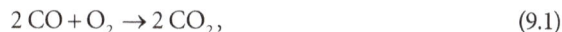

$$2\,CO + O_2 \rightarrow 2\,CO_2, \tag{9.1}$$

HC oxidation (shown on a unit carbon basis)

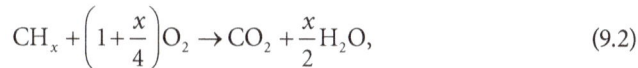

$$CH_x + \left(1 + \frac{x}{4}\right)O_2 \rightarrow CO_2 + \frac{x}{2}H_2O, \tag{9.2}$$

and NO reduction

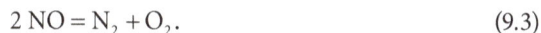

$$2\,NO = N_2 + O_2. \tag{9.3}$$

The redox process is depicted in **Figure 9.2**.

9.1.1 **Reduction and Oxidation Chemistry**

The essential function of an effective TWC is a high conversion of CO, HCs, and NOx at nearly stoichiometric exhaust conditions. The performance of the TWC is a strong function of AFR or λ. Oxidation of CO and HCs is limited when the exhaust is rich ($\lambda < 1$), creating a reducing condition. Likewise, the reduction of NOx is limited when the exhaust is lean ($\lambda > 1$), in an oxidizing condition. The narrow catalyst operating range, centered on the stoichiometric AFR ($\lambda = 1$), is where the conversions are highest and is often called the operating window for the catalyst, as shown in **Figure 9.3**. It would be very desirable to have a wide AFR operating window, with simultaneous redox reactions resulting in high conversions, as it would lessen the AFR control burden associated with the TWC.

The TWC catalyst typically includes a blend of platinum (Pt), palladium (Pd), and rhodium (Rh). The Pt and Pd provide the active sites for the oxidation reactions, where

FIGURE 9.2 A graphical depiction of redox (reduction-oxidation) chemistry. The molecule being oxidized acts as the reducing agent for the molecule being reduced, which acts as the oxidizing agent for the molecule being oxidized. The oxygen atoms move from the oxidizing agent to the reducing agent, and the free electrons move from the reducing agent to the oxidizing agent.

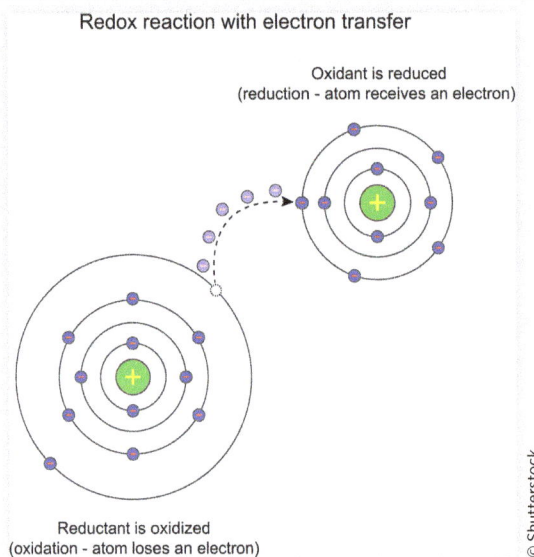

FIGURE 9.3 Three-way catalyst conversion with respect to λ, showing a high-conversion operating window near λ = 1 [9-10].

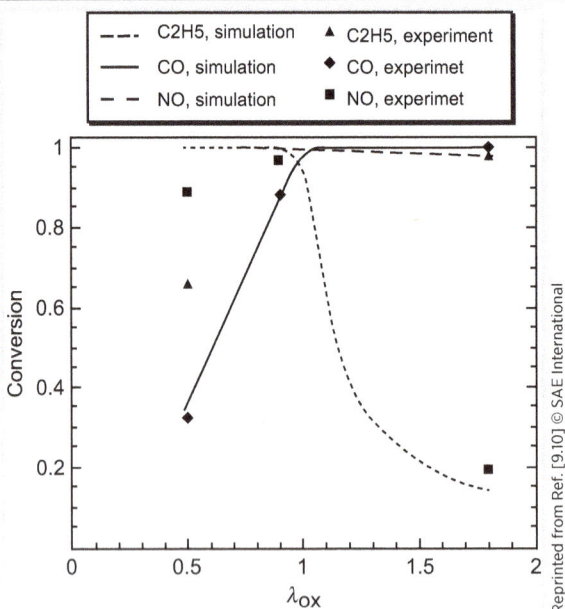

the CO is fully oxidized to CO_2 per the reaction in Eq. 9.1, and the HCs are oxidized to CO_2 and water per the reaction in Eq. 9.2. An additional oxidation reaction is

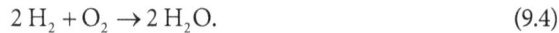

$$2\,H_2 + O_2 \rightarrow 2\,H_2O. \tag{9.4}$$

Rh provides the active sites for the reduction reactions, where NOx becomes N_2. Expanding the reduction reaction in Eq. 9.3 into a series of reaction options gives

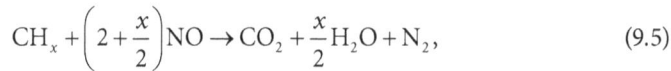

$$2\,NO + 2\,CO \rightarrow 2\,CO_2 + N_2, \tag{9.4}$$

$$CH_x + \left(2 + \frac{x}{2}\right)NO \rightarrow CO_2 + \frac{x}{2}H_2O + N_2, \tag{9.5}$$

$$2\,H_2 + 2\,NO \rightarrow 2\,H_2O + N_2, \tag{9.6}$$

and

$$2\,N_2O \rightarrow 2\,N_2 + O_2. \tag{9.7}$$

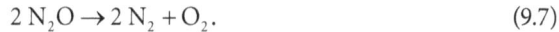

Additionally, there may be side reactions occurring that form desirable or undesirable products, such as

$$2\,NO + 5\,H_2 \rightarrow 2\,NH_3 + 2\,H_2O, \tag{9.8}$$

$$2\,NO + H_2 \rightarrow N_2O + H_2O, \tag{9.9}$$

and

$$2\,NH_3 \rightarrow N_2 + 3\,H_2. \tag{9.10}$$

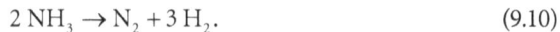

Rh also facilitates the water-gas shift (WGS) reaction,

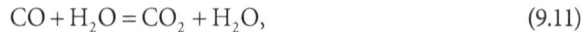

$$CO + H_2O = CO_2 + H_2O, \tag{9.11}$$

and steam reforming (shown on a unit carbon basis),

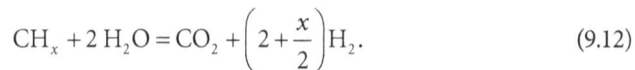

$$CH_x + 2\,H_2O = CO_2 + \left(2 + \frac{x}{2}\right)H_2. \tag{9.12}$$

The above reduction reactions occur while the engine operates in a narrow window around the stoichiometric AFR but they occur most efficiently when the exhaust is slightly richer than stoichiometric ($\lambda < 1$). Likewise, the oxidation reactions occur most efficiently when the exhaust is slightly leaner than stoichiometric ($\lambda < 1$). Therefore, the exhaust gas is controlled in a manner to create dithering or small-amplitude oscillations between fuel-rich and fuel-lean conditions. The desired products of these reactions are CO_2, H_2O, and N_2. Fortunately, at typical exhaust temperatures (approximately 500°C), these reaction products are thermodynamically favored. Therefore, the overall conversion efficiency of the TWC is determined both by the activity of the catalysts to promote the reactions and by the availability of oxidizing and reducing agents in the exhaust. Because oxidizing agents are in short supply during rich AFR, the CO, HCs, and hydrogen (H_2) in the exhaust are in competition for the limited amounts of O_2 and NO. Similarly, reducing agents are limited during lean AFR, and the O_2 and NO are in competition for the available CO, HCs, and H_2. As discussed in Section 9.3, some oxygen storage capacity (OSC) in the catalytic converter helps buffer the changes in demand as AFR changes.

The TWC is selective for several preferred reactions, including

- CO and HC oxidation instead of H_2 oxidation
- NO reduction by CO instead of H_2 oxidation or oxygen reduction
- NO reduction to N_2 instead of N_2O or NH_3

The WGS reaction shown in Eq. 9.11, and any steam reforming of the HCs, as shown in Eq. 9.12, may additionally contribute to CO and HC conversions under rich, reducing conditions. Note that there are sufficient concentrations of the reactants in the exhaust to completely react the CO, HCs, and NO over the entire AFR range, if the catalyst had ideal selectivity and could promote a subset of the reactions while suppressing the remainder. Therefore, there have been many efforts to develop more selective catalyst formulations, the topic of Section 9.2.

9.1.2 Warm-up Strategies

Vehicles fitted with a TWC have the lowest conversion efficiency and, therefore, emit the highest levels of pollution when they are first started, in a time period known as "cold-start." The duration of this cold-start period is dependent on how long it takes the catalyst to warm up to its light-off temperature. Improvements in the catalyst formulation, substrate, and warm-up strategy have brought this time down from minutes for the first TWC to tens of seconds for current systems.

Warm-up strategies are either passive or active, depending on whether or not external energy is used. Active systems, such as exhaust gas ignition requires the engine to run with a fuel-rich AFR to increase the exhaust concentrations of HCs, and then adds supplemental air to the exhaust to create overall near-stoichiometric conditions for combustion on the catalyst [9-11, 9-12]. Passive systems rely on management of the exhaust temperature by the engine. They include retarding the spark timing, increasing the engine speed, or early exhaust valve opening [9-13] to increase the enthalpy to the catalyst. Exhaust aftertreatment systems (EAS) with TWCs usually include a small close-coupled catalytic converter near the engine that can warm up quickly due to its lower thermal mass and higher exhaust inlet temperatures. A single small-volume TWC would give only limited conversion under most road-load conditions, and therefore a second, larger TWC is placed further downstream in the EAS. An example of an EAS with close-coupled and underfloor TWCs together is shown in Figure 7.16.

9.1.3 Transient Behavior

The TWC typically requires closed-loop control to rapidly cycle the AFR about the stoichiometric point ($\lambda = 1$), at a frequency of about 1 Hz [9-14]. This style of operation was determined to be required for the redox capability of the catalyst [9-15]. The TWC also contains OSC in the washcoat to adsorb excess oxygen during lean conditions and release it during rich conditions. This OSC extends the operating range under which the catalyst can selectively reduce NO [9-15]. This stored oxygen increases the conversion of CO and HCs when the exhaust is under slightly reducing conditions. In this way, the overall redox chemistry is better supported as the engine cycles the AFR between net oxidizing and net reducing conditions [9-16]. This strategy enables the stoichiometric redox reactions with the OSC component facilitating the catalytic reactions.

9.2 Catalyst Formulations

The choice of platinum group metals (PGM, also called precious or noble metals) for automotive catalytic converters is the result of three factors. First, PGM have the required activity for the high space velocities (SV) in automotive applications. Second, PGM resist poisoning by sulfur and other compounds present in gasoline and lubricants compared to base metal catalysts. Third, PGM have good stability during the hydrothermal aging that occurs during normal use [9-5]. The catalyst functions are described in Section 9.1.1, and the specific blend of Pt, Pd, and Rh depends on the relative costs of the materials and the details of the application.

The first TWC formulations used Pt and Rh to facilitate redox chemistry. Ceria (CeO_2) was also incorporated into the washcoat because it can oxidize further in the fuel-lean exhaust and thereby store oxygen. This OSC helps keep the catalyst surface concentrations near stoichiometric for the redox reactions and also helps stabilize the catalysts against hydrothermal aging. Further refinement of the catalyst formulation by substituting Pd for some of the Pt further improved the performance over a range of operating conditions. The application of Pd represented a significant breakthrough in TWC technology due to its superior catalytic activity for HC oxidation and its thermal stability [9-17]. Substrate optimizations for the geometric area, cell density, and thermal properties provided additional benefits to performance and operating range. In recent years, the cost of TWCs have been reduced by catalyst thrifting, leading to lower PGM loadings for a target pollutant conversion level.

9.2.1 Platinum

Platinum (Pt) is used in TWCs as the active site for the oxidation of CO and HCs. Its presence is particularly beneficial during warm-up. Air can be injected into the exhaust upstream of the TWC during warm-up so that it operates as an oxycat. The Pt contribution to the CO and HC conversions during warm-up becomes increasingly important as the Rh loading is decreased because Rh is susceptible to thermal damage [9-14].

In a dual catalyst or dual converter system where the TWC is followed by an oxycat, the oxycat usually contains Pt, Pd, or a Pt-Pd mixture. This oxycat is used to improve CO and HC oxidation. The Pt and Pd used in this kind of system contribute to the total PGM loading. The composition and distribution of the PGM in the TWC has been the subject of significant research in order to design a durable system, as discussed later in Section 9.6.

Platinum is also known to catalyze the reduction of NOx in an exhaust gas composition with stoichiometric levels of oxygen [9-1, 9-6, 9-18], however, this is likely only a minor contribution to the overall NOx conversion. Pt is not as effective as Rh at reducing NO in the exhaust gas with high CO concentrations, or when SO_2 is present [9-19, 9-20, 9-21]. In addition, Pt has a smaller AFR operating window than Rh where the selectivity of reducing NO to N_2 is favored. Pt also catalyzes the reduction of NO to undesirable byproducts, including NH_3 under reducing conditions or N_2O [9-22].

9.2.2 **Palladium**

Palladium, like Pt, is used to facilitate CO and HC oxidation in TWCs. Some Pd-only catalysts are in use in specific applications, such as pre-catalysts, or to promote CO and HC oxidation in the second (oxidation) catalyst used in dual catalyst systems.

There are a few significant issues with Pd including its susceptibility to sulfur poisoning, ability to alloy with Rh, inhibition of NO reduction selectivity in presence of HCs, and potential for detrimental interactions Pt or Rh [9-23]. The PGM interactions can be avoided by separating the catalytic materials into layers or zones, in order to provide each with its ideal environment within the exhaust, as described in Section 15.1.

9.2.3 **Rhodium**

The PGM most often used for the catalytic reduction of NOx in engine exhaust is Rh, which is the primary component for controlling NOx emissions in TWCs. Rh provides active sites to promote the reduction of NO to N_2 in TWCs [9-1, 9-15]. It is the preferred constituent due to its high selectivity for reducing NOx to elemental nitrogen with very little ammonia byproduct formation, provided that the AFR is close to stoichiometric. Due to its high selectivity for the NO reduction reaction, only a small amount of Rh (on the order of 1/10 the amount of Pt) is needed. Rh catalyzes the reduction of NO with either CO or H_2, though the reaction with H_2 is much faster at low exhaust temperatures [9-24]. Rh also strongly contributes to CO oxidation. Although both Pt and Pd simultaneously catalyze the oxidation of CO and HCs and reduction of NOx in the typical TWC operating window, they are not as effective as Rh [9-1].

Another important property of Rh in a TWC is its ability to catalyze the steam reformation of exhaust HCs, as shown in Eq. 9.12 [9-25, 9-26]. However, a major product of reforming is CO, which needs to be oxidized further. Rh also has high specific activity as an oxidation catalyst for CO, comparable to that of both Pt and Pd [9-27]. For this reason, Rh is believed to improve the low-temperature performance of oxycats. With its strong abilities for both oxidation and reduction, it may seem logical to consider an all-rhodium catalyst. However, Rh is two to six times as expensive as Pt or Pd, depending on commodity markets, and so the cost of the required loadings to obtain the oxidation performance preclude this option on cost grounds.

One potential drawback to Rh is its sensitivity to preconditioning in oxygen. It was observed that heating an Rh catalyst on alumina (Al_2O_3) in oxygen (calcining) at 500°C for 4 hours lead to a reversible loss in CO oxidation activity [9-28]. Calcination of Rh/Al_2O_3 catalysts at temperatures above 700°C further degrades the catalyst due to sintering—a process further described in Section 7.2.3.2.

9.2.4 **Compositions**

The first TWC typically used a combination of Pt and Rh catalysts. In the early 1990s, the typical Pt:Rh ratio in TWCs was approximately 5:1. This formulation gave the best tradeoff between performance and cost. Efforts to reduce the precious metal costs in the mid-1990s began to replace some or all of the Pt with Pd. Most of the oxycats (described in Chapter 8) deployed on vehicles between 1975 and 1980 used loadings between 1.77

and 2.47 g/L (50 to 70 g/ft³) with a 5:2 Pt:Pd ratio [9-29]. The GM pellet catalyst used for MY 1981–1982 vehicles used Pd with Pt and Rh. The literature indicates that the ranges of the loadings were as follows:

- Pd: 0 to 4.24 g/L (0–120 g/ft.³)
- Pt: 1.06 to 3.53 g/L (30 to 100 g/ft.³)
- Rh: 0.18 to 0.35 g/L (5 to 10 g/ft.³)

In most cases, the Pt:Pd ratio was greater than 1. During the development process, the importance of the position of the specific PGM within the catalyst was recognized and advanced impregnation techniques were developed to practically accomplish this.

Because PGM catalysts are expensive, engineers wish to reduce the catalyst loading to reduce cost while maintaining the desired performance. This catalyst thrifting process is challenging, as the effectiveness of PGM catalysts changes as they age from the mechanisms described in Section 1.3.2.2 and Section 9.7.

9.2.5 Alternatives to PGM

Since the deployment of the first oxycats, chemists and engineers have sought a base metal catalyst suitable for use in a TWC application as a replacement for expensive PGM catalysts. To date, this search for base metal substitutes has been unsuccessful. Base metal catalysts are not nearly as effective in typical exhaust gas compositions, SVs or temperatures. They are also susceptible to poisoning, particularly by sulfur [9-30]. Moreover, base metals have been shown to degrade PGM catalyst performance for oxycats [9-14, 9-31].

Transition metals have also been investigated as a means by which to reduce the noble metal loading. Molybdenum oxide (MoO_3), in combination with Pt on alumina (Pt/Al_2O_3), was investigated as a potential replacement for Rh. While it performed better than pure Pt/Al_2O_3 in selectivity for the reduction of NO to N_2 and was less susceptible to CO poisoning, it did not compare favorably to the Rh catalyst [9-32].

9.3 Oxygen Storage Capacity and Thermal Stability

The purpose of OSC is to buffer the lean-rich swings in exhaust gas conditions during normal vehicle operation. Its use pre-dates the first commercial application of TWCs [9-5, 9-15, 9-33]. Ceria (CeO_x) compounds were one of the first components to be developed for their oxygen storage and redox capability. Ceria has a combination of attractive properties as a storage material including redox cycling between cerium's valent ion states, easy impregnation onto Al_2O_3, compatibility with PGM catalysts, and good thermal stability [9-5, 9-34]. Ceria is incorporated to adsorb oxygen under lean, oxidizing conditions and release it during rich, reducing conditions and buffer the catalyst conditions to the stoichiometric point. This behavior widens the range in which the conversions of CO, HCs, and NOx are possible [9-34]. The total OSC is directly related to the amount of ceria present, though not all of it is available for reaction due to kinetic limitations during short engine transients [9-6]. Another benefit associated with ceria is the stabilization of the Pt on an Al_2O_3 washcoat against sintering [9-35]. Additionally, the ceria promotes the WGS reaction shown in Eq. 9.11. Heating the ceria-containing washcoats to about 900°C in the air reduces the OSC and can reduce the catalyst performance below that of one without OSC [9-35].

The introduction of OSC into the TWC created the need for more thermal stability because high ceria loadings can promote thermal deactivation of the catalysts. When lower emissions standards required the close-coupling of the TWC to mitigate cold-start emissions, this placement exacerbated durability issues with high ceria formulations, since the TWC was now operating in a hotter environment [9-5]. The solution to this problem was the stabilization of the ceria by another metal oxide such as zirconia (ZrO_2). Use of mixed oxides, such as $Al_2O_3/CeO_2/ZrO_2$, have been developed to improve thermal stability, have higher OSC, and lower light-off temperature as compared to a CeO_2-only or CeO_2-ZrO_2 formulations [9-36].

Ever since OSC was first added to TWCs, there has been a strong interest in developing compounds that have minimal negative interactions with other catalyst components. Ceria, while it benefits the oxidation reactions on the Pt and Pd catalysts, is not useful in the vicinity of Rh. Rh facilitates the reduction reactions, which are inhibited by available oxygen. However, it is possible to optimize the Pt or Pd environment to be different than the one for Rh, if the catalysts are divided into two or more layers, with their own promoter packages, as described in Section 15.1.

9.4 Control System

The first TWCs used a basic closed-loop emissions control system that employed an oxygen sensor, electronic control unit (ECU), fuel injector, and air pump [9-9, 9-37]. The exhaust oxygen sensor was used to indicate the state of the exhaust and whether or not there is excess oxygen present. The sensor acts as a switch for small changes in exhaust gas AFR (or λ) present in the exhaust, based on the presence or lack of oxygen as described in Section 6.1.3. The ECU receives the voltage signal from the oxygen sensor and sends the corresponding command to the fuel injector to slightly increase or decrease the fueling until the oxygen sensor once again detects that the exhaust AFR is back to stoichiometric ($\lambda = 1$). This simple system has the capability of controlling spark timing, idle speed, EGR, and the transmission converter clutch [9-14]. In addition, the oxygen sensor supports the on-board diagnostics functions, as discussed in Section 6.3. Thus, the oxygen sensor is used to recognize indicators of a catalyst malfunction and signal the driver, via the malfunction indicator lamp, that a problem exists and requires attention.

The inherent oscillations from a feedback control system in fuel injector command in a closed-loop system produce the well-known oscillations in the exhaust AFR that are often referred to as "dither." The specific amplitude and frequency of the oscillations depend on the physical characteristics of the engine, fuel injection system, and calibration. The typical oscillation frequency is on the order of 1 Hz. The closed-loop control system was necessitated by the low NOx emissions concentration thresholds and made more widespread by the continued development of accurate and low-cost sensors

[9-8, 9-38]. A more rigorous discussion of sensors, on-board diagnostics, and control strategies for exhaust aftertreatment devices and systems can be found in Chapter 6, Sensors and On-Board Diagnostics.

9.5 Substrates

A substrate, or catalyst support, is the solid, underlying material to which the catalyst is affixed. The substrate may have a high active surface area, itself. Since the activity of the catalyst occurs on the surface atoms, great effort is made to maximize the surface area of a catalytic material by distributing it over the support material to provide as much active area as possible. The substrate may be inert or as in the case of zeolites, it may participate in the catalytic reactions. Alumina (Al_2O_3) supports are chosen for PGM catalysts, due to their high surface area with advantageous pore structure, resistance to attrition, availability, and structural stability under typical automotive exhaust conditions [9-14, 9-39]. Factors that determine the substrate choice include size, cost, coatability, thermal properties and durability, and supplier availability.

9.5.1 Pellets

Among the first supports used for exhaust aftertreatment catalysts were alumina pellets [9-40], such as in the GM system shown in Figure 9.1. The alumina pellets are composed of thermally-stable transitional alumina phases stabilized by incorporating metal oxides [9-41], and were chosen because of their similarity to catalyst supports used in industrial chemical reactors. Support properties were improved to meet the need for improved converter performance in the late 1970s to increase the Brunauer–Emmett–Teller (BET) surface area and decrease the density of the supports [9-42]. The performance of the TWC on low-density supports was able to be improved, even while lowering the Rh loading by more than a factor of four [9-42]. This improvement, especially in cold start performance, was found to be due to the lower catalyst mass absorbing less of the exhaust energy when the support had lower density [9-43]. Further improvements were necessary to meet the 1980 emissions requirements, and this time catalyst efficiency was improved by moving to supports with the high geometric surface area. Increasing the surface area was accomplished by changing the shape of the pellets or reducing their size, to improve the surface area to volume ratio [9-42]. One significant issue with pellets is that the distribution of PGM catalysts in the pellets must be controlled in order to maintain the catalyst performance during use. The PGM catalysts should be within the first 150 µm of the outer edge of the pellet in order for them to be readily accessible to the reacting gases, but also to have sufficient sub-surface penetration to avoid rapid deactivation due to poisoning. However, this scheme is in competition with the desire to minimize PGM loadings [9-44, 9-45].

9.5.2 Monoliths

Thin-walled ceramic honeycomb monolith supports are now the most common substrates in use for TWCs and other catalytic converters. They are described in detail in Section 7.2.1.3.

Monoliths replaced pellets because of their superior durability under the vibrational loads encountered in the EAS.

9.6 Durability

The longer a catalyst is used, the more its performance degrades from its de-greened state because of catalyst aging. The catalysts in TWCs are subject to the typical aging and degradation mechanisms that are discussed in Section 7.2.3.2.

For TWCs, one of the leading catalyst poisoning species was lead (Pb) [9-46, 9-47, 9-48, 9-49]. Tetraethyl lead was used as a fuel additive to boost research octane number, reduce engine knock, and reduce engine wear. However, when an SI engine was operated with a leaded fuel (0.125 g Pb/L), the catalyst activity was found to decrease by roughly 25% after just 60 h of operation [9-46]. During rich operation, there is strong potential for the catalyst to reduce the SO_2 to hydrogen sulfide (H_2S). This reaction is often noticeable during the engine/catalyst warm-up, as it gives off a rotten egg type odor.

References

9-1.   Taylor, K.C., "Nitric Oxide Catalysis in Automotive Exhaust Systems," *Catalysis Reviews—Science and Engineering* **35**, no. 4 (1993): 457–481.

9-2.   Herz, R.K., "Dynamic Behavior of Automotive Catalysts. 1. Catalyst Oxidation and Reduction," *Industrial & Engineering Chemistry Product Research and Development* **20**, no. 3 (1981): 451–457.

9-3.   Environmental Protection Agency, "National Air Quality and Emissions Trends Report, 1989," Report No. EPA-450/4-91-003, 1991.

9-4.   Collins, N.R. and Twigg, M. V., "Three-Way Catalyst Emissions Control Technologies for Spark-Ignition Engines—Recent Trends and Future Developments," *Topics in Catalysis* **42**, no. 1-4 (2007): 323–332.

9-5.   Shelef, M. and McCabe, R.W., Twenty-Five Years after Introduction of Automotive Catalysts: What Next?," *Catalysis Today* **62**, no. 1 (2000): 35–50.

9-6.   Twigg, M.V., "Progress and Future Challenges in Controlling Automotive Exhaust Gas Emissions," *Applied Catalysis B: Environmental* **70**, no. 1-4 (2007): 2–15.

9-7.   Cooper, B., Evans, W., and Harrison, B., "Aspects of Automotive Catalyst Preparation, Performance and Durability," *Catalysis and Automotive Pollution Control* **30** (1987): 117–141.

9-8.   Zemke, B. and Gumbleton, J., "General Motors Progress Towards the Federal Research Objective Emission Levels," SAE Technical Paper 800398, 1980, https://doi.org/10.4271/800398.

9-9.   Hegedus, L.L. and Gumbleton, J.J., "Catalysts, Computers, and Cars: A Growing Symbiosis," *CHEMTECH;(United States)* **10**, no. 10 (1980): 403.

9-10.   Braun, J., Hauber, T., Többen, H., Zacke, P., Chatterjee, D., Deutschmann, O., and Warnatz, J., "Influence of Physical and Chemical Parameters on the Conversion Rate of a Catalytic Converter: A Numerical Simulation Study," SAE Technical Paper 2000-01-0211, 2000, https://doi.org/10.4271/2000-01-0211.

9-11. Brett, S.C., Eade, D., Hurley, R.G., Gregory, D., Collins, N.R., Morris, D., and Collingwood, I.T., "Evaluation of Catalysed Hydrocarbon Traps in the EGI System: Potential for Hydrocarbon Emissions Reduction," SAE Technical Paper 981417, 1998, https://doi.org/10.4271/981417.

9-12. Collins, N.R. and Twigg, M.V., "Three-Way Catalyst Emissions Control Technologies for Spark-Ignition Engines—Recent Trends and Future Developments," *Topics in Catalysis* **42**, no. 1-4 (2007): 323–332.

9-13. McCabe, R., "Cold-Start and Low Temperature Emissions Challenges," *CLEERS Workshop*, Dearborn, MI, 2014.

9-14. Taylor, K.C., "Automobile Catalytic Converters," editors: John R. Anderson, Michel Boudart *Catalysis* (Berlin: Springer, 1984), 119–170.

9-15. Gandhi, H., Piken, A., Shelef, M., and Delosh, R., "Laboratory Evaluation of Three-Way Catalysts," SAE Technical Paper 760201, 1976, https://doi.org/10.4271/760201.

9-16. Schlatter, J.C. and Mitchell, P.J., "Three-Way Catalyst Response to Transients," *Industrial & Engineering Chemistry Product Research and Development* **19**, no. 3 (1980): 288–293.

9-17. Wang, J., Chen, H., Hu, Z., Yao, M., and Li, Y., "A Review on the Pd-Based Three-Way Catalyst," *Catalysis Reviews* **57**, no. 1 (2015): 79–144.

9-18. Twigg, M.V., "Roles of Catalytic Oxidation in Control of Vehicle Exhaust Emissions," *Catalysis Today* **117**, no. 4 (2006): 407–418.

9-19. Summers, J.C. and Hegedus, L.L., "Effects of Platinum and Palladium Impregnation on the Performance and Durability of Automobile Exhaust Oxidizing Catalysts," *Journal of Catalysis* **51**, no. 2 (1978): 185–192.

9-20. Summers, J.C. and Hegedus, L.L., "Modes of Catalyst Deactivation in Stoichiometric Automobile Exhaust," *Industrial & Engineering Chemistry Product Research and Development* **18**, no. 4 (1979): 318–324.

9-21. Hegedus, L.L., Summers, J.C., Schlatter, J.C., and Baron, K., "Poison-Resistant Catalysts for the Simultaneous Control of Hydrocarbon, Carbon Monoxide, and Nitrogen Oxide Emissions," *Journal of Catalysis* **56**, no. 3 (1979): 321–335.

9-22. Otto, K. and Yao, H., "The Reduction of Nitric Oxide by Hydrogen over Ptγ-Al2O3 as a Function of Metal Loading," *Journal of Catalysis* **66**, no. 1 (1980): 229–236.

9-23. Summers, J.C. and Baron, K., "The Effects of SO2 on the Performance of Noble Metal Catalysts in Automobile Exhaust," *Journal of Catalysis* **57**, no. 3 (1979): 380–389.

9-24. Kobylinski, T.P. and Taylor, B.W., "The Catalytic Chemistry of Nitric Oxide: II. Reduction of Nitric Oxide over Noble Metal Catalysts," *Journal of Catalysis* **33**, no. 3 (1974): 376–384.

9-25. Gandhi, H., Piken, A., Stepien, H., Shelef, M., Delosh, R., and Heyde, M., "Evaluation of Three-Way Catalysts. Part II," SAE Technical Paper 770196, 1977, https://doi.org/10.4271/770196.

9-26. Schlatter, J.C., "Water-Gas Shift and Steam Reforming Reactions over a Rhodium Three-Way Catalyst," SAE Technical Paper 780199, 1978, https://doi.org/10.4271/780199.

9-27. Taylor, K.C., "Sulfur Storage on Automotive Catalysts," *Industrial & Engineering Chemistry Product Research and Development* **15**, no. 4 (1976): 264–268.

9-28. Schlatter, J., Taylor, K., and Sinkevitch, R., "The Behavior of Supported Rhodium in Catalyzing CO and NO Reactions," *Adv. in Catal. Chem. Symp*, Snowbird, UT, 1979.

9-29. Koltsakis, G.C. and Stamatelos, A.M., "Catalytic Automotive Exhaust Aftertreatment," *Progress in Energy and Combustion Science* **23**, no. 1 (1997): 1–39.

9-30. Barnes, G.J., "Catalysts for the Control of Automotive Pollutants," *A Comparison of Platinum and Base Metal Oxidation Catalysts* (ACS Publications, 1975), Chapter 7, 72-84, ISBN13: 9780841202191.

9-31. Kummer, J., "Oxidation of CO AND C2H4 by Base Metal-Catalysts Prepared on Honeycomb Supports," *Advances in Chemistry Series* **143** (1975): 178–192.

9-32. Gandhi, H., Yao, H., and Stepien, H., "Catalysis Under Transient Conditions," *The Use of Molybdenum in Automotive Three-Way Catalysts* (ACS Publications, 1982), Chapter 6, 143–162, ISBN13: 9780841206885.

9-33. Summers, J.C. and Ausen, S.A., "Interaction of Cerium Oxide with Noble Metals," *Journal of Catalysis* **58**, no. 1 (1979): 131–143.

9-34. Shelef, M., Graham, G.W., and McCabe, R.W., *Catalysis by Ceria and Related Materials* (London: Imperial College Press, 2002), 343–389.

9-35. Yao, Y.-F.Y., "Oxidation of Alkanes over Noble Metal Catalysts," *Industrial & Engineering Chemistry Product Research and Development* **19**, no. 3 (1980): 293–298.

9-36. Chen, B.H.-Y. and Chang, H.-L.R., "Development of Low Temperature Three-Way Catalysts for Future Fuel Efficient Vehicles," *Johnson Matthey Technology Review* **59**, no. 1 (2015): 64–67.

9-37. Pereira, C., Kim, G., and Hegedus, L., "A Novel Catalyst Geometry for Automobile Emission Control," *Catalysis Reviews Science and Engineering* **26**, no. 3-4 (1984): 503–523.

9-38. Cederquist, A.L., Devlin, S.S., Hart, D.L., and Moon, R., "The Microcomputer Based Engine Control System for the IIEC-2 Concept Car," SAE Technical Paper 790508, 1979, https://doi.org/10.4271/790508.

9-39. Osment, H.E., "Active Aluminas as Catalyst Supports for Treatment of Automotive Exhaust Emissions," SAE Technical Paper 730276, 1973, https://doi.org/10.4271/730276.

9-40. Summers, J. and Thompson, C., "Aging of Pelleted and Monolithic Automotive Emission Control Catalysts," SAE Technical Paper 800842, 1980, https://doi.org/10.4271/800842.

9-41. Gauguin, R., Graulier, M., and Papee, D., "Catalysts for the Control of Automotive Pollutants," *Advances in Chemistry Series* (American Chemical Society, 1975), ISBN13: 9780841202191.

9-42. Adomaitis, J.R., Smith, J.E., and Achey, D.E., "Improved Pelleted Catalyst Substrates for Automotive Emissions Control," SAE Technical Paper 800084, 1980, https://doi.org/10.4271/800084.

9-43. Oh, S.H., Cavendish, J.C., and Hegedus, L.L., "Mathematical Modeling of Catalytic Converter Lightoff: Single-Pellet Studies," *AIChE Journal* **26**, no. 6 (1980): 935–943.

9-44. Wei, J., "Catalysis for Motor Vehicle Emissions," *Advances in Catalysis*, Vol. **24** (Elsevier, 1975), 57–129.

9-45. Wei, J. and Becker, E.R., "The Optimum Distribution of Catalytic Material on Support Layers in Automotive Catalysis," *Advanced Chemical Series* **143** (1975): 116–132.

9-46. Williamson, W., Gandhi, H., Heyde, M.E., and Zawacki, G.A., "Deactivation of Three-Way Catalysts by Fuel Contaminants—Lead, Phosphorus and Sulfur," SAE Technical Paper 790942, 1979, https://doi.org/10.4271/790942.

9-47. Williamson, W.B., Lewis, D., Perry, J., and Gandhi, H.S., "Durability of Palladium Automotive Catalysts: Effects of Trace Lead Levels, Exhaust Composition, and Misfueling," *Industrial & Engineering Chemistry Product Research and Development* **23**, no. 4 (1984): 531–536.

9-48. Williamson, W.B., Stepien, H.K., Watkins, W.L., and Gandhi, H.S., "Poisoning of Platinum-Rhodium Automotive Three-Way Catalysts by Lead and Phosphorus," *Environmental Science & Technology* **13**, no. 9 (1979): 1109–1113.

9-49. Larese, C., Granados, M.L., Galisteo, F.C., Mariscal, R., and Fierro, J., "TWC Deactivation by Lead: A Study of the Rh/CeO2 System," *Applied Catalysis B: Environmental* **62**, no. 1-2 (2006): 132–143.

Lean NOx Traps

Breathe in that smog and feel lucky that only in L.A. will you glimpse a green sun or a brown moon. Forget that propaganda you've heard about clean air; demand oxygen you can see in all its glorious discoloration.

—John Waters, Crackpot: The Obsessions of John Waters

As discussed in Chapter 3, Emissions Control Regulations, nitrogen oxides (NOx) are one of the key criteria pollutants that must be managed for emissions compliance and to protect public health. The purpose of the lean NOx trap (LNT), also known as a NOx adsorber catalyst (NAC), NOx storage catalyst (NSC), or NOx storage and reduction (NSR) system, is to convert NOx into nitrogen gas (N_2) through a reduction reaction. This chapter should give the reader an understanding of the key LNT functions in the exhaust aftertreatment system (EAS), typical catalyst and washcoat choices, and possible failure modes.

The first publications on LNTs first appeared in 1995 [10-1], including work at Toyota by Iguchi et al. [10-2], and at Johnson Matthey and Daimler by Brogan et al. [10-3] and Bögner et al. [10-4]. Targeted applications of these works were light-duty lean-burn spark-ignition (SI) engines, although subsequent work on LNTs has focused on incorporating them into EAS for compression-ignition (CI) engines.

As the name suggests, the LNT adsorbs and holds NOx from the exhaust when the exhaust is in a net oxidizing state, such as from fuel-lean combustion where $\lambda > 1$. Note that λ is defined as the ratio of the air-fuel ratio (AFR) at the current operating point to the stoichiometric AFR for the given fuel. The LNT is then regenerated by desorbing the NOx and reducing it to N_2 during a fuel-rich combustion mode where $\lambda < 1$. The reducing agents are hydrocarbons (HC), carbon monoxide (CO), and hydrogen gas (H_2) that are generated by partial decomposition of the fuel in the engine under rich operating conditions. In their basic function, LNTs are similar to three-way catalysts (TWCs), described in Chapter 9, in that they use HC and CO to reduce NOx, and NOx and residual oxygen (O_2) to oxidize the HC and CO, but LNTs do these steps in sequence cyclically, instead of simultaneously.

From a practical view, LNTs are among the most complicated catalytic converters to implement in an EAS. In addition to the normal catalytic converter considerations, such as placement within the EAS, packaging, and thermal management, the LNT also requires cyclical changes in engine operation that are superimposed on the normal transient behavior of the internal combustion engine (ICE) in an application. An additional challenge is that the most effective catalyst for LNT function is platinum (Pt), which is expensive, and the LNT requires a substantial loading of Pt—approximately 3.5 g/L (100 g/ft³)—for good performance through the full useful life (FUL) of the engine. For reference, a TWC or diesel oxidation catalyst (DOC) might have 1.1–2.5 g/L (30 to 70 g/ft³) of Pt or similar catalyst, depending on the application. Some LNT catalyst formulations also add rhodium (Rh), which is even more expensive on a unit mass basis.

The main benefit of the LNT is that it uses a fluid, fuel, already onboard the vehicle as the reducing agent used to remove the NOx. Thus, the LNT does not require the additional components, such as the urea-water solution tank, dosing system, sensors, and on-board diagnostic functions, that urea-based selective catalytic reduction (SCR) systems do. Also, in-use compliance should be easier to guarantee, since the operator must supply fuel to make the engine work.

When used alone as the primary device for NOx conversion, an LNT monolith volume is typically 1.5–2.5 times the engine displacement. Thus, there are some smaller-engine applications where omitting the urea-water solution tank and dosing system will offset the additional catalyst cost of the LNT.

The LNT is intended for use on engines using fuel-lean combustion. The LNT works best in an EAS that usually experiences lean exhaust with $\lambda > 1$, but where a rich mode with $\lambda < 1$ is achievable. In this chapter, we present the normal function of an LNT, with a focus on the NOx storage and reduction performance; possible side reactions; typical catalysts; aging mechanisms; and other considerations for LNT design.

10.1 Lean NOx Trap (LNT) Function

The LNT requires the exhaust gas to cycle between a net oxidizing state from fuel-lean operation (the "lean period") and a net reducing state from fuel-rich operation (the "rich period") [10-5]. During the lean period of the cycle, the NOx is trapped in the LNT as nitrites (NO_2^-) or nitrates (NO_3^-) by chemisorption with the storage compound. The NOx trapping can be very efficient up until the storage compound becomes saturated. During the rich phase, the chemisorbed NOx is released and reduced to N_2. The NOx release also frees up the storage compound sites for the next lean period.

The LNT function then superimposes a periodic series of transient events over the normally transient operation of an ICE in a vehicle, thus complicating engine control. Nevertheless, the attractive features of the LNT are that the reducing agent is fuel, which is already onboard the vehicle for normal operation, and that the trapping of NOx during the lean period is nearly complete.

Challenges with LNT operation include the following:

- The overall duration of the lean-rich cycle
- The relative lengths of the lean and rich periods

- The minimum AFR or λ reached during the rich phase
- Competitive chemisorption of sulfur oxides (SOx) on the NOx storage sites
- Desulfation strategy
- Catalyst aging and degradation

The normal operation of an LNT is for it to cycle between a NOx adsorbing mode with lean (λ > 1) exhaust and a NOx desorption and reduction mode with rich (λ < 1) exhaust. A goal is to have smooth transitions from the one mode to the other. Maintaining the appropriate lean and rich conditions and then managing the transitions between them is governed by the control logic and calibration within the engine control unit (ECU). For example, the early work by Brogan et al. [10-3] used a cycle comprising 60 s of lean operation with λ = 1.35 followed by 0.3 s of rich operation with λ = 0.7. Bögner et al. [10-4] evaluated several cycle options, but still had relatively long stoichiometric or rich periods of 15 or 30 s compared to lean periods of 30, 60, or 90 s. Nevertheless, their focus was on characterizing the catalysts and not optimizing the cycle for best overall NOx conversion (x_{NOx}).

Compared to this work by Brogan et al. [10-3], most modern calibrations use a longer rich period of 4–10 s and make the exhaust less rich, with λ between 0.85 and 0.95 [10-1]. 60 s lean operation and 5 s rich is a typical cycle for an LNT. The longer rich duration is needed because LNTs have a larger NOx storage capacity than the early ones developed in 1995, and thus have more NOx to reduce to N_2. Also, boosted engines have a slower dynamic response making a rich period of under 1 s challenging. The rich λ is held closer to stoichiometric to mitigate fuel in oil dilution and smoke generation in the engine, which reduces PM loading and regeneration demands in the DPF. Bisaiji et al. [10-6, 10-7] at Toyota have described a fast LNT cycle, called DiAir, that comprises 6 s of lean operation and 1 s of rich. Fast cycling can improve NOx conversion and lower nitrous oxide (N_2O) yields [10-8, 10-9, 10-10, 10-11], but it requires an engine and associated actuators that can respond to this very dynamic set point trace in parallel with the normal transient engine demands. Muncrief et al. [10-12] found that shorter rich pulses with lower λ were more effective at high NOx conversion in the LNT than longer rich pulses with λ closer to 1. An example of the potential benefits of the fast LNT cycle is shown in **Figure 10.1** [10-11].

Because of its dynamic, if periodic, behavior during normal operation, it can be challenging to model and characterize an LNT well, especially during transient operation. One issue, described by Harold et al. [10-10, 10-13] at the University of Houston and by others [10-14], is that there are reaction fronts that propagate through the washcoat in each channel in the axial direction, especially during the transitions from lean to rich and back. For example, during the lean period, the LNT storage material fills up first in the front of the system. Additional NOx fills the storage sites further downstream as the lean period continues, as illustrated in **Figure 10.2** [10-15]. Over time, this leads to the front part of the LNT being very heavily used during normal operation and the back part, more lightly used. As the LNT ages, the active zone slowly migrates from the front face of the LNT monolith further into the monolith. In addition, the reaction fronts during the lean-to rich transition when the NOx is released and reduced also significantly influence the overall performance of the LNT, including the selectivity for N_2O formation, S_{N2O}.

FIGURE 10.1 Effects of (a) slow or (b) fast cycles in LNT performance on C_3H_6 conversion (circle), NH_3 selectivity (diamond), and CO yield (triangle, using unit carbon basis) at catalyst mid-point with an anaerobic rich feed condition with 3.5% H_2O (blue markers) or 0% (red markers). (Figure 6 in [10-11].)

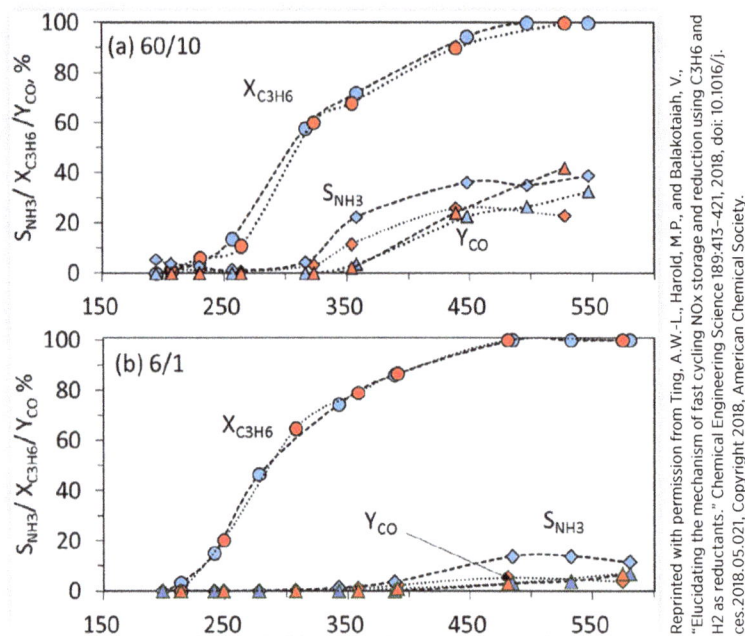

Reprinted with permission from Ting, A.W.-L., Harold, M.P., and Balakotaiah, V., "Elucidating the mechanism of fast cycling NOx storage and reduction using C3H6 and H2 as reductants." Chemical Engineering Science 189:413–421, 2018, doi: 10.1016/j.ces.2018.05.021, Copyright 2018, American Chemical Society.

FIGURE 10.2 Illustration of lean NOx trap function showing reaction fronts during (a) NOx trapping ($\lambda > 1$) and (b) NOx reduction ($\lambda < 1$) periods in the cycle. (Figure 2-5 in [10-15].)

•**(A) Sorption (Oxidizing Atmosphere): Period = 30–120 s**

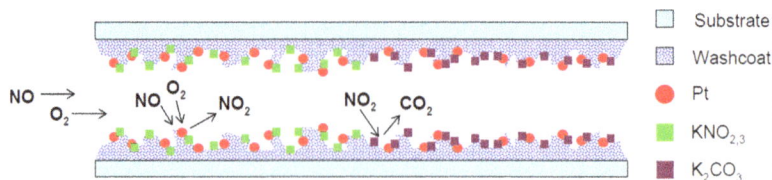

$2 NO + O_2 + Pt + K_2CO_3 \longrightarrow 2 KNO_{2,3} + Pt + CO_2$

•**(B) Regeneration (Reducing Atmosphere): Period = 1–10 s**

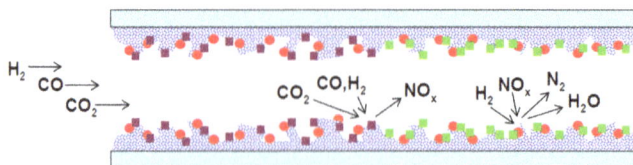

$2 KNO_{2,3} + Pt + H_2 + CO_2 \longrightarrow 2 KOH + N_2 + CO_2 + Pt + H_2O \longrightarrow K_2CO_3 + N_2 + 2 H_2O + Pt$

•**A-B-A-B-A-B Cycle Repeated**

There are several known challenges with LNT operation, especially during the transitions between the lean and rich periods. One is the "NOx puff," in which low but measurable quantities of nitrogen dioxide (NO_2), nitrogen monoxide (NO), and N_2O escape the LNT. This release can be enough to significantly influence the measured emissions from the engine and EAS. A second issue is ensuring good selectivity for N_2 instead of ammonia (NH_3) or N_2O. N_2O, a potent greenhouse gas, is regulated in many jurisdictions and needs to be minimized. NH_3 is also regulated on cycle, and so forming it may or may not be a problem depending on the overall EAS configuration. NH_3 formation is discussed in Section 10.1.3, and using that NH_3 in a downstream SCR system is discussed further in Section 15.5.1.

Choi et al. [10-16] found that the LNT can have some amount of axial back-mixing, that is, propagation upstream of a species by diffusion. This effect was observed by the consumption of H_2 by O_2 propagating upstream during the lean-to-rich transition.

The main advantage of LNTs is that they use the fuel already onboard the vehicle to reduce NOx species to N_2, which simplifies the in-use compliance by the operator compared to what is needed for a urea-based SCR system. LNTs can be cost-effective for smaller sized systems, where the urea storage tank and dosing system cost would outweigh the catalyst cost. LNTs also provide an advantage when exhaust temperatures are less than 200°C because they can start removing NOx from the exhaust gas before the SCR catalysts become active.

10.1.1 NOx Trapping during Lean Operation

NOx species are trapped in the LNT during lean operation by forming nitrates with the NOx storage compound in the LNT washcoat. The most common combination of catalyst, storage compound, and washcoat in the LNT are Pt and barium oxide (BaO) supported on γ-alumina (γ-Al_2O_3) [10-5]. The catalyst facilitates the NOx storage and reduction reactions, discussed in Section 10.2. The NOx storage compounds are part of the washcoat and are discussed in more detail in Section 10.3.

During lean operation, the NOx composition in the engine-out exhaust gas is typically about 90% NO and 10% NO_2. Thus, the first step for storing NOx during the LNT lean-rich cycle is to oxidize NO to NO_2 on the Pt catalyst sites within the LNT per the following reversible reaction:

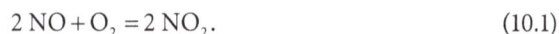

$$2\,NO + O_2 = 2\,NO_2. \tag{10.1}$$

The conversion of NO to NO_2 in Eq. 10.1 is limited by chemical kinetics at temperatures below 300°C or so, and limited by thermodynamics above 350°C or so, as shown in **Figure 10.3** [10-17]. Varying the oxygen concentration affects the thermodynamic equilibrium of Eq. 10.1. Decreasing the space velocity (SV) in the LNT system gives the reactants more time to undergo chemical reactions. Thus, with a lower SV, the NO and NO_2 will reach their thermodynamic equilibrium in Eq. 10.1 at a lower temperature than that shown in Figure 10.3. The NO oxidation reaction is inhibited by both NO and NO_2, and at lower temperatures, the rate of O_2 adsorption on the Pt catalytic sites limits the rate [10-18].

The next step in the NOx storage process is NOx chemisorption on the storage compound, which Epling et al. [10-5] the note is largely from the reaction of NO_2 with the alkali- or alkaline-earth storage compounds. This is especially true when

FIGURE 10.3 Steady-state NO$_2$:NOx ratios from experiments with varying feed conditions as shown. The first curve (16% O$_2$, 115 ppm NO$_2$) shows the equilibrium curve with respect to temperature[10-17].

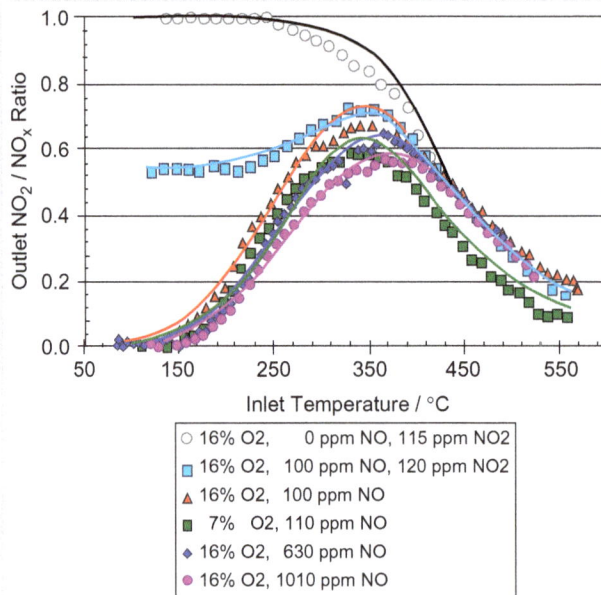

Reprinted from Ref. [10.17]. © SAE International

the LNT temperature is below 300°C [10-19]. BaO is the most common storage compound, as described in Section 10.3. In the presence of the combustion products carbon dioxide (CO$_2$) and water (H$_2$O), BaO reacts to form barium carbonate (BaCO$_3$) and barium hydroxide (Ba(OH)$_2$), respectively. Of these two, the BaCO$_3$ is more likely to form. The BaO or BaCO$_3$ then reacts with surface-bound NO$_2$ to form barium nitrate (Ba(NO$_3$)$_2$). The thermodynamic equilibrium of Eq. 10.1 is disrupted by the competing reactions of the NO$_2$ with the available storage compound within the washcoat. There are two main routes to form bound nitrate species, including the direct nitration route [10-20] of

$$2\,BaO + 2\,NO_2 + O_2 = 2\,Ba\left(NO_3\right)_2, \tag{10.2}$$

or the net disproportion reaction recommended by Nova et al. [10-21],

$$BaO + 3\,NO_2 \rightarrow Ba\left(NO_3\right)_2 + NO. \tag{10.3}$$

The NOx storage reactions, shown in Eqs. 10.2 and 10.3, remove NO$_2$ from the gas within the LNT monolith, which will drive the forward reaction in Eq. 10.1. Olsson et al. [10-22] describe how the NO in the exhaust gas needs to be oxidized to NO$_2$ following Eq. 10.1 so that the NOx may be trapped by the BaO in the washcoat as Ba(NO$_3$)$_2$ using the reaction shown in Eq. 10.2. Rafigh et al. [10-23] describe that the reaction Eq. 10.3 [10-13] reduces NO$_2$ to form BaO$_2$ as an intermediate.

Nova et al. [10-21] also found that increasing the number of Ba storage sites that neighbor Pt catalyst sites increase the overall NOx storage capacity. In addition, Olsson et al. [10-24, 10-25] have shown that some of the nitrate (NO$_3^-$) formed on the Pt catalyst spills over onto the γ-Al$_2$O$_3$ that constitutes the main part of the washcoat support instead of reacting with the NOx storage compound. This spillover effect supplements the NOx storage capacity of the LNT. Interestingly, Watling et al. [10-26] found that the NOx storage capacity was enhanced by having a higher NO concentration, which suggests that the NOx storage capacity is not strictly tied to the quantity of available barium (Ba). The O$_2$ concentration did not seem to affect the overall NOx storage capacity, though.

Olsson and colleagues [10-24, 10-25] have presented kinetic models for LNT chemistry that show a strong coverage dependency. Rafigh et al. [10-23] similarly observe that the NOx storage process seems to involve two different types of NOx storage sites within the LNT, each of which has its own rate of NOx storage. One is a small-capacity site that fills quickly, and the other is a large-capacity site that fills slowly during the lean period. These results can be seen in **Figure 10.4**, where the filling of the first type of storage site leads to a rapid breakthrough in LNT outlet NOx concentration, but the slow filling of the second site leads to a slower increase in NOx breakthrough. The feed gas contained 300 ppm NO and 10% O$_2$ [10-23]. The second type of storage site stores NO as barium nitrite (Ba(NO$_2$)$_2$), which then reacts to form Ba(NO$_3$)$_2$.

Epling et al. [10-27] note that BaO is the first barium species to form nitrates within the LNT during the lean, trapping mode. In the presence of CO$_2$ or H$_2$O, however, the BaO forms BaCO$_3$ or Ba(OH)$_2$, respectively, during the rich period. Their experimental data also suggest that there are at least two different sites for NOx storage within the catalyzed washcoat and that CO$_2$ influences one pathway while H$_2$O influences the other.

Because the NOx reduction process starts with NOx chemisorption on the LNT storage material, without the reducing event of rich operation, eventually the LNT saturates with NOx and the pollutant begins to pass through the LNT. Olsson et al. [10-22] identified that the NO in the exhaust needs to be oxidized to NO$_2$ on the Pt sites

FIGURE 10.4 Measured reactor outlet concentrations of NO (blue) and NO$_2$ (brown) from NOx storage experiment at 209°C. (Data from [10-23].)

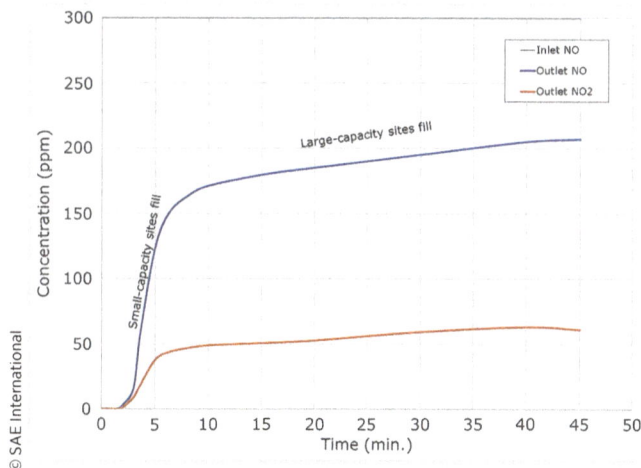

so that the NOx can be stored as $Ba(NO_3)_2$. Part of this process requires the sorbate precursor BaO to decompose as part of the nitrate formation.

10.1.2 NOx Reduction to Nitrogen

During the rich part of the cycle, the NO_2 is released from its storage compounds and then reduced to other nitrogen species. This NO_2 release comes from the decomposition of $Ba(NO_3)_2$ during the rich period, which is the reverse of Eq. 10.2. The NO_2 release appears to be promoted by the presence of CO_2 in the exhaust [10-28, 10-29] since that promotes $BaCO_3$ formation instead of BaO when the NO_2 is released,

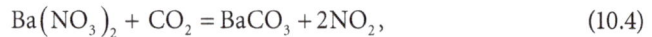

$$Ba(NO_3)_2 + CO_2 = BaCO_3 + 2NO_2, \tag{10.4}$$

Several reduction reactions are possible, depending on which reducing agents are available in the exhaust. Likely options include H_2, CO, or HC, which are typically represented by propene (C_3H_6), as shown below.

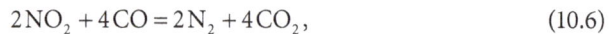

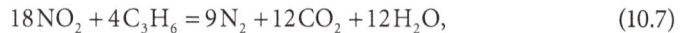

$$2NO_2 + 4H_2 = 2N_2 + 4H_2O, \tag{10.5}$$

$$2NO_2 + 4CO = 2N_2 + 4CO_2, \tag{10.6}$$

$$18NO_2 + 4C_3H_6 = 9N_2 + 12CO_2 + 12H_2O, \tag{10.7}$$

The most likely are the reactions in Eqs. 10.5 and 10.6, since both H_2 and CO, are formed both in the engine and on the Pt catalyst in the LNT through reforming of the available HC in the exhaust. In particular, H_2 seems to be particularly effective at NOx reduction [10-30], especially at lower temperatures below 250°C.

Eq. 10.7 uses propene (C_3H_6) as a model HC for NOx reduction, which is a typical practice. Other HC species could be used instead if they are known to react directly with NO_2 in the LNT. The reactions can be scaled to ensure appropriate stoichiometry by assuming that the HC present in fuel can be represented by $(CH_x)_n$, where $x \approx 1.9$ and n is between 3 and 16 for diesel exhaust.

While CO can directly reduce NO_2 as in Eq. 10.6, there is some evidence that in the presence of water in the exhaust, CO will form H_2 through the water-gas shift (WGS) reaction, and that H_2 will then reduce the NO_2. Larson et al. [10-31] also proposed the following two pathways for reducing NOx with CO:

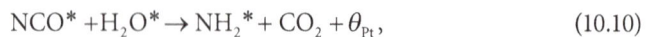

$$H_2O* + CO* = 2H* + CO_2, \quad \text{or} \tag{10.8}$$

$$N* + CO \rightarrow NCO* \quad \text{and} \tag{10.9}$$

$$NCO* + H_2O* \rightarrow NH_2* + CO_2 + \theta_{Pt}, \tag{10.10}$$

where θ_{Pt} is an open Pt surface site and "*" indicates a surface-bound species, here bound to the Pt catalyst in the LNT. The first pathway, shown in Eq. 10.8, uses CO in the WGS reaction to produce surface-bound hydrogen (H*), which then reduces the NOx to N_2

as in Eq. 10.5. The second pathway, shown in Eqs. 10.9 and 10.10, uses an isocyanate (NCO) intermediate to make a surface-bound amine group, NH_2*. In their overall reaction scheme, Larson et al. [10-31] suggest that this NH_2* group helps produce N_2O instead of N_2. **Figure 10.5** shows that N_2O and NH_3 can be produced by an LNT, especially at lower temperatures of 100°C–200°C, although these results are for steady-flow conditions at a given temperature.

Similarly, CO can also yield N_2O and NH_3 in an LNT under steady-flow conditions, as shown in **Figure 10.6**. There is some evidence [10-32] that the NH_3 formed in the LNT can be reacted with NO and NO_2 within the LNT to form N_2 through the SCR reactions,

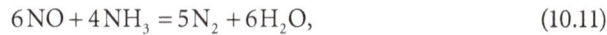

$$6NO + 4NH_3 = 5N_2 + 6H_2O, \qquad (10.11)$$

FIGURE 10.5 Concentrations of nitrogen-containing species exiting LNT with respect to temperature for 1:1 NO: H_2 feed [10-31].

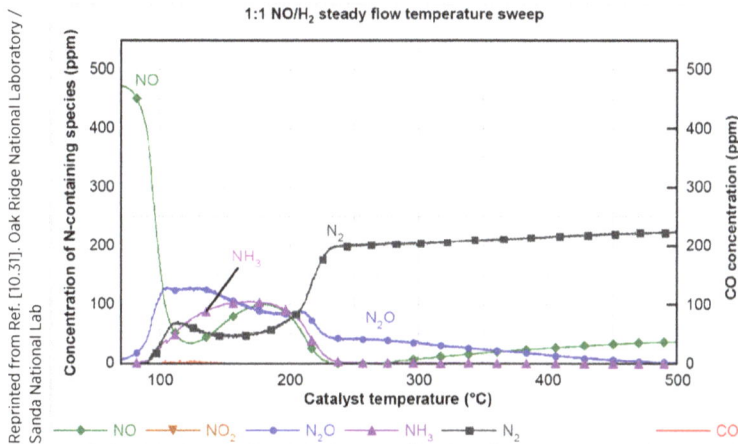

FIGURE 10.6 Concentrations of nitrogen-containing species exiting LNT with respect to temperature for 1:2.5 NO: CO feed [10-31].

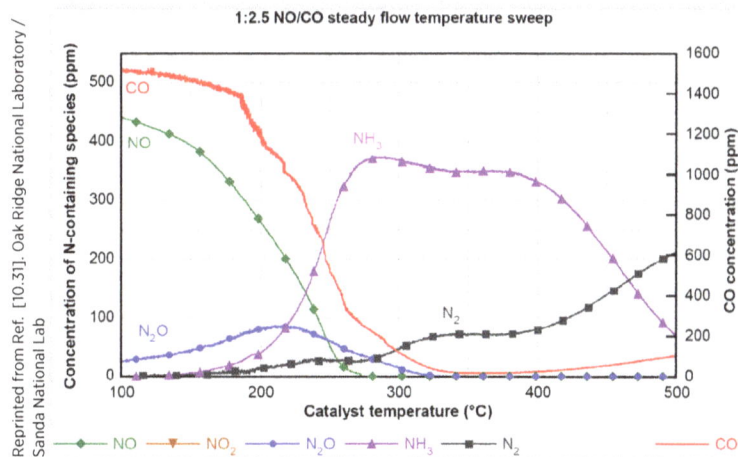

$$NO + NO_2 + 2\,NH_3 = 2\,N_2 + 3\,H_2O, \quad \text{and} \tag{10.12}$$

$$6\,NO_2 + 8\,NH_3 = 7\,N_2 + 12\,H_2. \tag{10.13}$$

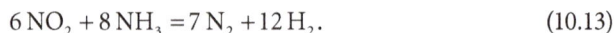

The SCR reactions are described in more detail in Chapter 12, Selective Catalytic Reduction Systems.

Epling et al. [10-27] looked at the effects of CO_2 and water on LNT performance with a system using Pt and Ba on an alumina (Al_2O_3) catalyst support. They found that both CO_2 and water lowered NOx conversion from about 80% to 85% to about 55% to 62%. Unfortunately, both compounds are products of combustion, and will always be present in engine exhaust.

During the rich phase of the operating cycle, the Pt catalyst sites are chemically reduced, making them able to reduce NOx to N_2 [10-27]. The experiments of Epling et al. [10-27] confirmed the proposed mechanism from Bögner et al. [10-4], which suggests that the nitrates decompose faster than the Pt sites reduce, leading to a release of NOx from the LNT.

10.1.3 Ammonia Generation

With a sufficiently long or "deep" rich period, NOx can be reduced to form NH_3 in the LNT instead of the preferred N_2. Usually, N_2 is the preferred product of NOx reduction, and it does form first in the LNT [10-33]. Thus, making NH_3 would usually be a waste of reducing agent, be it H_2, CO, or HC. There are, however, some circumstances in which it is useful to generate excess NH_3 in the LNT, such as for use in an SCR unit further downstream in the EAS. This combination of LNT and SCR system is one approach for passive SCR, and the combined system is described further in Section 15.5.

Both experimental and model results by Kumar et al. [10-13] of an extended rich period have found an increase in both NH_3 and H_2 concentrations leaving the sample LNT, as shown in **Figure 10.7**. The lean conditions used to fill the LNT included 500 ppm

FIGURE 10.7   Comparison of experimental and model results of LNT regeneration [10-13].

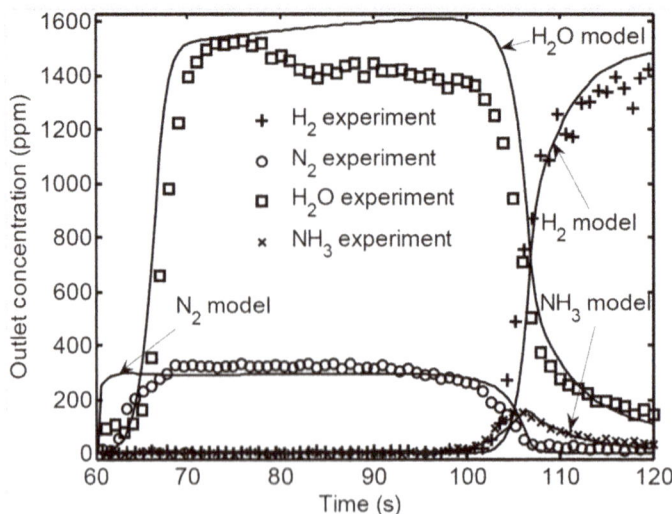

Reprinted from Catalysis Today, Vol. 147, Supplement, Divesh Bhatia, Robert D. Clayton, Michael P. Harold, Vemuri Balakotaiah, A global kinetic model for NO x storage and reduction on Pt/BaO/Al2O3 monolithic catalysts, S250-S256, Copyright 2009, with permission from Elsevier

NO and 5% O_2 in the feed gas. The rich conditions for this test used 1500 ppm H_2 with the balance being argon (Ar) for 100 s. The particular LNT sample had 2.70 wt% Pt and 14.6 wt% BaO in the catalyst. Similar work by Koči et al. [10-34] has determined that the NH_3 selectivity is highest when the LNT is around 200°C. At higher temperatures in the LNT, the NH_3 is used to reduce NOx to N_2 or is oxidized to NOx or both [10-35], depending on the relative amounts of NH_3 and NOx.

Wittka et al. [10-36, 10-37] have also looked at NH_3 generation for the purpose of an LNT–SCR combined system, and added Rh to the catalyst to help support the NH_3-forming reaction pathways. The Rh is important because it facilitates the WGS reaction, as in Eq. 10.8, which allows the CO to react with water to form surface-bound H* (or H_2), which is then used to make NH_3 [10-31, 10-37].

10.1.4 Additional Functions

One of the key functions needed in the LNT is the ability to purge sulfates from the BaO storage compound. The desulfation process is described in Section 10.4.

Because the LNT has a high Pt content, the LNT system can act as an oxidation catalyst and oxidize CO and HC when there is enough O_2 present in the exhaust. This oxidizing catalyst function is particularly useful in a modern EAS configuration that has the following units in series: LNT, diesel particulate filter (DPF), SCR system, and ammonia slip catalyst (ASC). Unfortunately, the NOx trapping function tends to inhibit the oxidation reactions [10-15].

Epling et al. [10-5] note that using an LNT as an oxidation catalyst means that exothermic reactions will occur in the LNT. These reactions can generate axial thermal gradients in the LNT system that affect the chemistry [10-10]. In addition, exothermic oxidation reactions can create transient temperature swings in the LNT [10-38], ranging from a modest 2°C–3°C at the downstream end of the LNT to a significant 50°C–100°C at the inlet end.

The catalyst in an LNT is similar to those used in a TWC, and one could consider using the LNT as a TWC when a lean SI engine is in a stoichiometric mode. As described in Chapter 9, however, modern TWC washcoats need significantly more oxygen storage capacity (OSC) than helps LNT function [10-5].

10.2 Catalyst Formulations

The most common catalysts used in LNTs are platinum group metals (PGM), which include Pt, palladium (Pd), and Rh. Of these, LNT catalysts are almost always only Pt because it is the most effective catalyst for the NOx storage and reduction reactions. Occasionally some Rh will be added, especially if the LNT is intended to have a higher selectivity for NH_3 since Rh promotes the WGS reaction that produces the H_2 used to make NH_3. Pd is generally not used as a catalyst in LNTs. The catalyst—Pt only or Pt with some Rh—is supported on a γ-Al_2O_3 washcoat that also includes the storage compound, which is usually BaO. The details of the storage compound and washcoat are discussed in Section 10.3.

The Pt catalyst needs to support the reduction and oxidation chemistry needed to make the LNT work. For example, NO needs to be oxidized to NO_2 during the lean period per the reaction in Eq. 10.1 so that it can react with the NOx storage compound, *e.g.*, BaO. Later, during the rich period, the catalyst needs to facilitate the reduction of released NO_2 or NO to N_2 gas, per the reactions in Eqs. 10.5, 10.6, and 10.7.

In one of the first papers on LNTs, Bögner et al. [10-4] prepared catalysts that used 1.45 g/L (41.1 g/ft^3) of Pt and 0.29 g/L (8.2 g/ft^3) of Rh. The washcoat support was a mix of Al_2O_3, ceria (CeO_2), and zirconia (ZrO_2). Wittka et al. [10-37] looked at LNT performance as part of an LNT–SCR combined system. They tested LNT units with a low PGM catalyst loading of 3.0 g/L (85 g/ft^3) and a high loading of 4.59 g/L (130 g/ft^3). Parks et al. [10-15] evaluated two LNT formulations that each had 3.53 g/L (100 g/ft^3) of Pt. Li et al. [10-9] tested a catalyst with 3.2 g/L (90 g/ft^3) of Pt and Rh in an 8:1 ratio. All of these loadings suggest that significant amounts of Pt are needed in a typical LNT device for passenger cars, much fewer vehicles with larger engines.

10.3 Other Washcoat Functionality and Substrates

Most LNTs use a γ-alumina (γ-Al_2O_3) support for the washcoat. The Pt and Rh catalyst, the NOx storage compound, and other constituents are then blended into the washcoat to provide the full functionality of the LNT. In addition to the dispersed compounds in the support, the γ-Al_2O_3 support itself seems to enhance the NOx storage capacity of the LNT because it can support some of the NO_3^- formed on the Pt catalyst [10-24, 10-25].

For example, Xu et al. [10-39] evaluated the performance of an LNT washcoat that relied on Al_2O_3 as the primary NOx storage compound, with 2 wt% to 3 wt% alkaline earth metals to enhance the performance. The alkaline earth metals used included magnesium (Mg), calcium (Ca), strontium (Sr), and Ba. These LNTs had a PGM catalyst loading of 3.53 g/L (100 g/ft^3) Pt and 0.71 g/L (20 g/ft^3) Rh, which is comparable to the catalyst loadings in commercial LNTs using BaO as the NOx storage compound.

In addition to the choice of catalyst, including type and loading, a major design consideration is the choice of NOx storage compound. The NOx storage compounds need to be strongly basic to react with the acidic species NO and NO_2. Iguchi et al. [10-2] evaluated several alkali metals and alkaline earth metals for their ability to facilitate NOx and HC conversion in the LNT. As shown in **Figure 10.8** [10-2], sodium (Na), potassium (K), and Ba oxides (Na_2O, K_2O, and BaO, respectively) had the best overall performance over the Japanese 10+15 mode cycle, which has since been superseded by other drive cycles as described in Section 4.1.7.

The challenge with Na_2O or K_2O as NOx storage compounds is that they are water-soluble, which is an issue for engine exhaust applications where there is a significant water vapor content in the exhaust. The risk is that water condenses in the LNT during a cold start or after engine shut-down and then dissolves and washes away the soluble NOx storage compounds. K_2O has been studied by some researchers because it is less soluble than Na_2O [10-40], but BaO is even less soluble and is, therefore, more

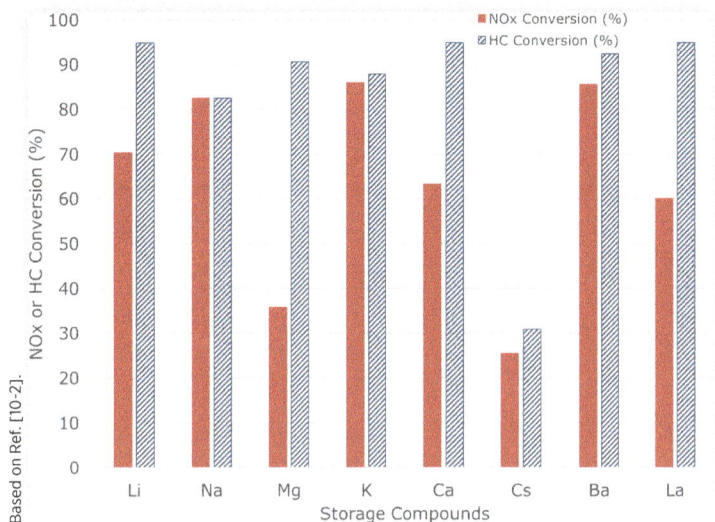

FIGURE 10.8 NOx storage component effect on NOx and HC conversion in an LNT on the Japanese 10+15 mode cycle [10-2].

stable in the LNT. Thus, BaO is the most common NOx storage compound used in commercial LNTs.

Note that in the presence of CO_2 or H_2O, BaO forms $BaCO_3$ or $Ba(OH)_2$, respectively. Since CO_2 and H_2O are present in engine exhaust as normal products of combustion, especially during the lean regeneration phase of the lean/rich cycle, it is unlikely that the Ba is present as BaO [10-27]. Nevertheless, most models of the NOx storage chemistry assume that the Ba is present as either BaO or $Ba(NO_3)_2$ with good fidelity to measured results. Lietti et al. [10-41] have found that BaO preferentially reacts to form $Ba(NO_3)_2$, followed by $Ba(OH)_2$ and $BaCO_3$. Disselkamp et al. [10-33] found that the $Ba(NO_3)_2$ forms a monolayer on the surface of the Al_2O_3 support with some larger bulk $Ba(NO_3)_2$ particles also present.

Ji et al. [10-42] created several model catalysts in order to evaluate the aged performance, but they also made a comparison to fresh catalyst performance. They varied the Pt, BaO, and CeO_2 or ceria-zirconia (CeO_2-ZrO_2) loadings in their catalysts to evaluate the performance of aged catalysts. The model LNT catalysts also had either a high level (20 g/ft^3 or 0.71 g/L) or a low level (10 g/ft^3 or 0.35g/L) of Rh. One of the functions Ji et al. [10-42] evaluated was the OSC of the catalyst and washcoat. The OSC comes from the CeO_2 or CeO_2-ZrO_2 in the washcoat; the ceria compounds were with lanthanum (La) in the γ-Al_2O_3 of the washcoat. Ji et al. [10-42] report that for fresh catalyst the OSC scales linearly with CeO_2 content. The CeO_2-ZrO_2 blend further enhanced OSC compared to the same quantity of CeO_2.

Rafigh et al. [10-23] evaluated the OSC from the CeO_2 in their LNT washcoat and determined that the OSC is a function of reactor temperature, as shown in **Figure 10.9**. The stored oxygen in the washcoat facilitates the WGS reaction, shown in Eq. 10.8.

Modern EAS, especially for light-duty applications, tend to have the LNT in the leading position. Thus, the substrates used for LNTs need to be robust enough to handle

FIGURE 10.9 Calculated oxygen storage capacity in moles per unit reactor volume [10-23].

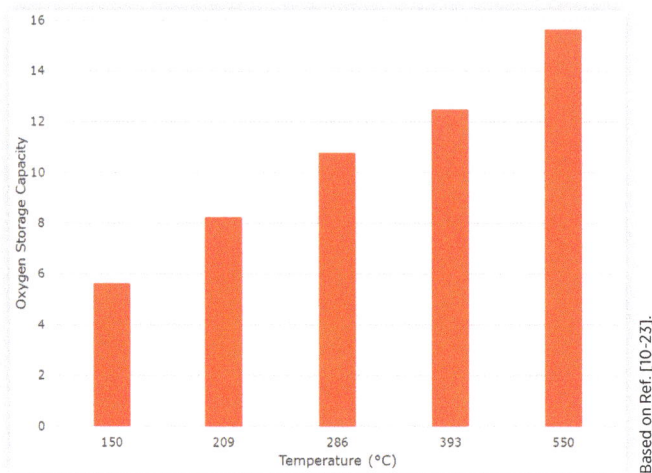

Based on Ref. [10-23].

the exhaust gas temperatures close to the engine or turbine outlet. These temperatures are expected to be 300°C–650°C for HDDE applications, and 150°C–300°C for LDDE applications. Light-duty SI engines may have hotter exhaust gas temperatures, although these will be mitigated by the calibration, which will move toward stoichiometric or fuel-rich conditions at high load to prevent knock. For these expected temperature ranges, cordierite substrates are suitable, although some close-coupled LNTs may use silicon carbide (SiC) substrates for improved thermal durability.

10.4 LNT Failure Modes

There are several key aging effects that degrade LNT performance over time. One aging mechanism is that the PGM catalysts Pt and Rh are prone to sintering within the washcoat as the LNT is exposed to high temperatures. Another key concern is the susceptibility of LNT catalysts and washcoat compounds to sulfur poisoning, especially the NOx storage compounds. Unfortunately, one of the ways to recover the LNT performance from sulfur poisoning is to put it through a high-temperature desulfation cycle, which leads to a trade-off between degradation from sulfur and degradation from sintering.

The catalyst sintering is caused by small Pt or Rh particles migrating within the washcoat at elevated temperatures and merging together. Thus, the catalyst mass stays the same, but the particles become larger which removes active surface area for the chemical reactions to occur. One of the consequences of the Pt or Rh catalyst sintering within the LNT is that the catalysts tend to withdraw from the BaO storage sites, which degrades the NOx storage and NOx reduction functions of the LNT. The sintering of the Pt seemed to be mitigated by the presence of CeO_2 in the overall catalyst formulation, which is consistent with the findings of Diwell et al. [10-43] for TWCs.

In addition, other compounds in the washcoat are susceptible to hydrothermal aging effects. For example, La-stabilized CeO_2 also appears to sinter during high-temperature events, although CeO_2-ZrO_2 seems to have better thermal stability. The NOx storage capacity also seems to be affected by aging [10-44, 10-45], by 35%–85% depending on the operating temperature and SV. The degradation in NOx storage appears to be greater at lower temperatures.

Ji et al. [10-42] used model catalysts to evaluate aging mechanisms for second-generation LNT catalyst formulations that, for example, contain CeO_2 or CeO_2-ZrO_2 in the washcoat. They varied the Pt, BaO, and CeO_2 or CeO_2-ZrO_2 loadings in their catalysts to evaluate the performance of aged catalysts. The CeO_2-ZrO_2 blend retained its OSC better than the washcoat with a similar loading of La-stabilized CeO_2 did. Also, the OSC from the CeO_2 helps the LNT catalysts retain good selectivity for N_2 and against NH_3 after aging, especially at higher temperatures of 350°C to 450°C [10-42].

The LNT's catalyst and storage compound are both very sensitive to sulfur oxides (SOx): sulfur dioxide (SO_2) and sulfur trioxide (SO_3). Engine exhaust typically only has SO_2, but SO_3 is readily formed from it in the presence of oxygen over Pt. The SOx species adsorb onto the Pt catalyst, and since they block active sites they impair the catalyst function, especially the oxidation reactions during the lean period [10-5]. Nevertheless, sulfur compounds are reversibly driven off within normal operating temperatures of 400°C–500°C, and so this is an easily managed issue.

The bigger problem is that SOx species react with BaO to form barium sulfate ($BaSO_4$), which is very thermally stable. $BaSO_4$ formation,

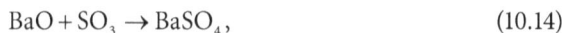

$$BaO + SO_3 \rightarrow BaSO_4,$$
(10.14)

leads to a corresponding reduction in the NOx storage capacity. The quantity of sulfur stored is proportional to the BaO content, and cumulative stored sulfur level is mildly affected by the sulfur composition or exposure time. The $BaSO_4$ forms first at the front face of the LNT and then keeps forming from the front to the back. Fridell et al. [10-47] found that SO_2 uptake more strongly degraded the NOx storage capability when it happened during the rich period than during the lean period. They hypothesize that this happens at BaO sites closer to the Pt catalyst during the rich period, and these sites are more active for the NOx trapping. Since the SOx in the exhaust comes from sulfur compounds in the fuel, lowering the sulfur content of the fuel is the main mitigation technique used. Mahzoul et al. [10-47] also found that the water vapor present in the exhaust helps mitigate $BaSO_4$ formation and facilitates desulfation. Similarly, the OSC from the CeO_2 content provides a modest benefit to this sulfur poisoning effect [10-5].

The adoption of diesel fuels with very low sulfur levels, such as ultra-low sulfur diesel (ULSD) in the US with fuel sulfur content at or below 15 ppm, was encouraged by the expectation of LNTs being adopted, in addition to the other benefits of lower SOx emissions. It is also helpful to use low-sulfur (or low SAPS) lubricant, although the sulfation from lubricant-derived sulfur is outweighed by that from fuel-derived sulfur.

Desulfation of the NOx storage compound typically requires temperatures over 650°C, and sometimes over 700°C, which means that each desulfation event hydrothermally ages the LNT catalyst. The desulfation event must last long enough to ensure that SOx is driven out of the whole LNT. A challenge of desulfation is that it is superimposed on the other operating requirements of the engine, and so a desulfation strategy needs to be robust to keep the LNT inlet temperature high while also meeting other demands.

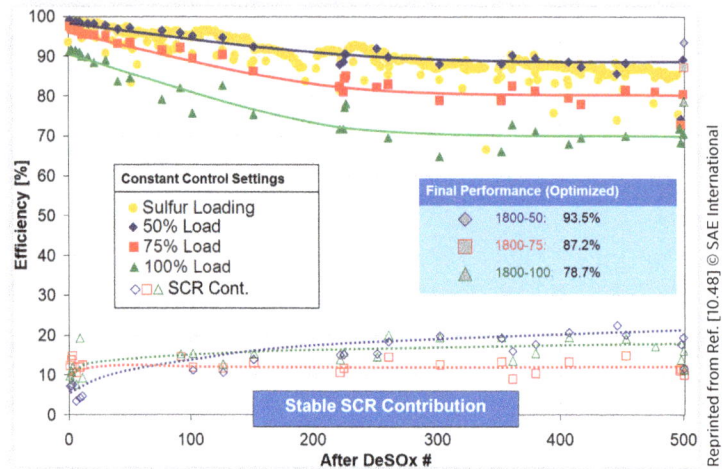

FIGURE 10.10 NOx conversion degradation in LNT after desulation cycles [10-48].

McCarthy et al. [10-48] conducted a series of sulfation–desulfation cycles on an LNT in a combined LNT–SCR system and found that the NOx conversion in the LNT dropped by 20%–30% from the de-greened catalyst performance after 200–300 cycles, as shown in **Figure 10.10**. However, after that point the LNT performance seems stable.

Another challenge for LNTs is that they have a tendency to form nitrous oxide (N_2O) during their normal operation. N_2O selectivity (S_{N2O}) can be mitigated by the formulation of the catalyst and the washcoat, but the transition from lean to rich during normal operation also influences S_{N2O} [10-49]. Another mitigation strategy is to use a downstream system in the EAS to either reduce the N_2O to N_2 or oxidize the N_2O back to NO or NO_2, preferably the former.

References

10-1.   Keenan, M., "Exhaust Emissions Control: 60 Years of Innovation and Development," SAE Technical Paper 2017-24-0120, 2017, https://doi.org/10.4271/2017-24-0120.

10-2.   Iguchi, S., Kihara, T., Harada, J., Tanaka, T., and Katoh, K., "NOx Storage Reduction 3-Way Catalyst for Lean Burn System," *16th International Vienna Motor Symposium*, Vienna, 1995.

10-3.   Brogan, M.S., Brisley, R.J., Walker, A.P., Webster, D.E. et al., "Evaluation of NOx Storage Catalysts as an Effective System for NOx Removal from the Exhaust Gas of Leanburn Gasoline Engines," SAE Technical Paper 952490, 1995, https://doi.org/10.4271/952490.

10-4.   Bögner, W., Krämer, M., Krutzsch, B., Pischinger, S., Voigtländer, D., Wenninger, G., Wirbeleit, F., Brogan, M.S., Brisley, R.J., and Webster, D.E., "Removal of Nitrogen Oxides from the Exhaust of a Lean-Tune Gasoline Engine," *Applied Catalysis B: Environmental* 7, no. 1–2 (1995): 153–171, https://doi.org/10.1016/0926-3373(95)00035-6.

10-5.   Epling, W.S., Campbell, L.E., Yezerets, A., Currier, N.W., and Parks, J.E., "Overview of the Fundamental Reactions and Degradation Mechanisms of NOx Storage/Reduction Catalysts," *Catalysis Reviews* **46**, no. 2 (2004): 163–245, https://doi.org/10.1081/CR-200031932.

10-6.   Bisaiji, Y., Yoshida, K., Inoue, M., Umemoto, K., and Fukuma, T., "Development of Di-Air - A New Diesel deNOx System by Adsorbed Intermediate Reductants," *SAE Int. J. Fuels Lubr.* **5**(1): 380–388, 2011, https://doi.org/10.4271/2011-01-2089.

10-7.   Bisaiji, Y., Yoshida, K., Inoue, M., Takagi, N., and Fukuma, T., "Reaction Mechanism Analysis of Di-Air-Contributions of Hydrocarbons and Intermediates," *SAE Int. J. Fuels Lubr.* **5**(3): 1310–1316, 2012, https://doi.org/10.4271/2012-01-1744.

10-8.   Chansai, S., Burch, R., Hardacre, C., and Naito, S., "Origin of Double Dinitrogen Release Feature during Fast Switching between Lean and Rich Cycles for NOx Storage Reduction Catalysts," *Journal of Catalysis* **317** (2014): 91–98, https://doi.org/10.1016/j.jcat.2014.05.020.

10-9.   Li, M., Zheng, Y., Luss, D., and Harold, M.P., "Impact of Rapid Cycling Strategy on Reductant Effectiveness during NO_x Storage and Reduction," *Emission Control Science and Technology* **3**, no. 3 (2017): 205–219, https://doi.org/10.1007/s40825-017-0071-5.

10-10. Ting, A.W.-L., Li, M., Harold, M.P., and Balakotaiah, V., "Fast Cycling in a Non-Isothermal Monolithic Lean NOx Trap Using H2 as Reductant: Experiments and Modeling," *Chemical Engineering Journal* **326** (2017): 419–435, https://doi.org/10.1016/j.cej.2017.05.002.

10-11. Ting, A.W.-L., Harold, M.P., and Balakotaiah, V., "Elucidating the Mechanism of Fast Cycling NOx Storage and Reduction Using C3H6 and H2 as Reductants," *Chemical Engineering Science* **189** (2018): 413–421, https://doi.org/10.1016/j.ces.2018.05.021.

10-12. Muncrief, R.L., Kabin, K.S., and Harold, M.P., "NOx Storage and Reduction with Propylene on Pt/BaO/Alumina," *AIChE Journal* **50**, no. 10 (2004): 2526–2540, https://doi.org/10.1002/aic.10208.

10-13. Kumar, A., Bhatia, D., Clayton, R., Xu, J., Harold, M.P., and Balakotiah, V., "Elucidating the Mechanism of NOx Storage and Reduction," *12th CLEERS Workshop*, Dearborn, MI, 2009.

10-14. Maurer, M., Holler, P., Zarl, S., Fortner, T., and Eichlseder, H., "Investigations of Lean NOx Trap (LNT) Regeneration Strategies for Diesel Engines," SAE Technical Paper 2017-24-0124, 2017, https://doi.org/10.4271/2017-24-0124.

10-15. Parks, J.E., Storey, J.M., Theiss, T.J., Ponnusamy, S., Ferguson, H.D., Williams, A.M. et al., "Lean NOx Trap Catalysis for Lean Natural Gas Engine Applications," report, September 1, 2007; [Tennessee]. (https://digital.library.unt.edu/ark:/67531/metadc896904/: accessed December 5, 2019), University of North Texas Libraries, UNT Digital Library, https://digital.library.unt.edu; crediting UNT Libraries Government Documents Department.

10-16. Choi, J.-S., Prikhodo, V., Chakravarthy, K., and Daw, C.S., "Assessing Monolith Length Effect on LNT Performance," *9th CLEERS Workshop*, Dearborn, MI, 2006.

10-17. Watling, T.C., Ahmadinejad, M., Țuțuianu, M., Johansson, Å., and Paterson, M.A.J., "Development and Validation of a Pt-Pd Diesel Oxidation Catalyst Model," SAE Technical Paper 2012-01-1286, 2012, https://doi.org/10.4271/2012-01-1286.

10-18. Bhatia, D., McCabe, R.W., Harold, M.P., and Balakotaiah, V., "Experimental and Kinetic Study of NO Oxidation on Model Pt Catalysts," *Journal of Catalysis* **266**, no. 1 (2009): 106–119, https://doi.org/10.1016/j.jcat.2009.05.020.

10-19. Al-Harbi, M. and Epling, W.S., "Investigating the Effect of NO versus NO_2 on the Performance of a Model NO_x Storage/Reduction Catalyst," *Catalysis Letters* **130**, no. 1-2 (2009): 121-129, https://doi.org/10.1007/s10562-009-9912-3.

10-20. Fridell, E., Skoglundh, M., Westerberg, B., Johansson, S., and Smedler, G., "NOx Storage in Barium-Containing Catalysts," *Journal of Catalysis* **183**, no. 2 (1999): 196–209, https://doi.org/10.1006/jcat.1999.2415.

10-21. Nova, I., Lietti, L., and Forzatti, P., "Mechanistic Aspects of the Reduction of Stored NOx over Pt-Ba/Al2O3 Lean NOx Trap Systems," *Catalysis Today* **136**, no. 1-2 (2008): 128–135, https://doi.org/10.1016/j.cattod.2008.01.006.

10-22. Olsson, L., Westerberg, B., Persson, H., Fridell, E., Skoglundh, M., and Andersson, B., "A Kinetic Study of Oxygen Adsorption/Desorption and NO Oxidation over Pt/Al_2O_3 Catalysts," *The Journal of Physical Chemistry B* **103**, no. 47 (1999): 10433–10439, https://doi.org/10.1021/jp9918757.

10-23. Rafigh, M., Dudgeon, R., Pihl, J., Daw, S., Blint, R., and Wahiduzzaman, S., "Development of a Global Kinetic Model for a Commercial Lean NOx Trap Automotive Catalyst Based on Laboratory Measurements," *Emission Control Science and Technology* **3**, no. 1 (2017): 73–92, https://doi.org/10.1007/s40825-016-0049-8.

10-24. Olsson, L., Persson, H., Fridell, E., Skoglundh, M., and Andersson, B., "A Kinetic Study of NO Oxidation and NOx Storage on Pt/Al_2O_3 and $Pt/BaO/Al_2O_3$," *The Journal of Physical Chemistry B* **105**, no. 29 (2001): 6895–6906, https://doi.org/10.1021/jp010324p.

10-25. Lindholm, A., Currier, N.W., Li, J., Yezerets, A., and Olsson, L., "Detailed Kinetic Modeling of NOx Storage and Reduction with Hydrogen as the Reducing Agent and in the Presence of CO_2 and H_2O over a Pt/Ba/Al Catalyst," *Journal of Catalysis* **258**, no. 1 (2008): 273-288, https://doi.org/10.1016/j.jcat.2008.06.022.

10-26. Watling, T.C., Bolton, P.D., and Swallow, D., "The Effect of NO and O_2 Concentration on NO_x Storage over a Lean NO_x trap: An Experimental and Modelling Study," *Chemical Engineering Science* **178** (2018): 312-323, https://doi.org/10.1016/j.ces.2017.12.007.

10-27. Epling, W.S., Campbell, G.C., and Parks, J.E., "The Effects of CO_2 and H_2O on the NOx Destruction Performance of a Model NOx Storage/Reduction Catalyst," *Catalysis Letters* **90**, no. 1-2 (2003): 45-56, https://doi.org/10.1023/A:1025864109922.

10-28. Balcon, S., Potvin, C., Salin, L., Tempère, J.F., and Djéga-Mariadassou, G., "Influence of CO_2 on Storage and Release of NO_x on Barium-Containing Catalyst," *Catalysis Letters* **60**, no. 1 (1999): 39-43, https://doi.org/10.1023/A:1019034402660.

10-29. Amberntsson, A., Persson, H., Engström, P., and Kasemo, B., "NO_x Release from a Noble Metal/BaO Catalyst: Dependence on Gas Composition," *Applied Catalysis B: Environmental* **31**, no. 1 (2001): 27-38, https://doi.org/10.1016/S0926-3373(00)00266-6.

10-30. Theis, J., Jen, H.-W., McCabe, R., Sharma, M., Balakotaiah, V., and Harold, M.P., "Reductive Elimination as a Mechanism for Purging a Lean NOx Trap," SAE Technical Paper 2006-01-1067, 2006, https://doi.org/10.4271/2006-01-1067.

10-31. Larson, R., Chakravarthy, K., Daw, C.S., and Pihl, J., "Modeling Kinetics of NH_3 and N_2O Formation in Lean NOx Traps," *9th CLEERS Workshop*, Dearborn, MI, 2006.

10-32. Clayton, R.D., Harold, M.P., and Balakotaiah, V., "Performance Features of Pt/BaO Lean NOx Trap with Hydrogen as Reductant," *AIChE Journal* **55**, no. 3 (2009): 687–700, https://doi.org/10.1002/aic.11710.

10-33. Disselkamp, R., Kim, D.H., Kwak, J.-H., Peden, C., Szanyi, J., and Tonkyn, R., "Fundamental Studies of NOx Adsorber Materials," *8th CLEERS Workshop*, Dearborn, MI, 2005.

10-34. Koči, P., Plát, F., Štěpánek, J., Bártová, Š., Marek, M., Kubíček, M., Schmeißer, V., Chatterjee, D., and Weibel, M., "Global Kinetic Model for the Regeneration of NO_x Storage Catalyst with CO, H_2 and C_3H_6 in the Presence of CO_2 and H_2O," *Catalysis Today* **147** (2009): S257-S264, https://doi.org/10.1016/j.cattod.2009.07.036.

10-35. Pihl, J., Parks, J.E., Toops, T.J., Daw, C.S., and Root, T., "Experimental Studies of N Species Selectivity during Regeneration of Lean NOx Traps," *9th CLEERS Workshop*, Dearborn, MI, 2006.

10-36. Wittka, T., Holderbaum, B., Maunula, T., and Weissner, M., "Development and Demonstration of LNT+SCR System for Passenger Car Diesel Applications," *SAE Int. J. Engines* **7**(3): 1269–1279, 2014, https://doi.org/10.4271/2014-01-1537.

10-37. Wittka, T., Holderbaum, B., Dittmann, P., and Pischinger, S., "Experimental Investigation of Combined LNT+ SCR Diesel Exhaust Aftertreatment," *Emission Control Science and Technology* **1**, no. 2 (2015): 167–182, https://doi.org/10.1007/s40825-015-0012-0.

10-38. Kabin, K.S., Muncrief, R.L., and Harold, M.P., "NOx Storage and Reduction on a Pt/BaO/Alumina Monolithic Storage Catalyst," *Catalysis Today* **96**, no. 1-2 (2004): 79–89, https://doi.org/10.1016/j.cattod.2004.05.008.

10-39. Xu, L., Graham, G., McCabe, R., and Hoard, J., "The Study of an Alumina-Based Lean NOx Trap (LNT) for Diesel," *9th CLEERS Workshop*, Dearborn, MI, 2006.

10-40. Toops, T.J., Smith, D.B., Epling, W.S., Parks, J.E., and Partridge, W.P., "Quantified NO_x Adsorption on Pt/K/gamma-Al_2O_3 and the Effects of CO_2 and H_2O," *Journal of Applied Catalysis B: Environmental* **58**, no. 3-4 (2005): 255–264, https://doi.org/10.1016/j.apcatb.2004.10.022.

10-41. Lietti, L., Forzatti, P., Nova, I., and Tronconi, E., "NOx Storage Reduction over Pt-Ba/γ-Al2O3 Catalyst," *Journal of Catalysis* **204**, no. 1 (2001): 175–191, https://doi.org/10.1006/jcat.2001.3370.

10-42. Ji, Y., Fisk, C., Easterling, V., Crocker, M., Choi, J.-S., and Partridge, W.P., "NOx Storage-Reduction Characteristics of LNT Catalysts Subjected to Simulated Road Aging," *12th CLEERS Workshop*, Dearborn, MI, 2009.

10-43. Diwell, A.F., Rajaram, R.R., Shaw, H.A., and Truex, T.J., "The Role of Ceria in Three-Way Catalysts," *Studies in Surface Science and Catalysis* **71** (1991): 139–152, https://doi.org/10.1016/S0167-2991(08)62975-4.

10-44. Ottinger, N.A., Nguyen, K., Bunting, B.G., Toops, T.J., and Howe, J., "Effect of High Temperature Lean/Rich Thermal Aging on NOx Storage and Reduction over a Fully-Formulated LNT," *12th CLEERS Workshop*, Dearborn, MI, 2009.

10-45. De Abreu Goes, J.E., Olsson, L., Berggrund, M., Kristoffersson, A., Gustafson, L., and Hicks, M., "Performance Studies and Correlation between Vehicle-and Rapid-Aged Commercial Lean NOx Trap Catalysts," *SAE Int. J. Engines* **10**, no. 4 (2017): 1613–1626, https://doi.org/10.4271/2017-01-0940.

10-46. Fridell, E., Persson, H., Olsson, L., Westerberg, B., Amberntsson, A., and Skoglundh, M., "Model Studies of NOx Storage and Sulphur Deactivation of NOx Storage Catalysts," *Topics in Catalysis* **16**, no. 1–4 (2001): 133–137, https://doi.org/10.1023/A:1016699202043.

10-47. Mahzoul, H., Gilot, P., Brilhac, J.-F., and Stanmore, B.R., "Reduction of NOx over a NOx-Trap Catalyst and the Regeneration Behaviour of Adsorbed SO_2," *Topics in Catalysis* **16**, no. 1-4 (2001): 293–298, https://doi.org/10.1023/A:1016600829785.

10-48. McCarthy, J.E., Jr., Korhumel, T.G., Jr., and Marougy, A.P., "Performance of a Fuel Reformer, LNT and SCR Aftertreament System Following 500 LNT Desulfation Events," *SAE Int. J. Commer. Veh.* **2**, no. 2 (2010): 34–44, https://doi.org/10.4271/2009-01-2835.

10-49. Lindholm, A., Currier, N.W., Fridell, E., Yezerets, A., and Olsson, L., "NOx Storage and Reduction over Pt Based Catalysts with Hydrogen as the Reducing Agent: Influence of H_2O and CO_2," *Applied Catalysis B: Environmental* **75**, no. 1–2 (2007): 78–87, https://doi.org/10.1016/j.apcatb.2007.03.008.

Particulate Filters

Though I spends me time in the ashes and smoke

In this 'ole wide world there's no 'appier bloke

—"Chim Chim Che-ree" from Mary Poppins

Concern for the environment and human health is resulting in increasingly rigorous regulation of engine particulate emissions by countries around the world, as discussed in Chapter 3, Emissions Control Regulations. At least currently, control of particulates from diesel-fueled compression-ignition (CI) engines is generally more challenging than for gasoline-fueled spark-ignition (SI) engines because CI engines tend to emit larger quantities of particulates in the regulated size/mass range. Since the early 2000s, particulate filters (PFs) have been integrated into the exhaust aftertreatment system (EAS) to mechanically filter particulate matter (PM) from the exhaust gas. This filtration reduces both the mass and number of particulates that are dispersed to the atmosphere. To achieve this reduction without harming the efficiency or gas exchange of the engine, engineers must traverse the delicate balance between particulate capture efficiency and pressure drop across the filter, which can cause a decline in engine performance. First introduced for CI engines, there is momentum building for them to be widely applied to SI engines in the near future, especially those with direct injection (DI) fuel systems.

This chapter discusses the following topics on PFs:

- Particulate matter
- PF function in the EAS
- Substrate materials
- Catalysts and washcoats
- Passive and active regeneration strategies
- Accumulation and removal of ash buildup in the filter.

11.1 **Particulate Matter**

Particulate matter (PM), also commonly called "soot," is primarily composed of the carbonaceous black smoke particles found in engine exhaust. Most often associated with heavy-duty diesel engines (HDDEs), PM is made by both SI and CI engines. Largely composed of unburned additives of lubrication oil and engine debris upstream, ash particulates possess a scale larger than the average exhaust particulates. Ash accumulation effects on PFs have attention as a result of the wide application of PFs and the rigorous PM/PN regulations. The effects of ash are presented in an extended special section, by C. Justin Kamp.

In contrast to gaseous exhaust emissions, particulates are not a well-defined chemical species and are often defined in terms of how they are measured in the exhaust. Thus, as discussed in Section 5.3.5, what is measured as PM may include additional solids, such as ash, or even liquid droplets present in the exhaust. Carbonaceous particulate formation occurs in-cylinder in a complex series of reactions in the combustion chamber. Particulate formation, including ash formation, is discussed in detail in Section 2.7.

As diesel particulate filter (DPF) technology has continued to evolve, it has become increasingly more important to understand the nature of the diesel particulate matter. It is critical to understand the reactivity of the PM for optimizing both the filtration and regeneration performance of DPFs over their in-use lifetime. As gasoline particulate filters (GPFs) continue to develop, a similar need is developing on the gasoline side—for gasoline particulates from SI engines with either port flow injection (PFI) or DI.

11.2 **Particulate Filter Functions**

The particulate filter has two main functions. First, it must mechanically remove the particulate from the exhaust, while allowing the exhaust gas to pass through. Second, it must periodically self-clean, or regenerate, by oxidizing the stored particulate matter to carbon dioxide (CO_2).

11.2.1 **Mechanical Filtration**

Particulate filters have been used to reduce particulate emissions from diesel exhaust for nearly two decades, with the most prevalent PF design being the ceramic wall-flow monolith depicted in **Figure 11.1**. This design is derived from the automotive three-way catalyst (TWC) flow-through monolith. By plugging alternate channels, the filter is transformed from flow-through to wall-flow, and the exhaust flow that enters through open channels, encounters the plugs on the back end, and is forced to flow through the porous walls of the ceramic substrate to be able to exit through a channel open on the downstream end. The particulates are trapped in and on the wall, creating partially clogged wall pores and a surface particulate filter cake. This strategy leads to high filtration efficiency and a high surface area to volume ratio, making it well suited to regulating particle emissions.

The mechanisms of filtration in particulate filters are based upon the separation of solid particles from gaseous fluid flow. This involves the passage of the gas through a

FIGURE 11.1   Illustration of wall flow filter.

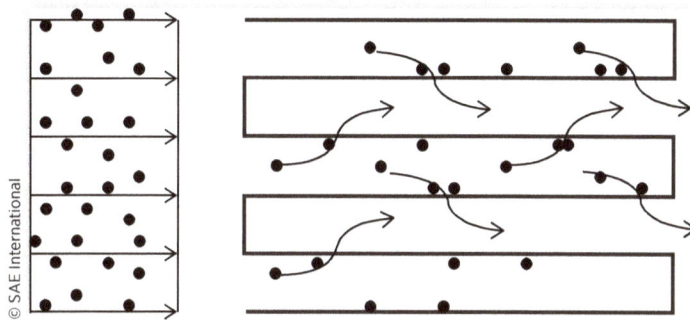

porous media, which the solid particles cannot pass through, thereby causing the reten-
tion of the particulates within the filter pores (deep bed filtration) and on the walls of
the filter media (surface filtering). The mean pore size of the filter media will determine
which of the two filtration mechanisms dominate. If the mean pore diameter is larger
than the mean diameter of the particulates, deep bed filtration will dominate until the
deposited particulates fill the pores by sieving. Once the pores have filled, there is a
transition to surface filtration. At this point, there is a layer of collected particles formed
on the surface of the walls of the filter, as shown in **Figure 11.2**. It is this "cake" of trapped
particulates that becomes the principal filtration mechanism for the filter media, and
this process is referred to as cake filtration.

The common ceramic wall-flow diesel particulate filters employ a combination of
deep bed and surface filtration. The initial period of filtration in a clean filter is dominated
by deep bed filtration, which has a large effect on the pressure drop across the filter, and
is responsible for the initial steep slope seen in the pressure drop versus operational time
plots, such as the one in **Figure 11.3**.

The overall deep bed filtration mechanism can be modeled as the sum of three
aerosol deposition mechanisms including (1) diffusional deposition, (2) inertial deposition, and (3) flow-line interception. These are depicted graphically in **Figure 11.4**. The
central gray sphere shown in the diagram represents a unit collector, the collecting body
of the filter. Particle-laden gas flows around the filter media, as indicated by the stream-
lines. Particulates are depicted as the small black spheres traveling with the exhaust gas
stream. *Diffusional deposition* depends on the Brownian motion of the soot particles

FIGURE 11.2   Illustration of deep bed filtration and cake (surface) filtration.

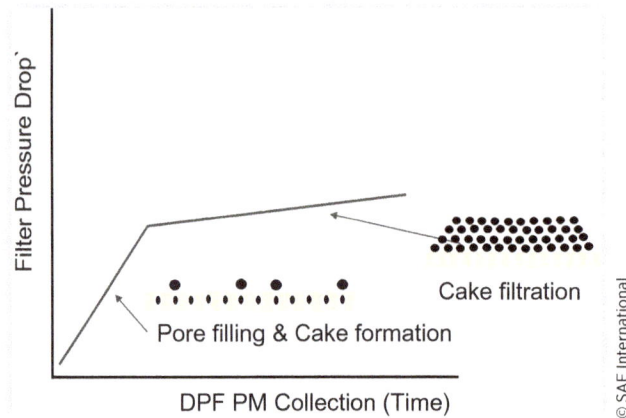

FIGURE 11.3 Example pressure drop with respect to DPF soot loading.

FIGURE 11.4 Illustrations of the deep bed filtration mechanisms. (a, top) Diffusional deposition by Brownian motion, (b, middle) Inertial deposition, also known as interception, and (c, bottom) Flow-line or direct interception of the particle by the substrate collector.

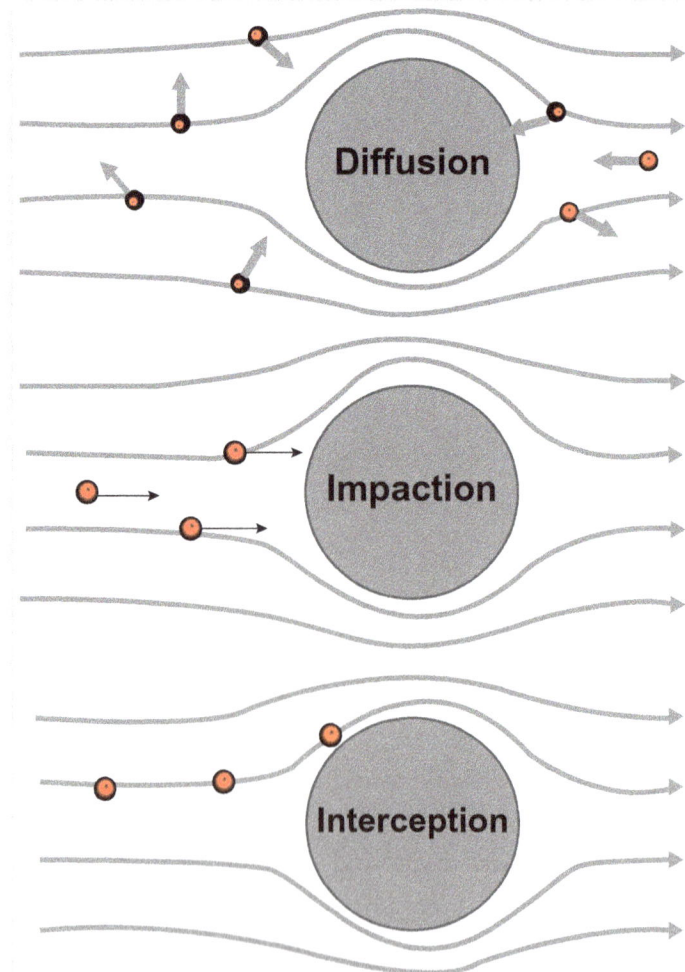

below 300 nm in diameter. These particles do not move uniformly along the gas stream-lines, but rather, they diffuse from the gas to the surface of the unit collector, where they are then collected. *Inertial deposition*, also referred to as *interception*, becomes an increas-ingly important term as particles increase in diameter and mass. As they approach the unit collector, the particles carried along in the gas follow the streamline, but may collide with the filter media obstruction due to their inertia. The collision of the particle with the unit collector results in the capture of the particle. *Flow-line interception*, or *direct impaction*, may occur when a gas streamline passes within a distance smaller than the particle's radius of the unit collector. In this case, a particle traveling along the gas streamline will impact the collector body and be collected without the influence of Brownian motion or inertia.

The deep bed filtration portion is characterized by lower filtration efficiency and lower overall pressure drop, as compared to the cake filtration regime. However, this filtration mechanism is responsible for the largest slope on the pressure drop versus time plot, as the collected particulates rapidly change the presented porosity of the filtration media, and therefore the ability of the gas to pass through easily. As the pore capacity is filled with particulates by the deep bed filtration mechanism, there is a transition to cake filtration as the particulates form a soot cake on the walls of the inlet channels. This transition is responsible for the knee in the pressure drop curve of Figure 11.3.

Cake filtration is highly efficient, as the collected particulates effectively form a packed bed filter on top of the filter wall and pre-filter the particulates from the gas flow. However, the formation of the cake is the cause of the relatively high pressure drop seen in Figure 11.3. Once the cake is formed, the rate of change of the pressure drop decreases and grows much less quickly than in the deep bed filtration regime.

11.2.2 Regeneration

The filter materials can only hold a fixed soot volume, and as the filter becomes filled (clogged) with particulates, they create an obstruction to the gas flow and an increased pressure drop across the device. This pressure drop creates an unfavorable backpressure against the engine exhaust. For these reasons, the filter must remove the soot either periodically or continuously by oxidizing the stored soot. This process is called active or passive regeneration, respectively. The temperature and composition of the exhaust gas are the key parameters that influence the regeneration process, especially to get the active regeneration event started. The type of regeneration process depends upon controlling the exhaust temperature entering the particulate filter and the oxygen (O_2) or nitrogen dioxide (NO_2) content of the gas.

11.2.2.1 Oxidation of Soot with Oxygen: Because diesel-fueled CI engines always have free O_2 present in the exhaust gas, oxygen-based regeneration should be a viable path to removing soot from the filter. The carbon-oxygen reaction has important roles in many industrial processes, most notably in coal and biomass power plants and gasification reactions. Thus, there already exists a tremendous body of previous work on solid carbon oxidation, of which there are several examples [11-1, 11-2, 11-3, 11-4, 11-5, 11-6, 11-7].

Despite several decades of research, solid carbon oxidation is still not fully understood and remains a very active area of continuing research.

The global reactions between solid carbon ($C_{(s)}$) and O_2 to form carbon monoxide (CO) and CO_2 are given by

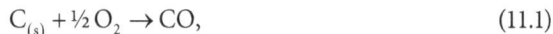

$$C_{(s)} + \tfrac{1}{2}O_2 \rightarrow CO, \tag{11.1}$$

and

$$C_{(s)} + O_2 \rightarrow CO_2. \tag{11.2}$$

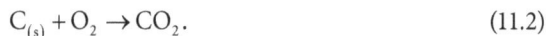

In addition, the direct oxidation of CO is a key contributor to the global carbon conversion

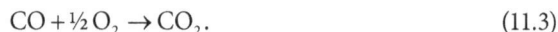

$$CO + \tfrac{1}{2}O_2 \rightarrow CO_2. \tag{11.3}$$

One of the chief barriers in establishing any kind of general kinetic mechanism is the huge variation in the microscopic properties of solid carbon, which depend heavily on the carbon source. These microscopic properties in turn have a large effect on the many elementary steps involved in the overall reaction including bulk gas and solid pore transport, adsorption, surface migration, surface reaction, and desorption [11-8]. Despite the lack of a general mechanism, several semi-global intrinsic kinetic mechanisms have been proposed in the literature, two of which are especially relevant here and are discussed below.

One important generalization about global oxidation rate that has arisen from previous studies is the three-zone model for carbon particle burnout [11-9, 11-10, 11-11], which takes into account the concurrent processes of boundary layer oxygen diffusion, intra-particle oxygen diffusion, and surface adsorption and chemical reaction. Work by investigators such as Essenhigh, Walker, and Smith [11-6, 11-9, 11-12, 11-13, 11-14] has firmly established the existence of three distinct regimes in which chemical reaction or one of the diffusion processes limits the overall reaction rate. This Three Zone Theory has been widely accepted and used to interpret data in the coal and char literature for over 30 years and is represented schematically in **Figure 11.5**.

Zone I burning occurs when chemical reactions at the carbon surface (including adsorption/desorption of reactants and/or products) limit the overall oxidation rate. Using reaction rates measured under these conditions, it is possible to determine the intrinsic activation energy (which still includes adsorption, desorption, and multiple elementary steps). In Zone II burning, solid phase diffusion of oxygen (e.g., via pores) contributes significantly to slowing the overall oxidation rate. The effective global activation energy under these conditions is roughly half that of the intrinsic activation energy. Zone III burning occurs when external diffusion of oxygen from the bulk gas to the particle becomes rate limiting. Here, the apparent activation energy is very small. Typically, particle burnout transitions from Zones I to III as the reaction (particle) temperature increases. Thus, it is possible to explore each type of behavior by selecting the temperature of experimental observation. In general, it is always prudent to consider the impact of all three types of processes when evaluating kinetic parameters.

FIGURE 11.5 Three Zone Theory rate controlling zones for heterogeneous oxidation reactions [11-1, 11-2, 11-7].

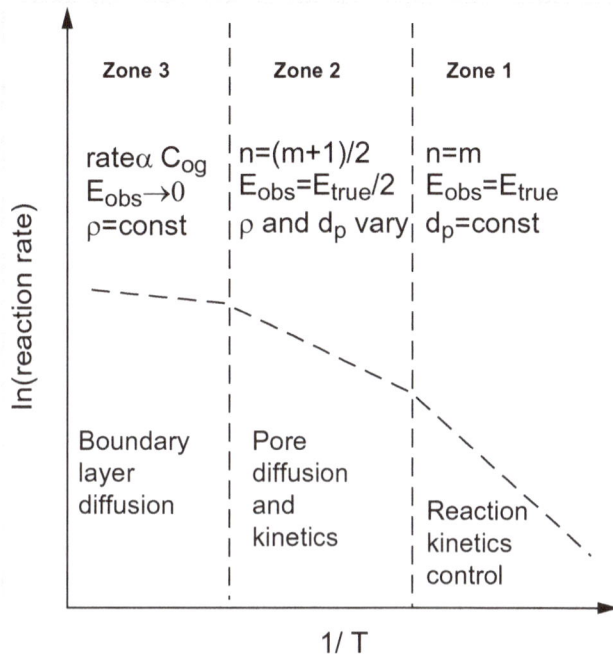

The global n^{th} order rate equation is often used to describe char oxidation at conditions typical of industrial boilers [11-15, 11-16, 11-17]. The most common form is

$$q_{rxn} = k_s P_{O,s}^n = A \cdot \exp\left(\frac{-E_{A,obs}}{RT_p}\right) \cdot P_{O,s}^n, \qquad (11.4)$$

which uses the Arrhenius rate expression discussed in Section 7.2.2. This simple equation has proved useful for modeling char oxidation at atmospheric pressure over small temperature ranges [11-12]. Due to its simplicity, this model is commonly used in comprehensive boiler combustion models. While this rate equation does not explicitly account for pore diffusion, these effects can be implicitly included in the pre-exponential factor A. This model has also been criticized for inadequately dealing with a wide range of experimental temperatures or high pressure data [11-18, 11-19]. However, it is acknowledged to be effective over smaller ranges in those critical parameters. An additional weakness is that this model cannot be extrapolated between Zones I and II; thus, the reaction order is often observed to vary with experimental conditions, between the limits of 0 and 1. Hurt [11-15] suggested that fits of experimental data with Eq. 11.4 can yield high fractional order due to surface heterogeneity. Simple surface reaction models that account for surface heterogeneity using power law functions have been proposed, but their use requires detailed surface characterization for the each fuel [11-20].

More detailed approaches for modeling carbon oxidation kinetics attempt to account for both the intrinsic reaction factors and pore diffusion effects. Intrinsic reaction models typically include explicit terms for Langmuir–Hinshelwood adsorption of oxygen and/or desorption of carbon dioxide [11-5]. Laurendaeau [11-21] suggested a way for including Langmuir–Hinshelwood effects in the intrinsic reaction rate, r, by:

$$r_{in}'''(C) = \frac{k_1 C}{1 + KC} \tag{11.5}$$

where C is the molar concentration of carbon and k_1 and K are kinetic parameters representing the rate velocity and the equilibrium constants, respectively.

A practical consequence of rate equations with a form like the above is that as oxygen concentration increases, the oxidation rate asymptotically approaches a maximum value due to the complete filling of all available adsorption sites on the carbon surface. Depending on the degree of surface site occupancy, it also means that the apparent reaction order in oxygen can vary between 0 and 1.

The speed at which the carbon particle oxidizes is dependent on both the reactivity and the rate at which the oxygen transfers from the bulk gas to the particle. This reactivity depends on the accessibility to oxygen within the pores of the particle and the velocity of the reaction between oxygen and the surface, also known as the intrinsic reactivity [11-2]. Intrinsic reactivity is defined as the reaction rate in the absence of mass transfer limitation. Pore diffusion effects in carbon oxidation are typically addressed with an effectiveness factor, η [11-22, 11-23, 11-24]. The effectiveness factor is the quantity by which the molar flux is multiplied by to account for diffusion resistance in the conversion process; it is the ratio of the actual reaction rate to the intrinsic rate [11-2]. The effectiveness factor is a function of the Thiele modulus, Φ [11-24], which is the dimensionless factor that relates the intrinsic rate of the chemical reaction in the absence of mass transfer resistance to the rate of diffusion for the particle. The Thiele modulus is defined as

$$\Phi = R_p \cdot \sqrt{D_{eff}} \tag{11.6}$$

where R_p is the radius of the particle and D_{eff} is the effective diffusivity, which depends on the diffusing gas and the nature of the pore structure. The effective diffusivity is the combination of the diffusive resistance through the bulk and the Knudsen diffusion, D_K, which dominates in narrow pores.

Diesel particulate oxidation by O_2 (as opposed to NO_2) has typically been considered as a first-order process (in carbon) with a temperature dependence that obeys the Arrhenius equation

$$r = k \cdot [C]^\alpha [O_2]^\beta \tag{11.7}$$

$$k = A \cdot \exp\left(\frac{-E_A}{RT}\right) \tag{11.8}$$

where r is the reaction rate in mol/min, k is the reaction rate velocity, A is the frequency factor, E_A is the activation energy in J/mol, R is the universal gas constant 8.314 J/mol·K, T is the absolute temperature in K, $[C]$ is the molar concentration of carbon, $[O_2]$ is the molar concentration of oxygen, and α and β are the reaction orders in carbon and oxygen, respectively.

Reported kinetic parameter values for diesel particulate oxidation by O_2 vary widely [11-25]. Activation energies for non-catalyzed oxidation range from 36 kJ/mol [11-26] all the way up to 170 kJ/mol [11-27, 11-28], close to the activation energy of graphite at

188 kJ/mol [11-29]. Reaction orders in carbon are reported to be ⅔, in the range representative of the shrinking core model, but this has yet to be proven accurate for diesel particulate [11-30]. To further complicate matters, reaction orders in oxygen have been reported in orders ranging from ½ [11-26] to 1 [11-30]. Thus, without direct experimental information, it would be impossible to reliably use the currently available kinetic data for estimating the oxidation rates of previously untested diesel particulates.

Early studies of the uncatalyzed oxidation of diesel particulate compared to flame particulate indicated that diesel particulate oxidation was more complex [11-30]. This study also proposed Printex (a commercially available carbon black) as a reference standard for oxidation reactivity [11-31, 11-32]. In these studies, the particulates (diesel and Printex) were milled to uniform size, but the SOF of the diesel particulate was not measured nor removed. Scattering in the data prevented determination of the kinetic parameters, and it was proposed that the volatile hydrocarbons present on the particulate were complicating factors.

Innovative work by Yezerets et al. [11-31, 11-33, 11-34, 11-35] began to clarify the diesel particulate reactivity to oxygen. In 2002, they established a flow reactor methodology for measuring particulate kinetic parameters, and this methodology is the basis for the kinetic measurements set out in this research proposal. Their work focused on the oxidation of devolatilized, heavy duty (HD) diesel particulate and a carbon black model particulate with 10% O_2. Two types of experiments were conducted: temperature-programmed oxidation (TPO) and a step-response technique. Experiments were carried out on samples that were first pretreated with temperature-programmed desorption (TPD) under inert gas to remove adsorbed hydrocarbons. The devolatilization was done to normalize the particulate samples, thereby eliminating the impact of the SOF. These methods are described in Chapter 5, Emissions Measurements.

Using the above experimental techniques, Yezerets et al. determined that there was a significant difference in the behavior of the diesel particulate and the carbon black, including very different burnout profiles [11-33]. The differences observed between the samples lead the researchers to believe that the behavior of the sample depends significantly on its properties, indicating that carbon black may not be an appropriate "reference particulate." Of particular interest here, Yezerets et al. observed an apparent increase in the activation energy as burnout progressed [11-31]. This correlation between activation energy and burnout was again confirmed by Yezerets et al. [11-34] in 2003, leading the authors to conjecture that the presence of some type of highly reactive surface species in the devolatilized particulate produced an initial boost in reactivity that diminished with time.

Additional work by this same group in 2005 [11-35, 11-36] with the oxidation pulse technique to minimize local temperature variations confirmed the reactivity changes in the early stages of burnout and quantified the amount of highly reactive particulate to be approximately 10%–25% of the initial sample [11-34]. The remainder of the particulate appeared to have relatively unchanging properties. Overall, the observations of this group indicated that the oxidation rate for their devolatilized HD particulate could be approximated by:

$$r = k \cdot [C]^\alpha [O_2]^\beta \qquad (11.9)$$

where the reaction velocity constant k at each was calculated by

$$k = \frac{r}{[C]^{\alpha}[O_2]^{\beta}} \tag{11.10}$$

and the activation energy, E_A, was found from the Arrhenius relationship shown in Eq. 11.8. The reaction was found to be first order in carbon, with $\alpha = 1$ and $\beta = 0.61 \pm 0.03$. The activation energy was found to be 137 ± 8.1 kJ/mol. However, it should be noted that the concentration of carbon is changing over the reaction, even at a steady temperature.

Stamelos [11-37, 11-38] and researchers studied the impact of the volatile hydrocarbons on the oxidation of diesel particulate in a DPF, but they did not attempt to estimate Arrhenius kinetic parameters. The relative percentage of SOF was measured by TGA for particulate generated by a range of engine conditions and a speed-load vs. volatiles map was created. Chemical analysis and identification of the compounds of the volatile fraction was not attempted. Unpredictable regeneration behavior in DPFs was correlated to low exhaust temperature conditions, which resulted in a relatively high volatile content.

Strzelec et al. extended the work to show that evaluating the reaction rate on the basis of total surface area normalizes the kinetics of particulate oxidation independent of its fuel source or combustion type, to a single activation energy [11-39]. Additionally, they showed that physical characterization, particularly surface area, of a particulate sample with burnout is the most important factor in determining its reactivity with oxygen. The heterogeneous nature of reaction depends on the instantaneous surface area available, and when that is accounted for, a single activation energy of 113 ± 6 kJ/mol was found and it was shown that the consumption of the particulate by O_2 combustion proceeds via Zone II burning mode, with external and internal pore diffusion oxidation [11-40].

11.2.2.2 Oxidation of Soot with Nitrogen Dioxide: In exhaust, NO_2 is also present and has been shown to have a favorable impact on DPF regeneration, often referred to as passive oxidation. The chemical reaction for oxidation of carbon by NO_2 is

$$C_s + 2\,NO_2 \rightarrow CO_2 + 2\,NO \tag{11.11}$$

NO_2 has been shown to oxidize particulate at lower temperature as compared to O_2 [11-41, 11-42, 11-43, 11-44, 11-45, 11-46, 11-47, 11-48, 11-49, 11-50, 11-51]. The reported magnitude of the soot oxidation rates for NO_2 is greater than for O_2, indicating that NO_2 is a stronger oxidant for promoting low temperature oxidation in the range of 200–500°C. Typically, NO_2 is 5%–15% of the total NOx in the raw engine exhaust, or approximately 50 ppm. However, the diesel oxidation catalyst (DOC) that is typically placed upstream of the DPF, containing Pt and Pd, oxidizes NO to NO_2 increasing the mixture ratio significantly.

NO_2 oxidation of diesel particulate has been shown to be considerably different than oxidation by O_2. The NO_2 oxidation is shown to follow the shrinking core prediction [11-52], which assumes a homogenously shrinking sphere of radius r where only the surface carbon is oxidizing. The prediction holds for both mass remaining during reaction and surface area. Therefore, it is evident that particulates oxidized by NO_2 follow a Zone I, surface only, burnout trajectory, whereas particulates oxidized by O_2 have been shown to be experiencing Zone II burning, which includes both surface and pore oxidation. This hypothesis seems reasonable given that the surface area for the

PM samples is greater than that of surface-only burning. Zone I burning is diffusion controlled and therefore leads to low activation energy values, as measured here. The low activation energy, specific surface area progression with extent oxidation, HR-TEM images, and difference plots of fringe length and tortuosity paint a consistent picture of lower reactivity for O_2, whereby it preferentially attacks highly curved lamella, which are more reactive due to bond strain, and short lamella, which have a higher proportion of more reactive edge sites. By contrast, NO_2 reacts indiscriminately, immediately upon contact with the surface, leading to the Zone I style burning that leads to the shrinking core type oxidation. Therein the difference plot of fringe length shows the NO_2 oxidized sample to have a higher proportion of long lamella (short ones have been consumed by O_2) and a lesser fraction of long lamella, as NO_2 broke them up (by virtue of basal plane attack) reflecting the aggressive oxidation by NO_2 where differences in carbon site reactivity, edge versus basal are insignificant. By contrast, the difference plot in tortuosity shows a preferential population of straighter lamella for O_2 oxidized particulates, as more curved lamella (represented by higher tortuosity) are preferentially oxidized by O_2 given their greater bond strain. The lower reactivity of O_2 is sensitive to this difference. By contrast, NO_2 is insensitive to this difference and hence this sample retains a larger proportion of highly curved lamella, despite their greater propensity toward oxidation.

11.2.2.3 **Active Regeneration:** Active regeneration refers to those thermal regeneration schemes in which the exhaust temperature has been purposely increased through a manipulated exothermic event. These may or may not include the use of a catalyst. Oxidation of carbonaceous particulates by O_2 requires exhaust temperatures in excess of 550°C–600°C, temperatures which are not encountered in practical operation, where exhaust temperatures into the DPF are more often in the 300°C–450°C range over the typical operating conditions. Light duty diesel applications may even have exhaust temperatures as low as 200°C–300°C.

For oxidation to occur, the exhaust temperature is generally increased in one of two active methods: (1) exhaust gas heating, or (2) use of a higher in-cylinder combustion temperature. Exhaust gas heating methods such as electric heaters, fuel burners, and microwave heating have been explored in the past to assist with the active regeneration strategies. Filters that do not include a catalyst need exhaust in the 600°C–700°C range to oxidize the stored particulate. The DOC can also be used upstream as a gas pre-heater for the DPF. A late cycle, or post fuel injection, or fuel sprayed directly into the DOC is oxidized creating an exotherm that serves to increase the exhaust gas temperature and enable DPF regeneration. For catalyzed filters, the exhaust gas temperature required is lower, and therefore, while they may still require an active strategy, it is at a lower fuel penalty than for an uncatalyzed filter.

11.2.2.4 **Passive Regeneration:** Passive regeneration does not require any kind of elevation of the exhaust temperature, but rather uses the normal exhaust gas temperature and NO_2 already present in the exhaust to oxidize the stored particulate matter. Since this strategy does not require any kind of manipulation of the exhaust gas conditions already existing, it is considered to be a passive strategy. Passive regeneration occurs, at least to some extent, in all particulate filters when NO_2 is present in the engine-out exhaust or formed over the oxidation catalyst. The advantages to the passive strategy are smaller fuel penalties associated with regeneration, decrease need for control, and fewer components. In addition, since the NO_2 oxidation process takes place at lower temperatures, it extends the operating window for regeneration in the particulate filter.

11.3 **Filter Materials**

11.3.1 **Substrates**

Particulate filter substrates provide the filtration surface. The filter configuration (porosity and shape factors) heavily influences the filtration efficiency and pressure drop, or backpressure applied to the engine. Design targets for the substrate include the following: durability, high filtration efficiency, low pressure drop, the ability to withstand exhaust temperatures, thermal shock resistance, compatibility with washcoating, mechanical strength, chemical (mainly oxidative) stability, soot loading capacity, ash loading capacity, and low cost. Generally, the shape is of a "honeycomb" monolith, having been derived from modifying flow-through substrates by plugging alternate channels, creating a look similar to a honeycomb, as seen in **Figure 11.6**.

The cell density of the substrate is measured in cells per square inch (cpsi), and the higher the cpsi, the narrower the individual channels are. Wall thickness of the channels is reported in mils, which is one thousandths of an inch. The porosity of the substrate is measured by mercury intrusion, and generally ranges from 50%–70%, so that gas can easily pass through the porous media.

Flow enters the open channels and proceeds through the wall due to the existence of the plug at the channel outlet. For this reason, particulate filter substrates are referred to as wall-flow. Ceramic wall-flow monoliths are the most commonly used filters.

FIGURE 11.6   Example of a honeycomb monolith.

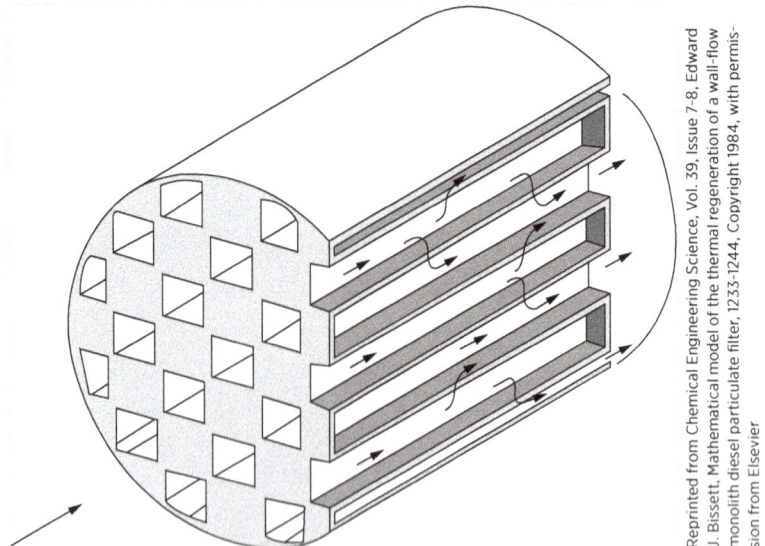

Reprinted from Chemical Engineering Science, Vol. 39, Issue 7-8, Edward J. Bissett, Mathematical model of the thermal regeneration of a wall-flow monolith diesel particulate filter, 1233-1244, Copyright 1984, with permission from Elsevier

11.3.1.1 **Cordierite:** Cordierite is a magnesium-alumino-silicate ceramic that can be manufactured by extrusion in a single brick, as shown in **Figure 11.7**.

The porosity of cordierite substrates is reaction-formed. Ground rice hulls are mixed into the ceramic matrix as a sacrificial templating material. After the monolith has been extruded, it is fired to harden the ceramic and to burn out the templating material, forming the pores. For this reason, the pores in cordierite substrates are neither uniform, nor homogeneous in size distribution.

Cordierite has a very low coefficient of thermal expansion and is resilient to the thermal cycling that can occur in the exhaust. It has good high-temperature resistance, up to 1,200°C, and good mechanical strengths at a low cost. For these reasons, the majority of particulate filters are made of cordierite.

11.3.1.2 **Silicon Carbide:** Silicon carbide substrates are made by sintering granules of SiC together. SiC has a higher melting point, as compared to cordierite, but it is not as thermally stable. Because of this, SiC substrates are manufactured in square segments that are then cemented together to create the substrate brick, as shown in **Figure 11.8**.

SiC has better thermal resistance than cordierite and is stable up to temperatures of about 1,800°C; however, it has a higher coefficient of thermal expansion, higher cost, and must be manufactured as segments. These drawbacks limit the deployment of SiC filters.

FIGURE 11.7 A cordierite particulate filter, which is continuous, as it was extruded as a single piece.

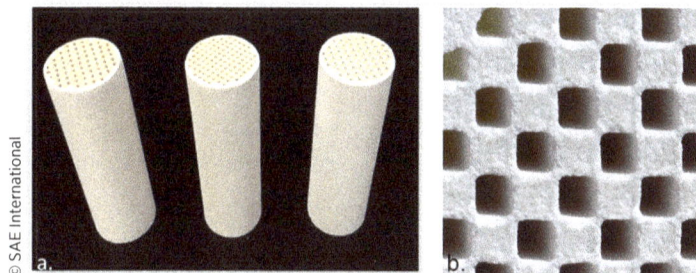

© SAE International

FIGURE 11.8 A SiC particulate filter. Note that the brick consists of segments that have been cemented together.

© SAE International

There are a few applications where SiC is preferable—as the risk of thermal shock for a filter increases with length. If the risk cannot be minimized by keeping the L/D ratio of the filter to a maximum of 1, and a long, thin filter is required, then SiC would be a better choice.

11.3.1.3 Other Materials: Other materials such as metal foams (most often nickel based), acicular mullite, and fiber filters have been proposed and manufactured. However, for reasons of cost, stability, manufacturability or filtration efficiency, they were not commercialized to the extent of cordierite or SiC. The vast majority of filter substrates are ceramic.

11.3.2 Catalysts and Washcoat

DPFs and GPFs are often catalytically coated to aid in the regeneration process and to convert any hydrocarbons remaining in the exhaust. Typically, the catalyst and washcoat used in the DPF is the same as the DOC, and the reader is directed to Section 8.2. Similarly, for GPFs, the catalyst and washcoat used is typically the same as the TWC, which is described in Section 9.2.

11.4 Diesel Particulate Filters

Diesel particulate filters (DPFs) haven been the dominant technology in place for particulate emissions control. DPFs function by mechanically filtering particulate matter from engine exhaust as the particle-laden gas stream passes through the wall of the porous filtration media. The gas is able to pass through the pores, but the particulates are captured in the pores and on the surface of the filter walls. This is especially important in the context of a varied and rapidly changing fuel supply, which will be discussed further in Chapter 16, Alternative Fuels. Diesel particulates are produced as micron and submicron scale aerosols during the combustion process from unburned fuel and lubricating oil residues. They are typically comprised of an elemental carbon (EC) substrate onto which partially combusted hydrocarbons are adsorbed. The amount of adsorbed hydrocarbons relative to elemental carbon is typically indicated in terms of soluble organic fraction (SOF), organic carbon fraction (OC), or volatile organic carbon (VOC), depending on how it is measured. The relative amounts of elemental C and hydrocarbons as well as the composition of the hydrocarbons are functions of the fuel, oil, the engine type, and the engine operation [11-53].

The most prevalent DPF design is the wall-flow monolith. This design is derived from the traditional three-way catalyst (TWC) flow-through monolith. By plugging alternate channels, the filter is transformed from flow-through to wall-flow, and the exhaust flow that enters through open channels, encounters the plugs on the back end, and is forced to exit through the porous walls of the ceramic substrate. This strategy leads to high filtration efficiency and a high surface area to volume ratio, making it well suited to regulating diesel particle emissions. Collection efficiencies greater than 90% have been reported in the literature [11-54, 11-55]. DPFs reduce tailpipe particulate emissions by mechanically filtering the diesel exhaust. The most prevalent DPF design is the wall-flow monolith, which is derived from the flow-through ceramic monoliths used for other types of catalytic exhaust after-treatment. DPF monoliths are unique in that alternate channels are plugged at opposite ends so that exhaust gases enter in one channel, pass through a porous wall, and then exit through a parallel channel. As a result, particulates are trapped in and on the wall, creating partially clogged wall pores and a surface a particulate cake. This strategy leads to high filtration efficiency and a high surface area to volume ratio, making it well suited to regulating diesel particle emissions.

Managing the trap particulate inventory is the greatest DPF control issue. As the inventory increases, backpressure on the engine also increases, which can adversely affect engine operation and fuel efficiency. To maintain acceptable engine operating conditions, the trap must be "regenerated" by oxidizing the stored particulate.

Regeneration can be accomplished by low temperature passive oxidation or active, catalyst-augmented oxidation, and/or high temperatures (>600 K). Ideally, under lean conditions, particulates are passively oxidized by NO_2 and excess oxygen in the exhaust gas [11-56] at typical exhaust temperatures, reducing the frequency of active regeneration events. When active regeneration is required, it is extremely important to be able to predict the rate of particulate oxidation to prevent excessive temperatures that can physically damage the DPF ceramic. It is also important to be able to accurately predict the level of preheating required to minimize excess fuel consumption. Both regeneration methods require implementation of some sort of model to guide the DPF control strategy.

DPF modeling efforts [11-57, 11-58] have focused on describing the filtration, gas flow, temperature distribution, and pressure drop during particulate accumulation and regeneration. Sub-models for particulate oxidation often assume simple, first-order heterogeneous reactions, with empirical parameters [11-58]. The high degree of variability in particulate oxidation behavior from one engine to the next and between different drive cycles makes it clear that an improved understanding of the fundamental processes involved in particulate oxidation is needed to meet the current needs for DPF modeling and controls. This is especially challenging now with the increasing market presence of biofuels. Thus, research into the impact of biofuels on particulate filtration and oxidation properties is essential in optimizing DPF technology [11-59].

11.5 Gasoline Particulate Filters

Ceramic wall-flow particulate filters have long been used to eliminate exhaust particulates. As one of the first wall-flow filters that are successfully commercialized, DPFs are well developed and widely applied on diesel vehicles for nearly two decades [11-54, 11-60, 11-61]. Their high filtration efficiency (>90%) earns them a major role in EAS for CI engines. The initial filtration efficiency an empty DPF is quite low at about 60%. This is far from a satisfied filtration performance and will not meet the diesel PM/PN regulations. However, a soot cake can quickly build up in the DPF inlet channels, which dramatically boost up the filtration efficiency to >95%. This soot cake has a lower permeability and porosity than DPF wall substrate and plays a predominate role in DPF particle filtration.

Despite the success of DPFs on diesel engines, they cannot be implemented on DI SI engines (also called gasoline DI or GDI engines) for two main reasons. First, PM emission from diesel engines is much higher than DI SI engines [11-60, 11-62, 11-63]. At the same time, diesel particulates are larger in size and help the rapid formation of the soot cake on the surface of DPF wall and achieve a high filtration efficiency [11-55, 11-64, 11-65]. Whereas the absence of soot cake due to the smaller size and lower concentration of DI SI particulates makes DPFs insufficient for DI SI engines aftertreatment [11-66]. Second, DI SI engines are more sensitive to back pressure [11-67]. DPFs with high overall pressure drop can choke the DI SI engines and lower the power output. Therefore, there is a need for specific GPFs with high filtration efficiency and low pressure drop to meet the ever-stringent PM and PN regulations.

GPFs have been developed from DPFs by making modifications, such as, reducing the filter material porosity, thinning the filter wall, and lowering the wall velocity. All these approaches effectively increase the filtration efficiency of particles from DI SI engines while keep the pressure drop low, an important aspect for DI SI engines. The major difficulty in combing these approaches in the GPF design is achieving balance between filtration efficiency and pressure drop.

11.6 **Ash by C. Justin Kamp**

While wall-flow filters, such as the DPF and the GPF are quite efficient at capturing soot, they are also efficient at capturing ash. Ash is comprised of incombustible materials including engine wear and corrosion debris, solids which may pass through the engine air filter, fuel impurities, alkali salts from engine coolant leaks and residuals of engine lubricant additives. Most of the ash trapped in the filter is lubricant-derived and is a result of the degradation of inorganic lubricant additives such as detergents (containing calcium and magnesium) and anti-wear (containing zinc, sulfur, and phosphorus). Note that lubricants are required to meet low sulfur, phosphorus, and volatility standards where ash accumulation in downstream particulate filters occurs slowly. Lubricant-derived ash leaves the engine as ultra-fine particles, in the 2–50 nm range, bound to soot particles, and grows into agglomerates of primary particles in the 0.5–2 µm range, as shown in **Figure 11.9**. During regeneration of the particle filter via oxidation of the carbonaceous soot particles, the inorganic, incombustible materials remain, as shown in Figure 11.9b. This remaining incombustible ash material forms a membrane on the filter surface, effectively clogging the filter substrate by way of filling substrate pores and by forming thick layers on the inlet channel walls, resulting in an increased filter pressure drop which negatively affects fuel economy. In this section, several ash-related topics will be discussed in more detail including ash characteristics, ash formation and accumulation, and the associated implications.

11.6.1 **Ash Characteristics**

11.6.1.1 **Ash Formation and Transport:** Ash exits the engine bound to soot as nanoscale particles. Ash particle growth, accumulation/agglomeration, and local

FIGURE 11.9 Images showing the various states of ash, beginning with (a) a TEM image of an engine-out ash precursor (approximately 5 nm in size), (b) a TEM image of a small, branched ash cluster, and (c) an SEM image of a DPF cross section showing an ash cake layer on the inlet channel surface. Adapted from [11-68, 11-69].

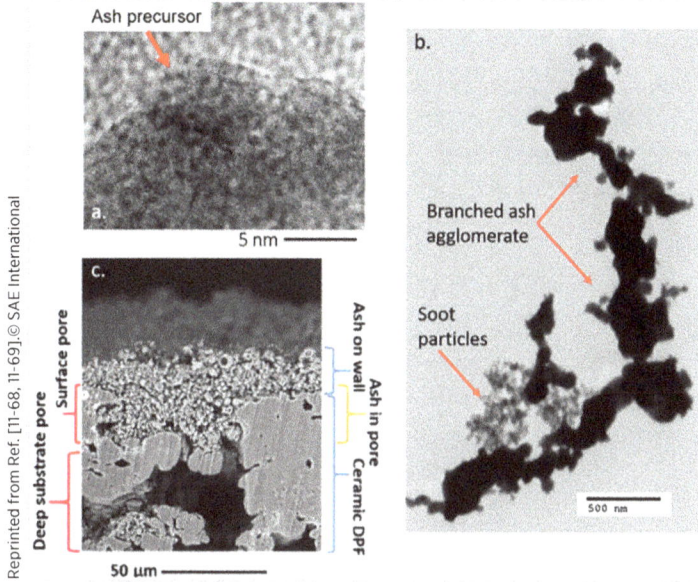

transport occurs within the particulate filter by way of a mechanism where ash follows the soot cake as it shrinks with oxidation during filter regeneration. This oxidation-induced local transport mechanism, as described in [11-68, 11-70, 11-71, 11-72], initially redistributes ash into small islands in and/or around filter surface pores, shown in **Figure 11.10a**, and eventually forms a uniform ash cake layer with increasing ash loading.

FIGURE 11.10 SEM images of the evolution of the ash cake layer on the inlet channel surface of a gasoline particulate filter, adapted from [11-68].

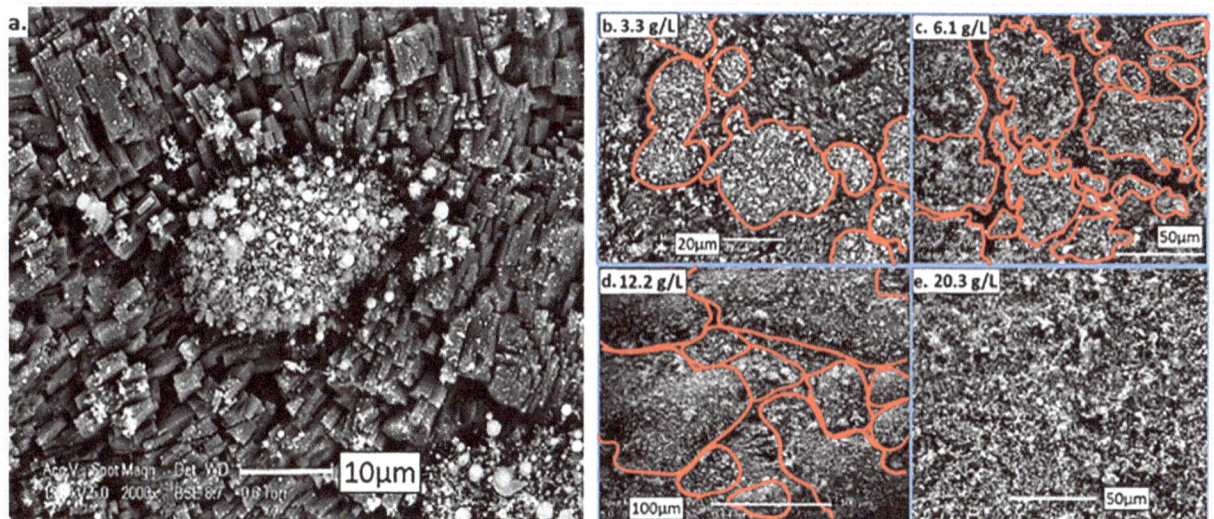

In addition, ash is transported within the filter via flow-induced transport, also described in [11-70, 11-72], whereby ash is pulled from the wall via shear force, re-entrained in the flow, and pushed to the back of the filter, forming the ash plug.

The result of the ash transport mechanisms, soot oxidation-induced and flow-induced, is a situation where ash can accumulate predominantly as a plug or as a cake layer, or most often as a combination of both. Generally, emissions control systems take this concept into consideration during design to minimize ash-related filter pressure drop; however, it is important to note that optimization in this context is unique to specific applications/exhaust flowrates/filter temperatures/substrates/etc. In addition, post-mortem, multi-scale analysis of the ash-loaded particulate filter (via nondestructive and destructive methods) provides useful information in regards to functionality and further optimization.

11.6.1.2 Ash Chemistry: Lubricant-derived ash, as stated previously, is an inorganic lubricant additive degradation product and typically is comprised of ionic crystalline salts including but not limited to the species of Ca, Mg, Zn, S, and P. Calcium and magnesium are typical detergent additives, while zinc, sulfur, and phosphorus are found in the ZDDP (zinc dithiophosphate) anti-wear additive. Some of the sulfur in ash also comes from the fuel. Lubricant-derived ash can be found in many forms with the following being most common: $CaSO_4$, $MgSO_4$ and $Zn_3(PO_4)_2/Zn_2P_2O_7$ [11-71, 11-72]. While ash chemistry can vary from the compounds listed above to oxides and dual cation species such as $CaZn_2(PO_4)_2$, $CaSO_4$ is typically the largest component of ash. Ash chemistry varies due to several reasons including lubricant additive formulation, specific engine parameters, and thermal environment in the aftertreatment system. Impurities in biofuels such as sodium or potassium can also be seen in ash species such as $Na_4Ca(SO_4)_3$ and $NaSO_4$. Ash composition is generally measured by a combination of several instruments including X-ray fluorescence (XRF), for elemental concentrations, X-ray diffraction (XRD) with Rietveld refinement, for species compound information, and inductively coupled plasma with atomic emission spectroscopy (ICP-AES), for precision elemental concentrations. Another form of ash chemistry worth mentioning is that of hydrated species. Hydrated ash has been found in gasoline particulate filters [11-68] and diesel applications which are subject to high humidity environments. The ash hydration cycle is semi-reversible and is seen in XRD data as $CaSO_4 \cdot nH_2O$, where $n = -0.1–2$.

11.6.1.3 Ash Morphology: Lubricant-derived ash has been found to leave the combustion chamber as ultra-fine particles in the 2–50 nm range, bound to soot particles as shown in **Figure 11.11**, and to form primary particles in the 0.5–2 μm range which agglomerate into porous networks. The lubricant additive chemistry, combustion conditions, and filter thermal environment as well as several other factors dictate the ash particle morphology and agglomerate characteristics. Primary ash particles range from spherical (typically Mg-based ash), to semi-spherical/angular (typically Ca-based ash) to branched, high-aspect ratio (typically Zn-based ash) morphology. The specific ash particle shapes and agglomeration characteristics are key to understanding ash-substrate

FIGURE 11.11   SEM images showing calcium-based (a) and zinc-based (b) ash and impact of ash morphology on ash packing and substrate interaction. The TEM images in the center show the morphological variations due to ash chemistry and are adapted from [11-73].

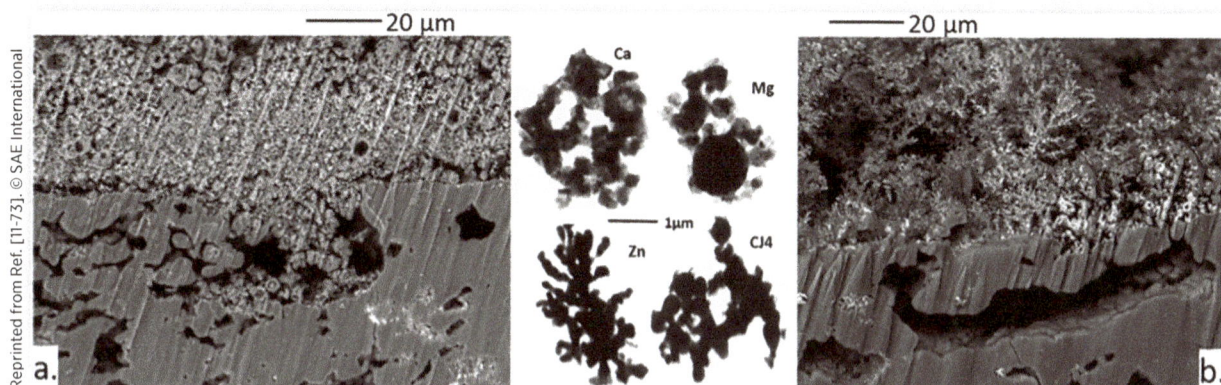

interactions as well as the overall impact of ash on the particulate filter functionality and pressure drop, as shown in Figure 11.11.

TEM images in Figure 11.11 (center) show small ash clusters where the ash chemistry, i.e., ash derived from specific lubricant additives, dictates particle and packing morphology. The SEM images in Figure 11.11 relate the chemistry-based ash morphology to agglomeration whereby calcium-based ash shown in Figure 11.11a packs together tightly with a lower permeability than the zinc-based ash shown in Figure 11.11b. It was found that for similar ash loadings, the calcium-based ash had a higher ash-related pressure drop than the zinc-based ash by almost a factor of three [11-74].

11.6.2 Impacts of Thermal Environment

Although the thermal environment in the particulate filter with active regeneration is designed to operate around 650°C, measurements have shown that short excursions (<1 min) to significantly higher temperatures (≈900°C–1,000°C) are not uncommon [11-75]. The thermal environment within the filter is an important detail as the chemistry and morphology of ash are subject to change with increasing temperature. High temperatures in general increase ash particle size, decrease ash porosity, increase ash density, increase the chemical interactions between ash and the substrate/catalyst, and promote oxide formation of the ash species. In-situ X-ray diffraction data and SEM images are shown in **Figure 11.12** to summarize chemical and morphological changes in ash due to temperature.

The in-situ XRD spectra excerpts shown in Figure 11.12 show that several significant changes occur to ash with increasing temperature. Higher temperatures result in (i) oxide formation, (ii) the instability of ash species with lower melting temperatures such as $Zn_3(PO_4)_2$ in comparison with much more stable species such as $CaSO_4$, and (iii) transition species indicative of chemical interaction between ash and the substrate, such as Zn_2SiO_4 [11-71]. In addition, the SEM images in Figure 11.2b-d reveal the severe structural changes present in ash due to temperature. Ash is shown to sinter and go from small, branched semi-spherical particles to large crystalline structures where the ash

FIGURE 11.12 An explanation of chemical and morphological changes in lubricant-derived ash due to temperature. (a) shows an excerpt of in-situ X-ray diffraction spectra for an ash sample undergoing heating from 25°C to 1,200°C, (b–d) show SEM images of ash samples which have seen progressively higher temperatures resulting in structural changes. Adapted from [11-71].

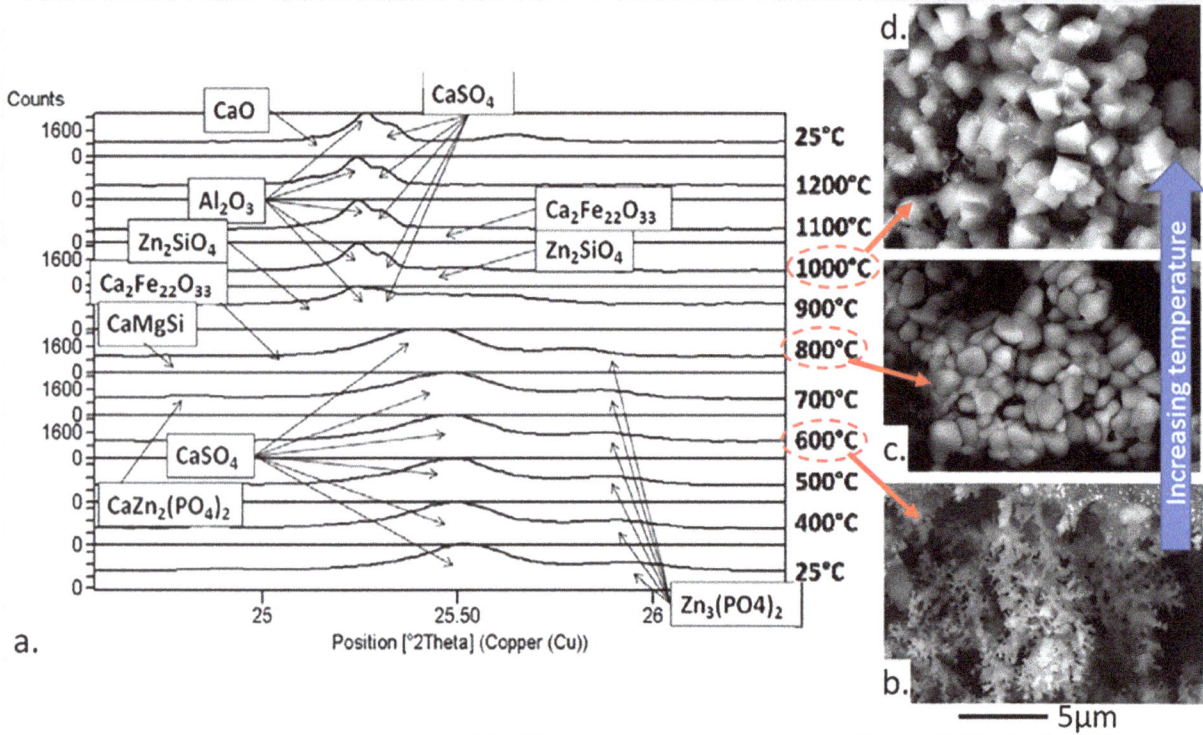

porosity transitions from 82%–96% (Figure 11.12b) to <40% (Figure 11.12d) and the ash density transitions from 0.2–0.5 g/cm^3 (b) to >1 g/cm^3 (d) when the ash has seen temperatures of 600°C (b) to 1,000°C (d), respectively. Furthermore, [11-71] found that as ash increases in density with temperature, it decreases in volume and can reduce the amount of ash clogging filter surface pores. However, as shown in Figure 11.12, high temperature ash densification is generally accompanied by chemical interactions with the substrate and is therefore not a viable filter cleaning method. In addition, ash sintering due to high temperatures has also been found to produce extreme particle growth in some cases where hollow particles can grow up to approximately 50 μm in size [11-76].

11.6.3 Ash Accumulation

The accumulation of lubricant-derived ash effectively reduces the available filtration area on the inlet channel surface as well as the inlet channel volume, resulting in an increase in the filter pressure drop. A 100% increase in filter pressure drop has been found with 1% sulfated ash over a 170–300k mile distance or with 1.6% sulfated ash 28–68k mile distance [11-77]. Interestingly, the ash loading alone has not been found to explain pressure drop [11-78], while pressure drop is strongly affected by ash characteristics such as particle morphology [11-71, 11-72, 11-79, 11-80, 11-81],

FIGURE 11.13   General ash accumulation observations including filter scale (a) and single channel scale (c) X-ray CT images, and SEM images of filter cross sections depicting ash presence on the inlet channel wall surface (b,d) and within the inlet channel surface pore (e).

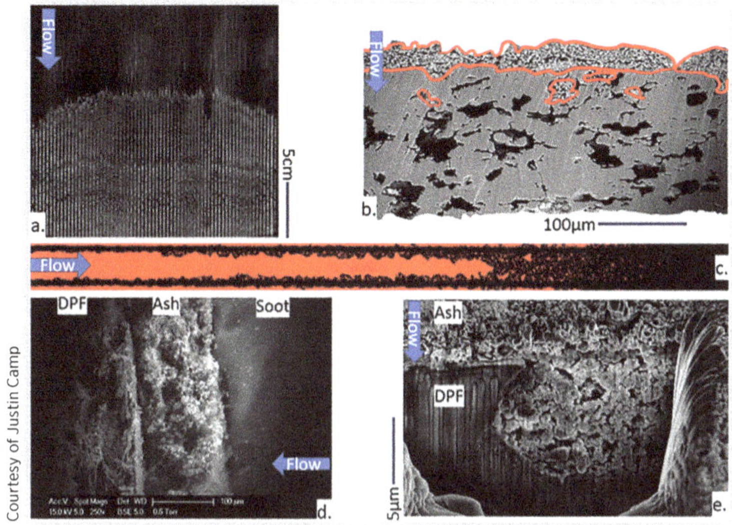

chemistry [11-72, 11-76, 11-80, 11-82], density [11-79, 11-83], and interaction with the substrate [11-71, 11-74, 11-76, 11-84, 11-85]. Ash accumulation in particulate filters results in a flow restriction-related fuel penalty of 1%–4% at 20 g/L of ash and 3%–7.6% at 60 g/L [11-75, 11-81, 11-86]. General ash accumulation observations at several specific length scales are given in **Figure 11.13**.

As shown in Figure 11.13, ash accumulates within the particle filter inlet channels and remains throughout the filter lifetime or until cleaned. Ash accumulates in the plug (outlet end of the inlet channels), as in (a), where the plug length generally grows with ash loading but not necessarily in a linear manner. Ash also accumulates on the inlet channel surface as the ash cake layer, as shown in (b-d), and within the inlet channel surface pores, as shown in (e). The ash plug effectively reduces the inlet channel volume, while the ash cake layer effectively reduces the filter surface area and creates a "membrane" on the inlet channel surface which grows in thickness with ash loading as it adds flow restriction to the exhaust passing through the filter.

11.6.4 Ash Abnormalities

Particulate filters in the field are subject to occasional ash-related abnormalities, some of which require either filter cleaning (cleaning is generally done every ~150k mi for non-problematic filters) or replacement due to excessive pressure drop and/or frequent regenerations. As mentioned previously, transitions can occur in ash chemistry such as in the cases of ash sintering and hydration. Other types of abnormalities include mid-channel ash deposits, as shown in **Figure 11.14b, c, e** and high density ash anchors, as shown in (f,g).

FIGURE 11.14 X-ray CT images of axial (a–e) and radial (f, g) particulate filter cross sections showing normal (a, d) and abnormal (b, c, e, f, g) ash accumulation.

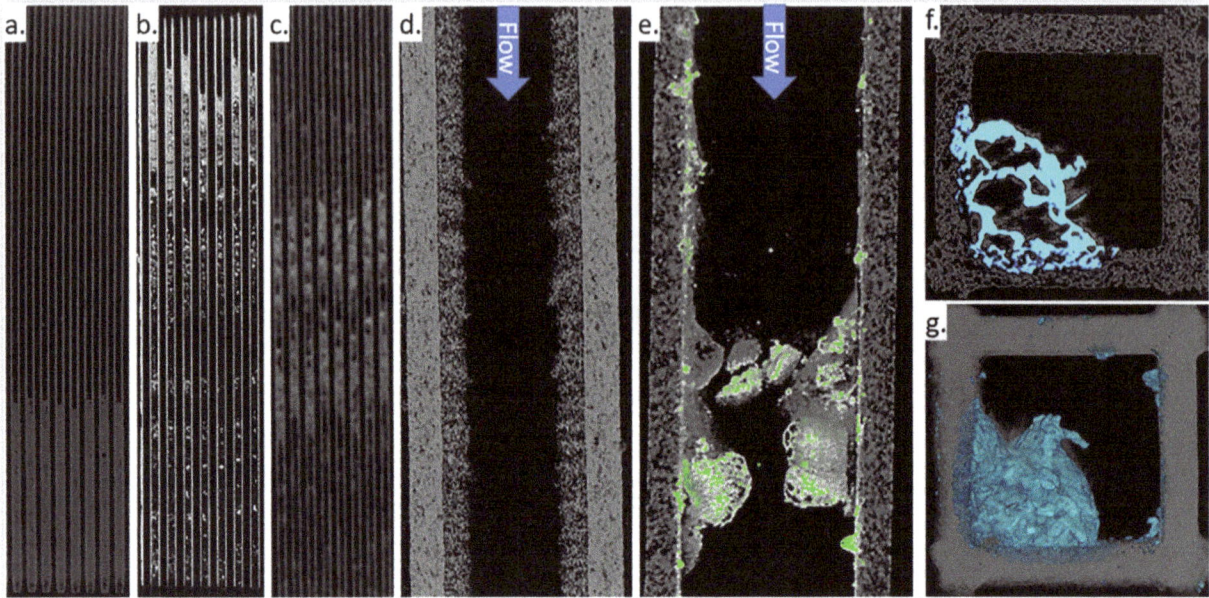

Courtesy of Justin Camp

The ideal ash accumulation scenario for most applications would be a thin cake layer of ash with a short, uniform ash plug, which would generally result in a relatively low ash-related pressure drop. In Figure 11.14b, c, e, f, g, various abnormalities are shown which significantly reduce the filter inlet channel volume and come from field-return filters which required replacement. Filters in (b,c) show excessive void space in the ash plugs, described by the "ash to gap ratio" in [11-69], which results in ash plug lengths which are far longer than expected for a given ash loading. The X-ray CT images in (f,g) show high density ash anchors whereby ash has undergone excessive sintering, turning into a near liquid state before penetrating almost the entire width of the filter wall before hardening into a glass-like, high density (>1 g/cm³) state. It has been found that these high-density ash anchors generally inhibit most ash cleaning methods.

11.6.5 Ash Implications

The discussion on lubricant-derived ash is an introduction to several characteristics of ash and ash accumulation within particulate filters. Ultimately, an understanding of ash is important due to the negative impact ash has on particulate filters including DPFs, GPFs, and SCRFs. Ash accumulation impacts the wall-flow filter by reducing inlet channel volume, reducing filtration surface area, reducing surface pore volume, all of which generally increase filter pressure drop. Excessive filter pressure drop can result in frequent filter regenerations or in the filter being pulled out of service. Ash-related filter pressure drop is typically discussed in 3 regimes including deep bed filtration, a transition point, and cake layer filtration. It has been shown in literature that for a filter with 33 g/L of ash, approximately 6% of the ash fills the substrate surface pores in the deep bed filtration phase and account for 40%–60% of the total ash-related pressure drop [11-87].

Thus, a fundamental understanding of ash-substrate interactions, especially ash accumulation within the filter wall in confined spaces is key to understanding ash-related pressure drop. [11-76] found through high resolution substrate cross sections that ash typically penetrates approximately 1 pore depth into the wall, or a depth of 30–50 μm. The pressure drop response of cake layer filtration is dictated in part by the permeability and packing density of the ash cake. An earlier study in this area found that in some circumstances, ash affects the depth penetration of soot into the filter pores and that a small amount of ash (12.5 g/L ash loading) can lower the total filter pressure drop as it prevents soot depth penetration while having a relatively low ash-related pressure drop increase [11-81].

The permeability of the ash layer can be obtained in several ways including physical measurement, discussed in [11-88], by CFD simulation through a 3D X-ray CT micronscale dataset, also described in [11-88], or by approximation. In addition, ash accumulates in the filter surface pores, reducing the applicability of this simple mathematical approach. Permeability values have been measured in numerous ways and given in the literature for this system where cordierite substrates have been found in the $0.95 \times 10^{-14} – 8 \times 10^{-13} \, \mathrm{m}^2$ range [11-65, 11-89, 11-90, 11-91, 11-92], $1.8–9.6 \times 10^{-14} \, \mathrm{m}^2$ for the ash layer [11-90, 11-92, 11-93, 11-94], and $3.2–3.3 \times 10^{-14} \, \mathrm{m}^2$ for the soot layer [11-55]. Ash permeability has been found to be function of ash chemistry and regeneration type (i.e., active or passive) [11-88]. An active area of ongoing practical research in this area is in the understanding of ash permeability and ash depth penetration by reducing the amount of material that penetrates the substrate pores or engineering the substrate to accommodate depth penetration while reducing the associated pressure drop.

References

11-1.   Essenhigh, R., "Fundamentals of Coal Combustion," *Chemistry of Coal Utilization*, M. Ellis (New York: Wiley Interscience, 1981), 1153–1311.

11-2.   Smith, I., "The Intrinsic Reactivity of Carbons to Oxygen," *Fuel* **57**, no. 7 (1978): 405–414.

11-3.   Marsh, H., "How Oxygen Molecules Gasify Carbons," *Special Publication of Chemical Society* **32** (1978): 133–174.

11-4.   Marsh, H., *Kinetics and Catalysis of Carbon Gassification, in Introduction to Carbon Science* (London: Butterworths, 1989).

11-5.   Essenhigh, R. and Misra, M., "Autocorrelations of Kinetic Parameters in Coal and Char Reactions," *Energy & Fuels* **4**, no. 2 (1990): 171–177.

11-6.   Essenhigh, R., "Combustion Characteristics of Carbon: Influence of the Zone I - Zone II Transition on Burnout in Pulverized Coal Flames," *Energy & Fuels* **13** (1999): 955–960.

11-7.   Smoot, L.D. and Smith, P., *Coal Combustion and Gassification* (New York: Plenum Press, 1985).

11-8.   Back, M., "The Kinetics of the Reaction of Carbon with Oxygen," *Canadian Journal of Chemistry* **75** (1997): 249–257.

11-9.   Walker, P., *Gas Reactions of Carbon* (New York: Academic Press, 1959).

11-10. Gray, D., Cogoli, J., and Essenhigh, R., "Problems in Pulverized Coal and Char Combustion," *Advances in Chemistry Series* **132**, no. 72 (1976).

11-11.   Essenhigh, R., Froberg, R., and Howard, J., "Combustion Behavior of Small Particles," *Industrial Engineering Chemistry* **57** (1965): 32–43.

11-12.   Smoot, D., "International Research Centers' Activities in Coal Combustion," *Progress in Energy and Combustion Science* **24** (1998): 409–501.

11-13.   Essenhigh, R., "Structure-Based Predictive Model for Char Combustion," Report, 1998, 1–20.

11-14.   Walker, P., "Carbon: An Old But New Material Revisited," *Carbon* **28**, no. 2/3 (1990): 262–279.

11-15.   Hurt, R. and Haynes, B., "On the Origin of Power-Law Kinetics in Carbon Oxidation," *Proceedings of the Combustion Institute* **30**, no. 2 (2005): 2161–2168.

11-16.   Hurt, R. and Gibbins, J., "Residual Carbon from Pulverized Coal Fired Boilers: 1. Size Distribution and Combustion Reactivity," *Fuel* **74**, no. 4 (1995): 471–480.

11-17.   Smith, P. and Smoot, L., "One-Dimensional Model for Pulverized Coal Combustion and Gasification," *Combustion Science and Technology* **23**, no. 1 (1980): 17–31, https://doi.org/10.1080/00102208008952393.

11-18.   Alvarez, T., Fuertes, A., Pis, J., and Ehrburger, P., "Influence of Coal Oxidation upon Char Gassification Reactivity," *Fuel* **74**, no. 5 (1995): 729–735.

11-19.   Essenhigh, R. and Mescher, A., "Influence of Pressure on the Combution Rate of Carbon," *Proceedings of the Combustion Institute* **26** (1996): 3085–3094.

11-20.   Zeng, D., *Effects of Pressure on Coal Pyrolysis at High Heating Rates and Char Combustion, in Chemical Engineering* (Provo: Brigham Young University, 2005).

11-21.   Laurendeau, N., "Heterogeneous Kinetics of Coal Char Gassification and Combustion," *Progress in Energy and Combustion Science* **4** (1978): 221–270.

11-22.   Bird, R., Stewart, W., and Lightfoot, E., *Transport Phenomena* (New York: John Wiley & Sons Inc., 1960).

11-23.   Vannice, M.A., *Kinetics of Catalytic Reactions* (New York: Springer Science + Business Media, 2005).

11-24.   Thiele, E., "Relation between Catalytic Activity and Size of Particle," *Industrial Engineering Chemistry* **31** (1939).

11-25.   Essenhigh, H., Shaw, D., Shu, X., and Misra, M., "Determination of Global Kinetics of Coal Volatiles Combustion," *International Combustion Symposium*, 1990.

11-26.   Stanmore, B., Brilhac, J., and Gilot, P., "The Oxidation of Soot: A Review of Experiments, Mechanisms and Models," *Carbon* **39**, no. 15 (2001): 2247–2268.

11-27.   Ciambelli, P., "Catalytic Combustion of Carbon Particles," *Catalysis Today* **27** (1996): 99–106.

11-28.   Lahaye, J., "Influence of Cerium Oxide on the Formation and Oxidation of Soot," *Combustion and Flame* **104** (1996): 199–207.

11-29.   Zaghib, K., Song, X., and Kinoshita, K., "Thermal Analysis of the Oxidation of Natural Graphite: Isothermal Kinetic Studies," *Thermochimica Acta* **371**, no. 1–2 (2001): 57-64.

11-30.   Neeft, J., "Kinetics of Oxidation of Diesel Soot," *Fuel* **76** (1997): 1129–1136.

11-31.   Yezerets, A., Currier, N., and Eadler, H., "Experimental Determination of the Kinetics of Diesel Soot Oxidation by O_2 - Modeling Consequences," SAE Technical Paper 2003-01-0833, 2003, https://doi.org/10.4271/2003-01-0833.

11-32.   Clague, A., Donnet, J., Wang, T., and Peng, J., "A Comparison of Diesel Engine Soot with Carbon Black," *Carbon* **37**, no. 10 (1999): 1553–1565.

11-33. Yezerets, A., Currier, N., Eadler, H., Popuri, S., and Suresh, A., "Quantitative Flow-Reactor Study of Diesel Soot Oxidation Processes," *SAE Technical Paper 2002-01-1684*, 2002, https://doi.org/10.4271/2002-01-1684.

11-34. Yezerets, A., Currier, N., Eadler, H., Suresh, A., Madden, P., and Branigin, M., "Investigation of the Oxidation Behavior of Diesel Particulate Matter," *Catalysis Today* **88**, no. 1–2 (2003): 17–25.

11-35. Yezerets, A., Currier, N., Kim, D., Eadler, H., Epling, W., and Peden, C., "Differential Kinetic Analysis of Diesel Particulate Matter (Soot) Oxidation by Oxygen Using a Step-Response Technique," *Applied Catalysis B, Environmental* **61**, no. 1–2 (2005): 120–129.

11-36. Yezerets, A., Currier, N., Epling, W., Kim, D., Peden, C., Muntean, G., Wang, C., Burton, S., and Vander Wal, R., "Towards Fuel Efficient DPF Systems: Understanding the Soot Oxidation Process," *DEER*, 2005.

11-37. Stratakis, G., Konstantas, G., and Stamatelos, A., "Experimental Investigation of the Role of Soot Volatile Organic Fraction in the Regeneration of Diesel Filters," *Proceedings of the Institution of Mechanical Engineers, Part D: Journal of Automobile Engineering* **217**, no. 4 (2003): 307–317.

11-38. Stamatelos, A. and Kandylas, I., "Modeling Catalytic Regeneration of Diesel Particulate Filters, Taking into Account Adsorbed Hydrocarbon Oxidation," *Industrial Engineering Chemistry* **38** (1999): 1866–1876.

11-39. Strzelec, A., Toops, T.J., and Daw, C.S., "Oxygen Reactivity of Devolatilized Diesel Engine Particulates from Conventional and Biodiesel Fuels," *Energy & Fuels* **27**, no. 7 (2013): 3944–3951.

11-40. Strzelec, A., Vander Wal, R.L., Lewis, S.A., Toops, T.J., and Daw, C.S., "Nanostructure and Burning Mode of Light-Duty Diesel Particulate with Conventional Diesel, Biodiesel, and Intermediate Blends," *International Journal of Engine Research* **18**, no. 5–6 (2017): 520–531, https://doi.org/10.1177/1468087416674414.

11-41. Leistner, K., Nicolle, A., and Da Costa, P., "Detailed Kinetic Analysis of Soot Oxidation by NO_2, NO, and $NO + O_2$," *The Journal of Physical Chemistry C* **116**, no. 7 (2012): 4642–4654, https://doi.org/10.1021/jp210312r.

11-42. Kandylas, I.P. and Koltsakis, G.C., "NO_2-Assisted Regeneration of Diesel Particulate Filters: A Modeling Study," *Industrial & Engineering Chemistry Research* **41**, no. 9 (2002): 2115–2123.

11-43. Jacquot, F., Logie, V., Brilhac, J.F., and Gilot, P., "Kinetics of the Oxidation of Carbon Black by NO_2: Influence of the Presence of Water and Oxygen," *Carbon* **40**, no. 3 (2002): 335–343.

11-44. Jung, J., Song, S., and Chun, K.M., "Characterization of Catalyzed Soot Oxidation with NO_2, NO and O_2 Using a Lab-Scale Flow Reactor System," *SAE Technical Paper 2008-01-0482*, 2008, https://doi.org/10.4271/2008-01-0482.

11-45. Kim, D., Lee, H.-S., Chun, K.M., Hwang, J.-H., Lee, K.S., and Chun, B.-H., "Comparison of Soot Oxidation by NO_2 Only and Plasma-Treated Gas Containing NO_2, O_2, and Hydrocarbons," *SAE Technical Paper 2002-01-2704*, 2002, https://doi.org/10.4271/2002-01-2704.

11-46. Lee, J.H., Lee, H.-S., Song, S., and Chun, K.M., "Experimental Investigation of Soot Oxidation Characteristic with NO_2 and O_2 Using a Flow Reactor Simulating DPF," *SAE Technical Paper 2007-01-1270*, 2007, https://doi.org/10.4271/2007-01-1270.

11-47. Setiabudi, A., Makkee, M., and Moulijn, J.A.J.A.C.B.E., "The Role of NO2 and O2 in the Accelerated Combustion of Soot in Diesel Exhaust Gases," *Applied Catalysis B: Environmental* **50**, no. 3 (2004): 185–194.

11-48. Shrivastava, M., Nguyen, A., Zheng, Z., Wu, H.-W., and Jung, H.S.J., "Kinetics of Soot Oxidation by NO_2," *Environmental Science and Technology* **44**, no. 12 (2010): 4796–4801.

11-49. Stanmore, B., Tschamber, V., and Brilhac, J.-F.J.F., "Oxidation of Carbon by NOx, with Particular Reference to NO_2 and N_2O," *Fuel* **87**, no. 2 (2008): 131–146.

11-50. Seong, H. and Choi, S., "Oxidation-Derived Maturing Process of Soot, Dependent on O_2-NO_2 Mixtures and Temperatures," *Carbon* **93** (2015): 1068–1076.

11-51. Tighe, C., Twigg, M., Hayhurst, A., Dennis, J.J.C., "The Kinetics of Oxidation of Diesel Soots by NO_2," *Combustion and Flame* **159**, no. 1 (2012): 77–90.

11-52. Strzelec, A., Vander Wal, R.L., Thompson, T.N., Toops, T.J., and Daw, C.S., "NO_2 Oxidation Reactivity and Burning Mode of Diesel Particulates," *Topics in Catalysis* **59**, no. 8–9 (2016): 686–694.

11-53. Sluder, S., Wagner, R., Storey, J., and Lewis, S., "Implications of Particulate and Precursor Compounds Formed During High-Efficiency Clean Combustion in a Diesel Engine," SAE Technical Paper 2005-01-3844, 2005, https://doi.org/10.4271/2005-01-3844.

11-54. Konstandopolous, A. and Johnson, J., "Wall-Flow Diesel Particulate Filters - Their Pressure Drop and Collection Efficiency," SAE SAE Technical Paper 890405, 1989, https://doi.org/10.4271/890405.

11-55. Konstandopoulos, A.G., Kostoglou, M., Skaperdas, E., Papaioannou, E., Zarvalis, D., and Kladopoulou, E., "Fundamental Studies of Diesel Particulate Filters: Transient Loading, Regeneration and Aging," SAE Technical Paper 2000-01-1016, 2000, https://doi.org/10.4271/2000-01-1016.

11-56. Nakatani, K., Hirota, S., Takeshima, S., Itoh, K., Tanaka, T., and Dohmae, K., "Simultaneous PM and NOx Reduction System for Diesel Engines," SAE Technical Paper 2002-01-0957, 2002, https://doi.org/10.4271/2002-01-0957.

11-57. Konstandopolous, A., Skaperdas, E., and Masoudi, M., "Inertial Contributions to the Pressure Drop of Diesel Particulate Filters," *SAE Paper Number 2001-01-0909*, 2001, https://doi.org/10.4271/2001-01-0909.

11-58. Huynh, C., Johnson, J., Yang, S., Bagley, S., and Warner, J., "A One-Dimensional Computational Model for Studying the Filtration and Regeneration Characteristics of a Catalyzed Wall-Flow Diesel Particulate Filter," SAE Technical Paper 2003-01-0841, 2003, https://doi.org/10.4271/2003-01-0841.

11-59. Strzelec, A., *Kinetic Model Development for the Combustion of Particulate Matter from Conventional and Soy Methyl Ester Diesel Fuels.* (Madison, Wisconsin: University of Wisconsin-Madison, 2009), 168.

11-60. Johnson, T., "Vehicular Emissions in Review," *SAE Int. J. Engines* **9**, no. 2 (2016): 1258–1275, https://doi.org 10.4271/2016-01-0919.

11-61. Bissett, E.J., "Mathematical Model of the Thermal Regeneration of a Wall-Flow Monolith Diesel Particulate Filter," *Chemical Engineering Science* **39**, no. 7-8 (1984): 1233–1244.

11-62. Maricq, M., "How are Emissions of Nuclei Mode Particles Affected by Emission Control," *HEI-Conference*, Portland, OR, May 2009.

11-63. Johnson, T. and Joshi, A., "Review of Vehicle Engine Efficiency and Emissions," SAE Technical Paper 2017-01-0907, 2017, https://doi.org/10.4271/2017-01-0907.

11-64. Konstandopoulos, A.G., Skaperdas, E., and Masoudi, M., "Inertial Contributions to the Pressure Drop of Diesel Particulate Filters," SAE Technical Paper 2001-01-0909, 2001, https://doi.org/10.4271/2001-01-0909.

11-65. Konstandopoulos, A.G., "Flow Resistance Descriptors for Diesel Particulate Filters: Definitions, Measurements and Testing," SAE Technical Paper 2003-01-0846, 2003, https://doi.org/10.4271/2003-01-0846.

11-66. Sheppard, J., *Size-Dependent Filtration of Cakeless Diesel Particulate Filters in Mechanical Engineering* (College Station, TX: Texas A&M University, 2014).

11-67. Dalla Nora, M., Lanzanova, T.D.M., and Zhao, H., "Effects of Valve Timing, Valve Lift and Exhaust Backpressure on Performance and Gas Exchanging of a Two-Stroke GDI Engine with Overhead Valves," *Energy Conversion and Management* **123** (2016): 71–83.

11-68. Lambert, C.K., Chanko, T., Jagner, M., Hangas, J., Liu, X., Pakko, J., and Kamp, C.J., "Analysis of Ash in Low Mileage, Rapid Aged, and High Mileage Gasoline Exhaust Particle Filters," SAE Technical Paper 2017-01-0930, 2017, https://doi.org/10.4271/2017-01-0930.

11-69. Kamp, C.J., Bagi, S., and Wang, Y., "Phenomenological Investigations of Mid-Channel Ash Deposit Formation and Characteristics in Diesel Particulate Filters," SAE Technical Paper 2019-01-0973, 2019, https://doi.org/10.4271/2019-01-0973.

11-70. Sappok, A., Govani, I., Kamp, C., Wang, Y., and Wong, V., "In-Situ Optical Analysis of Ash Formation and Transport in Diesel Particulate Filters during Active and Passive DPF Regeneration Processes," *SAE Int. J. Fuels Lubr.* **6**, no. 2 (2013): 336–349, (2013) https://doi.org/10.4271/2013-01-0519.

11-71. Sappok, A., Kamp, C., and Wong, V., "Sensitivity Analysis of Ash Packing and Distribution in Diesel Particulate Filters to Transient Changes in Exhaust Conditions," *SAE Int. J. Fuels Lubr.* **5**, no. 2 (2012): 733–750, https://doi.org/10.4271/2012-01-1093.

11-72. Bagi, S., Bowker, R., and Andrew, R., "Understanding Chemical Composition and Phase Transitions of Ash from Field Returned DPF Units and Their Correlation with Filter Operating Conditions," *SAE Int. J. Fuels Lubr.* **9**, no. 1 (2016): 239–259, https://doi.org/10.4271/2016-01-0898.

11-73. Kamp, C.J., Sappok, A., Wang, Y., Bryk, W., Rubin, A., and Wong, V., "Direct Measurements of Soot/Ash Affinity in the Diesel Particulate Filter by Atomic Force Microscopy and Implications for Ash Accumulation and DPF Degradation," *SAE Int. J. Fuels Lubr.* **7**, no. 1 (2014): 307–316, https://doi.org/10.4271/2014-01-1486.

11-74. Sappok, A., Rodriguez, R., and Wong, V., "Characteristics and Effects of Lubricant Additive Chemistry on Ash Properties Impacting Diesel Particulate Filter Service Life," *SAE Int. J. Fuels Lubr.* **3**, no. 1 (2010): 705–722, https://doi.org/10.4271/2010-01-1213.

11-75. Aravelli, K. and Heibel, A., "Improved Lifetime Pressure Drop Management for Robust Cordierite (RC) Filters with Asymmetric Cell Technology (ACT), SAE Technical Paper 2007-01-0920, 2007, https://doi.org/10.4271/2007-01-0920

11-76. Kamp, C.J., Sappok, A., and Wong, V., "Soot and Ash Deposition Characteristics at the Catalyst-Substrate Interface and Intra-Layer Interactions in Aged Diesel Particulate Filters Illustrated Using Focused Ion Beam (fib) Milling," *SAE Int. J. Fuels Lubr.* **5**, no. 2 (2012): 696–710, https://doi.org/10.4271/2012-01-0836.

11-77. Bodek, K.M. and Wong, V.V., "The Effects of Sulfated Ash, Phosphorus and Sulfur on Diesel Aftertreatment Systems-A Review," SAE Technical Paper 2007-01-1922, 2007, https://doi.org/10.4271/2007-01-1922

11-78. Bardasz, E.A., Cowling, S., Panesar, A., Durham, J., and Tadrous, T.N., "Effects of Lubricant Derived Chemistries on Performance of the Catalyzed Diesel Particulate Filters," SAE Technical Paper 2005-01-2168, 2005, https://doi.org/10.4271/2005-01-2168.

11-79. Bagi, S., Singh, N., and Andrew, R., "Investigation into Ash from Field Returned DPF Units: Composition, Distribution, Cleaning Ability and DPF Performance Recovery," SAE Technical Paper 2016-01-0928, 2016, https://doi.org/10.4271/2016-01-0928.

11-80. Sappok, A., Santiago, M., Vianna, T., and Wong, V.W., "Characteristics and Effects of Ash Accumulation on Diesel Particulate Filter Performance: Rapidly Aged and Field Aged Results," SAE Technical Paper 2009-01-1086, 2009, https://doi.org/10.4271/2009-01-1086.

11-81. Sappok, A. and Wong, V., "Ash Effects on Diesel Particulate Filter Pressure Drop Sensitivity to Soot and Implications for Regeneration Frequency and DPF Control," *SAE Int. J. Fuels Lubr.* **3**, no. 1 (2010): 380–396, https://doi.org/10.4271/2010-01-0811.

11-82. Sharma, V., Bagi, S., Patel, M., Aderniran, O., and Aswath, P.B., "Influence of Engine Age on Morphology and Chemistry of Diesel Soot Extracted from Crankcase Oil," *Energy & Fuels* **30**, no. 3 (2016): 2276–2284.

11-83. Monahan, G.J., *Characterizing Ash Properties and Effects on Diesel Particulate Filter Flow Resistance* (Massachusetts Institute of Technology, 2016).

11-84. Sappok, A., Wang, Y., Wang, R.-Q., Kamp, C., and Wong, V., "Theoretical and Experimental Analysis of Ash Accumulation and Mobility in Ceramic Exhaust Particulate Filters and Potential for Improved Ash Management," *SAE Int. J. Fuels Lubr.* **7**, no. 2 (2014): 511–524, https://doi.org/10.4271/2014-01-1517.

11-85. Sappok, A.G. and Wong, V.W., "Detailed Chemical and Physical Characterization of Ash Species in Diesel Exhaust Entering Aftertreatment Systems," SAE Technical Paper 2007-01-0318, 2007, https://doi.org/10.4271/2007-01-0318.

11-86. George, S. and Heibel, A., "Next Generation Cordierite Thin Wall DPF for Improved Pressure Drop and Lifetime Pressure Drop Solution," SAE Technical Paper 2016-01-0940, 2016, https://doi.org/10.4271/2016-01-0940.

11-87. Sappok, A. and Wong, V.W., "Lubricant-Derived Ash Properties and Their Effects on Diesel Particulate Filter Pressure-Drop Performance," *Journal of Engineering for Gas Turbines and Power* **133**, no. 3 (2011): 032805.

11-88. Kamp, C.J., Zhang, S., Bagi, S., Wong, V., Monahan, G., Sappok, A., and Wang, Y., "Ash Permeability Determination in the Diesel Particulate Filter from Ultra-High Resolution 3D X-Ray Imaging and Image-Based Direct Numerical Simulations," *SAE Int. J. Fuels Lubr.* **10**, no. 2 (2017): 608–618, https://doi.org/10.4271/2017-01-0927.

11-89. Iwata, H., Konstandopoulos, A., Nakamura, K., Kasuga, T., Ogyu, K., and Ohno, K., "Durability of Filtration Layers Integrated into Diesel Particulate Filters," SAE Technical Paper 2013-01-0837, 2013, https://doi.org/10.4271/2013-01-0837.

11-90. Nuszkowski, J., Thompson, G.J., Moles, N., Chiaramonte, M., and Hu, J., "Pressure Drop and Cleaning of In-Use Ash Loaded Diesel Particulate Filters," SAE Technical Paper 2006-01-3256, 2006, https://doi.org/10.4271/2006-01-3256.

11-91. Bouteiller, B., Bardon, S., Briot, A., Girot, P., Gleize, V., and Higelin, P., "One Dimensional Backpressure Model for Asymmetrical Cells DPF," SAE Technical Paper 2007-01-0045, 2007, https://doi.org/10.4271/2007-01-0045.

11-92. Millet, C.-N., Ménégazzi, P., Martin, B., Colas, H., and Bourgeois, C., "Modeling of Diesel Particulate Filter Regeneration: Effect of Fuel-Borne Catalyst," SAE Technical Paper 2002-01-2786, 2002, https://doi.org/10.4271/2002-01-2786.

11-93. Fujii, S. and Asako, T., "Design Optimization of Non-Catalyzed DPF from Viewpoint of Back Pressure in Ash Loading State," SAE Technical Paper 2011-01-2091, 2011, https://doi.org/10.4271/2011-01-2091.

11-94. Zarvalis, D., Lorentzou, S., and Konstandopoulos, A.G., "A Methodology for the Fast Evaluation of the Effect of Ash Aging on the Diesel Particulate Filter Performance," SAE Technical Paper 2009-01-0630, 2009, https://doi.org/10.4271/2009-01-0630.

Selective Catalytic Reduction Systems

Ray "Bones" Barboni: And what about the sun, does it ever shine around here, or is this smog around all the time?
Limo Driver with Sign: They say the smog is the reason we have such beautiful sunsets.
Ray "Bones" Barboni: That's what they say, huh? What a bunch of [B.S.].

—Elmore Leonard, Get Shorty

Removing nitrogen oxides (NOx) emissions from the exhaust of fuel-lean engines is particularly challenging due to lower exhaust temperature and relatively high oxygen (O_2) concentrations. The three-way catalyst (TWC), discussed in Chapter 9, is not effective at reducing NOx under these net oxidizing conditions. Therefore, an alternative technology is needed for NOx conversion in the exhaust aftertreatment system (EAS). While lean NOx traps (LNTs), described in Chapter 10, have been implemented in production for light-duty applications, LNT costs scale quickly with respect to engine size, and the system is susceptible to rapid aging. Selective catalytic reduction (SCR) systems, by contrast, are currently the most widely applied method for removing NOx from compression-ignition (CI) engines. The SCR system reduces NOx, specifically nitrogen monoxide (NO) and nitrogen dioxide (NO_2), by reacting it with ammonia (NH_3) over a catalyst to produce nitrogen gas (N_2) and water (H_2O). The decomposition of urea ($CO(NH_2)_2$) to generate the ammonia used in the SCR system also makes carbon dioxide (CO_2).

SCR was first developed for the reduction of NOx in the exhaust of stationary industrial applications, including engines used for power generations, and has been well proven in this market for decades. Later, SCR was adapted for mobile internal combustion engines—first for on-road engines, and later, non-road. SCR was first used in power plants in the late 1970s in Japan, followed by deployment in Europe in the mid-1980s. SCR systems were introduced in the United States first for gas turbines [12-1] in the 1990s, then further refined and developed for NOx control from coal-fired power plants [12-2]. Additional common applications for

SCR included removing NOx from the exhaust from heaters used in plants and refineries, and from boilers in the chemical processing industry and municipal waste plants [12-3, 12-4]. The introduction of ammonia- or urea-based SCR systems for NOx control from mobile applications began with marine engines [12-5]. The large displacement volumes and quasi-steady-state operation of marine engines were similar to stationary generators, which made the adaptation of the SCR system straightforward. Ships equipped with ammonia SCR systems were able to convert 92% of the NOx in the exhaust.

Work on developing vehicular SCR systems dates back to the 1980s [12-6, 12-7, 12-8], when the technology began to be adapted for engines. Application of SCR to vehicles requires overcoming several issues, such as the introduction of the reductant, reduction in the presence of excess O_2, thermal management, and active control strategies [12-9, 12-10]. The first commercial urea-SCR system in a vehicle was introduced in late 2004 by Nissan Diesel on its Quon heavy-duty truck [12-11]. In 2007, the US Environmental Protection Agency (EPA) proposed a NOx limit of 0.2g/bhp·h be achieved in model year (MY) 2010, which pushed the development of SCR systems for heavy-duty diesel engines (HDDE). SCR was subsequently implemented in light-duty EAS in the following years. As of the writing of this book, urea SCR is the technology of choice for meeting the NOx limits for heavy-duty truck and bus engines by the majority of manufacturers. For light-duty vehicles, SCR was introduced in some EPA Tier-2 vehicles, while others used LNTs. By the mid-2010s, most of the Tier-2 vehicles with LNTs were converted to urea SCR. In the European Union (EU), SCR was introduced on some Euro 5 models and was more widely in use on Euro 6 vehicles. SCR was introduced in non-road CI engines to meet the US Tier-4i or EU stage-IIIB emission standards, discussed in Chapter 3, Emissions Control Regulations.

A modern SCR system comprises several physical and chemical processes, as follows:

- The introduction of urea-water solution (UWS) to the exhaust gas
- The decomposition of the urea into the active species for reaction, ammonia
- The adsorption of the ammonia on the catalyst surface
- The reaction of NO and NO_2 with ammonia on the catalyst surface to form N_2 and water
- Side reactions that form byproducts

The urea is delivered in a urea-water solution (UWS) known as diesel exhaust fluid (DEF) in the US and AdBlue in the EU. The process steps above, along with their thermodynamic, diffusion, and kinetic limits, are described in detail in this chapter.

In addition to the device-specific issues identified earlier, application to the vehicle platforms required overcoming additional barriers related to the UWS. These barriers included the infrastructure for urea distribution, development of dosing methods and strategies, and catalyst optimization for NOx conversion. Some regulatory bodies, notably EPA, were initially skeptical about using SCR to meet emissions limits. Concerns included ensuring that the UWS is available together with diesel fuel throughout a nationwide distribution network (availability of infrastructure for urea) and that it would always be replenished in a timely manner by vehicle operators with the correct UWS. Ultimately, SCR proved to be a more robust emission technology than the main alternative, LNTs, and has been widely used in all types of mobile CI engines since.

One concern when using SCR is the unintentional emissions of an excess reductant, ammonia, which is also called ammonia slip. To avoid these potential ammonia emissions, an ammonia slip catalyst (ASC) may be required downstream of the SCR. ASCs are described further in Chapter 13.

Due to the continuing development work on the catalysts, materials, and control strategy, current production SCR systems can achieve greater than 98% NOx conversion. This makes SCR the preferred catalyst technology for reducing engine-out NOx to the levels mandated by emissions regulations.

12.1 Selective Catalytic Reduction Functions

The SCR system is able to selectively reduce NOx despite the O_2-rich (fuel-lean) exhaust environment by reacting with ammonia over a catalyst. The direct use of ammonia has been pragmatically ruled out, due to the safety concerns of carrying aqueous or gaseous ammonia on board the vehicle. Thus, ammonia is indirectly delivered by using the UWS, urea in an aqueous solution, as the ammonia source. The UWS is sprayed into the exhaust pipe upstream of the SCR system, where it decomposes into ammonia and CO_2. The ammonia reacts with the NOx on the catalytic sites in the SCR. Specifically, an aqueous solution of urea (32.5% by weight) is injected into the exhaust gas stream using an injection system, similar in nature to a diesel fuel injector. A high-level illustration of an SCR catalytic converter with an engine leading to an exhaust pipe, an upstream oxidation catalyst, a urea dosing system, a mixer, and a urea SCR catalyst is shown in **Figure 12.1**. Other examples of an EAS with SCR system are provided in Section 7.5, including the EcoDiesel EAS in Figure 7.18, the Duramax EAS in Figure 7.20, the Power Stroke EAS in Figure 7.23, and the Scania DC13 EAS illustrated in Figure 7.29.

12.1.1 Thermal Decomposition of Urea to Ammonia

The necessary first step following the UWS injection is the evaporation of the aqueous urea solution droplets. Once evaporated the urea can decompose into ammonia in a

FIGURE 12.1   Illustration of SCR system [12-12].

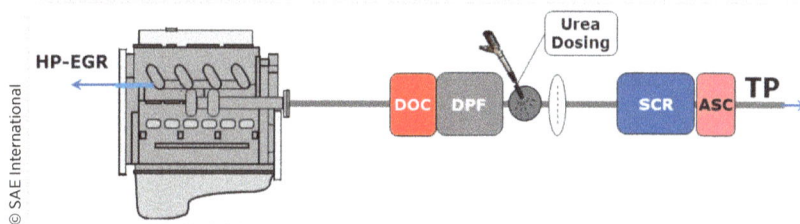

process called thermolysis [12-13]. This evaporation often leads to the formation of molten or solid urea, as shown in Eq. 12.1, which then sublimates into the gas phase.

$$CO(NH_2)_{2,aqueous} \rightarrow CO(NH_2)_{2,gas} + n\,H_2O \tag{12.1}$$

Based on exhaust gas temperatures, the thermolysis of urea to ammonia takes place in the exhaust pipe and forms equimolar amounts of ammonia and isocyanic acid (HNCO),

$$CO(NH_2)_2 \rightarrow NH_3 + HNCO. \tag{12.2}$$

While HNCO is very stable in the gas phase, it hydrolyzes easily on the surface of the SCR catalyst, forming ammonia and CO_2 per

$$HNCO + H_2O \rightarrow NH_3 + CO_2. \tag{12.3}$$

The thermolysis of urea, shown in Eq. 12.2, is thermodynamically limited [12-13] and endothermic. At exhaust temperatures below 200°C, there is a risk that the urea decomposition is inadequate, as the thermolysis reaction kinetics slow down. For sustained operation, the minimum required temperature for urea droplet evaporation is 170°C. If the exhaust gas temperature or wall surface temperature stays below 200°C, there is a risk of deposit formation. These deposits occur when a thin film of solid urea is formed on the exhaust pipe surface or front face of the SCR monolith. Deposited urea is less likely to undergo thermolysis and hydrolysis, thus affecting urea usage, ammonia slip, and NOx conversion in the SCR system. Evaporation of urea from the hot exhaust pipe wall cools the wall and increases the risk of melamine deposit formation, which can glaze the exhaust pipe wall [12-10].

The injector configuration and control of the UWS spray is of great importance in the SCR system, as it correlates with the concentration and uniformity of the ammonia delivery to the catalyst. Based on urea droplet evaporation, film formation, and heat transfer in the pipe wall, dynamic models have been developed to control the urea solution spray in such a manner as to enhance the performance of the urea SCR catalyst [12-14]. Additionally, static mixers are sometimes installed upstream of the SCR catalyst to homogenize the radial distribution of the exhaust gas flow and urea concentration in the exhaust pipe. They can also help heat the pipe wall by forcing warmer gas to the wall.

Several computational fluid dynamics (CFD) analyses have been done to investigate the effect of mixer design on the uniformity index, a measure of how homogeneous the ammonia concentration is across the cross-section of the exhaust pipe, as well as the overall NOx conversion in the SCR catalyst [12-15, 12-16]. The hydrolysis of HNCO at temperatures below 350°C is incomplete in the exhaust pipe upstream of the SCR catalyst and is completed only on the acidic sites of the catalyst [12-17, 12-18, 12-19].

12.1.2 Reduction Reactions

The three chemical reactions for NOx reduction are referred to as the "standard," "slow," and "fast" SCR reactions. All three reactions reduce NOx to N_2, just with different reactant stoichiometries and at varying rates, as shown in **Figure 12.2** below. All three processes are governed by two limits that affect the rate of NOx conversion: mass transfer and kinetic. The mass transfer limits affecting how quickly NOx and ammonia can get to the catalyst

FIGURE 12.2   Schematics of the SCR reactions.

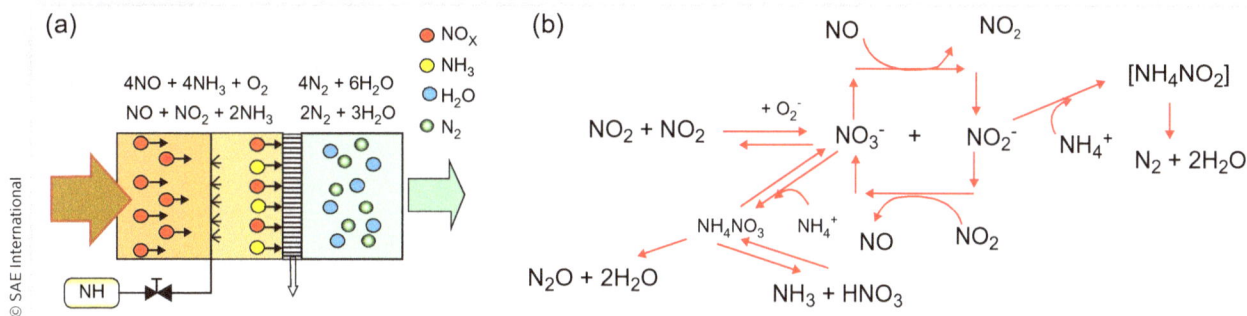

surface are the same for all three SCR reactions. These limits vary with catalyst type and configuration [12-20]. Where they differ is in their kinetic limits, which are determined by the rates of reaction, where the rate is itself a function of reactant concentration and the catalyst. Although vanadia-based catalysts were the first employed, nowadays iron (Fe) and copper (Cu) zeolites are typical. The catalysts are discussed in detail in Section 12.4. Another key factor in these reactions is the ratio of NO to NO_2. The fast SCR reaction, shown in Eq. 12.6, has a stoichiometric ratio of 1:1 for $NO:NO_2$. Therefore, the overall rate of NOx conversion is enhanced by an upstream oxidation catalyst capable of converting some of the NO to NO_2 to get closer to a 1:1 ratio. However, complete oxidation of the NOx is not desirable, as the reaction rate with the only NO_2 is the slowest, as seen in Eq. 12.5.

The standard SCR reaction only consumes NO and is given by

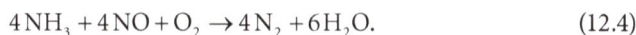

$$4\,NH_3 + 4\,NO + O_2 \rightarrow 4\,N_2 + 6\,H_2O. \tag{12.4}$$

This reaction represents a stoichiometric, 1:1 molar ratio between the ammonia and NO with some additional O_2 consumption in the process. By contrast, the SCR reaction that only consumes NO_2 is much slower, especially over a vanadia catalyst and is given by

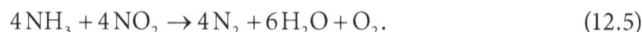

$$4\,NH_3 + 4\,NO_2 \rightarrow 4\,N_2 + 6\,H_2O + O_2. \tag{12.5}$$

This slower reaction rate earns it the name "slow SCR reaction" [12-21]. This reaction is far less dominant since the NO_2 concentration is rarely higher than the NO concentration, even with an oxidation catalyst creating NO_2 from NO.

When both NO and NO_2 are present, the reaction is relatively fast and is therefore called the "fast SCR reaction." This reaction consumes both NO and NO_2 with an overall 1:1 stoichiometric molar ratio between ammonia and the combined NOx species, per

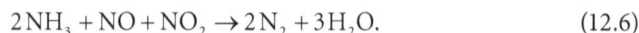

$$2\,NH_3 + NO + NO_2 \rightarrow 2\,N_2 + 3\,H_2O. \tag{12.6}$$

This reaction does not consume any additional exhaust O_2. The fast SCR reaction is one of the two preferred reactions on Fe/zeolite catalysts for maximum NOx conversion, as discussed in Section 12.4. The reaction rate for the fast SCR reaction in Eq. 12.6 is much faster [12-22, 12-23, 12-24] than the standard SCR reaction in Eq. 12.4 [12-10, 12-25, 12-26, 12-27].

An oxidation catalyst upstream of the SCR catalyst can strongly influence the rate of NOx consumption by forming NO_2 and moving the $NO:NO_2$ ratio closer to 1:1. The additional NO_2 increases the portion of total NOx consumed by the fast SCR reaction, which leads to better NOx conversion and less ammonia slip, even at higher space velocities.

The mechanisms of the ammonia SCR reactions have been studied extensively since the early 2000s. The surface SCR reactions are complex and the role of the intermediate nitrate species have been shown to have important roles, particularly for the fast SCR reaction. Several studies have indicated that the rapid decomposition of ammonium nitrate (NH_4NO_2) is important to the SCR surface-catalyzed nitrogen formation process [12-28, 12-29]. This reaction, also used commercially to prepare chemically pure nitrogen, is

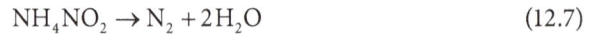

$$NH_4NO_2 \rightarrow N_2 + 2H_2O \tag{12.7}$$

At higher temperatures, the ammonium nitrate can instead decompose to undesirable emissions species, as shown in Eq. 12.12 in Section 12.1.3.2.

12.1.3 Formation of Secondary Emissions and Byproducts

12.1.3.1 **Ammonia Slip:** The ammonia emissions from SCR are known as ammonia slip. These ammonia emissions occur if there is an excess reductant that does not react with NOx over the catalyst. For example, higher ammonia to NOx ratios make slip more likely. The standard and fast SCR reactions shown in Eqs. 12.4 and 12.6, respectively, both have equimolar (1:1) ammonia to NOx ratios (ANR). Ratios higher than 1:1 will increase the ammonia slip [12-30, 12-31]. In practical application, the ANR is generally between 0.9 and 1.1, to minimize ammonia slip while achieving a satisfactory NOx conversion level.

Figure 12.3 shows clear trends for both ammonia slip and NOx conversion as a function of temperature. ammonia slip decreases with increasing temperature, whereas NOx conversion may either increase or decrease, depending on the catalyst formulation. These phenomena will be further described in Section 12.4.

FIGURE 12.3   Ammonia slip and NOx conversion as a function of NH_3: NOx ratio. Data from [12-32].

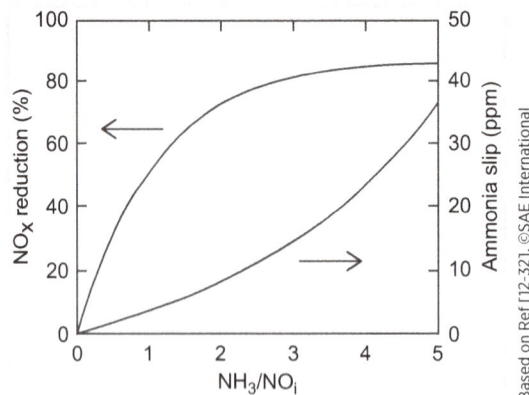

12.1.3.2 **Byproduct Formation:** There are several additional reactions that may occur in the SCR that can create undesirable byproducts. Ammonia can be directly oxidized into N_2 via

$$4\,NH_3 + 3\,O_2 \rightarrow 2\,N_2 + 6\,H_2O, \tag{12.8}$$

N_2O via

$$2\,NH_3 + 2\,O_2 \rightarrow N_2O + 3\,H_2O, \tag{12.9}$$

or even NO via

$$4\,NH_3 + 5\,O_2 \rightarrow 4\,NO + 6\,H_2O, \tag{12.10}$$

instead of converting NOx into N_2. The nitrous oxide (N_2O) formed in Eq. 12.9 is particularly problematic as it is an extremely powerful oxidizer and therefore a major scavenger of ozone O_3 in the stratosphere. Also, N_2O has 300 times the global warming potential of CO_2, as discussed in Section 1.4 of Chapter 1, Introduction. The oxidation to NO shown in Eq. 12.10 is typically only of concern at exhausting temperatures greater than 500°C. Unfortunately, in this reaction, the ammonia adds to the tailpipe NOx instead of removing it.

At temperatures below 200°C, the ammonia can react with NO_2 to produce ammonium nitrate (NH_4NO_3) [12-33], by

$$2\,NH_3 + 2\,NO_2 + H_2O \rightarrow NH_4NO_3 + NH_4NO_2. \tag{12.11}$$

Ammonium nitrate can precipitate out of the gas as a white solid crystal, which is highly soluble in water. Also, ammonium nitrate can readily form an explosive mixture when mixed with hydrocarbons (HC) from the fuel or oil. For these reasons, it is best to avoid ammonium nitrate formation by keeping the exhaust temperature greater than 200°C, or by controlling the gas concentration of ammonia to keep it out of the necessary reaction stoichiometry. At high exhaust temperatures, the ammonium nitrate bound on the catalyst surface can decompose forming N_2O [12-29]. This reaction is commercially used to manufacture N_2O,

$$NH_4NO_3 \rightarrow N_2O + 2\,H_2O \tag{12.12}$$

12.1.4 Warm-up Strategies and Other Operational Considerations

Although all exhaust aftertreatment catalysts need to reach their respective light-off temperatures before they function, this is especially important for SCR systems. This is because the urea decomposition, described in Section 12.1.1, and the catalytic function, described in Section 12.1.2, both have a minimum operating temperature of about 170°C to 200°C. The main warm-up strategies for SCR are based on increasing the available exhaust enthalpy, as discussed in Section 7.3.2.1.

One strategy is to place a close-coupled SCR system (ccSCR) at the turbine outlet [12-34, 12-35]. While beneficial as a warm-up strategy, this adds complexity to the hardware because the EAS then requires a second urea injector, SCR system, and ASC. In addition, it also requires modification of the controls software, since the ECU needs to manage two SCR units and the DPF in between them [12-34].

12.2 **Urea Processing**

Ammonia is an excellent chemical reducing agent for use in SCR. However, handling ammonia is problematic due to its caustic nature and toxicity. The ideal reductant for use in SCR would be non-toxic, safe to handle, easy to transport, abundant, and inexpensive. There is no reductant species that fits these criteria that can be directly used in the SCR. However, commercially available, nontoxic urea meets the majority of the criteria and can easily decompose into ammonia. An aqueous urea solution is stable and compatible with the vehicle storage requirements for mobile SCR applications. Though urea is quite soluble in water (even at concentrations greater than 50 wt%), the UWS used in SCR systems is 32.5 wt% urea in balance water, which forms a eutectic solution. The benefit of the eutectic concentration is that as the UWS changes phase from liquid to solid and back, the concentration of the liquid solution and the corresponding ice all remain the same at 32.5 wt%. The consideration of the urea concentration in the solid phase is important because the freezing point of the UWS is –11°C (12°F). Because this temperature is within the normal operating window of a vehicle, especially in northern climates, a thermal management strategy is essential to proper urea-SCR system function.

One of the main challenges of implementing SCR systems is processing the UWS within the EAS to ensure the proper function of the SCR unit. Fundamentally, the UWS dosing system should be designed to provide a uniform concentration of ammonia in the gas phase over the front face of the SCR monoliths. To achieve that end, the following process steps need to happen in the urea dosing system and in the part of the EAS that starts with the urea injector and ends with the SCR unit:

1. The UWS is maintained in the liquid state and the reservoir quantity is monitored.

2. The UWS is injected or sprayed into the exhaust gas with a minimum amount of wall wetting.

3. The UWS evaporates within the exhaust gas before the gas enters the SCR monolith.

4. The urea decomposes into ammonia and CO_2 in the humid exhaust gas, as described in Section 12.1.1.

5. The ammonia formed is uniformly distributed within the exhaust gas flow field.

6. The exhaust gas is evenly distributed over the front face of the SCR monolith.

This section describes the operating strategy and typical hardware needed (urea tank, injector controls, and mixer) for successful operation. Potential failure modes in the urea processing system are also discussed.

12.2.1 **Urea Injection System**

The SCR function requires precise control of the ammonia concentration in the exhaust [12-36]. The ammonia concentration is dependent on the thermal decomposition of the UWS, which is in turn dependent on the dosing into the exhaust. Therefore, underdosing the UWS will result in insufficient NOx conversion to meet the regulatory requirement because there is insufficient ammonia present. Overdosing, injecting too much UWS, will lead to undesirable ammonia or byproduct emissions, as described in Section 12.1.3. Preventing this ammonia slip is one reason that an ASC, described in Chapter 13, maybe required when using SCR on a vehicle.

The main purpose of the urea injection system is to monitor and manage the stored UWS and deliver it to the exhaust gas as the dosing strategy requires [12-37]. The system includes the following components: UWS storage tank, tank heater, UWS pump, delivery and return lines, injector unit, and dosing control unit. An example of a urea injection system diagram for the Scania DC13 engine and EAS is shown in Figure 7.29, in Section 7.5.5. The urea injection system will also include the sensors described in Chapter 6, Sensors and On-Board Diagnostics, that ensure proper function of this part of the system.

In order to inject the solution into the exhaust, the UWS must be pumped from its storage tank and sprayed through an atomizing nozzle, similar in nature to a fuel injector [12-38]. One challenge of the urea injection system is that the design range for ambient temperature is typically −40°C (−40°F) to +40°C (104°F), which encompasses the freezing point of the eutectic UWS, −11°C (12°F). Thus, the storage tank usually includes an integrated heater element that starts melting the UWS when the engine is started. The storage tank may also include a heat exchanger that uses warm engine coolant to keep the urea solution liquid, as shown in Figure 7.29. One advantage of the solution being at its eutectic concentration is that the urea concentration in the liquid equals that in the solid portion. Thus, the whole tank does not need to be thawed before dosing can begin.

In addition, dosing devices to enhance the low-temperature performance of SCR have been developed. These typically include a heated bypass line that uses an electric heater to promote urea thermolysis, as shown in Eq. 12.2. The feed also includes a catalyst to further hydrolyze HNCO into ammonia and CO_2, per Eq. 12.3. The device is activated only when the exhaust gas temperature falls below the urea evaporation and decomposition temperature of about 180°C.

An additional challenge for the engine and EAS manufacturer is ensuring that the UWS storage tank does not run dry and that the operator has put the correct fluid into the tank. Thus, the storage tank includes a level sensor and a UWS quality sensor, both described in Chapter 6, Sensors and On-Board Diagnostics, that help ensure that the correct fluid at the correct concentration was put in.

Lastly, the UWS needs to be delivered into the exhaust gas flow through an injector. Aspects of the injector design include the target UWS flow rates, delivery pressure, nozzle design, and spray pattern, and injector orientation within the EAS [12-39]. As shown in **Figure 12.4**, there are several options for the injector orientation within the exhaust pipe. The preferred approach is to inject the solution symmetrically into the flow field, with the injector either aiming upstream or downstream, but packaging constraints usually limit the engineers' ability to achieve that orientation. It is typical to use CFD to assess how the UWS spray interacts with the exhaust gas flow at key operating conditions [12-16]. A significant degree of mixing of the ammonia and exhaust gas is essential for overall performance.

FIGURE 12.4   Urea solution injector concepts.

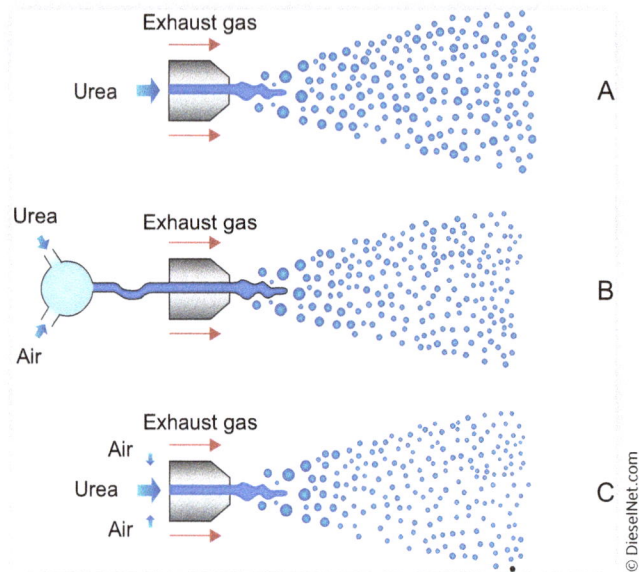

Another consideration is the range of exhaust flows into which the injector sprays the urea solution. Since most injectors control the flow rate through pulse width modulation (PWM), the UWS flow during each pulse will reach its maximum. Thus, there is a risk of urea-based deposits forming because the spray hits the wall when the exhaust gas flow rate is near its minimum and cannot entrain the droplets into the flow.

One of the other SCR system design considerations is how much UWS to carry onboard the vehicle. The length of time that a tank of UWS lasts depends on a number of factors, including the vehicle type and operating parameters. The typical range of UWS consumption rate is about 1%–5% of the fuel consumption rate [12-11, 12-40, 12-41], depending on the engine size, duty cycle, and the NOx conversion needed in the SCR system. Thus, for heavy-duty vehicles with high fuel and UWS consumption rates, the UWS tank is sized so that it will be refilled approximately every other fuel fill. However, for light-duty vehicles (LDVs), refilling the UWS tank is seen as a maintenance item, so the tanks are sized large enough that they are refilled on an oil change interval of 5,000 mi. (8,047 km) to 15,000 mi (24,140 km). This design guideline could mean packaging a tank of about 26.5 L (7 gal.) to about 90 L (23 gal.) on the LDV.

There have been several attempts to carry the urea supply on-board as a solid, instead of using UWS [12-42]. There are several advantages to solid urea, as opposed to the UWS. The first advantage is that solid urea requires less volume for the reductant storage tank, which is a key consideration in vehicles with constrained packaging. Alternatively, using solid urea would enable a longer driving range for a given storage volume. A second advantage is that by having the solid urea directly sublimate, there would be no exhaust cooling effect associated with the evaporation of water from the aqueous solution. Third, with no water, there is no concern for the UWS freezing. These potential advantages aside, all of the solid urea dosing systems that have been proposed are much more complex than the UWS dosing system, which is a major barrier to their adoption.

12.2.2 Mixing Volume

The SCR subsystem needs a certain amount of volume between the UWS injection point and the front face of the SCR monolith. This volume, which includes the exhaust pipe and the SCR inlet volume, needs to provide enough residence time to allow the urea to decompose to ammonia—first by thermolysis (Eq. 12.2), then by hydrolysis (Eq. 12.3)—before it enters the SCR monolith. For many HDDE, the guideline is to include about a meter (about 40 in.) between the urea injector and the transition to the SCR inlet volume.

Another function of the mixing volume is to promote the radial dispersion of the urea or ammonia across the cross-section of the gas flow. The flow field entering the injection point and through to the SCR monolith front face strongly influences the degree of dispersion. A static mixer, described in Section 12.2.3, is often included to expedite the mixing process. The uniformity index (UI) is the standard figure of merit to evaluate how evenly distributed the gas flow or the ammonia concentration is over the front face of the SCR monolith. The UI for the gas flow is defined as follows:

$$UI = 1 - \frac{1}{2A\overline{v_z}} \int_0^{2\pi} \int_0^1 \left(v_z\left(r^*,\theta\right) - \overline{v_z} \right) dr^* d\theta, \tag{12.13}$$

where r^* is the dimensionless radius, r/R, θ is the angular position, $v_z(r^*,\theta)$ is the axial gas velocity as a function of position in the cross-section of the SCR inlet face, $\overline{v_z}$ is the average gas flow velocity over the SCR inlet face, and A is the cross-sectional area. A perfectly uniform flow has a UI of 1.0. Similarly, we can define the UI for concentration as follows:

$$UI = 1 - \frac{1}{2A\overline{c_{NH_3}}} \int_0^{2\pi} \int_0^1 \left(c_{NH_3}\left(r^*,\theta\right) - \overline{c_{NH_3}} \right) dr^* d\theta, \tag{12.14}$$

where $c_{NH3}(r^*,\theta)$ is the ammonia concentration as a function of position and $\overline{c_{NH_3}}$ is the average ammonia concentration.

12.2.3 Static Mixers

The purpose of the static mixer is to provide a uniform ammonia concentration in exhaust gas so that the entire catalyst volume is utilized for the SCR reaction and avoid ammonia bypass. Static mixers are often installed between the UWS injector and the SCR monolith to improve the dispersion of the UWS droplets in the exhaust without adding exhaust pipe length. By directing the bulk exhaust gas flow from the center of the pipe to the walls, static mixers also increase the wall temperature. Packaging space is always a concern in EAS design and the shortest possible distance between the injector and SCR system. A smaller mixing length also helps mitigate heat loss from the exhaust gas. The longer droplet residence time provides an additional opportunity for the urea decomposition process to generate ammonia before the exhaust stream reaches the inlet face of the SCR monolith. It is desirable to evenly distribute the ammonia across the cross-section of the gas flow, for optimal catalyst performance. The heat transfer to the exhaust pipe walls is increased via flow turbulence originating from the vanes.

The static mixers can also serve a second purpose if they are used as injection targets for the UWS. Urea crystals will form on the mixer element as the H_2O from the urea solution evaporates, but then the urea crystals will sublimate into the exhaust. Because the mixer is hotter than the wall, these urea crystals should sublimate and decompose more readily than if formed on the exhaust pipe wall. However, this approach is not without risks, as the urea deposits on the mixer could end up blocking the exhaust gas flow if they steadily accumulate.

While static mixers are effective in evenly distributing the ammonia across the cross-section of the gas flow, they introduce a flow resistance, and so increase the pressure drop of the EAS. This back pressure is one of the main challenges of using the mixer, as it can lead to changes in engine performance, emissions speciation and concentration, and a fuel consumption penalty. The typical value of maximum acceptable backpressure is approximately 12 kPa at 600°C, which often presents a trade-off consideration for the ammonia UI.

12.2.4 **Failure Modes**

Failure modes for the urea injection system include deposits of urea crystals, polyurea oligomers, and ammonium nitrate crystals. As discussed in Section 12.2.3, sometimes the urea will finish decomposing on the front face of the SCR monolith or in the entrance zone of the channels. However, these urea deposits can also block the exhaust gas flow in the exhaust pipe or on the front face of the SCR monolith. Urea crystal deposits are highly likely when the exhaust gas stays below 230°C. During warm-up, urea dosing can start even when the gas at the urea injector reaches 150°C, but on the expectation that hotter gas will follow and help remove any deposits generated during warm-up. As described in Section 12.4, the typical SCR catalysts used for on-road and non-road engines tend to have low activity below 150°C, so the low-temperature threshold is well matched for the two key parts of the SCR subsystem.

At temperatures below 200°C, the urea does not effectively decompose into ammonia following the reaction shown in Eq. 12.2. Partial decomposition of urea has been shown to produce several byproducts such as biuret, cyanuric acid, ammelide, and ammeline [12-43, 12-44]. These byproducts can form solid deposits in the exhaust system and can lower NOx conversion due to insufficient reductant concentrations delivered to the catalyst [12-45, 12-46]. Further, these deposits may face plugging the substrate, occluding flow and adding to the backpressure penalty. Though the temperature is a major factor in urea deposit formation, both the type of catalyst and its surface area may also contribute, as described in Section 12.4 on catalyst formulations and Section 12.5 on substrates.

12.3 **Alternatives to Urea**

UWS is the dominant delivery method for the ammonia reductant used in SCR, with good availability in the US, EU, and Japan. Nevertheless, there have been several alternatives considered to overcome challenges with UWS. Some notable drawbacks to UWS are that it is corrosive, has the high creeping capability, crystallizes easily, and must be free of impurities. In addition, the evaporation of the water can decrease the exhaust gas temperature by 10°C–15°C, which is especially problematic during low-temperature operation. To overcome these difficulties, solid reductants such as metal ammines and ammonium salts have been proposed. The major advantage of such solid reductants is the increased ammonia quantity that may be carried in a given volume. Some of the salts have also been shown to have better low-temperature performance for dealing with cold start emissions. Also, because the ammonia is directly released in a solid to gas transition process, it mitigates wall film and deposit formation. Additional alternatives to the UWS delivery include alternative (direct) precursors of the ammonia reductant, as well as an alternate HC reductant. These alternatives are discussed in more detail in this section.

12.3.1 **Direct Ammonia Dosing Systems**

Safety issues preclude the use of liquid or gaseous ammonia in a pressure vessel. Ammonia vapor can irritate the human respiratory tract at as little as 50 ppm and becomes toxic at high concentrations. Ammonia has a flammability range of 15%–28% in air. However, alternative ammonia direct dosing strategies have been developed. These strategies include systems with the controlled decomposition of a storage material to release ammonia. Several materials have been investigated as alternative sources of direct ammonia injection into the exhaust including solid urea, guanidinium formate, methanamide, metal amines, and ammonium salts. Because solid urea has a decomposition temperature above 125°C and both the guanidinium formate and methanamide are also aqueous solutions they were deemed to be too complex as replacements. Therefore, only the metal ammines (trade name AdAmmine™) and ammonium salts (specifically ammonium carbamate) are considered in this text.

One significant issue with introducing gaseous ammonia directly into the exhaust is the stratification of the mixture. The SCR catalyst requires a uniform distribution of ammonia to the front face of the catalyst to maximize the NOx conversion efficiency over the entire catalyst brick. This is difficult to achieve without sufficient mixing length but can be aided by the use of a static mixer, as described in Section 12.2.3.

12.3.1.1 **Amminex AdAmmine™ Cartridge (Metal Ammines):** The ammonia capacity of the urea solution limits the service interval during vehicle operation and creates the requirement of infrastructure for an additional automotive fluid at fuel stations. Ammonia storage in the solid-state is safer because it reduces the ammonia volatility as compared to liquid ammonia. To be a competitive storage method, solid ammonia storage must be low-cost and have a high volumetric capacity or storage density. In addition, the stored ammonia must be able to be released dynamically upon demand. Ammonia can be adsorbed in a solid inorganic salt matrix as a metal ammine complex [12-47, 12-48]. Low-cost metal halides such as magnesium hexamine chloride

$(Mg(NH_3)_6Cl_2)$ can store more than 50% ammonia by weight. The preparation of such a hexamine salt is done by adsorption and is an exothermic process. Likewise, the release of ammonia from the ammine complex is endothermic, and will, similar to urea decomposition, require energy from the exhaust gas to occur. The magnesium hexamine chloride releases the stored ammonia between 120°C and 350°C.

For automotive applications, powdered materials with low bulk density are not well-suited for on-board storage or introduction into the exhaust system. Amminex A/S has demonstrated that they can produce a solid unit "cartridge" with a density similar to that of a non-porous single crystal [12-48]. The cartridges have been designed to have a pore structure that will let the ammonia escape without a complex diffusion path during its release [12-49]. A 100 g AdAmmine cartridge holds 51.7 g of ammonia [12-48]. To get the equivalent ammonia mass from UWS, one needs 280 g of the solution, which favors the AdAmmine. Thus, to generate the 5 kg of ammonia that is needed for a 30,000 km service interval, one needs about 10 kg or 8 L of AdAmmine compared to about 28 kg or 26 L of UWS. The AdAmmine setup with the cartridge, controllers, and heater is shown in **Figure 12.5** below.

As shown in Figure 12.5, with the Amminex AdAmmine system, there is no need for an injector oriented at the entry to the exhaust pipe. Since there is no water and the ammonia is directly desorbing from the matrix, there is also no need for a mixing section in the EAS. A single tube with multiple exit points can release ammonia at multiple positions across the cross-section of the exhaust pipe to enable mixing.

The appeal of the ASDS system is its simplicity, where the reductant is made available on demand as the ammonia is desorbed from a high-capacity solid storage material. The European Commission Joint Research Center evaluated the ASDS for retrofit purposes and released a study showing that the technology was able to achieve NOx conversions from 25% on the cold-start cycle to 82% for the hot-start cycle in the laboratory and on-road, even for the high emitting Euro 6b diesel vehicles [12-50].

FIGURE 12.5 AdAmmine Ammonia Storage and Delivery System (ASDS) [12-48].

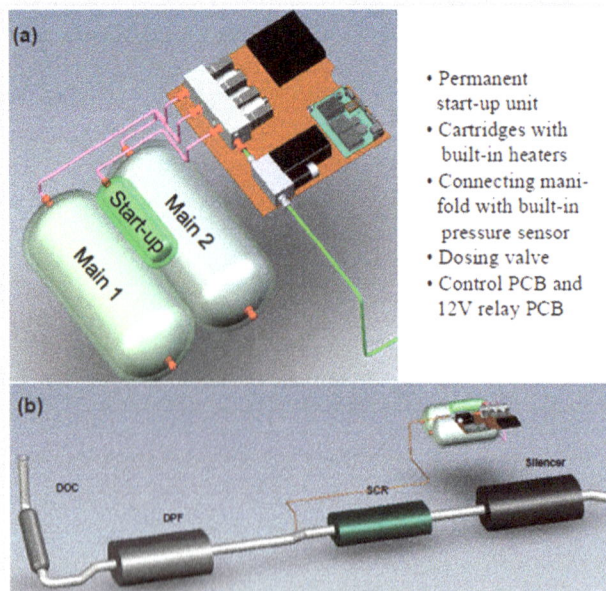

12.3.1.2 **Ammonium Salts:** Ammonium salts are a potential source of ammonia for use in SCR systems. The higher ammonia densities of the salts compared to urea solutions, would allow the use of smaller packaging for use onboard the vehicle, similar to the metal ammines discussed above. Unlike in urea systems, where there is uncertainty in the amount that is converted to ammonia under various exhaust conditions, gaseous injection of ammonia may enable the use of SCR catalysts at low temperatures, even after aging [12-51]. When heated, the ammonium salts decompose into ammonia, leaving no solid residue. One such salt, ammonium carbamate ($NH_4[H_2NCO_2]$), has been demonstrated to have a range of desirable properties for use as an ammonia source for SCR [12-52]. The global equilibrium decomposition of ammonium carbamate is shown in Eq. 12.15. More details on the precise mechanism of the decomposition can be found in [12-53].

$$NH_4COONH_2 = 2\,NH_3 + CO_2 \qquad (12.15)$$

When the ammonium carbamate is heated in a closed system, it decomposes until the equilibrium vapor pressures of the ammonia and CO_2 are reached. Once the system achieves equilibrium, the forward reaction ceases unless the temperature is increased or the product gases are allowed to leave. When the system cools, the process goes in the reverse direction, reforming the original ammonium salt.

Although an ammonium salt system requires less volume to store the ammonia, it requires significant additional volume for heaters and pressure containment [12-52].

12.3.2 Passive Ammonia SCR

Passive ammonia SCR is not a fully passive catalytic converter, such as the systems described in Chapter 14, Passive Aftertreatment Systems. Instead, passive ammonia SCR systems take advantage of the tendency of LNTs or TWCs to form ammonia. Passive ammonia SCR configurations are described in detail in Section 15.5.

12.3.3 Hydrocarbon SCR (HC-SCR)

The SCR of NOx using hydrocarbons (HC-SCR) has been extensively studied as a potential alternative to ammonia or urea-based SCR [12-54, 12-55, 12-56, 12-57], due to the ready availability of fuel-based reductant carried on-board the vehicle. In fact, the HCs already present in diesel exhaust are an easy source of reductant. Due to the selectivity of the reaction, increasing concentrations of HC in the exhaust increases the NOx conversion rate. However, there is a limited supply of exhaust HC as compared to the concentration of NOx and therefore, enrichment of the exhaust with additional fuel is required. Enrichment can be achieved by either a late fuel injection in-cylinder or by separate injection of fuel into the EAS upstream of the catalyst.

While HC-SCR catalysts that make use of the normally occurring HC in the exhaust are passive in nature, those that use HC enrichment are considered to be active systems. These systems require the use of a liquid dosing injection system (an injector, controller, and sensors), but do not require an additional storage tank for a separate liquid reductant. Hydrocarbon enrichment necessarily imparts a fuel consumption penalty since the fuel sprayed into the exhaust system does not add to the motive power of the vehicle. In addition, enrichment may require the use of an extra oxidation catalyst to convert any HC that slip through the HC-SCR. For these reasons, and due to the high efficacy of the urea-SCR systems, HC-SCR has not seen mass deployment.

12.4 **Catalyst Formulations**

A number of base metal and precious metal catalysts, paired with zeolite, alumina (Al_2O_3), silica-alumina (SiO_2-Al_2O_3), titania (TiO_2), and zirconia (ZrO_2) washcoats, have been shown to have lean NOx reduction capability [12-58]. However, the majority of these systems were eliminated as NOx reduction catalysts due to their limited NOx conversion efficiency or required performance temperature. Today, there are three families of SCR catalyst materials that are used commercially in-vehicle applications: vanadia (V_2O_5), iron zeolite (Fe/zeolite), and copper zeolite (Cu/zeolite) [12-59].

The base metal exchanged zeolite catalysts of interest for SCR applications are Fe/zeolite and Cu/zeolite catalysts, which are found to be more thermally stable than vanadia catalyst formulations are at typical diesel exhaust temperatures. Typical temperature ranges over which these SCR catalysts and platinum (Pt) are effective at promoting the SCR reactions are shown in **Table 12.1**.

The zeolite materials exhibit higher NOx conversions at high temperatures without losing catalytic activity. A comparison of the SCR performance of a vanadium catalyst (V-SCR, using V_2O_5), iron zeolite (Fe-SCR, using Fe/zeolite) and copper zeolite (Cu-SCR, using Cu/zeolite) with respect to NO_2:NOx ratio is shown in **Figure 12.6**.

One limitation with the base-metal exchanged zeolites is that they facilitate forming N_2O at higher NO_2:NOx ratios. However, the selective oxidation of ammonia to N_2 is dominant above 350°C. Important features of NOx conversion on zeolites are discussed below by comparing the two most common base-metal zeolite catalysts - iron (Fe) and copper (Cu) zeolites.

TABLE 12.1 Effective temperature ranges for SCR catalysts.

Catalyst	Temperature Range (°C)
Platinum (Pt)	175–275
Vanadia (V_2O_5)	300–450
Copper or iron zeolite	300–600
Base metal	250–425

FIGURE 12.6 NOx conversion with respect to NO_2:NOx ratio for several SCR catalysts at 200°C.

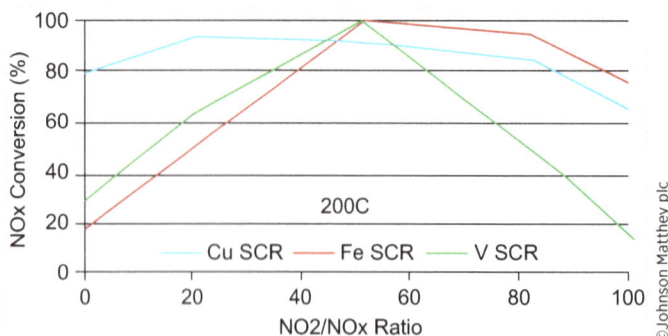

©Johnson Matthey plc

12.4.1 Vanadia Catalysts

Vanadia is the cheapest of the SCR catalyst materials and was the original SCR catalyst. When the ammonia SCR technology was first introduced to treat Japanese power plant emissions, researchers investigated the use of vanadium pentoxide (V_2O_5) catalysts for reducing NO with ammonia [12-60, 12-61]. When it was adapted for vehicle applications nearly 20 years later, HDDEs had exhaust temperatures that were high enough to use these same vanadia catalysts. An early example of treating exhaust gas with an SCR unit was published by Jacob et al. [12-62], in which the exhaust of a 6-cylinder, 14.6 L diesel engine was treated with a DPF and SCR catalyst. This system used a 4:1 ratio of SCR catalyst volume to engine displacement. Extra ammonia was added to ensure NOx conversion of approximately 90% in the SCR during operation. Vanadia-titania (V_2O_5-TiO_2) catalysts have also been used in stationary sources. Typical catalyst compositions are 1%–3% V_2O_5 and 10% tungsten trioxide (WO_3) on a high surface area titania washcoat. Detailed analysis of NOx reduction kinetics with ammonia on vanadia catalysts for SCR applications is discussed by Li et al. [12-63].

One significant issue is that at exhaust gas temperatures greater than 400°C, vanadia catalysts facilitate forming N_2O [12-64]. One of the possible reaction mechanisms leading to the formation of N_2O is given by

$$4\,NH_3 + 4\,NO + 3O_2 \rightarrow 4\,N_2O + 6H_2O. \qquad (12.16)$$

Additionally, vanadia can become vapor and leave the catalytic converter when it is heated above 550°C. Vanadia is poisonous, and the threshold temperature is easily reached in engine exhaust, especially if the SCR system is downstream of the DPF. Thus EPA has given vanadia SCR systems extra scrutiny to ensure that the systems do not present a public health hazard, and so this catalyst is not preferred.

Another disadvantage for the development of the vanadia catalysts is their temperature range for operation. It has been shown that the temperatures on light-duty diesel engines (LDDEs) are too low for effective operation and that the vanadia has poor durability when exposed to temperatures over 550°C. Also, the vanadia catalyst facilitates ammonium nitrate deposit formation when exhaust temperatures are below 200°C. The following ammonium nitrate formation reaction requires NO_2 as the oxidizing reactant, and is assumed to occur at higher NO_2 concentrations in the exhaust:

$$2\,NH_3 + 2\,NO_2 \rightarrow NH_4NO_3 + N_2 + H_2O \qquad (12.17)$$

Zeolite catalysts do not facilitate the reaction in Eq. 12.17. For these reasons, base-metal zeolite formulation catalysts were introduced as an alternative to meet the emission regulations. These catalysts were able to operate in a markedly lower temperature window, and therefore were much more appropriate for vehicle applications.

12.4.2 Iron Zeolites

The first base-metal zeolite catalyst used for SCR applications was iron (Fe) zeolite because they are reasonably stable with good high-temperature performance. Fe/zeolites support the direct oxidation of ammonia to N_2 with a selectivity of almost 100% until approximately 600°C. At higher temperatures, the selectivity toward N_2 formation decreases with increases in NO formation [12-65]. They also exhibit a higher ammonia storage capacity at lower temperatures than vanadia catalysts, as both the Lewis and Brønsted acid sites on the Fe/zeolite catalyst bind ammonia at temperatures up to 350°C; whereas only Brønsted acid sites bind ammonia at higher temperatures.

Apart from the global reactions that occur both on the vanadium and zeolite catalysts, they are prone to individual side reactions in a narrow temperature range.

Fe/zeolites exhibit a distinct NO oxidation capability. The reversible NO oxidation to NO_2 on active zeolites is given by

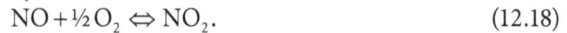

$$NO + \tfrac{1}{2}O_2 \Leftrightarrow NO_2. \tag{12.18}$$

On the Fe/zeolite catalyst, the reaction rate of the slow SCR reaction involving NO_2, shown in Eq. 12.5, is comparable to the rate of the fast SCR reaction involving NO and NO_2, shown in Eq. 12.6, between 300°C and 450°C. The Fe/zeolite has a distinct NO oxidation capability. It can oxidize up to 40% of the NO present at 350°C [12-46, 12-65, 12-66, 12-67]. Therefore, the NO oxidation reaction has been incorporated into the reaction kinetics scheme for Fe/zeolite-based SCR models.

Overall, Fe/zeolite has the best high-temperature performance, with good thermal durability and sulfur tolerance, and only minimal HC poisoning or parasitic ammonia oxidation. Fe/zeolite does not promote undesirable byproducts, unlike the potential for dioxin over Cu/zeolites, described in Section 12.4.3. However, Fe/zeolites have the slowest reaction rate for the standard SCR reaction at low temperatures, and the NO_2 concentration in the exhaust must be managed to improve low-temperature performance.

12.4.3 Copper Zeolites

Copper zeolites (Cu/zeolites) store twice the amount of ammonia than Fe/zeolites do at all temperatures [12-68]. They also have the best low-temperature performance, as they are more active below 350°C than either vanadia or Fe/zeolite catalysts. The higher activity at temperatures below 350°C makes Cu/zeolites a better choice for LDDE applications. In contrast, Fe/zeolites are more active at temperatures greater than 350°C and are thermally stable until 650°C. Thus, Fe/zeolites are an option for HDDE applications. This difference can also be observed in **Figure 12.7**, where a comparison of NOx and ammonia conversions is made between the Cu/zeolite and Fe/zeolite catalysts.

FIGURE 12.7   NOx and ammonia conversion and N_2O formation with respect to temperature for Cu/zeolite and Fe/zeolite SCR catalysts [12-68]

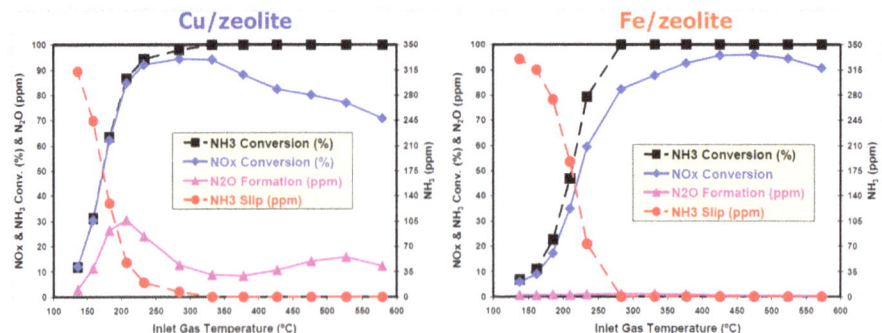

Reprinted from Applied Catalysis B: Environmental 66(3-4):208-216, Kröcher, O., Devadas, M., Elsener, M., Wokaun, A., Söger, N., Pfeifer, M., Demel, Y., and Mussmann, L., "Investigation of the selective catalytic reduction of NO by NH3 on Fe-ZSM5 monolith catalysts.", Copyright 2006, with permission from Elsevier

One significant issue with Cu/zeolite is its sensitivity to sulfur poisoning, which requires occasional high-temperature (>550°C) desulfation to thermally desorb the poison. EPA has also expressed concern about the potential to form dioxin over the copper sites, but this concern is mitigated by the use of smaller zeolite pore sizes to exclude HC from the pores. The significant improvements made to Cu/zeolites in the last decade, such as controlling the pore sizes, have made it the preferred catalyst for high performing NOx conversion systems.

12.5 Substrates

SCR system substrates are usually the same materials used for other EAS subsystems. Thus, cordierite and silicon carbide (SiC) are typical choices, and the specific choice may depend on the expected operating temperatures that the SCR system will see. It is possible to extrude monoliths made entirely of zeolites, though, so for the Cu/zeolite and Fe/zeolite SCR catalysts, it is possible to create a monolith that is also inherently made of just the catalyst. This approach has the potential to save a processing step, although one may still want to coat a zeolite monolith with a catalytic washcoat to make sure that only as much catalyst as can be used is supplied.

References

12-1.   Klimstra, J., "Catalytic Converters for Natural Gas Fueled Engines—A Measurement and Control Problem," SAE Technical Paper 872165, 1987, https://doi.org/10.4271/872165.

12-2.   Radojevic, M., "Reduction of Nitrogen Oxides in Flue Gases," *Nitrogen, the Confer-Ns* (Amsterdam: Elsevier, 1998), 685–689, https://doi.org/10.1016/B978-0-08-043201-4.50097-6.

12-3.   Cobb, D., Glatch, L., Ruud, J., and Snyder, S., "Application of Selective Catalytic Reduction (SCR) Technology for NOx Reduction from Refinery Combustion Sources," *Environmental Progress* **10**, no. 1 (1991): 49–59.

12-4.   Luck, F. and Roiron, J., "Selective Catalytic Reduction of NOx Emitted by Nitric Acid Plants," *Catalysis Today* **4**, no. 2 (1989): 205–218.

12-5.   Lamas, M.I. and Rodríguez, C.G., "Emissions from Marine Engines and NOx Reduction Methods," *Journal of Maritime Research* **9**, no. 1 (2012): 77–81.

12-6.   Bosch, H. and Janssen, F., *Catalytic Reduction of Nitrogen Oxides: A Review on the Fundamentals and Technology* (Amsterdam: Elsevier, 1988), 7.

12-7.   Held, W., Koenig, A., Richter, T., and Puppe, L., "Catalytic NOx Reduction in Net Oxidizing Exhaust Gas," SAE Technical Paper 900496, 1990, https://doi.org/10.4271/900496.

12-8.   Caton, J. and Siebers, D., "Reduction of Nitrogen Oxides in Engine Exhaust Gases by the Addition of Cyanuric Acid," *ASME Journal of Enginering for Gas Turbines Power* **111** (1989): 387–393.

12-9.   Koebel, M., Madia, G., and Elsener, M., "Selective Catalytic Reduction of NO and NO_2 at Low Temperatures," *Catalysis Today* **73**, no. 3–4 (2002): 239–247.

12-10. Nova, I. and Tronconi, E., eds., *Urea-SCR Technology for deNOx After Treatment of Diesel Exhausts* (New York: Springer, 2014), https://doi.org/10.1007/978-1-4899-8071-7.

12-11. Hirata, K., Masaki, N., Ueno, H., and Akagawa, H., "Development of Urea-SCR System for Heavy-Duty Commercial Vehicles," SAE Technical Paper 2005-01-1860, 2005, https://doi.org/10.4271/2005-01-1860.

12-12. Kasab, J.J., de Monte, M., Hadl, K., Noll, H., Mannsberger, S., Graf, G., Theissl, H., and Arnberger, A., "Using Close-Coupled SCR to Meet Ultra-Low NOx Requirements," in *2019 CLEERS*, Ann Arbor, MI, 2019.

12-13. Ebrahimian, V., Nicolle, A., and Habchi, C., "Detailed Modeling of the Evaporation and Thermal Decomposition of Urea-Water Solution in SCR Systems," *AIChE Journal* **58**, no. 7 (2012): 1998–2009.

12-14. Malmberg, S., Votsmeier, M., Gieshoff, J., Söger, N., Mußmann, L., Schuler, A., and Drochner, A., "Dynamic Phenomena of SCR-Catalysts Containing Fe-Exchanged Zeolites–Experiments and Computer Simulations," *Topics in Catalysis* **42**, no. 1 (2007): 33–36.

12-15. Sjövall, H., Olsson, L., Fridell, E., and Blint, R.J., "Selective Catalytic Reduction of NOx with NH_3 over Cu-ZSM-5-The Effect of Changing the Gas Composition," *Applied Catalysis B: Environmental* **64**, no. 3–4 (2006): 180–188.

12-16. Drennan, S., Kumar, G., Quan, S., and Wang, M., "Application of Automatic Meshing to Urea-Water Injection Simulation for Engine Aftertreatment," SAE Technical Paper 2015-01-1057, 2015, https://doi.org/10.4271/2015-01-1057.

12-17. Devarakonda, M., Tonkyn, R., Tran, D., Lee, J.H., and Herling, D., "Modeling Competitive Adsorption in Urea-SCR Catalysts for Effective Low Temperature NOx Control," *ASME 2010 International Mechanical Engineering Congress and Exposition*, American Society of Mechanical Engineers, 2010

12-18. Devarakonda, M., Tonkyn, R., and Lee, J., "Modeling Species Inhibition and Competitive Adsorption in Urea-SCR Catalysts," *SAE Int. J. Fuels Lubr.* **5**(2): 867–874, 2012, https://doi.org/10.4271/2012-01-1295.

12-19. Cavataio, G., Girard, J., Patterson, J., Montreuil, C., Cheng, Y., and Lambert, C., "Performance Characterization of Cu/Zeolite and Fe/Zeolite Catalysts," *Cross-cut Lean Exhaust Emissions Reduction Simulations (CLEERS) Workshop*, Dearborn, MI, 2007.

12-20. Metkar, P.S., Balakotaiah, V., and Harold, M.P., "Experimental Study of Mass Transfer Limitations in Fe-and Cu-Zeolite-Based NH_3-SCR Monolithic Catalysts," *Chemical Engineering Science* **66**, no. 21 (2011): 5192–5203.

12-21. Kowatari, T., Hamada, Y., Amou, K., Hamada, I., Funabashi, H., Takakura, T., Nakagome, K., "A Study of a New Aftertreatment System (1): A New Dosing Device for Enhancing Low Temperature Performance of Urea-SCR," SAE Technical Paper 2006-01-0642, 2006, https://doi.org/10.4271/2006-01-0642.

12-22. Nova, I., Ciardelli, C., Tronconi, E., Chatterjee, D., and Bandl-Konrad, B., "NH_3-NO/NO_2 Chemistry over V-Based Catalysts and Its Role in the Mechanism of the Fast SCR Reaction," *Catalysis Today* **114**, no. 1 (2006): 3–12.

12-23. Nova, I., Ciardelli, C., Tronconi, E., Chatterjee, D., and Weibel, M., "NH_3-NO/NO_2 SCR for Diesel Exhausts after Treatment: Mechanism and Modelling of a Catalytic Converter," *Topics in Catalysis* **42**, no. 1 (2007): 43–46.

12-24. Chatterjee, D., Burkhardt, T., Weibel, M., Tronconi, E., Nova, I., and Ciardelli, C., "Numerical Simulation of NO/NO_2/NH_3 Reactions on SCR-Catalytic Converters: Model Development and Applications," SAE Technical Paper 2006-01-0468, 2006, https://doi.org/10.4271/2006-01-0468.

12-25. Chatterjee, D., Burkhardt, T., Weibel, M., Nova, I., Grossale, A., and Tronconi, E., "Numerical Simulations of Zeolite and V-Based SCR Catalytic Converters," SAE Technical Paper 2007-01-1136, 2007, https://doi.org/10.4271/2007-01-1136.

12-26. Colombo, M., Nova, I., and Tronconi, E., "Detailed Kinetic Modeling of the NH_3-NO/NO_2 SCR Reactions over a Commercial Cu-Zeolite Catalyst for Diesel Exhausts after Treatment," *Catalysis Today* **197**, no. 1 (2012): 243–255.

12-27. Tronconi, E., Nova, I., Ciardelli, C., Chatterjee, D., and Weibel, M., "Redox Features in the Catalytic Mechanism of the "standard" and "fast" NH_3-SCR of NOx over a V-Based Catalyst Investigated by Dynamic Methods," *Journal of Catalysis* **245**, no. 1 (2007): 1–10.

12-28. Ciardelli, C., Nova, I., Tronconi, E., Chatterjee, D., and Bandl-Konrad, B., "A 'Nitrate Route' for the Low Temperature "Fast SCR" Reaction over a V_2O_5–WO_3/TiO_2 Commercial Catalyst," *Chemical Communications* **23** (2004): 2718–2719.

12-29. Grossale, A., Nova, I., and Tronconi, E., "Role of Nitrate Species in the "NO_2-SCR" Mechanism over a Commercial Fe-zeolite Catalyst for SCR Mobile Applications," *Catalysis Letters* **130**, no. 3–4 (2009): 525–531.

12-30. Ito, E., Hultermans, R., Lugt, P., Burgers, M., Van Bekkum, H., and Van den Bleek, C., "Selective Reduction of NOx with Ammonia over Cerium Exchanged Zeolite Catalysts: Towards a Solution for an Ammonia Slip Problem," *Studies in Surface Science and Catalysis* (Amsterdam: Elsevier, 1995), 661–673.

12-31. York, A.P., Watling, T.C., Cox, J.P., Jones, I.Z., Walker, A.P., Blakeman, P.G., Ilkenhans, T., Allansson, R., and Lavenius, M., "Modeling an Ammonia SCR DeNOx Catalyst: Model Development and Validation," SAE Technical Paper 2004-01-0155, 2004, https://doi.org/10.4271/2004-01-0155.

12-32. Heck, R.M. and Farrauto, R.J., *Catalytic Air Pollution Control: Commercial Technology* (New York: Van Nostrand Reinhold, 1995).

12-33. Khair, M.K. and Majewski, W.A., *Diesel Emissions and Their Control* (Warrendale, PA: SAE International, 2006).

12-34. de Monte, M., Hadl, K., Noll, H., and Mannsberger, S., "SCR Control Strategies with Multiple Reduction Devices for Lowest NOx Emissions," *SAE Heavy-Duty Diesel Emissions Control Symposium*, 2018.

12-35. Harris, T.M., McPherson, K., Rezaei, R., Kovacs, D., Rauch, H., and Huang, Y., "Modeling of Close-Coupled SCR Concepts to Meet Future Cold Start Requirements for Heavy-Duty Engines," SAE Technical Paper 2019-01-0984, 2019, https://doi.org/10.4271/2019-01-0984.

12-36. Gabrielsson, P.L., "Urea-SCR in Automotive Applications," *Topics in Catalysis* **28**, no. 1–4 (2004): 177–184.

12-37. Devarakonda, M., Parker, G., Johnson, J.H., Strots, V., and Santhanam, S., "Model-Based Estimation and Control System Development in a Urea-SCR aftertreatment System," *SAE Int. J. Fuels Lubr.* **1**, no. 1 (2009): 646–661, https://doi.org/10.4271/2008-01-1324.

12-38. Gieshoff, J., Pfeifer, M., Schäfer-Sindlinger, A., Spurk, P., Garr, G., Leprince, T., and Crocker, M., "Advanced Urea SCR Catalysts for Automotive Applications," SAE Technical Paper 2001-01-0514, 2001, https://doi.org/10.4271/2001-01-0514.

12-39. Gekas, I., Gabrielsson, P., Johansen, K., Nyengaard, L., and Lund, T., "Urea-SCR Catalyst System Selection for Fuel and PM Optimized Engines and a Demonstration of a Novel Urea Injection System," SAE Technical Paper 2002-01-0289, 2002, https://doi.org/10.4271/2002-01-2089.

12-40. Seher, D.H., Reichelt, M., and Wickert, S., "Control Strategy for NOx-Emission Reduction with SCR," SAE Technical Paper 2003-01-3362, 2003, https://doi.org/10.4271/2003-01-3362.

12-41. Shost, M., Noetzel, J., Wu, M.-C., Sugiarto, T., Bordewyk, T., Fulks, G., and Fisher, G.B., "Monitoring, Feedback and Control of Urea SCR Dosing Systems for NOx Reduction: Utilizing an Embedded Model and Ammonia Sensing," SAE Technical Paper 2008-01-1325, 2008, https://doi.org/10.4271/2008-01-1325.

12-42. Koebel, M., Elsener, M., and Madia, G., "Recent Advances in the Development of Urea-SCR for Automotive Applications," SAE Technical Paper 2001-01-3625, 2001, https://doi.org/10.4271/2001-01-3625.

12-43. Xu, L., Watkins, W., Snow, R., Graham, G., McCabe, R., Lambert, C., and Carter III, R., "Laboratory and Engine Study of Urea-Related Deposits in Diesel Urea-SCR Aftertreatment Systems," SAE Technical Paper 2007-01-1582, 2007, https://doi.org/10.4271/2007-01-1582.

12-44. Schaber, P.M., Colson, J., Higgins, S., Thielen, D., Anspach, B., and Brauer, J., "Thermal Decomposition (pyrolysis) of Urea in an Open Reaction Vessel," *Thermochimica Acta* **424**, no. 1–2 (2004): 131–142.

12-45. Kowatari, T., Hamada, Y., Amou, K., Hamada, I., Funabashi, H., Takakura, T., and Nakagome, K., "A Study of a New Aftertreatment System (1): A New Dosing Device for Enhancing Low Temperature Performance of Urea-SCR," SAE Technical Paper 2006-01-0642, 2006, https://doi.org/10.4271/2006-01-0642.

12-46. Nishioka, A., Sukegawa, Y., Katogi, K., Mamada, H., Kowatari, T., Mukai, T., and Yokota, H., "A Study of a New Aftertreatment System (2): Control of Urea Solution Spray for Urea-SCR," SAE Technical Paper 2006-01-0644, 2006, https://doi.org/10.4271/2006-01-0644.

12-47. Walker, G., *Solid-State Hydrogen Storage: Materials and Chemistry* (Amsterdam: Elsevier, 2008).

12-48. Johannessen, T., Schmidt, H., Svagin, J., Johansen, J., Oechsle, J., and Bradley, R., "Ammonia Storage and Delivery Systems for Automotive NOx Aftertreatment," *SAE Technical Number 2008-01-1027*, 2008, https://doi.org/10.4271/2008-01-1027.

12-49. Hummelshøj, J.S., Sørensen, R.Z., Kustova, M.Y., Johannessen, T., Nørskov, J.K., and Christensen, C.H., "Generation of Nanopores during Desorption of NH_3 from $Mg(NH_3)_6Cl_2$," *Journal of the American Chemical Society* **128**, no. 1 (2006): 16–17.

12-50. Giechaskiel, B., Suarez-Bertoa, R., Lahde, T., Clairotte, M., Carriero, M., Bonnel, P., and Maggiore, M., "Emissions of a Euro 6b Diesel Passenger Car Retrofitted with a Solid Ammonia Reduction System," *Atmosphere* **10**, no. 4 (2019): 180

12-51. Cavataio, G., Jen, H.-W., Warner, J.R., Girard, J.W., Kim, J.Y., and Lambert, C.K., "Enhanced Durability of a Cu/Zeolite Based SCR Catalyst," *SAE Int. J. Fuels Lubr.* **1**(1): 477–487, 2009, https://doi.org/10.4271/2008-01-1025.

12-52. Fulks, G., Fisher, G.B., Rahmoeller, K., Wu, M.-C., D'Herde, E., and Tan, J., "A Review of Solid Materials as Alternative Ammonia Sources for Lean NOx Reduction with SCR," SAE Technical Paper 2009-01-0907, 2009, https://doi.org/10.4271/2009-01-0907.

12-53. Ramachandran, B., Halpern, A.M., and Glendening, E.D., "Kinetics and Mechanism of the Reversible Dissociation of Ammonium Carbamate: Involvement of Carbamic Acid, *The Journal of Physical Chemistry A* **102**, no. 22 (1998): 3934–3941.

12-54. Burch, R., Breen, J., and Meunier, F., "A Review of the Selective Reduction of NOx with Hydrocarbons under Lean-Burn Conditions with Non-Zeolitic Oxide and Platinum Group Metal Catalysts," *Applied Catalysis B: Environmental* **39**, no. 4 (2002): 283–303.

12-55. Amiridis, M.D., Zhang, T., and Farrauto, R.J., "Selective Catalytic rReduction of Nitric Oxide by Hydrocarbons," *Applied Catalysis B: Environmental* **10**, no. 1–3 (1996): 203–227.

12-56. Schmieg, S.J., Blint, R.J., and Deng, L., "Control Strategy for the Removal of NOx from Diesel Engine Exhaust Using Hydrocarbon Selective Catalytic Reduction," *SAE Int. J. Fuels Lubr.* **1**(1): 1540–1552, 2009, https://doi.org/10.4271/2008-01-2486.

12-57. Thomas, J.F., Lewis, S.A., Bunting, B.G., Storey, J.M., Graves, R.L., and Park, P.W., "Hydrocarbon Selective Catalytic Reduction Using a Silver-Alumina Catalyst with Light Alcohols and Other Reductants," SAE Technical Paper 2005-01-1082, 2005, https://doi.org/10.4271/2005-01-1082.

12-58. Smedler, G., Ahlström, G., Fredholm, S., Frost, J., Lööf, P., Marsh, P., Walker, A., and Winterborn, D., "High Performance Diesel Catalysts for Europe beyond 1996," *SAE Paper Number 950750*, 1995, https://doi.org/10.4271/950750.

12-59. Walker, A.P., "Current and Future Trends in Catalyst-Based Emission Control System Design," *SAE Heavy Duty Diesel Emission Control Symposium*, Gothenburg, Sweden, 2012.

12-60. Inomata, M., Miyamoto, A., and Murakami, Y., "Mechanism of the Reaction of NO and ammonia on Vanadium Oxide Catalyst in the Presence of Oxygen under the Dilute Gas Condition," *Journal of Catalysis* **62**, no. 1 (1980): 140–148, 10.1016/0021-9517(80)90429-7.

12-61. Inomata, M., Miyamoto, A., Ui, T., Kobayashi, K., and Murakami, Y., "Activities of Vanadium Pentoxide/Titanium dioxide and Vanadium Pentoxide/Aluminum Oxide Catalysts for the Reaction of Nitric Oxide and Ammonia in the Presence of Oxygen," *Industrial & Engineering Chemistry Product Research and Development* **21**, no. 3 (1982): 424–428, doi:10.1021/i300007a014.

12-62. Jacob, E., Emmerling, G., Döring, A., Graf, U., Harris, M., and Hupfeld, B., "Reduction of NOx from HD Diesel Engines with Urea SCR Compact Systems (Controlled Diesel Catalyst)," *19th International Vienna Motor Symposium*, Vienna, 1998, 10.13140/2.1.3927.9368.

12-63. Li, J., Chang, H., Ma, L., Hao, J., and Yang, R.T., "Low-Temperature Selective Catalytic Reduction of NOx with NH_3 over Metal Oxide and Zeolite Catalysts—A Review," *Catalysis Today* **175**, no. 1 (2011): 147–156.

12-64. Kondratenko, V.A., Bentrup, U., Richter, M., Hansen, T.W., and Kondratenko, E.V., "Mechanistic Aspects of N_2O and N_2 Formation in NO Reduction by NH_3 over Ag/Al_2O_3: The Effect of O_2 and H_2," *Applied Catalysis B: Environmental* **84**, no. 3-4 (2008): 497–504.

12-65. Devadas, M., Kröcher, O., Elsener, M., Wokaun, A., Söger, N., Pfeifer, M., Demel, Y., and Mussmann, L., "Influence of NO_2 on the Selective Catalytic Reduction of NO with Ammonia over Fe-ZSM5," *Applied Catalysis B: Environmental* **67**, no. 3–4 (2006): 187–196.

12-66. Xu, L., McCabe, R.W., and Hammerle, R.H., "NOx Self-Inhibition in Selective Catalytic Reduction with Urea (Ammonia) over a Cu-Zeolite Catalyst in Diesel Exhaust," *Applied Catalysis B: Environmental* **39**, no. 1 (2002): 51–63.

12-67. Kröcher, O., Devadas, M., Elsener, M., Wokaun, A., Söger, N., Pfeifer, M., Demel, Y., and Mussmann, L., "Investigation of the Selective Catalytic Reduction of NO by ammonia on Fe-ZSM5 Monolith Catalysts," *Applied Catalysis B: Environmental* **66**, no. 3–4 (2006): 208–216.

12-68. Cavataio, G., Girrd, J., Patterson, J.E., Montreuil, C., Cheng, Y., and Lambert, C.K., "Performance Characterization of Cu/Zeolite and Fe/Zeolite Catalysts for the SCR of NOx," in *10th CLEERS Workshop*, Dearborn, MI, 2007.

CHAPTER 13

Ammonia Slip Catalysts

Slip slidin' away
Slip slidin' away
You know the nearer your destination
The more you're slip slidin' away

—Paul Simon, "Slip Slidin' Away"

Ammonia slip catalysts (ASC) supplement the function of selective catalytic reduction (SCR) system in an exhaust aftertreatment system (EAS). ASC are known by many names including ammonia oxidation catalysts (AOC), AmOx catalysts, or ammonia clean-up catalysts (CUC). The purpose of the ASC is to remove any residual ammonia (NH_3) from the exhaust gas leaving the SCR system. As shown in **Figure 13.1**, where the ASC is labeled "NH3," and in the examples in Section 7.5, the SCR system and ASC are positioned after all the other EAS components. Thus, the ASC is the last catalytic converter in the EAS before the exhaust gas exits the tailpipe. From reading this chapter, one will learn about the main functions of the ASC, potential side reactions, typical configurations, and materials choices.

As discussed in Chapter 12, Selective Catalytic Reduction Systems, the ammonia intended to reduce nitrogen oxides (NOx) to nitrogen gas (N_2) on the SCR catalysts may exit the SCR unit instead. This ammonia slip typically arises from one of the three causes, as discussed by Hünnekes, et al. [13-2]. Ammonia slip is a problem both because it is regulated and because humans can smell it at a concentration as low as 25 ppm. Ammonia gas can cause respiratory distress when the concentration is above 100 ppm, but that level would be unusual from a normally functioning EAS. The regulatory target in the United States (US) for on-road applications is 10 ppm average at the tailpipe of an engine and EAS at its full useful life as measured over the applicable regulatory transient cycle: either FTP-75 or heavy-duty FTP transient cycle (HD FTP) (see Section 4.1.1 or Section 4.2.4, respectively). Likewise, China requires that non

FIGURE 13.1 Engine and EAS architecture, showing ASC (as "NH$_3$") with other catalytic converters and sensors. (Figure 2 in [13-1].)

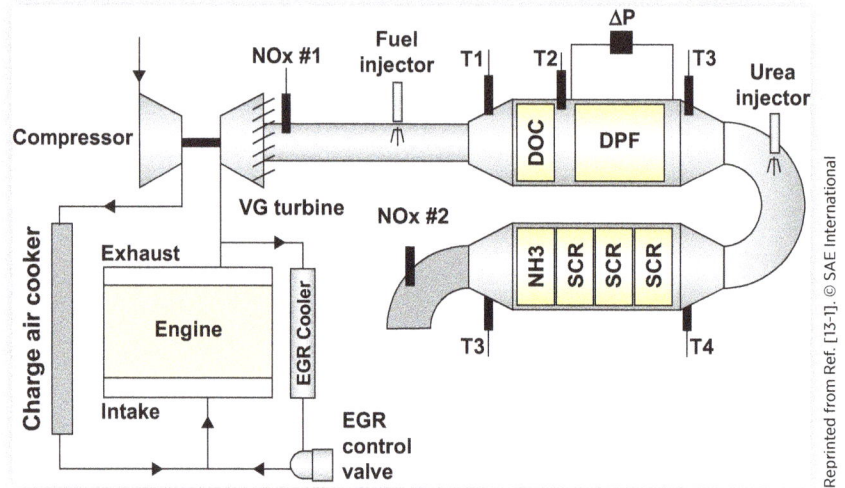

Reprinted from Ref. [13-1]. © SAE International

road applications emit not more than 25 ppm ammonia on average over the relevant test cycles, especially the Non Road Transient Cycle described in Section 4.2.7.

Ammonia slip has several causes, which are described below. First, the SCR reaction may not consume all of the ammonia supplied [13-2]. For example, the SCR reaction may be relatively slow, leading to a low conversion of ammonia and NOx in the SCR system. Alternatively, excess ammonia may be supplied to the SCR unit compared to the stoichiometric quantity needed to reduce NOx to N$_2$, that is, the ammonia to NOx ratio (ANR) is greater than one. This excess ammonia will pass through the SCR unit unreacted.

Second, the ammonia storage capacity of the SCR catalyst is a function of temperature, and the capacity decreases as the SCR temperature increases. Thus, if there is a high temperature excursion in the SCR system, it can cause ammonia to desorb from the catalyst and exit the SCR system with the exhaust gas [13-2].

Third, the incomplete conversion of injected urea to ammonia upstream of the SCR unit can generate ammonia slip [13-2]. This slip can occur either because the urea finishes decomposing downstream of the SCR unit, or because urea-based deposits such as urea crystals later decompose to form ammonia in excess of what the SCR dosing controller demands.

There are various strategies to mitigate ammonia slip, including modifying the SCR catalyst to improve NOx conversion; improving the distribution of ammonia over the SCR monolith inlet face; and tightening control over the ANR through model-based, predictive control for the urea solution dosing strategy. Depending upon the SCR system placement within the EAS, active thermal management strategies may also be needed.

According to Girard et al. [13-3], one of the reasons for mitigating ammonia slip from the SCR is the risk that it could form nitrous oxide (N$_2$O) on an oxidation catalyst downstream. For example, as shown in the schematic of the Power Stroke™ EAS in Figure 7.23, Ford places the SCR system upstream of the catalyzed diesel particulate filter (CDPF), which is able to oxidize ammonia to N$_2$O, nitrogen monoxide (NO), or nitrogen dioxide (NO$_2$). Nonetheless, it is counter-productive for the ASC to generate NO or NO_2, much less N_2O, since it is part of a system designed to remove NOx from the exhaust [13-4].

Girard et al. [13-3] noted that "the greatest concern over NH_3 slip occurs at low temperatures [under 200°C] during cold starts, and at moderate exhaust temperatures (250°C–350°C) with both moderate and very high space velocities." Below 350°C the catalysts are selective for N_2O instead of N_2, and above 350°C the catalysts are at a risk of oxidizing ammonia to NOx.

There has been some discussion about whether an ASC device needs to be included in the EAS at all. For example, Georgiadis et al. [13-5] discuss an approach to meet Euro 6 standards with a combined SCR and diesel particulate filter (SDPF) system* that excludes an ASC. As discussed in Section 3.2.3, the European Union (EU) tightened the on-cycle tailpipe NOx limits from the earlier Euro 5 standard. Euro 6 has forced the inclusion of SCR systems and introduces the Real Driving Emissions (RDE) standards, which are described in Section 4.1.10. The ASC-free EAS configuration requires tight dosing control of the urea-water solution so that the ammonia and NOx conversions are high in the SCR system. Georgiadis et al. [13-5] implemented an adaptive dosing control algorithm to demonstrate how one could mitigate ammonia emissions at the tailpipe. This approach has been demonstrated, but it is not clear that the approach is robust enough yet to meet regulatory requirements in markets other than the EU.

By contrast, Sukumar et al. [13-6] noted that an advantage of adding the ASC downstream of the SCR unit is that it supports a more aggressive ammonia dosing strategy used to improve NOx conversion in the SCR, since any residual ammonia will be converted before being released into the atmosphere. The challenge, then, is to maximize the selectivity for N_2 in the ASC as part of an optimized SCR-ASC system that minimizes ammonia and NOx emissions.

13.1 Ammonia Slip Catalyst Function

The purpose of the ASC is to minimize ammonia slip by converting any ammonia leaving the tailpipe into N_2 and water (H_2O). It does so by reacting ammonia with residual oxygen (O_2) in the exhaust gas. Because the ASC promotes oxidation, it will help oxidize any residual carbon monoxide (CO) that may enter the ASC. This section covers the main function of the ASC within the EAS; side reactions, both desirable and otherwise; and other considerations.

13.1.1 Oxidizing Ammonia

The primary function of the ASC is to remove ammonia from the exhaust gas leaving an SCR unit. This function is especially important when the SCR unit is otherwise the last part of the EAS before the tailpipe, but it is also useful in EAS configurations that include a smaller, close-coupled SCR unit near the exhaust manifold or turbine outlet.

The preferred ammonia oxidation reaction is

$$4\,NH_3 + 3O_2 = 2N_2 + 6H_2O, \tag{13.1}$$

* SCR-DPF, or SDPF, systems are discussed in Section 15.2, Combined SCR and DPF.

in which the ammonia is converted to N_2 and water. The direct oxidation of ammonia per Eq. 13.1 is not the most likely pathway to N_2, despite being thermodynamically feasible. Several potential side reactions are possible instead, including the following that produce the potent greenhouse gas (GHG) N_2O or the criteria pollutant NO from the ammonia:

$$2NH_3 + 2O_2 = N_2O + 3H_2O \qquad (13.2)$$

or

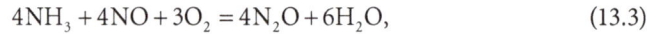

$$4NH_3 + 4NO + 3O_2 = 4N_2O + 6H_2O, \qquad (13.3)$$

and

$$4NH_3 + 5O_2 = 4NO + 6H_2O. \qquad (13.4)$$

All of the oxidation reactions are thermodynamically favorable, as the oxidation of ammonia to N_2 in Eq. 13.1 has a Gibbs free energy change of −327 kJ per mole of ammonia oxidized; the oxidation to N_2O in Eq. 13.2 or 13.3, −275 kJ/mol; and the oxidation to NO in Eq. 13.4, −241 kJ/mol [13-7]. Smith et al. [13-8] did bench reactor experiments that suggest that N_2O is formed via Eq. 13.3, as shown in **Figure 13.2**, and Shrestha et al. [13-4] likewise indicate that N_2O is formed by a reaction between N atoms and NO molecules that are adsorbed onto the surfaces of Pt crystallites. In the experiment in Smith et al. [13-8], the space velocity (SV) is 100,000 h^{-1}, the ammonia inlet concentration is 100 ppm, and the NO is 50 ppm.

A good ASC option will have high catalytic activity and good selectivity for N_2 across the range of operating states—temperature, composition, and flow—present in the SCR system outlet flow [13-3, 13-7]. It will also need to have a good durability and be robust to high temperature excursions while being exposed to the water and carbon dioxide (CO_2) of diesel exhaust.

Platinum (Pt) supported on alumina (Al_2O_3) is very effective at oxidizing ammonia, but has poor selectivity for N_2, and instead tends to form N_2O and NO per the reactions in Eqs. 13.3 and 13.4 [13-7]. The surface-bound N species come from dissociated, surface-bound ammonia. Therefore, Pt/Al_2O_3 on its own is not the preferred choice for treating ammonia slip.

FIGURE 13.2 Ammonia slip catalyst functional evaluation. (Figure 1 in [13-8].)

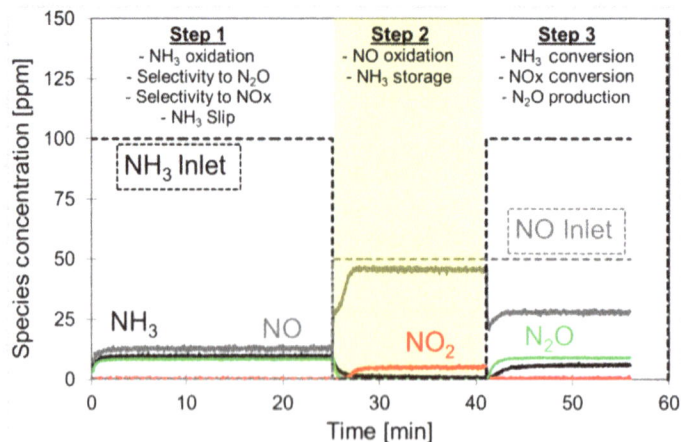

Girard et al. [13-3] did bench reactor experiments to evaluate ammonia oxidation performance over a typical Pt oxidation catalyst for a CDPF. They found that this typical oxidation catalyst had a moderate selectivity for N_2O over N_2 below 350°C, and a very high selectivity for NOx overall above that temperature. Lowering the Pt catalyst loading did not significantly improve the N_2 yield. Their work reinforces the hypothesis that N_2O forms from a mixture of NO and N on a Pt oxidation catalyst.

Kamasamudram et al. [13-7] also did bench reactor experiments to evaluate ammonia oxidation over an ASC as well as over a diesel oxidation catalyst (DOC) sample monolith and an SCR sample monolith. These experiments were under steady-state conditions at several temperatures, and compared ammonia conversion (x_{NH3}) and the relative yields of N_2O and N_2, which are labeled "selectivity" in **Figure 13.3**. For the results shown in Figure 13.3, the reaction conditions were as follows [13-7]:

- DOC: inlet $y_{NH3} \approx 100$ ppm and SV = 100,000 h^{-1}
- ASC: inlet $y_{NH3} \approx 100$ ppm and SV = 100,000 h^{-1}
- SCR: inlet $y_{NH3} \approx 200$ ppm and SV = 40,000 h^{-1}

The DOC, with its relatively high platinum group metal (PGM) catalyst loading, strongly favors NO formation as expected, as shown in **Figure 13.3(b)**. The SCR catalyst, which is an iron-exchanged zeolite (Fe/zeolite) catalyst with no PGM content, has a high selectivity for N_2 and low selectivity for N_2O when it oxidizes ammonia, even above

FIGURE 13.3   Steady-state ammonia conversion and product yield (labeled selectivity) over DOC, ASC, and SCR catalysts. (Figure 2 in [13-7].)

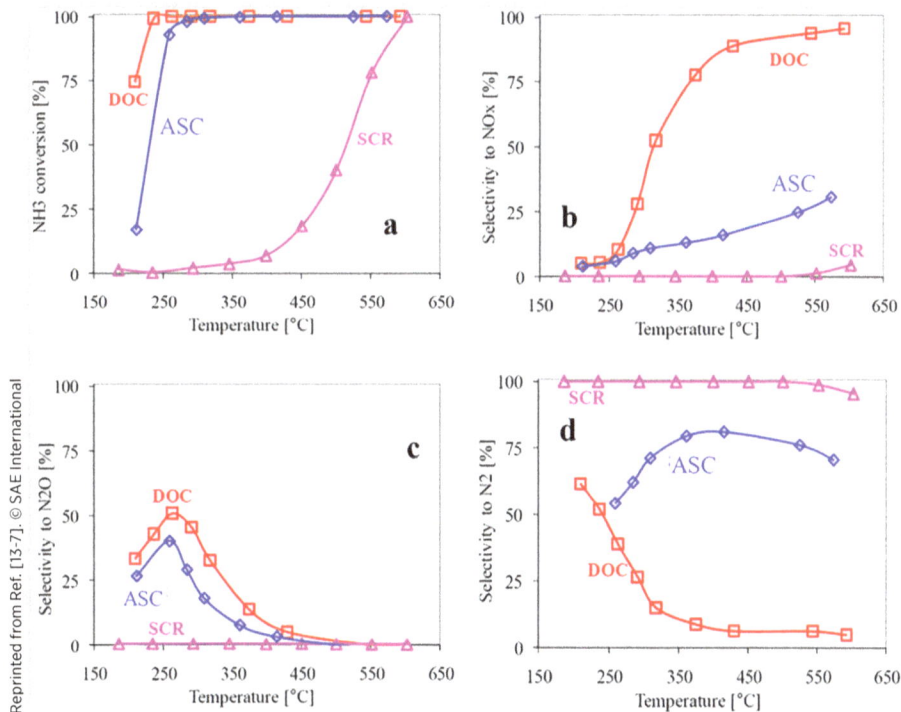

FIGURE 13.4 Ammonia storage on DOC, ASC, and SCR catalysts. (Figure 6 in [13-7].)

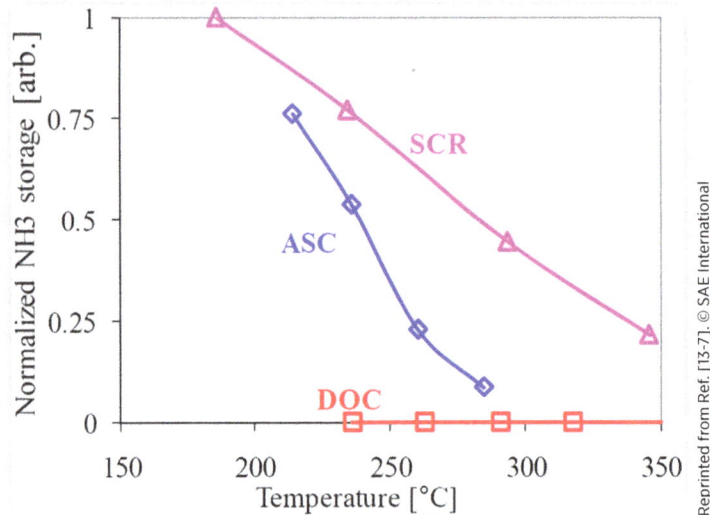

500°C or so. This selectivity for N_2 arises from the residual ammonia stored on the SCR catalyst—some of the ammonia is oxidized to NOx on the SCR catalyst, but then the NOx reacts with the stored ammonia to form N_2 per the SCR reactions [13-7].* The PGM loading in the ASC was "substantially lower than a DOC." The Fe/zeolite in the ASC unit can also to store a modest amount of ammonia, as shown in **Figure 13.4**; this stored ammonia improves the selectivity for N_2 over NOx compared to the DOC.

To get the best selectivity for N_2, modern ASC use a bi-functional catalyst that combines an oxidation catalyst, such as Pt/Al_2O_3, with a reduction catalyst, such as the copper-exchanged zeolite (Cu/zeolite) or Fe/zeolite used for SCR [13-4]. The details of the bi-functional ASC configuration are described in Sections 13.2 and 13.3, since it affects both the catalyst choice and how they are placed within the ASC unit.

Commercially available ASC in use around 2007 typically used Pt/Al_2O_3 oxidation catalysts only. They were found to have a high ammonia conversion when above 225°C at low SV of 80,000 h^{-1}, and had good selectivity for N_2 [13-3]. At higher SV or at lower temperatures, however, the ammonia conversion was lower, and thus improving the conversion was therefore a target of the work by Girard et al. [13-3].

The ammonia oxidation reaction on the Cu/zeolite catalyst Cu/SSZ-13 appears to be consistent with the two-site model. It is not a single reaction step consistent with an Arrhenius reaction rate [13-9]. The results by Olsson et al. [13-9], indicate that at low temperatures, ammonia is stored and then oxidized on the S1 storage sites in the Cu/SSZ-13 structure, and at high temperatures, on the S2 storage sites. Zheng et al. [13-10] also evaluated ammonia oxidation on Cu/SSZ-13. In addition, the ammonia conversion over Cu/SSZ-13 appeared inhibited by the presence of propene (C_3H_6), as the temperature for the high conversion zone is raised by 50°C when propene is present, as shown in **Figure 13.5**. The presence of hydrogen gas (H_2) and CO did not seem to significantly influence the rate of ammonia oxidation over this SCR catalyst. For the results in Figure 13.5 the propene concentration is 500 ppm, whereas

---

* The SCR reactions are discussed in detail in Chapter 12.

FIGURE 13.5 Ammonia conversion by oxidation over Cu/SSZ-13 in the presence of 500 ppm C_3H_6 or 1 vol% CO + 1 vol% H_2. (Figure 5 in [13-10].)

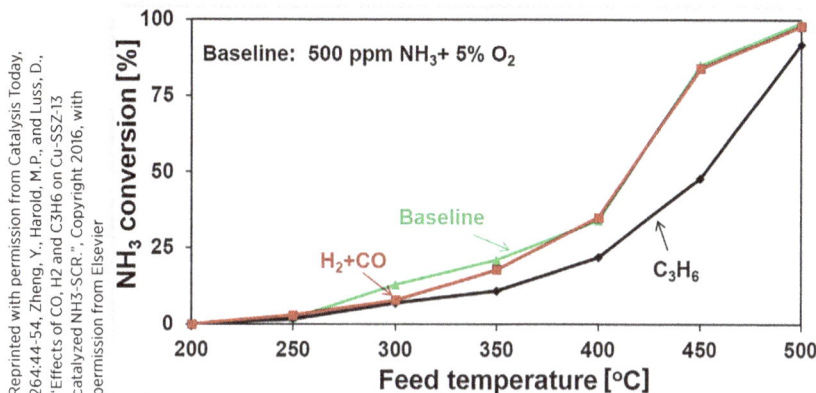

the H_2 and CO are both at 1 vol%. Although ammonia oxidation is an undesirable side reaction over an SCR catalyst, in the ASC context it is a feature because the ammonia oxidation shows a high selectivity for N_2.

13.1.2 Other Operational Considerations

The ASC is typically the very last catalytic converter in the EAS. Therefore, its warm-up profile is affected by the thermal mass of the full, preceding EAS. It is also difficult to control independently, since it immediately follows the SCR system in the EAS, either in its own monolith or as a zone coating on the final SCR monolith. A bi-functional ASC catalyst that combines both ammonia oxidation and SCR function to improve the selectivity for N_2 (S_{N_2}) must be sufficiently hot for the catalyst to light off and sustain the desired chemistry. Similarly, in configurations where the ASC follows an SDPF unit, the ASC must also be able to handle the hot exhaust during filter regeneration events. Note that the ASC typically has a small overall volume of about half the engine displacement.

To improve the system performance, model predictive control (MPC) has been investigated by Krejza et al. [13-11] to manage the SCR and ASC in an EAS for light-duty vehicles (LDV) subject to Euro 6 and RDE limits. The MPC is used to minimize the ammonia slip by continuously estimating the capacity of the ammonia reservoir in the SCR system, how much ammonia is currently stored there, and the estimated demand based on the current engine operating point.

13.2 Catalyst Formulations

ASC typically use PGM catalysts, especially Pt/Al_2O_3[13-8], for the oxidation function. Modern ASC often combine the oxidation catalyst with an SCR catalyst, such as Fe/zeolite or Cu/zeolite, to improve the selectivity for N_2[13-8]. As shown by Girard et al. [13-3] there can be some variability in the performance of an ASC. The selectivity of N_2 over N_2O, NO, or NO_2 is a particularly important metric for ASC catalysts, and one that is subject to some variability based on the catalyst formulation and related factors.

Olsson et al. [13-9] performed experiments to support the developing models of ammonia oxidation over SCR catalysts, especially Cu/SSZ-13. These experiments spanned the catalyst temperature range of 150°C–600°C with the following reactant concentrations: 400 ppm ammonia, 8 vol% O_2, and 5 vol% H_2O. As mentioned in Section 13.1.1, the results suggest that the ammonia oxidation reaction is not one single step consistent with an Arrhenius reaction rate but is consistent with a two-site model. Their results indicate that at low temperatures, ammonia is stored and then oxidized on the S1 storage sites within the Cu/SSZ-13 structure. The S1 sites lie within the six-membered rings in the SSZ-13 structure. At high temperatures, ammonia is stored and oxidized on S2 storage sites, which includes active sites within the cages or in a copper oxide (Cu_xO_y) form. The S1 and S2 storage sites include both copper and Brønsted acid sites that bind the ammonia, although the binding is stronger on the S2 sites. For this two-site model, the ammonia oxidation reaction on the S1 sites is estimated to have an activation energy E_{S1} of 72.7 kJ/mol, and on the S2 sites, an activation energy E_{S2} of 195.0 kJ/mol.

Shrestha et al. [13-4] investigated the selectivity of an ASC that used a bi-functional catalyst that combines a Pt/γ-Al_2O_3 oxidation catalyst with an Fe/ZSM-5 catalyst for SCR. As discussed above, Pt/γ-Al_2O_3 tends to form NOx instead of N_2 when catalyzing ammonia oxidation. Using the Fe/ZSM-5 means that the ASC catalyst can use the NOx formed on the Pt/γ-Al_2O_3 catalyst in an SCR reaction with the ammonia, which greatly improves the selectivity for N_2 in the ASC. In their study Shrestha et al. [13-4] placed a washcoat layer of Pt/γ-Al_2O_3 containing 10 g/ft³ Pt on the monolith and then placed a washcoat with Cu/zeolite or Fe/zeolite over it. The intent is for some of the ammonia entering the ASC to adsorb on the SCR catalyst and for some to diffuse through to the Pt/γ-Al_2O_3 layer, where it will form NOx and react with the stored ammonia as it diffuses back into the main channel.

Increasing the SV to 285,000 h^{-1} from 66,000 h^{-1} caused the asymptotic ammonia conversion at high temperature to drop from 100%. Shrestha et al. [13-4] believe that the ammonia is the rate-limiting reactant, and therefore at high SV the mass transport phenomena do not provide enough time for the ammonia to get to the catalyst and react thoroughly.

Shrestha et al. [13-4] evaluated both a dual-layer, bi-functional catalyst and a mixed bi-functional catalyst. They report some benefits from the mixed catalyst, especially with reducing the yield of N_2O from the ASC at lower temperatures, although the two variants are otherwise broadly similar in performance. Fe/ZSM-5 has a low activity for reducing N_2O at these lower temperatures, leading to the results shown in **Figure 13.6**.

Later, Shrestha et al. [13-12] looked at a bi-functional mixed or dual-layer Cu/SSZ-13 and Pt/γ-Al_2O_3 catalyst. Here, the chabazite SSZ-13 has smaller pores than the ZSM-5 zeolite described in the earlier work [13-4]. The Cu/SSZ-13 [13-12] appears to have a better selectivity for N_2 and for N_2O than the earlier Cu/ZSM-5 catalyst [13-4].

In Shrestha et al. [13-12] they used roughly 2.7 g/ft³ (0.095 g/L) of Pt on γ-Al_2O_3 washcoat and found that the Pt/γ-Al_2O_3 had to be milled separately from the Cu/SSZ-13 so that the Pt did not leach into the zeolite cages and becomes catalytically inaccessible. The lower Pt loading was used to reflect trends with productions systems, as the earlier work [13-4] had used a much higher Pt loading of about 10 g/ft³ (0.35 g/L).

Shrestha et al. [13-12] found that using 2.7 g Pt/ft³ (0.095 g/L) on γ-Al_2O_3 alone simply converted the ammonia to NOx, with a high selectivity for N_2O in the temperature

FIGURE 13.6 Desired and side-product yields from mixed bi-functional catalyst at 66,000 h^{-1} SV, with feed gas composition of 200 ppm NH_3, 5% O_2, 2.5% H_2O, and 2% CO_2. (Figure 5(b) in [13-4].)

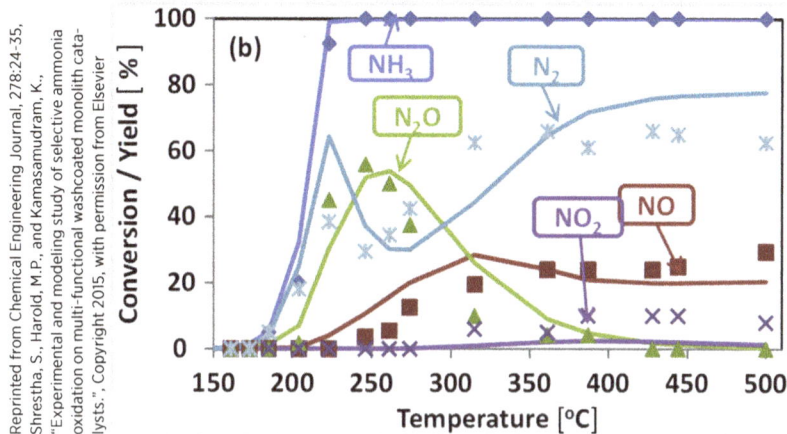

range of 220°C to nearly 400°C. Using a bi-functional ASC catalyst that combined 2.7 g Pt/ft^3 on γ-Al_2O_3 with one of several Cu/SSZ-13 layers did not significantly change the ammonia conversion but did improve the selectivity for N_2. Adding Cu/SSZ-13 to the ASC catalyst decreased the selectivity for N_2O, S_{N_2O}, from about 45% at 250°C with Pt/γ-Al_2O_3 catalyst only to about 35% at 250°C. The temperature range with $S_{N_2O} \geq 25\%$ also decreased from 220°C to 360°C to 220°C to 290°C by adding Cu/SSZ-13. One limiting factor for ammonia conversion to N_2 is that the Cu/SSZ-13 layer may impede the diffusion of ammonia into the Pt/γ-Al_2O_3 layer underneath. These trends held at the higher SV of 265,000 h^{-1}, although the overall ammonia conversion decreased from 100% down to 60% to 90% at high temperature based on the specific catalyst formulation.

Shrestha et al. [13-4, 13-12] evaluated the higher SV since the ASC tends to be a relatively small system compared to the SCR volume. The higher SV accentuates the mass transfer limitations on the effective reaction rate. As with the combination of Pt/γ-Al_2O_3 and Fe/ZSM-5 catalysts, the combination of Pt/γ-Al_2O_3 and Cu/SSZ-13 shows slightly lower N_2O selectivity when the catalysts are intermixed in one thicker layer than when the Cu/SSZ-13 layer is coated over the Pt/γ-Al_2O_3 layer, but it is a very limited effect, as shown in **Figure 13.7**. In the case shown in Figure 13.7, the Cu content was increased to 5.71 wt% from the 1.5 wt% or 3.0 wt% Cu used in the rest of the study [13-12]. When an equimolar mixture of ammonia and NO are fed to the reactor, S_{N_2O} increases below 200°C. As the SCR reactions become active, however, the peak in S_{N_2O} at 250°C is decreased.

Based on these results of the dual-layer and mixed catalysts, Shrestha et al. [13-12] ended up formulating and testing a catalyst washcoat configuration in which a diluted Cu/SSZ-13 layer was placed over a washcoat layer that mixed Cu/SSZ-13 and Pt/γ-Al_2O_3. In this way, the mass transfer resistance to the SCR reaction could be minimized. This configuration also shields the Pt from direct exposure to the bulk gas, so that NO formed by ammonia oxidation has a chance to be reduced to N_2 on the SCR catalyst before it escapes into the bulk gas in the monolith channel.

Therefore, the choice of catalysts and the formulation control both the ammonia conversion and the selectivity for N_2 over undesirable side products such as N_2O or NO.

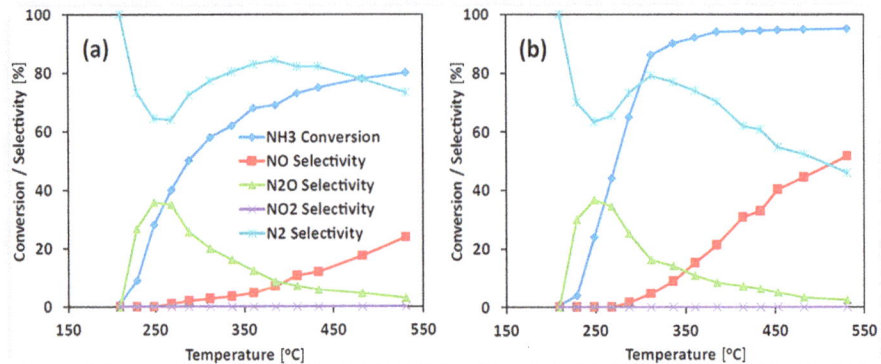

FIGURE 13.7 Desired and side-product yields from (a) layered or (b) mixed bi-functional catalyst at 265,000 h^{-1} SV, with feed gas composition of 500 ppm NH_3, 5% O_2, 2.5% H_2O, and 2% CO_2. (Figure 10 in [13-12].)

Reprinted from Catalysis Today, 267:130–144, Shrestha, S., Harold, M.P., Kamasamudram, K., Kumar, A., Olsson, L., and Leistner, K., "Selective oxidation of ammonia to nitrogen on bi-functional Cu–SSZ-13 and Pt/Al2O3 monolith catalyst.", Copyright 2016, with permission from Elsevier

13.3 Other Washcoat Functionality

The catalyst formulation used will determine the fundamental nature of the washcoat. For the bi-functional ASC that includes both SCR and oxidation catalysts, one approach is to have a dual-layer washcoat. For this approach, an SCR washcoat layer is coated over an oxidation catalyst (Pt/Al$_2$O$_3$) layer in the monolith [13-6], as shown in **Figure 13.8**. Here, the SCR and ammonia oxidation reactions take place in the in their respective layers, but the NOx formed by the ammonia oxidation diffuses back to react with ammonia in the SCR layer, thereby improving the overall selectivity for N$_2$.

Shrestha et al. [13-4] investigated the selectivity of an ASC that used a bi-functional catalyst that combines a Pt/γ-Al$_2$O$_3$ oxidation catalyst with an Fe/ZSM-5 catalyst for SCR. As discussed above, Pt/γ-Al$_2$O$_3$ favors NOx formation instead of N$_2$ when catalyzing ammonia oxidation. Using the Fe/ZSM-5 means that the ASC catalyst can use the NOx formed on the Pt/γ-Al$_2$O$_3$ catalyst in an SCR reaction with ammonia diffusing in from

FIGURE 13.8 Schematic of dual-layer washcoat for bi-functional ammonia slip catalyst.

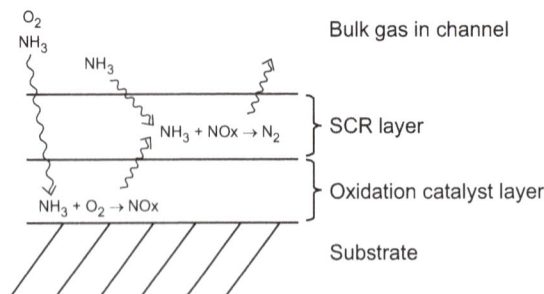

the channel, which greatly improves the selectivity for N_2 in the ASC. In their study, Shrestha et al. [13-4] placed a washcoat layer of Pt/γ-Al$_2$O$_3$ containing 10 g/ft^3 Pt on the monolith and then placed a washcoat with Cu or Fe zeolite over it. The intent was for some of the ammonia entering the ASC to adsorb on the SCR catalyst and for some to diffuse through to the Pt/γ-Al$_2$O$_3$ layer, where it forms NOx and reacts with the stored ammonia as it diffuses back into the main channel.

Shrestha et al. [13-4] did experiments to estimate the rates of ammonia diffusion through a Fe/ZSM-5 washcoat layer and the underlying Pt/γ-Al$_2$O$_3$ washcoat layer, as well as to confirm the kinetics models for the SCR and ammonia oxidation reactions. It is important to understand the diffusion of ammonia and its products through the washcoat, as the effective rate of reaction can be limited by the mass transfer.

Sukumar et al. [13-6] described modeling and simulation results from using a Cu-based ASC as part of a dual-layered SCR–ASC unit. This study included experiments to validate the models of the chemical kinetics. These experiments used a similar protocol to that used by Kamasamudram et al. [13-7] to evaluate ASC performance as the inlet gas composition is changed. In modeling the ASC, then, an important factor is coupling the diffusion of species through the washcoat layers and their chemistry.

Some ASC are zone coated at the back end of an SCR monolith, in which case the ASC is on the substrate used for the SCR. But even when the ASC is on a separate monolith, the substrate used for the ASC is usually the same as that for the preceding SCR monolith. Thus, the monolith material is usually cordierite for heavy-duty applications. For light-duty applications, the monolith may be cordierite, silicon carbide (SiC), metal foil, or some other material that is matched to the expected exhaust temperature range at the ASC location.

In a typical EAS configuration the ASC is the last catalytic converter before the tailpipe exit, which means that the exhaust is at its coolest and cordierite would be a logical choice. If the SCR–ASC combination were further upstream, for example, in a close-coupled position upstream of the particulate filter, then a higher-temperature material such as SiC would be more appropriate.

13.4 Ammonia Slip Catalyst Failure Modes

The ASC needs good activity over the full useful life of the EAS for the target application. Some mechanisms degrade the catalytic activity including sulfur poisoning and hydrothermal aging (HTA). In addition, there is a risk that the ASC could promote the formation of N_2O.

Sulfur poisoning, in which sulfur dioxide (SO_2) or sulfur trioxide (SO_3) from either the fuel or lubricant interacts with the catalyst surface, is a primary degradation mechanism for ASC. The sulfated catalyst does not promote ammonia oxidation to N_2 as designed, but may instead promote the alternative reactions from Eqs. 13.2, 13.3, and 13.4. The details of the sulfur poisoning depend upon the catalyst used. The catalysts and the effects of sulfation are described in Section 13.2, Catalyst Formulations. Sulfur poisoning is particularly an issue with Cu/zeolite catalysts that provide the SCR function [13-13]. Fortunately, the sulfur poisoning can be reversed with suitable catalyst heating to drive the sulfur oxides (SOx) back off the catalyst surface [13-7].

Smith et al. [13-8] studied the effects of SOx on the performance of an ASC with a Cu/zeolite layer placed over a Pt/Al_2O_3 layer in the washcoat, and how that performance recovers after a desulfation event. For this ASC configuration, the catalyst showed the most SO_2 desorption at 700°C, which suggests that the whole EAS will need to be strongly heated to remove SOx. On the other hand, they showed that there was only a small difference in ammonia oxidation with respect to adsorbed sulfate quantity around 280°C–330°C, and a negligible difference otherwise in conditions that were not particularly challenging for ammonia oxidation. It appears that the main effect of adsorbed sulfate is to promote the formation of the side products NO_2, NO, and N_2O. This change in performance is assumed to result from the sulfation of the Cu/zeolite SCR catalyst and not from poisoning the Pt/Al_2O_3[13-8]. Smith et al. [13-8] also evaluated the effect of sulfation on the NOx conversion of an ASC containing Cu/zeolite as shown in **Figure 13.9**. The adsorbed SOx significantly impairs the NOx conversion, especially below 350°C, where it is believed that the ammonia oxidation is primarily on the Pt/Al_2O_3[13-8].

HTA of the ASC is also possible and will degrade performance. The likely aging mechanism is Pt sintering in the washcoat. The Cu/zeolite or Fe/zeolite performance is also affected by HTA [13-13]. In the specific case of the Cu/SSZ-13 catalyst investigated by Luo et al. [13-13] the desorption spectra suggested that HTA might convert one Cu site into another, since the relative peaks change but the overall ammonia storage capacity remains constant. As shown in **Figure 13.10**, the aging behavior of Cu/SSZ-13 shows a shift from the high-temperature site to the low-temperature site after aging at 600°C. By contrast, another Cu/zeolite SCR catalyst, Cu/SAPO-34 is relatively unaffected by aging at 600°C. HTA of the ASC at 650°C degraded the catalyst's performance by reducing

FIGURE 13.9 Effect of sulfur poisoning on NOx conversion by ammonia over Cu/zeolite ASC. Inlet NH_3 concentration is 100 ppm, inlet NOx, 50 ppm [13-8].

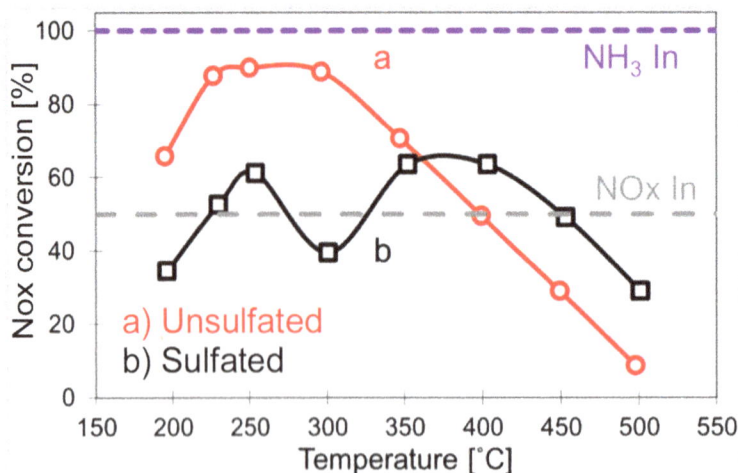

FIGURE 13.10 Ammonia temperature programmed desorption after aging at 600°C for (a) Cu/SSZ-13 and (b) Cu/SAPO-34. (Figure 1 in [13-13].)

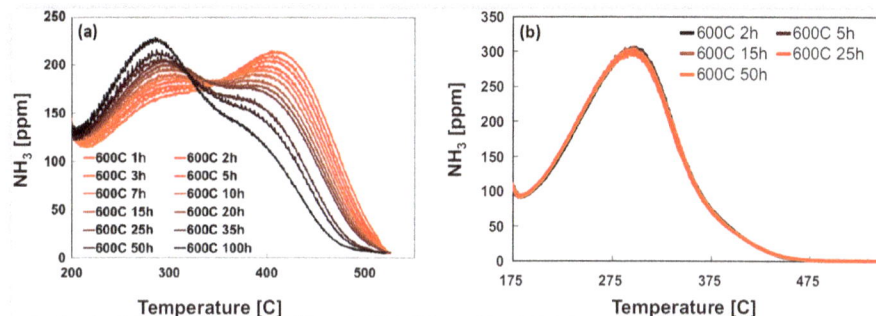

Reprinted from Journal of Catalysis, 348:291-299, Luo, J., Gao, F., Kamasamudram, K., Currier, N., Peden, C.H.F., and Yezerets, A., "New insights into Cu/SSZ-13 SCR catalysts acidity. Part I: Nature of acidic sites probed by NH3 titration.", Copyright 2017, with permission from Elsevier

the selectivity for N_2 [13-7]. HTA appears to degrade the ASC performance at a slower rate than sulfur poisoning does [13-8].

Shrestha et al. [13-4] saw relatively high selectivity for N_2O in their bi-functional ASC, especially in the temperature range of 210°C–350°C. This N_2O release may come from the use of Fe/ZSM-5 instead of a Cu/zeolite catalyst, which are generally more active at these lower temperatures than Fe/zeolite catalysts. Shrestha et al. [13-4] noted that Fe/ZMS-5 has a very low activity for the SCR reactions below 300°C, although at these temperatures the Pt/γ-Al$_2$O$_3$ can promote the SCR reactions. Hünnekes et al. [13-2] studied the effect of feed composition on ASC performance with a Pt loading of 5 g/ft^3 and found that the presence of NO increases the low-temperature selectivity to N_2O, which is not a desirable outcome.

References

13-1. Charlton, S., Dollmeyer, T., and Grana, T., "Meeting the US Heavy-Duty EPA 2010 Standards and Providing Increased Value for the Customer," *SAE Int. J. Commer. Veh.* **3**, no. 1 (2010): 101–110, https://doi.org/10.4271/2010-01-1934.

13-2. Hünnekes, E.V., Heijden, P.V.D., and Patchett, J.A., "Ammonia Oxidation Catalysts for Mobile SCR Systems," SAE Technical Paper 2006-01-0640, 2006, https://doi.org/10.4271/2006-01-0640.

13-3. Girard, J.W., Cavataio, G., and Lambert, C.K., "The Influence of Ammonia Slip Catalysts on Ammonia, N_2O and NO_X Emissions for Diesel Engines," SAE Technical Paper 2007-01-1572, 2007, https://doi.org/10.4271/2007-01-1572.

13-4. Shrestha, S., Harold, M.P., and Kamasamudram, K., "Experimental and Modeling Study of Selective Ammonia Oxidation on Multi-Functional Washcoated Monolith Catalysts," *Chemical Engineering Journal* **278** (2015): 24–35, https://doi.org/10.1016/j.cej.2015.01.015.

13-5. Georgiadis, E., Kudo, T., Herrmann, O., Uchiyama, K., and Hagen, J., "Real Driving Emission Efficiency Potential of SDPF Systems without an Ammonia Slip Catalyst," SAE Technical Paper 2017-01-0913, 2017, https://doi.org/10.4271/2017-01-0913.

13-6. Sukumar, B., Dai, J., Johansson, A., Markatou, P., Ahmadinejad, M., Watling, T., Ranganath, B., Nande, A., and Szailer, T., "Modeling of Dual Layer Ammonia Slip

Catalysts (ASC), SAE Technical Paper 2012-01-1294, 2012, https://doi.org/10.4271/2012-01-1294.

13-7.   Kamasamudram, K., Yezerets, A., Chen, X., Currier, N., Castagnola, M., and Chen, H.-Y., "New Insights into Reaction Mechanism of Selective Catalytic Ammonia Oxidation Technology for Diesel Aftertreatment Applications," *SAE Int. J. Engines* **4**(1): 1810–1821, 2011, https://doi.org/10.4271/2011-01-1314.

13-8.   Smith, M.A., Kamasamudram, K., Szailer, T., Kumar, A., and Yezerets, A., "Impact of Sulfur-Oxides on the Ammonia Slip Catalyst Performance, SAE Technical Paper 2014-01-1545, 2014, https://doi.org/10.4271/2014-01-1545.

13-9.   Olsson, L., Wijayanti, K., Leistner, K., Kumar, A., Joshi, S.Y., Kamasamudram, K., Currier, N.W., and Yezerets, A., "A Multi-Site Kinetic Model for NH_3-SCR over Cu/SSZ-13," *Applied Catalysis B: Environmental* **174-175** (2015): 212–224, https://doi.org/10.1016/j.apcatb.2015.02.037.

13-10. Zheng, Y., Harold, M.P., and Luss, D., "Effects of CO, H_2 and C_3H_6 on Cu-SSZ-13 Catalyzed NH_3-SCR," *Catalysis Today* **264** (2016): 44–54, https://doi.org/10.1016/j.cattod.2015.06.028.

13-11. Krejza, P., Pekar, J., Figura, J., Lansky, L.,von Wissel, D., and Zhang, T., "Cascade MPC Approach to Automotive SCR Multi-Brick Systems," SAE Technical Paper 2017-01-0936, 2017, https://doi.org/10.4271/2017-01-0936.

13-12. Shrestha, S., Harold, M.P., Kamasamudram, K., Kumar, A., Olsson, L., and Leistner, K., "Selective Oxidation of Ammonia to Nitrogen on Bi-Functional Cu–SSZ-13 and Pt/Al_2O_3 Monolith Catalyst," *Catalysis Today* **267** (2016): 130–144, https://doi.org/10.1016/j.cattod.2015.11.035.

13-13. Luo, J., Gao, F., Kamasamudram, K., Currier, N., Peden, C.H.F., and Yezerets, A., "New Insights into Cu/SSZ-13 SCR Catalyst Acidity. Part I: Nature of Acidic Sites Probed by NH_3 Titration," *Journal of Catalysis* **348** (2017): 291–299, https://doi.org/10.1016/j.jcat.2017.02.025.

Passive Aftertreatment Systems

I had learned that science is a rewarding, active process of discovery, not the passive absorption of what others had discovered.

—Harold E. Varmus [14-1]

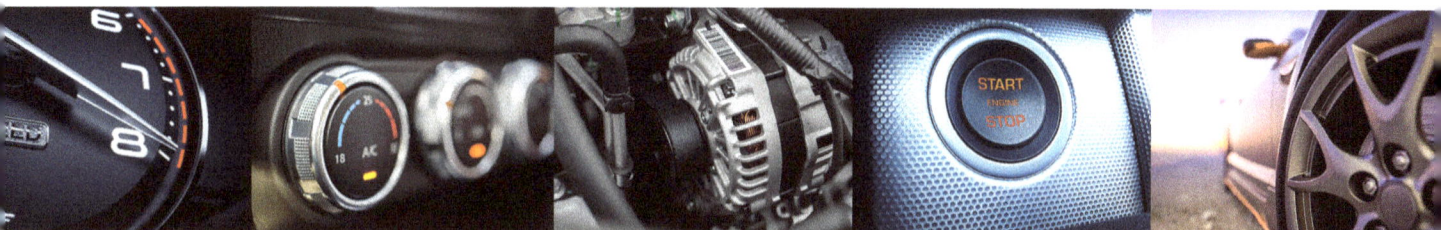

One of the main challenges encountered when developing an exhaust aftertreatment system (EAS) for regulatory compliance is that many of the regulatory test cycles described in Chapter 4, Emissions Testing Protocols, start with the engine and EAS at an ambient temperature of 20°C (68°F) to 30°C (86°F). The warm-up and keep-warm strategies described in Section 7.3.2 are necessary to get the catalytic converters within the EAS above their light-off temperatures so that they convert gaseous pollutants, such as nitrogen oxides (NOx) and non-methane hydrocarbons (NMHCs). Nevertheless, during warm-up, the exhaust gas passes through the EAS mostly untreated, and up to 90% of on cycle emissions can be attributed to the cold start period [14-2].

A related, additional challenge is that there can be situations in normal use where the temperature of the exhaust gas entering the EAS drops below the point where the EAS functions well, which also increases in-use emissions under those conditions. This challenge already exists for heavy-duty diesel engines (HDDEs), and Chen [14-3] indicates that it may become more prevalent as internal combustion efficiency improvements reduce exhaust gas temperatures for both spark-ignition and compression-ignition engines.

These challenges motivate an interest in improved thermal management strategies; in finding catalysts that light-off temperature and convert criteria pollutants at lower temperatures, as described elsewhere in this book; and in passive aftertreatment systems. Common features of passive aftertreatment systems include the trapping and release behavior with respect to temperature, and the need for a purge strategy during normal operation to reset the system for the next cold start. A passive aftertreatment system collects its target pollutant when the exhaust gas temperature is below a given threshold, and then releases the pollutant above that threshold temperature. The system generally does not contribute to the conversion of the target pollutant itself, but delays the flow of pollutant until such time that it can be chemically converted further downstream in the EAS. After reading this chapter, one will learn about both passive NOx adsorbers (PNA) and hydrocarbon (HC) traps, these are two types of passive aftertreatment systems currently in development or preproduction.

Depending on the application and the temperature thresholds for the trapping and release from the passive aftertreatment system, there is a risk that the passive system will not empty fully before the next cold exhaust event. Over time, this could lead to the passive system being saturated and unable to trap further pollutant during either a cold start or low-temperature operation. The EAS control scheme will likely need to model the quantity of pollutant stored within the passive aftertreatment system both to monitor and to actively manage its fill state.

Another challenge for passive aftertreatment systems is that they are likely to experience some of the hottest temperatures in the EAS during normal use as they are generally placed upstream of the active EAS systems, and are therefore close to the exhaust manifold or turbine outlet. Thus, these systems must be robust against hydrothermal aging (HTA) at high temperature (up to 800°C–850°C) so that they maintain their function over the regulated full useful life (FUL) of the application.

Specific details on PNA and HC traps are provided in their respective sections below. These details include a description of the normal function of the passive aftertreatment systems; typical catalyst, washcoat, and substrate materials used; and failure modes.

14.1 Passive NOx Adsorbers

One of the main challenges in removing NOx from exhaust gas is that the catalysts used for NOx reduction are generally not active when the catalyst is below 200°C, which then affects both on-cycle and in-use emissions. This limitation on NOx conversion performance and ways to mitigate it are described in more detail in earlier chapters, including the following:

- Chapter 7 for thermal management of catalysts
- Chapter 9 for three-way catalysts (TWCs)
- Chapter 10 for lean NOx traps (LNTs)
- Chapter 12 for selective catalytic reduction (SCR) systems

Although the thermal management strategies discussed in Section 7.3.2 are broadly useful for NOx control, there has also been research to address this issue since the 1990s on PNA. The goal of the research has been to develop materials that will trap NOx at low temperatures and then release it from the PNA once the catalytic NOx reduction system is warmed-up and capable of high NOx conversion. For example, Descorme et al. [14-4] evaluated low-temperature nitrogen monoxide (NO) adsorption at room temperature (*ca.* 25°C or 77°F) on materials containing palladium (Pd). Around the same time, Cordatos and Gorte [14-5] also investigated NO adsorption onto model catalysts using Pd on ceria (CeO$_x$, usually CeO$_2$), Pd/ceria.

The NOx adsorption and desorption reactions may be represented using NO as

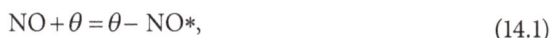

$$NO + \theta = \theta - NO*, \qquad (14.1)$$

where θ is a surface site for adsorption- and the asterisk indicates a surface-bound species. Good PNA materials favor the forward (adsorption) reaction in Eq. 14.1 at temperatures below 200°C and favor the reverse (desorption) reaction above 200°C. The adsorption of NO is strongly favored, and therefore the adsorption–desorption process is not a simple equilibrium process that is governed by the system temperature. Instead, Pihl et al. [14-6] hypothesize that the adsorption–desorption process is actually an active chemical cycle. Note also that the storage capacity of the PNA is governed both by the number of surface sites and by the equilibrium between the adsorption and desorption reactions. NOx molecules are therefore always to bind and release from the storage material even when the total stored is constant.

14.1.1 Passive NOx Adsorber Function

The function of the PNA is to trap and store as much of the NO entering the EAS as possible during cold exhaust events, and then to release that NO once the system temperature has gone above the release temperature threshold. That is, the PNA should favor the forward (adsorption) reaction in Eq.14.1 when the catalytic converter in the EAS for NOx removal is not above its light-off temperature. Once the catalytic converter is hot enough to convert NOx to nitrogen (N_2), the PNA should favor the reverse (desorption) reaction. The challenge is finding a material where the NO release temperature is high enough to ensure NOx conversion in the catalytic converter downstream, but not so high that the desorption temperature is difficult to reach and sustain. Most of the researchers are therefore targeting PNA NO release temperatures of 200°C–250°C, as most NOx reduction catalysts light-off around 200°C [14-7].

The placement of the PNA within the EAS also requires some thought. The PNA needs to be upstream of the urea doser for the SCR system, as the PNA contains Pd. This Pd will promote the oxidation of the ammonia (NH_3) from the urea solution into N_2 or NOx, which will thereby decrease the net NOx conversion of the EAS. The PNA also needs to be far enough upstream so that it warms up to its minimum temperature for adsorbing NOx under typical exhaust gas conditions. On the other hand, if the PNA is too far upstream of the SCR or similar catalytic converter for NOx removal, there is a risk that the NO will desorb from the PNA before the downstream catalysts have reached their light-off temperature. Work at Johnson Matthey and Cummins by Chilumukuru et al. [14-8] has examined EAS configurations for light-duty diesel engines (LDDEs), with an emphasis on emissions performance over the FTP-75 cycle, covered in detail in Section 4.1.1. For example, Chilumukuru et al. [14-8] describe placing a combined PNA and HC trap upstream of a catalytic converter that has an SCR catalyst on a diesel particulate filter (DPF) substrate.* Similarly, Henry et al. [14-9] recommend placing a low-temperature SCR system immediately downstream of the PNA so that when the PNA reaches the temperature when NOx starts desorbing—roughly from 180°C to 220°C—it can be reduced right away.

---

* The function of a combined SCR-DPF, or SDPF, system is described further in Section 15.2.

The diesel Cold Start Catalyst™ (dCSC™) strategy from Johnson Matthey [14-3] combines a PNA, HC trap, and diesel oxidation catalyst (DOC) to improve cold start emissions. The dCSC™ catalytic converter replaces the DOC in the EAS for a diesel engine. The PNA described by Chen et al. [14-10] uses a Pd/ceria PNA material, although they also investigated Pd-exchanged zeolites (Pd/zeolite).

The work of Henry et al. [14-9] provides an example of using a PNA to support meeting Tier 2 bin 2 requirements for NOx with an LDDE. For the engine they were studying, a Cummins ISF 2.8 LDDE with EAS assumed to contain DOC, DPF, and SCR system, the NOx emissions during the first 200s of FTP-75 exceeded the NOx budget for the complete 1874s drive cycle. This excess in NOx emissions happened because the base engine and EAS needed about 170s for the exhaust gas to warm up the SCR catalyst to its light-off temperature of 200°C. Therefore, Henry et al. [14-9] wanted to reduce the tailpipe NOx emissions during Bag 1 of the FTP-75 cycle, which covers the warm-up period on this cycle. To meet the overall drive cycle target they wanted the tailpipe NOx emissions during the first 200s of FTP-75 cut in half. The PNA used by Henry et al. [14-9] has its peak NOx storage capacity at 160°C, and from roughly 130°C–180°C the NOx storage capacity is at least 75% of that peak. However, the NOx storage capacity of the PNA at 200°C is half of the peak capacity at 160°C.

PNA are also being considered to support future requirements, such as the ultra-low NOx standard for HDDE currently under development and described in Sections 3.3.1.1 and 3.3.2. The Southwest Research Institute (SwRI) [14-11, 14-12] is evaluating the ability of future engine and EAS configurations to meet this standard. The preferred configuration for their future work, shown in **Figure 14.1**, uses a PNA upstream of the main part of the EAS in conjunction with a Volvo MD13TC 13 L HDDE similar to the Volvo D13 described in Section 7.5.5 [14-12].

Similarly, Theis and Lambert [14-7] have developed a test protocol for materials samples that simulates FTP-75 Bags 1 and 2 followed by US06, which they applied to the Ford Super Duty® truck with the Power Stroke® diesel engine, a version of which is described further in Section 7.5.3. FTP-75 and US06 are described in Sections 4.1.1 and 4.1.3, respectively. The Ford test protocol runs for 33 min, and has a maximum exhaust gas temperature of 250°C during the FTP-75 portion, and 350°C during the US06 portion. The bench reactor system warmed up more slowly on this test protocol than the real PNA system would on the vehicle.

D.H. Kim and colleagues [14-13] studied NO release from a PNA material that used Pd/ceria with some platinum (Pt) added to improve the thermal stability of the material. Within this Pd/ceria material, the NOx adsorbs onto the ceria, and the Pd activates the adsorption sites. During a temperature-programmed desorption (TPD) test they found two

FIGURE 14.1 Exhaust aftertreatment system to support ultra-low NOx performance, including passive NOx adsorber [14-12].

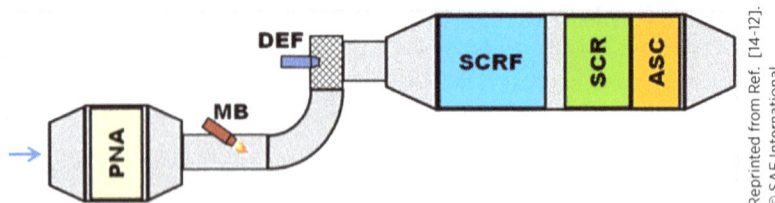

Reprinted from Ref. [14-12].
© SAE International

NO desorption peaks. The first, or alpha, desorption peak occurs around 300°C and is associated with the Pd in the material. The second, or beta, desorption peak occurs around 430°C and is associated with the Pt. The alpha peak is preferred, since the beta peak temperature is too high to be consistently reached, especially in a modern HDDE application. If the PNA material is given more time to adsorb NOx, the alpha peak release temperature drops slightly, to 290°C, and reflects a larger stored quantity that is slower to desorb from the PNA.

Crocker and colleagues at the University of Kentucky have also studied Pt and Pd alloys supported on a ceria-zirconia (CeO_2–ZrO_2) mixture containing 20% CeO_2 [14-14, 14-15]. They conducted bench studies of various materials while adsorbing NO at 120°C in the presence of water, oxygen (O_2), and carbon dioxide (CO_2), and then desorbing NO in a TPD experiment. They also found two NOx desorption peaks from the Pt–Pd sites, with the first (alpha) peak at 220°C or so, and the second (beta) at 380°C or so. As described earlier [14-13], the alpha peak is associated with the Pd sites, and the beta peak appears, the Pt sites. Presumably the choice of support for the Pt–Pd alloy also influences the peak desorption temperatures. Both desorption peaks shifted to lower temperatures after the materials were aged.

Khivantsev et al. [14-16] at the Pacific Northwest National Laboratories investigated the adsorption and desorption performance of a PNA using Pd/SSZ-13, because it has a release temperature of 310°C and a competitive NO storage capacity. This release temperature is attractive because a downstream NOx reduction catalytic converter will be active by the time the NO desorbs. In the experiments the Pd/SSZ-13 was exposed to a gas flow at 100°C that contained 200 ppm NO, and the team measured how much NO was adsorbed. Once the Pd/SSZ-13 sample was saturated, a TPD was started, again with 200 ppm NO in the gas flow, and the NO desorbed was also measured. Their experiments determined that the storage capacity of the Pd/SSZ-13 sample is about 0.43 mol NO per mol Pd, which is competitive with other Pd/zeolite materials, but which is less than full utilization of the Pd. Khivantsev et al. [14-16] hypothesized that not all of the Pd is available to be used for NO adsorption.

Pihl et al. [14-6] described work done at the Oak Ridge National Laboratory in which they studied a PNA material with 1.8 (g/L) of Pd (50 g/ft.³) in Pd/ZSM-5. This PNA material was blended into a washcoat that was placed onto a 400 cpsi (cell/in²; 62 cell/cm²) monolith sample. The monolith sample was exposed to feed gas at an inlet temperature of 100°C that included 200 pm NO and 200 pm carbon monoxide (CO) until it saturated with NO. Then, a TPD sequence was started with feed gas containing 200 ppm CO. The NO was observed to release from the PNA starting around 250°C, with the peak desorption rate at 340°C. Pihl et al. [14-6] also observed that during the NO adsorption phase, CO was released from the PNA material. To make their calculations, they measured the quantity of NO adsorbed and the quantity desorbed. The amounts measured were equivalent given the measurement errors, although the quantity desorbed usually appeared slightly larger than the quantity adsorbed.

Of course, exhaust gas is a complex mixture that includes, among other species, O_2, CO, CO_2, and water (H_2O), all of which can influence PNA performance. CO_2 and O_2 have a slight effect on the NOx adsorption performance of Pd/ZSM-5 [14-6], although the presence of CO appears to aid NOx adsorption on Pd/SSZ-13 [14-13]. Water in the exhaust gas appears to inhibit NO adsorption and increase CO oxidation on Pd/zeolite PNA materials, as shown in **Figure 14.2** [14-6]. Similarly, Liu et al. [14-17] found that water

FIGURE 14.2   Water decreases NO uptake and increases CO oxidation on Pd/ZSM-5 [14-6].

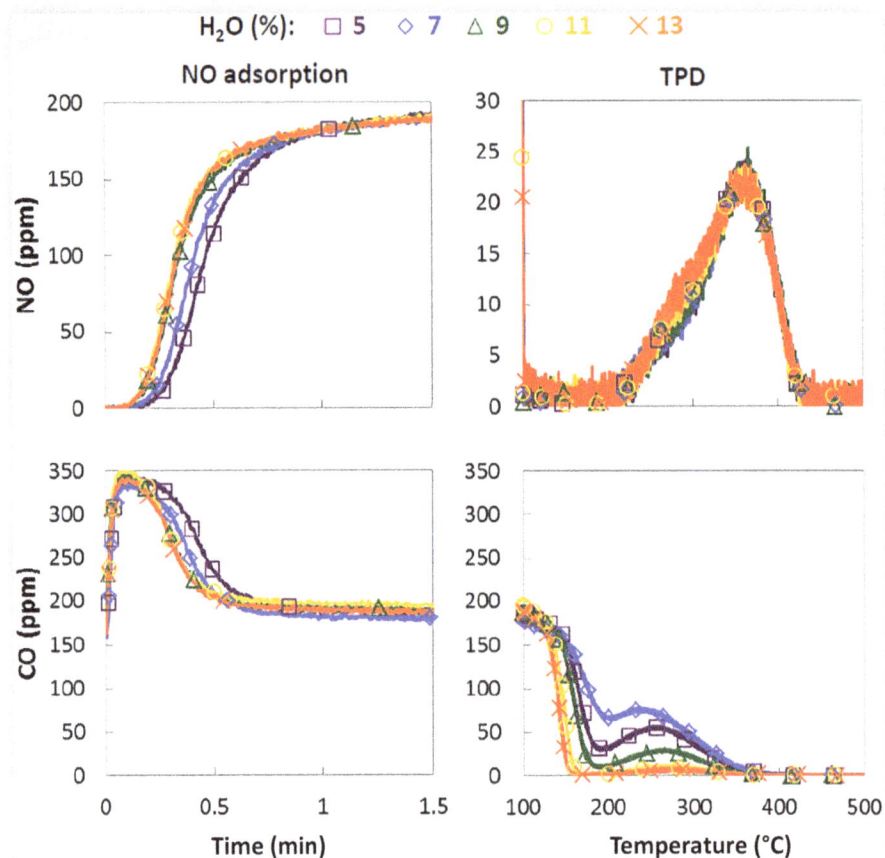

inhibited NOx adsorption below about 110°C, and Theis and Lambert [14-7], below 100°C. On the other hand, Chen [14-3] and Liu et al. [14-17] found that water did not affect NO adsorption on Pd^{2+} sites, which suggests that water blocks NO adsorption onto the acid sites in the zeolite. Water also appears to push NO desorption from Pd/CHA to significantly higher temperatures, from 150°C to 350°C, as shown in **Figure 14.3**, which is actually a more useful desorption temperature.

The NO concentration in the feed does not appear to affect the total quantity of NO trapped, although higher concentrations fill the PNA material faster. The stored NO on the Pd/ZSM-5 also inhibits CO oxidation. By varying the feed gas temperature, Pihl et al. [14-6] determined that the NO storage capacity of Pd/ZSM-5 is highest in the range of 125°C–150°C, as shown in **Figure 14.4**. Above 150°C, the NO desorption rate increases and starts to limit the net quantity of NO stored.

Theis and Lambert [14-7] at Ford have studied PNA using Pd/CZO, which is Pd on a mix of 75% ceria and 25% zirconia. Their focus was on the NOx adsorption performance below 200°C. They found that the NOx storage capacity on the Pd/CZO was enhanced by reductants, such as ethene (C_2H_4), CO, hydrogen (H_2), and CO+H_2, as shown in **Figure 14.5**. These tests were run on a bench reactor system with a feed gas containing 140 ppm NO, 5% H_2O, 5% CO_2, and 10% O_2, with the balance being N_2.

FIGURE 14.3 Water increases NO desorption temperature on Pd/CHA [14-17].

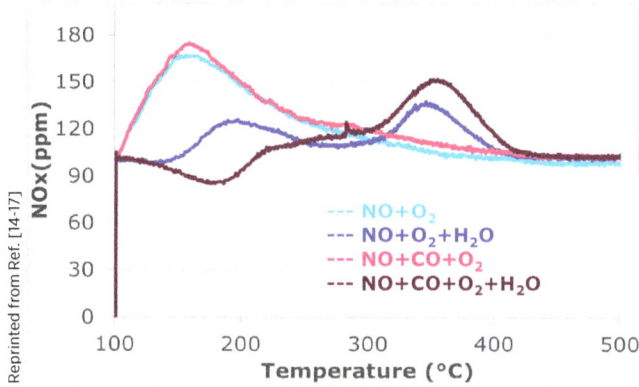

FIGURE 14.4 Best NO storage occurs when Pd/ZSM-5 passive NOx adsorber is between 125°C and 150°C [14-6].

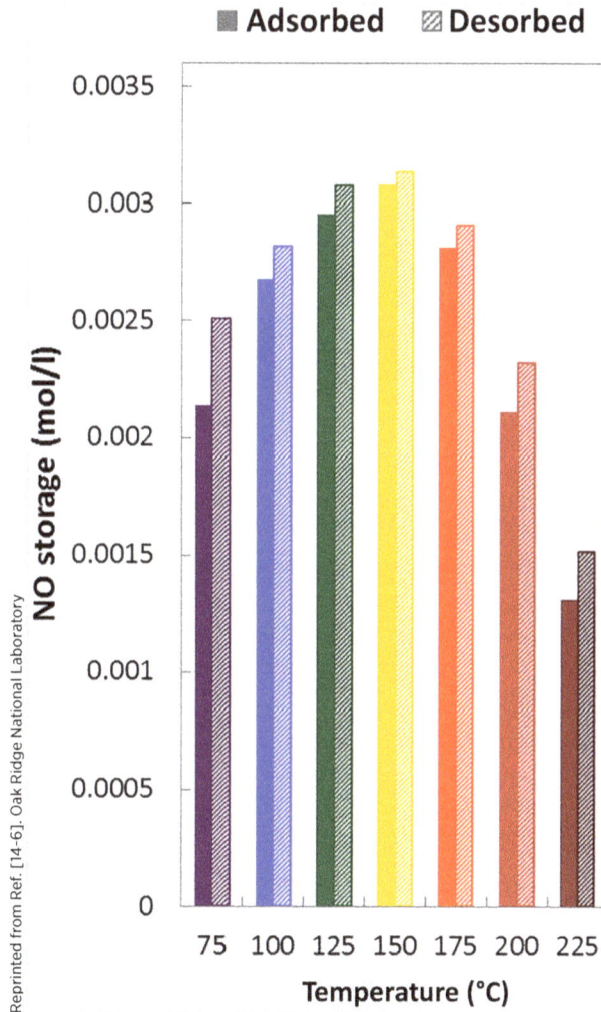

FIGURE 14.5 Transient tests on Pd/CZO with 140 ppm NO and various reducing agents, including C_2H_4, CO, H_2, and CO+H_2, in 5% H_2O, 5% CO_2, and 10% O_2 [14-7].

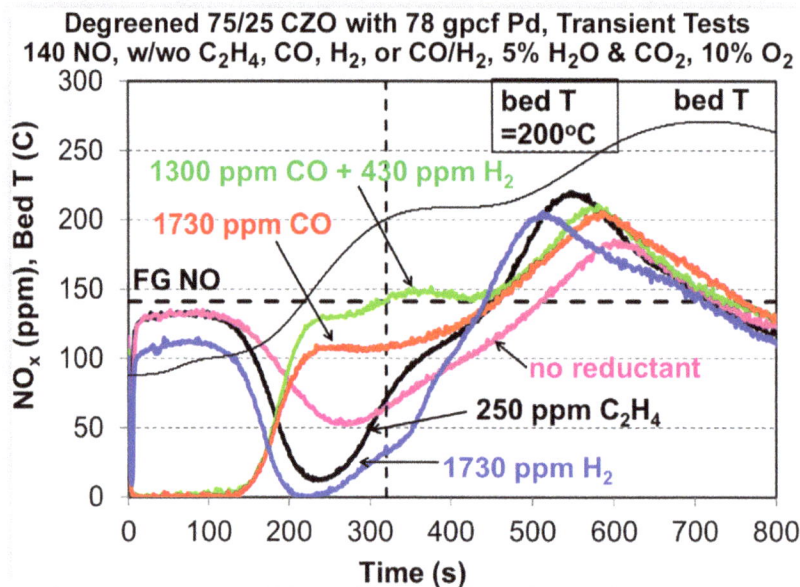

Theis and Lambert [14-7] found that the reductants, however, seem to enhance NOx storage by promoting the formation of chemical intermediates. For example, their DRIFTS reactor system detected a peak corresponding to isocyanate (NCO) when NO and CO were supplied to a Pd/ceria PNA sample. Unfortunately, the carbonates present generate DRIFTS peaks that mask the presence of HC–NO intermediates. The reductants and other species could also participate in nitrogen chemistry. For example, some of the adsorbed NO left the PNA as NO_2, and the presence of reductants means that some of the NO can be reduced to N_2. By contrast, Lardinois et al. [14-18] observed that CO seems to facilitate NO adsorption on Pd/CHA by forming a [Pd–(NO)–(CO)]²⁺ complex within the zeolite.

14.1.2 Catalyst Formulations

Palladium (Pd) is the catalyst of choice for PNA systems, and it is typically placed on either a ceria or zeolite support. For example, Cordatos and Gorte [14-5] investigated NO adsorption onto Pd/ceria model catalysts.

Likewise, Descorme et al. [14-4] evaluated Pd-containing materials such as Pd/ZSM-5 made from H/ZSM-5 zeolite exchanged with Pd tetrammine complex ([Pd(NH₃)₄]²⁺). They found that NO could be adsorbed by reducing Pd^{2+} to Pd^+, which led to the subsequent storage of nitrogen dioxide (NO_2) on the zeolite. Pihl et al. [14-6] hypothesize that the active sites for NO adsorption are coordinated Pd^{2+} ions within the ZSM-5 structure. Descorme et al. [14-4] used infrared absorption to determine that the Pd–NO complex in the zeolite structure also coordinated* with water and NO_2. Phil et al. [14-6] also found that the Pd oxidizes CO to CO_2.

---

* When chemical species are coordinated, it means that they are loosely bound together because of electrostatic forces.

TABLE 14.1 Comparison between dCSC™ and DOC [14-3].

System type	PGM mix (Pt:Pd)	PGM loading (g/ft.³)	Volume ratio (Cat:Engine)
DOC	4:1	115	2.0:3.0
dCSC™	8:13	152	2.0:3.0

Researchers at Johnson Matthey have been active in developing and evaluating PNA materials. Chen et al. [14-19] for example, have found several Pd/zeolite compounds that can efficiently adsorb NO below 200°C. Infrared spectra of and CO adsorption experiments on these Pd/zeolite PNA compounds suggest that Pd is located at the zeolites' exchange sites [14-3], and that the highly dispersed Pd atoms provide the low-temperature NO storage. Chen et al. [14-19] also found that the zeolite structure significantly affects NO storage capacity and desorption temperatures. The Pd sites in Pd/zeolites appear to adsorb NO directly, whereas in Pd/ceria compounds the NO must be converted to nitrates (NO_3^-) first to be stored.

The PNA described by Chen et al. [14-10] uses a Pd/ceria PNA material, although they also investigated Pd/zeolite materials. The catalyst mixture and loadings of the Johnson Matthey dCSC™ system are shown in **Table 14.1** [14-3]. The main challenge with the Pd/ceria material is that it has poor stability during HTA. Nevertheless, the NOx storage of the dCSC™ recovers after regeneration at 350°C or so [14-10], as shown in **Figure 14.6**.

Chen [14-3] indicated that Pd/zeolite PNA materials show high NO storage capacity until they saturate. The type of zeolite used for the PNA material also influences both the NO storage capacity and the NO bonding strength, which in turn affects the NOx release temperature.

Kim and other researchers looked at a PNA material that consists of highly dispersed Pd on a zeolite support [14-13]. This Pd/zeolite material offers improved NOx storage capability and improved resistance to sulfur dioxide (SO_2) poisoning [14-20]. Of the many zeolites available, Kim et al. [14-13] found that SSZ-13 chabazite (CHA) is a promising zeolite support for PNA applications. They varied the Pd content in Pd/SSZ-13 from 0 wt% to 5 wt%, and found that 2 wt% Pd in SSZ-13 provides the best NOx storage

FIGURE 14.6   dCSC™ NOx storage regeneration is effective at 350°C [14-10].

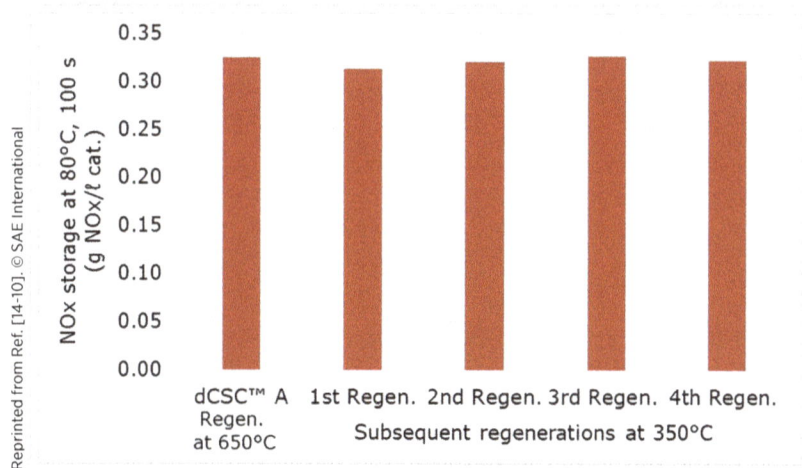

capability. Their work suggests that for <3 wt% Pd, most of the Pd is present as palladium oxide (PdO), but for >3 wt% Pd, HTA tends to promote sintering of the Pd and the formation of Pd–O–Pd clusters within the support material. Interestingly, at lower Pd concentrations, HTA appears to promote Pd dispersion and to produce Pd ions within Pd/SSZ-13. This latter feature is important, as NOx adsorbs onto ionic Pd sites within Pd/SSZ-13, unlike Pd/ceria, where the adsorption is on the ceria.

Khivantsev et al. [14-16] also investigated the effect of varying Pd concentration within Pd/SSZ-13. In the range of 0 wt% to 2 wt% Pd in SSZ-13, they found that the NO storage capacity is directly proportional to the Pd content, and that 2 wt% Pd has about 90% of the Pd available for NOx storage. However, the trend in NO storage capacity does not extend linearly from 2 wt% Pd to 5 wt% Pd. Lee et al. [14-21] also looked at Pd/ZSM-5 zeolite support, and found that 2 wt% Pd provided good performance, especially when the silicon (Si) to aluminum (Al) molar ratio (SAR) in the ZSM-5 was around 30:1.

Lee et al. [14-21] investigated the ability of Pd/ZSM-5 to adsorb NO at 120°C. The Pd/ZSM-5 material performs better when oxidized at 750°C than at 500°C. They found that the higher temperature promotes the formation of atomically dispersed Pd within the ZSM-5 structure, whereas at the lower temperature PdO agglomerates formed instead. The study results suggest that the dispersed Pd on the ZSM-5 is more responsible for chemisorptive NO adsorption than PdO clusters when the system is around 120°C. Lee et al. [14-21] also looked at varying the Pd loading and SAR of the ZSM-5, and their results indicate that there is an optimum molar ratio of Pd to Al within ZSM-5, estimated at 1:4. This result suggests that the Al in the ZSM-5 structure helps disperse the ionic Pd within the structure, and that increasing both together increases the overall NO storage capacity within this PNA material.

Zheng et al. [14-20] also evaluated various zeolite-based PNA materials, including Pd on beta zeolite (BEA), ZSM-5, and SSZ-13. They investigated the state of the Pd within these materials, including the oxidation states of the Pd and its dispersion and distribution within the zeolite structures. Zheng et al. [14-20] found that there were multiple Pd species within their materials, including Pd in the zeolites' cationic sites and PdO_x particles on the zeolite surfaces. They also found that the smaller sizes of the pores within SSZ-13 meant that more of the Pd was deposited on the surface, whereas in ZSM-5, more of the Pd was dispersed internally. Their assessment is that the primary NO adsorption sites are Pd(I) and PdO clusters, which are formed by reducing the cationic Pd(II) sites or the PdO_2 surface clusters within the zeolites. These adsorption sites can be reduced by CO or NO, depending on the exhaust gas composition. The presence of water promotes NO_2 desorption on these materials, which limits the NOx storage capacity of the material.

PNA performance can be affected by different ways of preparing the sample, including exchanging the H⁺ ions in the SSZ-13 with other ions, such as alkali metals and ammonium ion (NH_4^+), before adding the Pd [14-16]. Khivantsev et al. [14-16] found that the most active PNA material used incipient wetness impregnation to place 1 wt% Pd in NH_4/SSZ-13 with a SAR of 6:1. This Pd/SSZ-13 material appeared to have about 87% of the Pd available for NO storage. The experiments and related materials characterization work suggest that atomically dispersed Pd is the active trapping species. Moreover, it is Pd(I) that is bound to NO in the following chemisorption reaction:

$$Pd(II) + NO \rightarrow Pd(I) + NO^+ \tag{14.2}$$

The palladium carbonyl ($[Pd(CO)_2]^{2+}$) concentration in the PNA material increases with CO exposure from the feed gas. In the presence of water and CO, $[Pd–(CO)–(NO)]^{2+}$ complexes can also form within the PNA material.

Crocker and colleagues at the University of Kentucky are working with a partner group at the University of California-Berkeley to model the Pd within the zeolite structure [14-18]. The modeling results suggest that the Pd^{2+}, Pd^+, and Pd^0 characteristics are mostly independent of the neighboring Al within the zeolite.

14.1.3 Washcoat and Substrate

More than in most other EAS, the washcoat in a PNA strongly affects the function of the overall device. For example, the choice of ceria or zeolite support influences the Pd state and distribution and where the NOx adsorbs, as discussed in Sections 14.1.1 and 14.1.2.

Ceria in the washcoat not only influences the Pd in a PNA, but can also directly affect NOx storage capacity. For example, Cordatos and Gorte [14-5] found that NO reduces to N_2 or nitrous oxide (N_2O) on Pd/ceria. They also found that chemically reducing the ceria support in the Pt/ceria material increased the amount of adsorbed NO. This effect on performance was not seen on Pd/α-Al_2O_3, suggesting that the oxidation state of the ceria is responsible for the change in NO adsorption capacity.

Jones et al. [14-22] also looked at blending 5%–20% praseodymium oxide (Pr_2O_3) in CeO_2 to make a support for Pt, Pd, or both. They found that the Pt-Pd mix on an 80% CeO_2–20% Pr_2O_3 mixed oxide support worked best, including the benefits to overall NOx storage capacity and thermal stability. Nevertheless, the mixed oxide thermal stability was worse than that of a Pd/zeolite PNA material.

Zeolite supports have therefore become a more common area of focus in recent years. Bench reactor studies to assess NOx trapping and release properties of PNA materials under various feed compositions by Zheng et al. [14-20] found that the zeolite structure affects both where the Pd adsorption sites are located and the NO mass transport to and from the internal adsorption sites. For example, the smaller sizes of SSZ-13 pores means that more of the Pd is deposited on the surface of the zeolite, whereas in ZSM-5, which has larger pores, more of the Pd is dispersed internally [14-20].

Lardinois, Crocker et al. [14-18] have been focusing on stoichiometric operating conditions for combined PNA and HC trap systems for use with spark-ignited gasoline engines, again to improve cold start NOx control. They have found that BEA zeolite is a promising material, but the structure of BEA has nine different Al sites, making it very complex to understand. Thus, they also made a CHA PNA material to have a simpler zeolite model system to study. The commercial BEA Lardinois et al. [14-18] used had an SAR of 11:1, and the CHA, an SAR of 14:1. The Pd in the Pd/CHA was dispersed in particles generally <2 nm in diameter, albeit with some clusters over 20 nm in diameter.

Lee et al. [14-21] found that Pd/ZSM-5 needed a higher temperature—over 750°C—to activate; however, HTA at this temperature decreased the NOx adsorption capacity of the material. Thus more work is needed to understand the appropriate temperature to prepare the PNA for use.

PNA substrates are usually the same materials used for other EAS subsystems. Thus, cordierite and silicon carbide (SiC) are typical choices. As the NOx storage capacity is proportional to the zeolite content, although, some PNA substrates are extruded zeolites, which saves a process step and effectively makes the substrate the washcoat.

14.1.4 Failure Modes

PNA are subject to the typical failure modes that affect other EAS components, including sulfur poisoning and HTA. There are some distinct modes as well, which generally involve deactivating the catalyst by reducing the Pd ions in the PNA to a metallic form.

The introduction of ultra-low sulfur diesel in the United States and elsewhere has helped mitigate sulfur oxides (SOx) in the exhaust, but SOx can also come from the lubricant. SOx interfere with the adsorption of NOx on PNA materials, as they do with other materials used to manage NOx. In the case of PNA, SO_2 blocks the Pd sites for adsorption in Pd/ceria [14-13]. Moreover, the NOx storage function of Pd/ceria does not fully recover after a desulfation cycle above 600°C. Thus, despite the promise shown by Pd/ceria, it is too vulnerable to SOx poisoning to be a useful production PNA material. By contrast, Chen [14-3] shows that most of the performance of Pd/zeolite materials is recovered after desulfation and that these materials are stable after multiple sulfation-de-sulfation cycles, making them robust to SOx poisoning. These results are confirmed by Liu et al. [14-17], where various sulfur loadings of 0.5–2.0 g sulfur (S) per liter of PNA were imposed on a Pd/CHA PNA material, but this sulfur had a negligible effect on the NOx storage capacity. Moreover, they found that sulfur both delays the release of NOx to higher temperatures and inhibits CO oxidation on Pd/CHA.

Lee et al. [14-13] indicate that the best conditions for HTA to rapidly age the Pd/SSZ-13 PNA material to a FUL state are in the range of 700°C–750°C for about 5 h. Sodium or potassium ions present in zeolite that are residuals from the starting material appear to inhibit the formation of Pd ions in the structure by binding to the Al sites within the SSZ-13 structure. Khivantsev et al. [14-16] evaluated their HTA protocol, which used 10% water in a gas feed at 750°C for 16 h. They saw a 10%–20% decrease in NO storage performance after aging, which was less severe than the effect seen by Lee et al. [14-13]. Khivantsev et al. [14-16] examined the material, and found that HTA promotes the formation of PdO clusters within the washcoat, and that these PdO clusters are not active for NOx trapping. Given enough time, the HTA also de-aluminates the SSZ-13, which also reduces the NOx trapping function, since the Al sites within the SSZ-13 zeolite support the Pd^{2+} ions. Lee et al. [14-13] find that HTA above 750°C promotes de-alumination of the SSZ-13, as does extended aging of 10–15 h at 750°C. Ji et al. [14-14] also found that aging appears to deconstruct the Pt–Pd alloy, although the catalysts remain near each other on the CeO_2–ZrO_2 support after aging, as shown in **Figure 14.7**.

Lee et al. [14-21] examined vacuum treatment of PNA materials, and found that the Pd in Pd/ZSM-5 becomes fully metallic from this treatment. Only 34% of the Pd on Pd/SSZ-13 becomes metallic, which suggests that the main difference is the stability of the Pd^{2+} within the zeolite structure.

FIGURE 14.7 Aging of Pt-Pd alloy on CeO$_2$-ZrO$_2$ support [14-14].

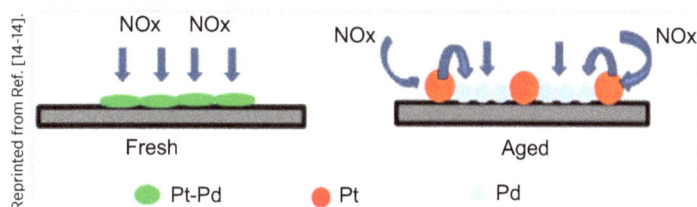

Lee et al. [14-13] evaluated the effect of reductive aging of Pd/zeolite catalyst by CO or hydrogen (H$_2$). This aging mechanism appears to form Pd metal within the material, which does not store NO. Reductive aging by CO above 500°C strongly decreased the NO adsorption capacity by promoting the formation of Pd metal in the PNA material. H$_2$ exposure had a lesser reductive aging effect, although Lardinois et al. [14-18] reduced a Pd/BEA PNA material with H$_2$ at 500°C, and the process formed Pd metal. CO has a stronger effect than H$_2$ because palladium carbonyl complexes (Pd(CO)$_n$) are volatile and therefore mobile within the structure of SSZ-13 [14-13]. As the CO is removed, the Pd forms metallic clusters from sintering. Oxidation with air can convert some of the Pd metal back into active species, but Pd/BEA does not recover its original storage capacity after the reduction–oxidation cycle. Pd/CHA also has its Pd reduced to metal when reacted with H$_2$ at 300°C, and after subsequent oxidation it is able to recover some of its adsorption performance, but not all [14-18].

Theis and Lambert [14-7] found that reactions with NO and reductants such as C$_2$H$_4$, CO, H$_2$, or CO+H$_2$ convert Pd^{2+} ions to PdO$_x$ over time, which degrades the PNA trapping performance. With some careful experimentation, they have determined that some PdO is reduced during each test, and that this reduction also degrades the PNA performance over time. To support this hypothesis, Theis and Lambert [14-7] found that there was very little reaction between the PdO$_x$ and reducing agent during the initial storage period, which contradicts the more common hypothesis that CO or similar reduces a PdO$_x$ site that subsequently stores NOx. Instead, when NO and the reducing agent are introduced together, they can form a chemical intermediate on the PdO$_x$ site that then promotes the reduction of that PdO$_x$ site. Moreover, they found that NOx storage decreased from test to test, even as more PdO$_x$ sites were reduced. Therefore, for practical applications, a PNA designed for approximately stoichiometric exhaust gas will need to be operated to protect the Pd/zeolite materials from rich transients, especially ones with H$_2$ present, to mitigate the reduction of Pd ions in the Pd/zeolite materials [14-18].

14.2 Hydrocarbon Traps

The purpose of the HC trap is to absorb the HC emissions released from the engine during the cold start period, and then release them as the exhaust temperature increases to the point where the HC species can be converted in the EAS [14-23]. Work on HC traps dates back to the early 1990s, such as that described by Hochmuth et al. [14-24]. HC traps use an adsorbent material, typically a zeolite, to provide the HC trap function by chemisorption on the zeolite of the HC species that flow through the HC trap system. As with the PNA described in Section 14.1, HC traps are intended to work during both on-cycle and in-use conditions. This section describes the function, materials, and potential issues of HC traps.

14.2.1 Hydrocarbon Trap Function

It is essential that the HC trap adsorbs HC at low temperatures, and releases (desorbs) the HC into the EAS for subsequent conversion, as noted by Lafyatis et al. [14-25]. The HC trap sizing needs to balance HC storage capacity and breakthrough during low temperature operation with how the HC trap system affects overall EAS system warm-up. Adding the HC trap places a thermal mass in the EAS that will increase the time to get to the light-off temperature in the downstream EAS subsystem that converts HC into CO_2 and water, such as the TWC or DOC. In a typical light-duty spark-ignition (SI) gasoline engine, the lead or close-coupled TWC takes about 100 s to reach its light-off temperature during the warm-up phase.

An additional challenge of managing the combination of the HC trap and the HC oxidizing catalyst downstream of it is that the desorption of HC from the HC trap may not match well with the quantity of oxygen stored in or available to the oxidation catalyst [14-23]. The details of oxygen storage in an oxidation catalyst are covered in Chapter 8, and in a TWC, in Chapter 9.

Although experiments can evaluate the effectiveness of a candidate HC trap material for a given HC species, real engine exhaust contains a range of HC species. Therefore, the HC trap performance must generalize to the most common HC species expected. Lupescu [14-23] observes that methane (CH_4), alkanes, ethene, propene (C_3H_6), and acetaldehyde (CH_3CHO) are not expected to trap well, whereas ethanol (C_2H_5OH), larger alkenes, and aromatics, such as benzene (C_6H_6) or toluene ($C_6H_5CH_3$) should store easily within an HC trap.

Kyriakidou et al. [14-26] and Binder et al. [14-27] have found that Pd/ZSM-5 shows the best overall trapping capacity for both HC and NOx, as shown in **Figure 14.8**.

FIGURE 14.8   Propene and NO storage capacity with respect to trap material [14-27].

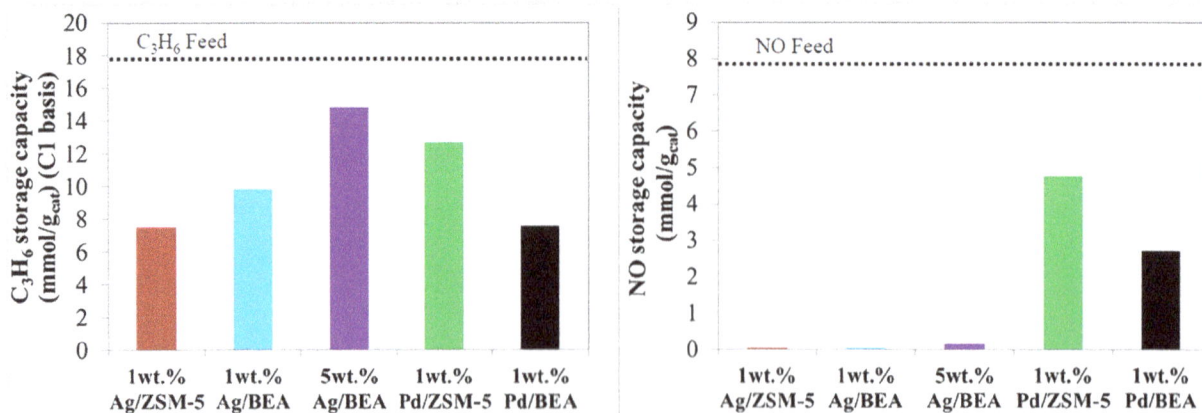

The 5 wt% silver (Ag) in BEA is better at trapping propene than the 1 wt% Pd in ZSM-5, but the Ag/BEA material does not trap a significant quantity of NO, and thus a separate PNA unit would be needed.

In addition, Burke et al. [14-28] and Iliyas et al. [14-29] have found that water competes for HC adsorption sites within the zeolite structure of the HC trap. Adding the HC trap alone or with a water trap will increase the time to get to the light-off temperature in the downstream EAS subsystem that converts HC into CO_2 and water.

As discussed in Section 14.1, Chilumukuru et al. [14-8] describe placing a combined PNA and HC trap upstream of an SCR–DPF, or SDPF, unit. This EAS configuration is intended to help an LDV with LDDE meet the Tier 3 bin 30 or SULEV30 regulatory targets described in Sections 3.2.1 and 3.2.2, respectively.

Another approach is to add HC trap functionality to a TWC to improve the low temperature emissions control performance during cold start [14-30]. One of the challenges noted is that HC traps often start releasing HC before the TWC has reached its light-off temperature of about 250°C, which limits the overall on-cycle conversion of HC. Future internal combustion engines are expected to be more efficient, which means that the exhaust gas temperature entering the EAS will be cooler on average and thereby increase the time to catalyst light-off temperature. For their advanced HC trap, Chang et al. [14-30] used an extruded zeolite substrate with 300 cpsi (47 cells/cm²), which has about double the HC trap material (4 g/in.³, or 0.24 g/cm³) compared to the conventional HC trap with about 2 g/in.³ (0.12 g/cm³) of zeolite washcoat on a ceramic substrate. Comparisons of the HC removal performance of the conventional and advanced HC traps combined with a low-temperature TWC, shown in **Figure 14.9**, indicate that either HC trap improves the low temperature HC emissions. Over the 30-s interval of the test, the inlet conditions were about 1700 ppm HC and 80°C gas temperature. In these conditions the conventional HC trap collects about 70% of the HC fed to it, and the advanced HC trap, about 85%.

On the subsequent warm-up of these materials, there is little difference in the HC release between the conventional and advanced HC traps, as shown in **Figure 14.10**.

FIGURE 14.9   Comparison of HC removal performance over 30 s with inlet gas temperature of 80°C and inlet HC concentration of 1700 ppm [14-30].

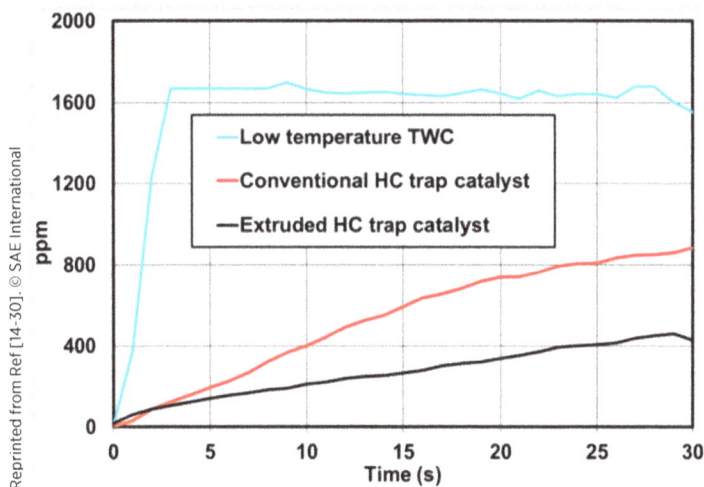

FIGURE 14.10 Comparison of HC release during temperature ramp from 80°C to 650°C [14-30].

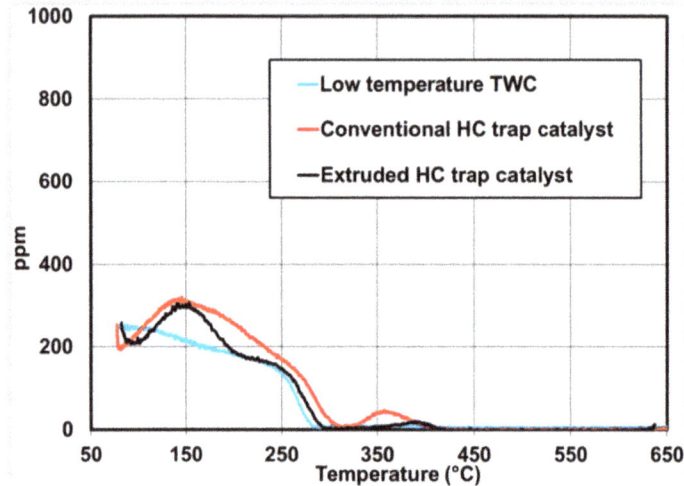

Reprinted from Ref. [14-30]. © SAE International

The HC traps have a release peak around 150°C that is not present with the low-temperature TWC. On-vehicle tests suggest that that the HC trap can reduce NMHC emissions over Bag 1 of the FTP-75 cycle by 5 mg/mi. from 22 mg/mi. to 17 mg/mi. The HC trap has no effect on NOx emissions.

14.2.2 Catalyst Formulations

As with the PNA, the main function of the HC trap is not catalytic, although Pd is the main trapping constituent, and it can have other catalytic activity. Weibenga et al. [14-31] studied several mixtures of Pd and rhodium (Rh) on one of two substrates, which are described in Table 14.2. The samples with the cordierite substrate had a TWC washcoat layered over an HC trap layer, where the Pd was included in both washcoat layers. The extruded zeolite substrate samples had a TWC washcoat layer only. In these samples, the Pd was only placed in the TWC washcoat; it was not exchanged into the zeolite. For both substrates, the Rh was only placed in the TWC washcoat layer.

Weibenga et al. [14-31] used bench reactor tests to assess how well these HC trap samples adsorbed the HC species propene, iso-pentane (C_5H_{12}), iso-octane (C_8H_{18}), and m-xylene (C_8H_{10}). These species were chosen to represent partly burned and unburned fuel species typical of gasoline exhaust. The zeolite structure appeared stable up to 900°C,

TABLE 14.2 Comparison of HC trap samples [14-31].

Substrate	Cell density (cpsi)	Wall thickness (mm)	Pd (g/ft.³)	Rh (g/ft.³)
Cordierite	400	0.102	10	2
			50	2
			100	2
Extruded zeolite	400	0.279	48	2
			68	2
			98	2

although the thermal aging degraded the effectiveness of the Pd to trap the smaller HC species. The m-xylene desorbs from these materials at a high enough temperature that it is oxidized by the TWC before it leaves the system.

14.2.3 Washcoat and Substrate

HC traps typically use a zeolite washcoat, because the pore size in the zeolite structure can be chosen to be about the same size as the HC molecules of interest, as illustrated in **Figure 14.11** [14-32].

Pore size affects the performance of an HC trap because the HC molecules adsorb to sites inside the pores, as shown in Figure 14.11. Experimental work by Czaplewski et al. [14-33] looked at the effect of pore size on adsorption using two different species, propene, and toluene. Their work indicated that the maximum adsorption by the trapping material occurs when the pore diameter is approximately 0.1 nm greater than the diameter of the hydrocarbon molecule, as shown in **Figure 14.12**. Czaplewski et al. [14-33] also found

FIGURE 14.11   Comparison of beta zeolite (BEA) structure to propene and toluene (1 Å = 0.1 nm) molecules [14-32].

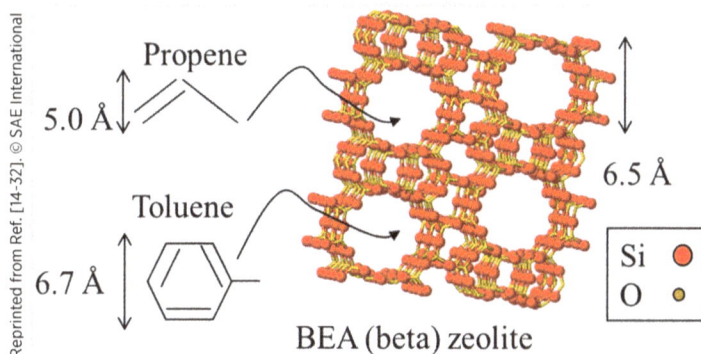

Reprinted from Ref. [14-32]. © SAE International

FIGURE 14.12   Effect of zeolite pore size on propene and toluene adsorption [14-33].

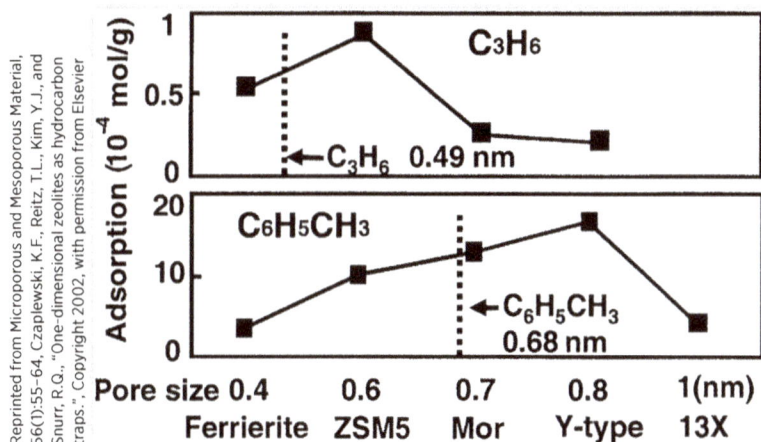

Reprinted from Microporous and Mesoporous Material, 56(1):55–64, Czaplewski, K.F., Reitz, T.L., Kim, Y.J., and Snurr, R.Q., "One-dimensional zeolites as hydrocarbon traps," Copyright 2002, with permission from Elsevier

that a three-dimensional porous zeolite structure did not adsorb HC species as well as a one-dimensional zeolite pore structure did.

There has been work toward finding an optimal zeolite pore size for an HC trap application [14-34]. The study showed that the zeolite HC trap material could not adsorb small HC molecules, specifically ethene. Smaller HC species are harder to trap and harder to oxidize than larger species, since there are fewer carbon–carbon bonds to split during catalytic combustion. Nevertheless, there may be ways to optimize the pore size for the set of HC species expected from the engine during the cold start period and similar operation with cool exhaust. A given zeolite material has only one characteristic pore size, and so the optimum may still fail to trap certain HC species. If a wide range of species is expected, a blend of zeolite HC trap materials could be zone coated or layered onto the monolith, as discussed in Section 15.1, Complex Catalyst Coatings, to provide the best overall capture of HC.

As with other EAS component washcoats, additional components can be mixed with the zeolites to enhance certain properties of the HC trapping material. For example, adding Ag to zeolite HC traps can delay desorption of HC from the material by increasing the temperature range over which desorption occurs and by releasing less HC at any given temperature as it increases [14-35]. Kyriakidou et al. [14-26] found that adding even 1 wt% Ag to BEA zeolite significantly improved its ability to trap propene at low temperature in the presence of water, and 5 wt% Ag improved performance further, as shown in **Figure 14.13**. It appears that 5 wt% Ag$^+$ in BEA promotes propene oxidation, as CO_2, ethene, and formaldehyde (CH_2O) are all released from the HC trap material [14-26]. A risk of using Ag/zeolite is that the Ag$^+$ can become volatile and deposit in the TWC or DOC downstream, where it reacts with the Pd and other metallic catalysts and thereby degrades the catalytic function of the downstream unit [14-35]. Because of this risk with Ag, work remains to find a suitable additive to the HC trap zeolite to

FIGURE 14.13   Propene trapping performance with respect to silver content [14-26].

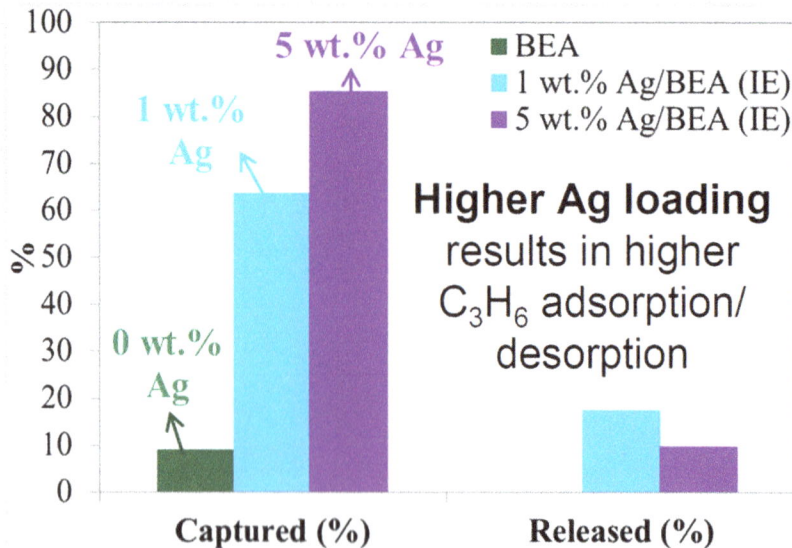

mitigate HC release from the trap before the TWC or similar EAS can support high conversion of HC.

Increasing the zeolite loading in an HC trap can increase the HC storage capacity [14-32]. With the higher zeolite loading, there are more storage sites for HC within the trapping material, which means that the HC coverage is more uniform within the HC trap during operation. This coverage improves the trapping efficiency as well as the HC storage capacity.

These HC trap zeolite materials are supported on a monolith substrate, which is usually cordierite or silicon carbide (SiC), depending on the engine and application. As with PNA systems, though, the HC storage capacity is proportional to the zeolite content, and so an extruded zeolite substrate can be used to significantly increase the storage capacity in the HC trap [14-32].

14.2.4 Failure Modes

The quantity of HC adsorbed by the HC trap is affected by the composition of the exhaust gas. For example, some experiments suggest that water may decrease the storage capacity by adsorbing to the sites that trap HC species. This competition between HC and water for zeolite sites, which are hydrophilic, is believed to come from water being polar whereas most HC species are not [14-36]. Lafyatis et al. [14-25] examined one approach to mitigate this competition for trapping sites in the zeolite, which is to use a water trap upstream of the HC trap. The water trap would remove water from the exhaust gas and thereby expose the HC trap to HC species only. The water trap would need to hold enough water during the warm-up period that the water that did break through had only a slight influence on the HC trap performance. The water trap would also need to release the trapped water back into the exhaust once the EAS was warm enough to release and convert the HC from the HC trap into CO_2 and water. Given that the exhaust gas leaving the engine has about 10% water in it, a water trap that removes all of the water vapor during the warm-up period is probably not feasible.

Aging is a significant concern for HC traps, especially HTA over 700°C [14-2]. The deterioration can lead to the HC trap being useless. Hiramoto et al. [14-37] investigated how to improve the thermal stability of a HC trap zeolite material, and **Figure 14.14** shows how the improved zeolite crystal structure starts collapsing at about 800°C, which is 50°C higher than the 750°C at which the baseline HC trap material does. The thermal durability of the HC trap can be improved by picking a different material with similar chemical characteristics, or mitigated by ensuring that the operating conditions do not exceed the temperature at which aging accelerates.

Another concern is that HC traps may facilitate the formation of N_2O in the EAS [14-38]. The N_2O concentration seems to vary with the Pt or Pd content of the TWC or DOC and also to be affected by the zeolite material chosen for the HC trap. It appears that N_2O concentration peaks when the exhaust is between 150°C and 225°C, as shown in **Figure 14.15**. The peak in N_2O formation increases in temperature and decreases in magnitude as the Pt:Pd ratio decreases, although above 200°C the selectivity results suggest that N_2O formation is governed by thermodynamics in that

---

FIGURE 14.14   Effect of hydrothermal aging on HC trap zeolite structure [14-37].

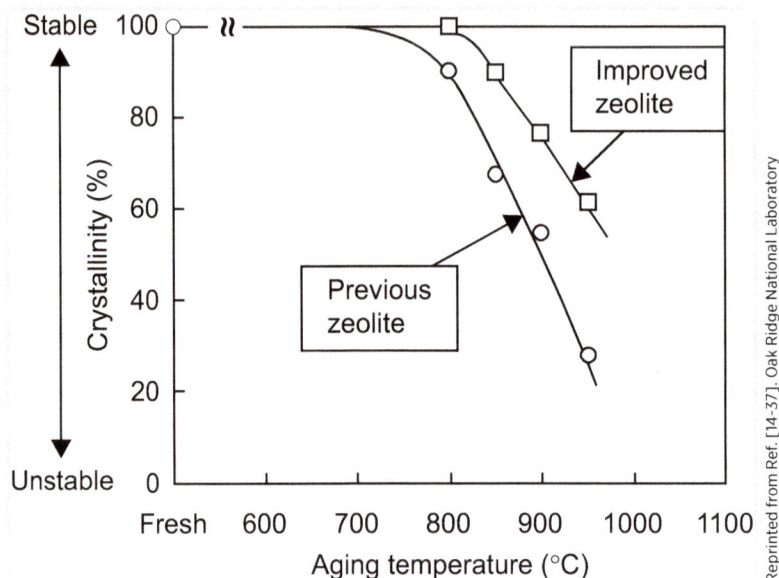

FIGURE 14.15 N_2O formation from zeolite HC trap [14-38].

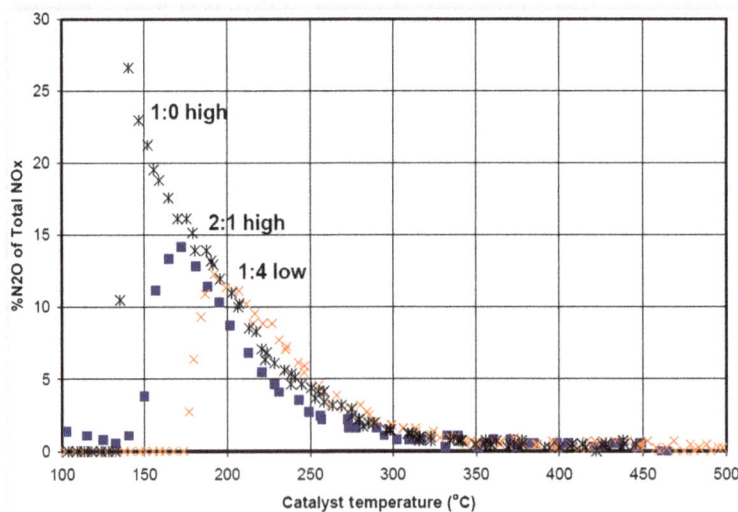

---

regime. Another solution to reduce N_2O formation could be to place a PNA upstream of the HC trap. This EAS configuration could let the temperature of the HC trap exceed the N_2O temperature window; nevertheless, there is still the risk of N_2O formation because NOx and HC will be present in the HC trap once the PNA warms up and starts releasing N_2O. By contrast, Chen et al. [14-10] found that the complete dCSC™ system tended to suppress N_2O emissions, compared to a standard DOC, as shown in **Figure 14.16**.

FIGURE 14.16   N$_2$O formation and NOx storage by DOC and dCSC™ [14-10].

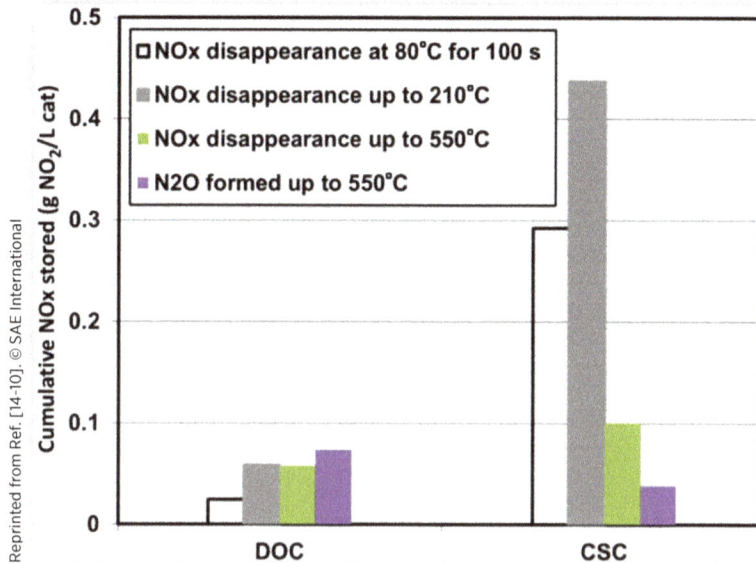

References

14-1.   Varmus, H., "The Writing Life of Nobel Prize-Winning Scientist Harold Varmus," *The Washington Post* (Washington, D.C: The Washington Post Co., 2010).

14-2.   Lupescu, J.A., Chanko, T.B., Richert, J.F., and DeVries, J.E., "Treatment of Vehicle Emissions from the Combustion of E85 and Gasoline with Catalyzed Hydrocarbon Traps," *SAE Int. J. Fuels Lubr.* **2**(1): 485–496, 2009, https://doi.org/10.4271/2009-01-1080.

14-3.   Chen, H.-Y., "Zeolite Supported Pd Catalysts for Low Temperature NO and HC Storage," *2016 CLEERS*, Ann Arbor, MI, 2016.

14-4.   Descorme, C., Gélin, P., Primet, M., and Lécuyer, C., "Infrared Study of Nitrogen Monoxide Adsorption on Palladium Ion-Exchanged ZSM-5 Catalysts," *Catalysis Letters* **41**, no. 3-4 (1996): 133–138, https://doi.org/10.1007/BF00811479.

14-5.   Cordatos, H. and Gorte, R., "CO, NO, and H$_2$ Adsorption on Ceria-Supported Pd," *Journal of Catalysis* **159**, no. 1 (1996): 112–118, https://doi.org/10.1006/jcat.1996.0070.

14-6.   Pihl, J. and Majumdar, S.S., "NO Adsorption and Desorption Phenomena on a Pd-Exchanged Zeolite Passive NOx Adsorber," *2018 CLEERS Workshop*, Ann Arbor, MI, 2018.

14-7.   Theis, J.R. and Lambert, C.K., "Effect of Reductants on the NOx Storage and Release Performance of Pd/CZO Low Temperature NOx Adsorbers," *2018 CLEERS Workshop*, Ann Arbor, MI, 2018.

14-8.   Chilumukuru, K., Gupta, A., Ruth, M., Cunningham, M., Kothand araman, G., Cumaranatunge, L., and Hess, H., "Aftertreatment Architecture and Control Methodologies for Future Light Duty Diesel Emission Regulations," *SAE Int. J. Engines* **10**: 1580–1587, 2017, https://doi.org/10.4271/2017-01-0911.

14-9.   Henry, C., Lengenderfer, D., Yezerets, A., Ruth, M., Chen, H.-Y., Hess, H., and Naseri, M., "Passive Catalytic Approach to Low Temperature NOx Emission Abatement," *DEER*, Detroit, MI, 2011.

14-10. Chen, H.-Y., Mulla, S., Weigert, E., Camm, K., Ballinger, T., Cox, J., and Blakeman, P., "Cold Start Concept (CSC™) A Novel Catalyst for Cold Start Emission Control," *SAE Int. J. Fuels Lubr.* **6**(2): 372–381, 2013, https://doi.org/10.4271/2013-01-0535.

14-11. Sharp, C., "Low NOx Diesel Engine Program Update and Next Steps," *2017 CLEERS Workshop*, Ann Arbor, MI, 2017.

14-12. Sharp, C., Webb, C.C., Neely, G., Carter, M., Yoon, S., and Henry, C., "Achieving Ultra Low NOX Emissions Levels with a 2017 Heavy-Duty On-Highway TC Diesel Engine and an Advanced Technology Emissions System - Thermal Management Strategies," *SAE Int. J. Engines* **10**(4): 1697–1712, 2017, https://doi.org/10.4271/2017-01-0954.

14-13. Lee, J., Ryou, Y., Kim, Y., Hwang, S., Lee, H., Kim, C.H., and Kim, D.H., "Characteristics of Pd-Based Passive NOx Adsorbers (PNA) for Cold Start Application," *2018 CLEERS Workshop*, Ann Arbor, MI, 2018.

14-14. Ji, Y., Bai, S., Xu, D., Crocker, M., Darab, J., and Harris, D., "Pt- and Pd-Promoted Ce-Zr Mixed Oxide for Low Temperature NOx Adsorber Applications," *2016 CLEERS*, Ann Arbor, MI, 2016.

14-15. Ji, Y., Xu, D., Bai, S., Graham, U., Crocker, M., Chen, B., Shi, C., Harris, D., Scapens, D., and Darab, J., "Pt- and Pd-Promoted CeO_2–ZrO_2 for Passive NOx Adsorber Applications," *Industrial & Engineering Chemistry Research* **56**, no. 4 (2017): 111–125, https://doi.org/10.1021/acs.iecr.6b03793.

14-16. Khivantsev, K., Jaegers, N., Kovarik, L., Wang, Y., Gao, F., and Szanyi, J., "Toward Rational Synthesis and Molecular Level Understand ing of Pd/Zeolite Passive NOx Adsorbers (PNA)," *2018 CLEERS Workshop*, Ann Arbor, MI, 2018.

14-17. Liu, D., Mantarosie, L., Islam, H., Novák, V., Sarwar, M., Chen, H.-Y., Collier, J.E., and Thompsett, D., "An "in-situ" Infrared and XAS Study on NO Adsorption on Pd/Zeolite under Complex Gas Feed," *2017 CLEERS*, Ann Arbor, MI, 2017.

14-18. Lardinois, T., Bates, J., Gounder, R., Pace, R.B., Ji, Y., Crocker, M., Van der Mynsbrugge, J., and Bell, A., "Pd/Zeolite Passive HC/NOx Adsorbers," *2018 CLEERS Workshop*, Ann Arbor, MI, 2018.

14-19. Chen, H.-Y., Collier, J.E., Liu, D., Mantarosie, L., Durán-Martín, D., Novák, V., Rajaram, R.R., and Thompsett, D., "Low Temperature NO Storage of Zeolite Supported Pd for Low Temperature Diesel Engine Emission Control," *Catalysis Letters* **146**, no. 9 (2016): 1706–1711, https://doi.org/10.1007/s10562-016-1794-6.

14-20. Zheng, Y., Kovarik, L., Engelhard, M.H., Wang, Y., Wang, Y., Gao, F., and Szanyi, J., "Low-Temperature Pd/Zeolite Passive NO_x Adsorbers: Structure, Performance, and Adsorption Chemistry," *The Journal of Physical Chemistry* **121**, no. 29 (2017): 15793–15803, https://doi.org/10.1021/acs.jpcc.7b04312.

14-21. Lee, J., Ryou, Y., Cho, S.J., Lee, H., Kim, C.H., and Kim, D.H., "Investigation of the Active Sites and Optimum Pd/Al of Pd/ZSM–5 Passive NO Adsorbers for the Cold-Start Application: Evidence of Isolated-Pd Species Obtained after a High-Temperature Thermal Treatment," *Applied Catalysis B: Environmental* **226** (2018): 71–82, https://doi.org/10.1016/j.apcatb.2017.12.031.

14-22. Jones, S., Ji, Y., and Crocker, M., "CeO_2-M_2O_3 Passive NOx Adsorbers for Cold Start Applications," *2016 CLEERS*, Ann Arbor, MI, 2016 .

14-23. Lupescu, J., "Overview of Automotive Zeolite HC Trap, Challenges for Gasoline Fuel and Current Research Areas," *2015 CLEERS*, Dearborn, MI, 2015.

14-24. Hochmuth, J.K., Burk, P.L., Tolentino, C., and Mignano, M.J., "Hydrocarbon Traps for Controlling Cold Start Emissions," SAE Technical Paper 930739 1993, https://doi.org/10.4271/930739.

14-25. Lafyatis, D.S., Ansell, G.P., Bennett, S.C., Frost, J.C., Millington, P.J., Rajaram, R.R., Walker, A.P., and Ballinger, T.H., "Ambient Temperature Light-off for Automobile Emission Control," *Applied Catalysis B: Environmental* **18**, no. 1–2 (1998): 123–135, https://doi.org/10.1016/S0926-3373(98)00032-0.

14-26. Kyriakidou, E.A., Choi, J.-S., Toops, T.J., and Parks II, J.E., "A Comparative Study of ZSM-5 and BEA-Zeolites for Hydrocarbon Trap Applications under Cold-start Conditions," *2016 CLEERS*, Ann Arbor, MI, 2016.

14-27. Binder, A.J., Kyriakidou, E.A., Toops, T.J., and Parks II, J.E., "Approaching the 150°C Challenge with Passive Trapping Materials and Highly Active Oxidation Catalysts," *2017 CLEERS*, Ann Arbor, MI, 2017.

14-28. Burke, N., Trimm, D., and Howe, R.F., "The Effect of Silica: Alumina Ratio and Hydrothermal Ageing on the Adsorption Characteristics of BEA Zeolites for Cold Start Emission Control," *Applied Catalysis B: Environmental* **46**, no. 1 (2003): 97–104, https://doi.org/10.1016/S0926-3373(03)00181-4.

14-29. Iliyas, A., Zahedi-Niaki, H., and Eić, M., "One-Dimensional Molecular Sieves for Hydrocarbon Cold-Start Emission Control: Influence of Water and CO_2," *Applied Catalysis A: General* **382**, no. 2 (2010): 213–219, https://doi.org/10.1016/j.apcata.2010.04.048.

14-30. Chang, H.-L., Chen, H.-Y., Koo, K., Rieck, J., and Blakeman, P., "Gasoline Cold Start Concept (gCSC™) Technology for Low Temperature Emission Control," *SAE Int. J. Fuels Lubr.* **7**(2): 480–488, 2014, https://doi.org/10.4271/2014-01-1509.

14-31. Wiebenga, M.H., Oh, S.H., and Qi, G., "Cold-Start Emission Reduction Potential and Limitations of Commercial Passive Hydrocarbon Adsorbers," *Emission Control Science and Technology* **3**, no. 1 (2016): 47–58, https://doi.org/10.1007/s40825-016-0052-0.

14-32. Nunan, J., Lupescu, J., Denison, G., Ball, D., and Moser, D., "HC Traps for Gasoline and Ethanol Applications," *SAE Int. J. Fuels Lubr.* **6**(2): 430–449, 2013, https://doi.org/10.4271/2013-01-1297.

14-33. Czaplewski, K.F., Reitz, T.L., Kim, Y.J., and Snurr, R.Q., "One-Dimensional Zeolites as Hydrocarbon Traps," *Microporous and Mesoporous Materials* **56**, no. 1 (2002): 55–64, https://doi.org/10.1016/S1387-1811(02)00441-9.

14-34. Kanazawa, T., "Development of Hydrocarbon Adsorbents, Oxygen Storage Materials for Three-Way Catalysts and NOx Storage-Reduction Catalyst," *Catalysis Today* **96**, no. 3 (2004): 171–177, https://doi.org/10.1016/j.cattod.2004.06.119.

14-35. Murakami, K., Tominaga, S., Hamada, I., Nagayama, T., Kijima, Y., Katougi, K., and Nakagawa, S., "Development of a High Performance Catalyzed Hydrocarbon Trap Using Ag-Zeolite," SAE Technical Paper 2004-01-1275, 2004, https://doi.org/10.4271/2004-01-1275.

14-36. Mukai, K., Kanesaka, H., Akama, H., and Ikeda, T., "Adsorption and Desorption Characteristics of the Adsorber to Control the HC Emission from a Gasoline Engine," SAE Technical Paper 2004-01-2983, 2004, https://doi.org/10.4271/2004-01-2983.

14-37. Hiramoto, Y., Takaya, M., Yamamoto, S., and Okada, A., "Development of a New HC-Adsorption Three-Way Catalyst System for Partial-ZEV Performance," SAE Technical Paper 2003-01-1861, 2003, https://doi.org/10.4271/2003-01-1861.

14-38. Lambert, C., Dobson, D., Gierczak, C., Guo, G., Ura, J., and Warner, J., "Nitrous Oxide Emissions from a Medium-Duty Diesel Truck Exhaust System," *SAE Int. J. Powertrains* **3**(1): 4–25, 2014, https://doi.org/10.1504/IJPT.2014.059410.

Combined Aftertreatment Systems

Alone we can do so little; together we can do so much.

—Helen Keller

Engineers have several constraints on their ability to design exhaust aftertreatment systems (EAS) for engines and vehicles, including EAS component cost, mass, flow restriction, and volume. The attraction of combining emissions control functions into a single catalytic converter in the EAS is to mitigate these constraints. For example, the combined system can have improved thermal management, especially for warm-up, since what had been the downstream catalytic converter is now closer to the start of the EAS. The combined system can also be smaller and lighter than two separate catalytic converters normally are. Nevertheless, the combination of systems is rarely as simple as combining or layering catalytic washcoats, or putting a catalytic washcoat on a particulate filter substrate.

Each of the constituent catalytic converters described in this chapter has been described individually earlier in this book, including the following:

- Thermal management of catalytic converters in Chapter 7
- Oxidation catalysts in Chapter 8
- Three-way catalysts (TWCs) in Chapter 9
- Lean NOx traps (LNTs) in Chapter 10
- Diesel particulate filters (DPFs) in Chapter 11
- Gasoline particulate filters (GPFs) in Chapter 11
- Selective catalytic reduction (SCR) systems in Chapter 12
- Ammonia slip catalysts (ASCs) in Chapter 13
- Hydrocarbon traps (HC traps) in Chapter 14

After reading this chapter, the engineer should better understand the rationale for certain combinations of EAS subsystems. Combining catalytic functions within one monolith through washcoat placement is discussed. Some combinations within one catalytic converter are described, including SCR on DPF; TWC on GPF; and HC trap and TWC. Lastly, combinations of catalytic converters to achieve passive SCR are described, including LNT+SCR and TWC+SCR.

15.1 **Complex Catalyst Coatings**

One way to combine functions within a single monolith is to apply multiple catalytic washcoats within the same monolith. The two main options are as follows:

- Axial zone coating, where one section along the length of a monolith channel has one catalyst, and another section, a second catalyst
- Layered zone coating, where one catalyst washcoat is placed over another to provide a blended function within one unit

Some specific applications of each zone coating approach are also discussed.

15.1.1 **Axial Zone Coating**

Axial zone coating is the process by which a specific catalytic washcoat is placed on only one axial section of a monolith support. It is a common way to incorporate several functions into one EAS unit. The process is relatively simple, since the monolith can be partly or fully dipped into the washcoat slurry to coat the relevant sections, as illustrated in **Figure 15.1**.

As mentioned by Girard et al. [15-1], a common example of axial zone coating in EAS is placing the ASC function on the downstream end of the final SCR system monolith. In this case, the SCR can be extended all the way through the monolith channel, or stopped short of where the ASC washcoat starts.

Another example is using axial zone coating of SCR washcoats within a monolith. One example documented by Stewart et al. [15-2, 15-3] shows that two axial zones of a washcoat containing Cu/SSZ-13 SCR catalyst were applied from the upstream and downstream ends of a DPF monolith, which led to an axial section where the washcoat was particularly thick, which is a waste of material. Villamaina et al. [15-4] studied an

FIGURE 15.1   Illustration of axial zone coating within an SCR system.

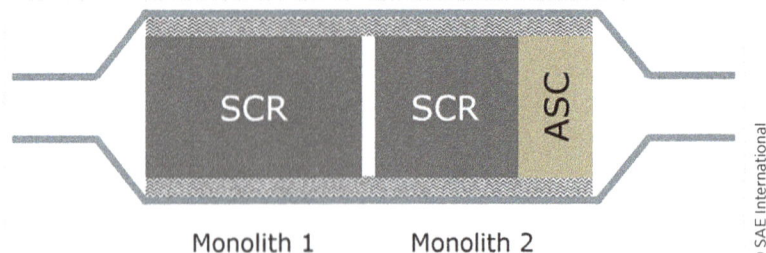

Monolith 1   Monolith 2

© SAE International

SCR in which the monolith had a copper (Cu) zeolite (Cu/zeolite) catalyst only, an iron (Fe) zeolite (Fe/zeolite) catalyst only, an axial zone-coated washcoat in which the Cu/zeolite preceded the Fe/zeolite, and an axial zone-coated washcoat in which the Fe/zeolite preceded the Cu/zeolite. As shown in **Figure 15.2**, the blend and order of the catalysts affected both the NO conversion and the nitrous oxide (N_2O) selectivity. In this case, having the Fe/zeolite come before the Cu/zeolite led to the best overall results.

FIGURE 15.2   Steady-state NO conversions (a) and N_2O concentrations (b) obtained in the standard SCR reaction. Space velocity is 75,000 h^{-1}, NH_3 = 500 ppm, NO = 500 ppm, O_2 = 5 vol.%, and H_2O = 5 vol.%. (Figure 1 in [15-4].)

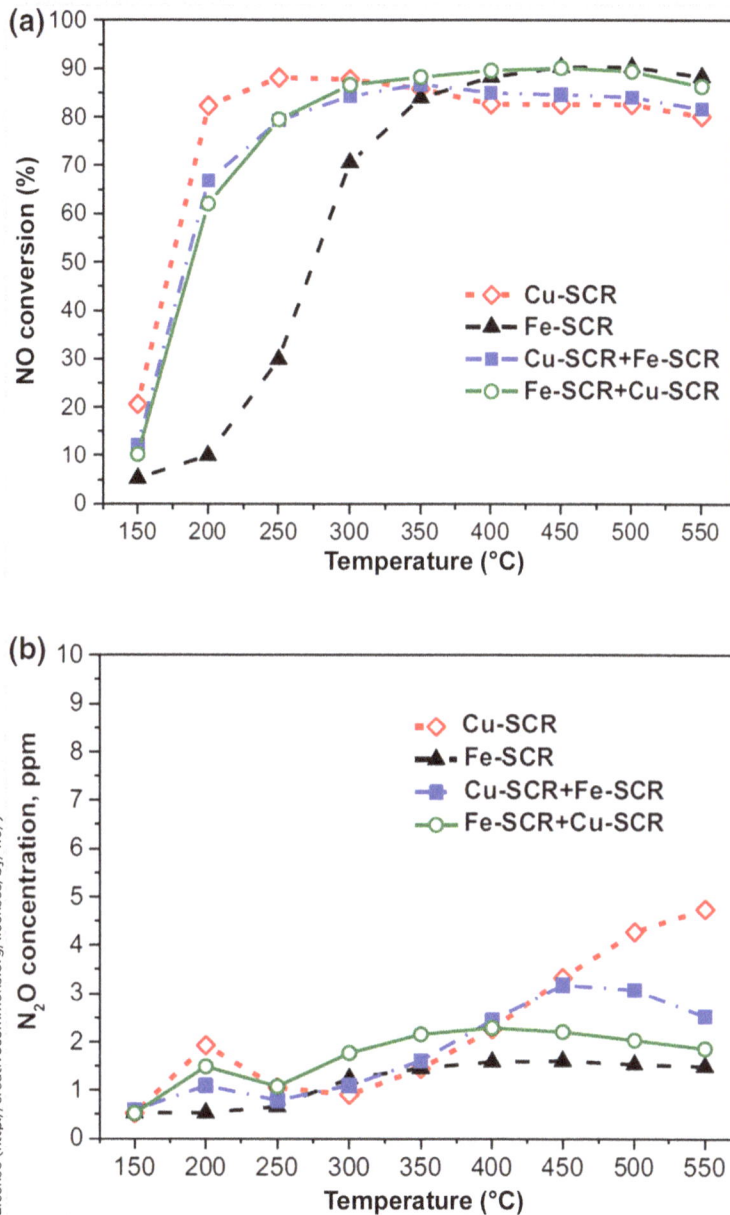

15.1.2 Layered Washcoats

Another approach to incorporating multiple catalytic functions within a monolith is to layer the washcoats over each other. This approach is used when blending the catalysts together into a single washcoat layer may cause one function to interfere with another.

Modern ASCs represent an example of an EAS system with a layered washcoat. As described in detail in Chapter 13, a Fe/zeolite SCR washcoat is layered over a platinum on alumina (Pt/Al_2O_3) ammonia (NH_3) oxidation layer, as shown in Figure 13.8. Some of the NH_3 entering the ASC diffuses through the upper washcoat layer to the oxidation catalyst where it is converted to nitrogen monoxide (NO) and nitrogen dioxide (NO_2). The nitrogen oxide (NOx) species then diffuse back through the SCR layer, where they react with the rest of the NH_3 to form nitrogen gas (N_2) and water (H_2O).

Another example is placing a layer of oxidation catalyst underneath an SCR washcoat in an SCR system. The oxidation catalyst layer allows the NO and NO_2 to re-equilibrate as they are consumed through the oxidation-reduction reaction,

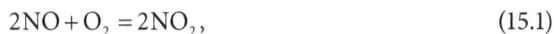

$$2NO + O_2 = 2NO_2, \tag{15.1}$$

which improves overall NOx conversion in the SCR system. Likewise, as described in Section 15.2, the SCR function in an SDPF is often layered with an oxidation catalyst washcoat, as the oxidation catalyst helps manage the particulate loading within the SDPF system. In some cases, the SCR layer is applied from the downstream side of the filter wall, with the oxidation catalyst on the upstream side.

A last example is the combination of TWC and hydrocarbon (HC) trap in different catalytic layers. As described in Section 15.4, the HC trap layer can be placed underneath the TWC so that as the HC is released from the HC trap layer it can react in the TWC layer.

15.2 Combined Selective Catalytic Reduction System and Diesel Particulate Filter

SCR systems and DPF are both proven technologies to remove NOx and particulate emissions, respectively, from diesel engine exhaust gas. One approach to decrease the overall cost and volume of the EAS for diesel engines is to integrate both of these functions into a single system, the SCR–DPF or SDPF [15-5]. The SDPF is also known by various trade names, including SCRF® from Johnson Matthey and EMPRO™ SCR.2F from BASF Catalysts.

Combining these systems for simultaneous NOx and particulate matter (PM) or particulate number (PN) control is a relatively new development. Some of the first system results were presented by Patchett and SantaMaria [15-6] from Engelhard (now BASF Catalysts). Nevertheless, Keenan [15-7] notes that SDPF has become a mainstream technology for light-duty diesel engine (LDDE) vehicle applications in Europe, not least because the overall package can be optimized to place the system closer to the turbine outlet than would be possible with the SCR and DPF units separated.

The EAS with SDPF needs to show the following four functions:

1. Conversion of NO to NO_2 to support the fast SCR reaction and passive soot regeneration
2. NOx reduction to form N_2
3. PM filtration and oxidation
4. NH_3 slip control

Of these functions, the NO to NO_2 conversion, from the forward reaction of Eq. 15.1, is one of the primary functions of the diesel oxidation catalyst (DOC) upstream of the SCR and DPF. But in the SDPF configuration, there is competition for NO_2 between the fast SCR reactions in Eq. 12.6 that convert NOx to N_2,

$$2NH_3 + NO + NO_2 \rightarrow 4N_2 + 3H_2O, \quad (12.6)$$

and the passive soot oxidation reaction in the DPF that removes carbonaceous soot ($C_{(s)}$) from the filter,

$$C_{(s)} + 2NO_2 \rightarrow CO_2 + 2NO. \quad (15.2)$$

When the DPF is a separate unit, it can also have an oxidation catalyst loaded into it to facilitate further oxidation of the NO back into NO_2 so that it can participate in the passive soot oxidation reaction (Eq. 15.2) multiple times before it leaves the DPF. However, the SDPF cannot have an oxidation catalyst because the catalyst facilitates the oxidation of NH_3 to NOx in competition with the desired SCR reactions.

Thus, because NO_2 plays an important role in NOx conversion in a flow-through SCR catalyst and in the passive regeneration of PM, NO_2 management is a key factor in the integrated system, irrespective of the ordering of the catalysts. This challenge has led to some SDPF configurations where the SCR washcoat is applied from the downstream side of the filter wall to help ensure that NO_2 encounters the soot in the filter before it is reduced. This configuration supports both of the needed SDPF functions: NOx conversion and PM trapping and removal. NH_3 slip resulting from high temperature excursions is expected to be handled by a downstream ASC, as described in Chapter 13.

Lee et al. [15-8] performed some of the earliest work on SDPF systems. They evaluated a combined Cu/zeolite SCR and wall-flow DPF system on a Chevy Silverado pick-up truck equipped with a prototype 4.9-L, 6-cylinder LDDE. The SDPF, called a two-way SCR/DPF catalyst by Lee et al. [15-8], was in an EAS downstream of a 0.85-ℓ close-coupled DOC and a larger, 2.3-L under-floor DOC. They evaluated how well the SPDF could simultaneously control NOx and PM. Most of the catalysts used were aged to simulate low-mileage and high-mileage conditions resulting from the high temperatures caused by DPF regeneration. The NOx reduction performance of the Cu/zeolite SDPF was found to be comparable with that of the standard Cu/zeolite flow-through SCR over the FTP-75 and US06 drive cycles, as described in Sections 4.1.1 and 4.1.3, respectively. The NOx conversion was independent of the level of soot loading in the filter, up to 5 g/L. Although

FIGURE 15.3 Cold flow tests indicating that the additional flow restriction from the SCR washcoat can be mitigated [15-9].

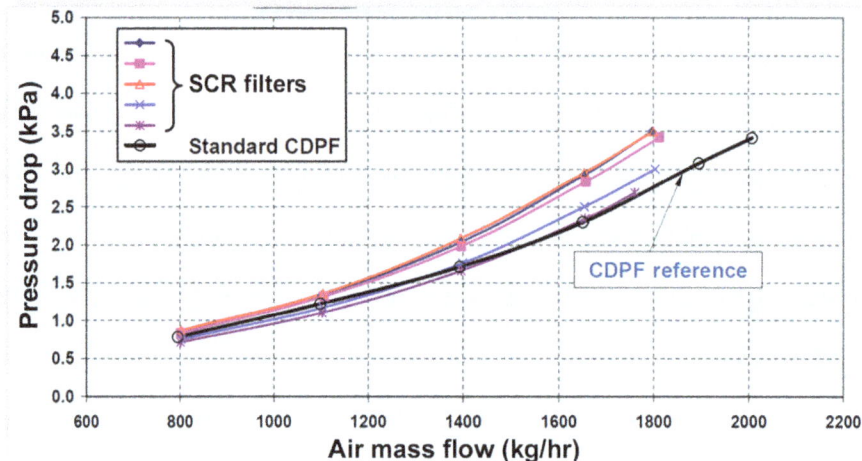

Reprinted from Ref. [15.9]. US Department of Energy

the filter was successfully regenerated by in-cylinder post injections, the multiple filter regenerations led to hydrothermal aging (HTA) of the oxidation and SCR catalysts, which led to significant performance degradation. The NOx conversion was also low during DPF regeneration because the NH_3 oxidized to NO at high temperatures, and was not available for the SCR reactions.

Oladipo et al. [15-9] discussed an SDPF as part of an EAS for a light–heavy-duty diesel engine (LHDDE). They evaluated the SDPF performance using both NH_3 and urea $[CO(NH_2)_2]$ as the reducing agent. The engine was the General Motors Duramax, a 6.6-L V-8 LHDDE meeting the 2004 EPA emission standards. The EAS included a DOC and an SDPF, where the DOC was a 5.1-L system with diameter of 7.5 in. (19 cm) and length of 7 in. (18 cm). The SDPF used a 17-L substrate with a diameter of 10.5 in. (26.7 cm) and a length of 12 in. (30.5 cm), and it underwent HTA for 48 h at 650°C for the NOx conversion study. Cold flow test results indicated that the SCR washcoat can be optimized to minimize the pressure drop penalty from adding it to the DPF wall, as shown in **Figure 15.3**, although these results do not include the pressure drop benefit of removing the SCR system and associated piping from the EAS. Note also that the comparison in Figure 15.3 is to a standard catalyzed DPF (CDPF).

Oladipo et al. [15-9] reported a peak NOx conversion of 95% from the SDPF. As shown in **Figure 15.4**, the NOx conversion was found to be comparable when either NH_3 or urea is used as reductants individually at an exhaust gas temperature of 250°C and an NH_3 to NOx ratio (ANR) of 1:1.

They also identified that urea decomposition, hydrolysis, and mixing were not limiting factors for NOx conversion. One possible reason for this observation might be the long mixing length between the injection points for urea or NH_3 and the SDPF front face in the EAS tested. A 4-h steady-state test with a space velocity (SV) of 21,000 h^{-1}, ANR of 1.0, and SDPF inlet gas temperature of 250°C caused high soot load to accumulate in the filter, which resulted in a 12% drop in NOx conversion. Experiments conducted on a flow-through SCR catalyst and a version of the SDPF after HTA for 50 h at 750°C indicate that the flow-through SCR catalyst has better NOx conversion across the target

FIGURE 15.4 NOx conversion and NH$_3$ slip at ANR = 1:1 using NH$_3$ or the comparable quantity of urea [15-9].

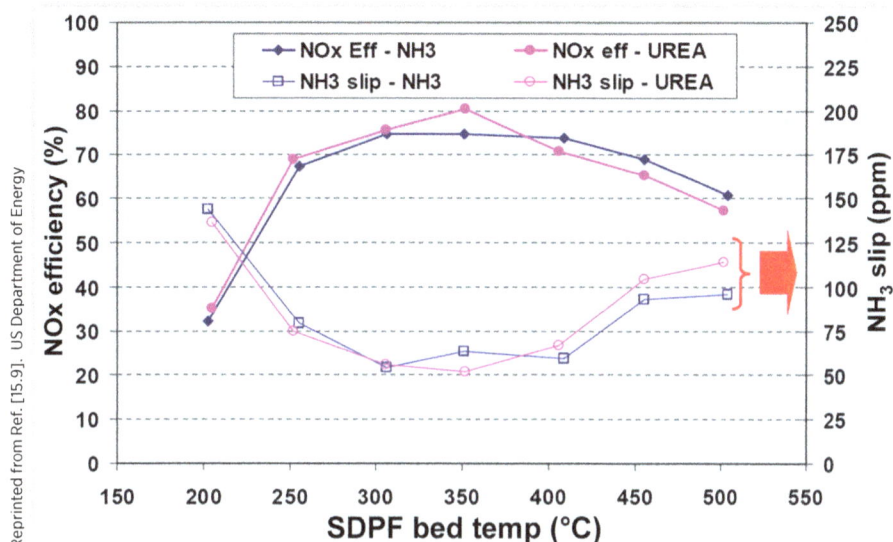

range of operating temperatures considered, 200°C–500°C. However, the SDPF shows a corresponding increase in NH$_3$ slip as well.

Devarakonda et al. [15-10] and Rappé [15-11, 15-12] have described work at the Pacific Northwest National Laboratory on SDPF systems. For example, an important consideration is balancing the SCR catalyst washcoat loading for good NOx conversion with the pressure drop across the SDPF unit. In addition, a design goal was to maximize both NOx conversion and passive soot oxidation per Eq. 15.2. An ultra-high porosity substrate made from cordierite was used to minimize the inherent flow restriction of the DPF substrate. Then, the SCR washcoat application was refined to minimize the effect of the washcoat on the DPF pressure drop. One approach is to place the SCR washcoat on the downstream side of the filter wall so that the NO$_2$ coming into the filter can help oxidize the trapped soot, as discussed above.

Several SCR catalyst formulations have been studied, including vanadium-based V$_2$O$_5$/WO$_3$/TiO$_2$ SCR catalysts and base-metal exchanged zeolites, especially Fe/zeolite and Cu/zeolite catalysts. In addition to the more general benefits of zeolite-based SCR catalysts, discussed in Chapter 12, they are also better suited to the SDPF application. Fe/zeolites have certain advantages such as high NOx conversion at the high-temperature range relevant to heavy-duty diesel engine (HDDE) applications, 300°C–650°C; direct NH$_3$ oxidation to N$_2$; and NO oxidation to NO$_2$. Fe/zeolites show this unique oxidation capability which can be used for passive regeneration of soot in the SDPF without a precious-metal catalyst that might otherwise oxidize NH$_3$. Cu/zeolites have high NOx conversion in the temperature range relevant to LDDE, 150°C–300°C, but they lose their activity at higher temperatures. Cu/zeolites can also form N$_2$O when there is a high NO$_2$ concentration in the exhaust gas. N$_2$O is a regulated, potent greenhouse gas (GHG) and needs to be minimized.

Chilumkuru et al. [15-13] described work on the Johnson Matthey diesel Cold Start Concept (dCSC™) catalyst for an advanced EAS. The dCSC™ system places a passive

NOx adsorber (PNA) and HC trap in a close-coupled position that replaces the DOC, and then has an SDPF followed by a secondary SCR system in the underfloor position. They found that this EAS configuration could meet the EPA Tier 3 bin 30 standards, as described in Chapter 3. The PNA is described in Section 14.1. In follow-up work, Cumaranatunge [15-14] described how the SDPF combination affected soot removal from the filter, especially during active regeneration. Cumaranatunge [15-14] reported that during active regeneration of the particulate filter in the SDPF, there is no competition for NO_2, because the equilibrium of the NO oxidation reaction in Eq. 15.1 strongly favors NO instead of NO_2 >500°C, which is the temperature at which the collected soot ignites. However, in general, the SDPF requires more time at a given temperature than a catalyzed DPF does to remove a given quantity of soot from the filter. It is believed that the presence of even a small amount of NO_2 is the main factor promoting soot oxidation, because the catalyst present in the catalyzed DPF helps convert the NO generated in the reaction in Eq. 15.2 back into NO_2. One advantage of the SDPF, though, is that there is considerably less NO_2 exiting the filter unit, as shown in **Figure 15.5**.

As of this writing, Sharp et al. [15-15] are evaluating future EAS configurations for the California Air Resources Board (CARB) to assess the feasibility of the ultra-low NOx standard for HDDE described in Sections 3.3.1.1 and 3.3.2. They are using an advanced EAS with a Volvo MD13TC HDDE, which is comparable to the Volvo D13 described in Section 7.5.5. Their preferred configuration for evaluation uses an SDPF, labeled an "SCRF" in **Figure 15.6**, to help manage NOx emissions.

Two SDPF scenarios have been explored with the following three objectives:

1. To determine the contribution of NO_2 to overall NOx conversion and the passive regeneration of PM in the integrated system

FIGURE 15.5 Catalyzed DPF ("CSF") releasing more NO_2 than SDPF ("SCRF®") during active regeneration at 550°C and 600°C [15-14].

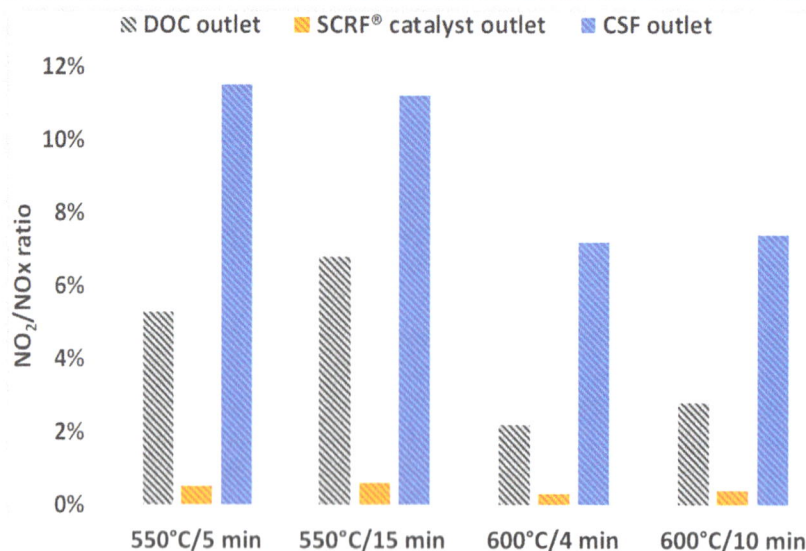

FIGURE 15.6 EAS to support ultra-low NOx performance, including SDPF (SCRF) [15-15].

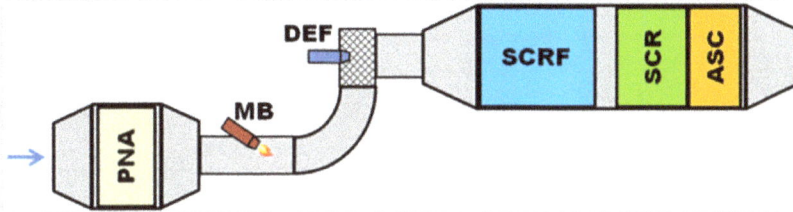

Reprinted from Ref [15-15]. © SAE International

2. To evaluate the integrated system and improve fuel economy by eliminating frequent active regenerations (through passive regeneration of PM by NO_2)

3. To determine if a DOC can be avoided, thus reducing the overall cost of the integrated system

The two scenarios include the following:

1. Uncatalyzed DPF coated with Fe/zeolite only

2. Catalyzed DPF with Fe/zeolite

The first option helps determine the NO oxidation capability of the Fe/zeolite catalyst and assesses if the NO_2 produced is enough to regenerate the PM in the filter. The second option enables evaluation of the CO and HC oxidation capabilities of the oxidation catalyst in the SDPF during active regeneration, and helps evaluate the NO oxidation capability when compared to the case when an oxidation catalyst is not used. If the evaluation determines that an oxidation catalyst is necessary in the SDPF, options for zone coating the SCR and oxidation catalysts on the DPF substrate will need to be explored, including the relative washcoat amounts of each zone.

Gao et al. [15-2, 15-3] characterized SCR washcoats that were applied to DPF substrates. In a medium-resolution X-ray computed tomography (CT) scan, a transition to a thicker washcoat layer was seen, as shown in **Figure 15.7**.

FIGURE 15.7 CT scan of SCR washcoat loading distribution over a DPF substrate suggesting that washcoat is not evenly loaded [15-3].

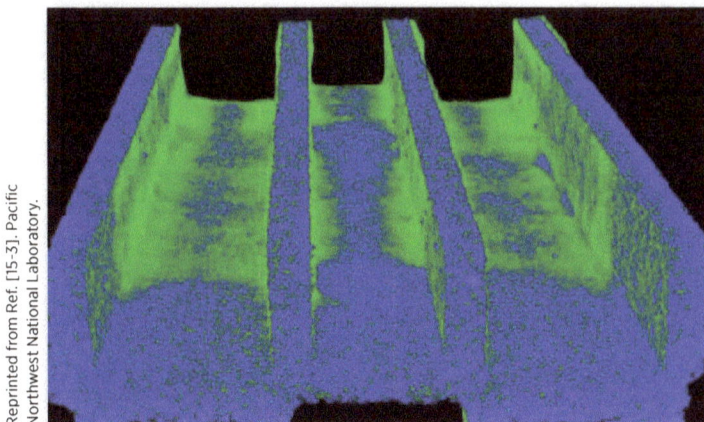

Reprinted from Ref. [15-3]. Pacific Northwest National Laboratory.

FIGURE 15.8 High-resolution CT scan of SDPF cross-section showing that SCR washcoat fills the porous filter walls [15-3].

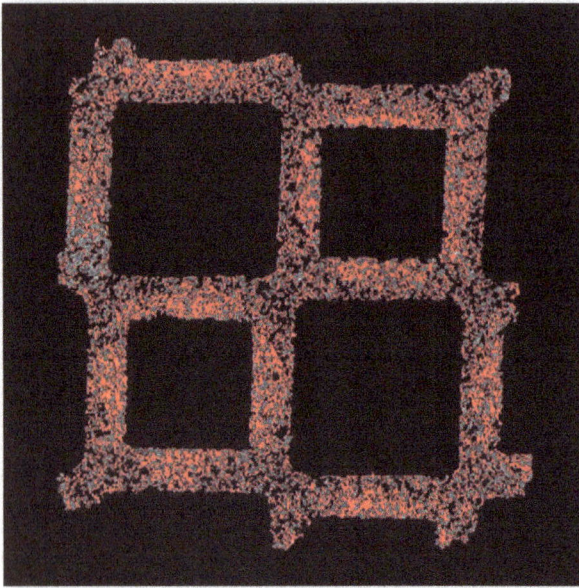

Reprinted from Ref. [15-3]. Pacific Northwest National Laboratory.

FIGURE 15.9 High-resolution CT scan of SDPF cross-section showing that thicker SCR washcoat fills and coats the porous filter walls [15-3].

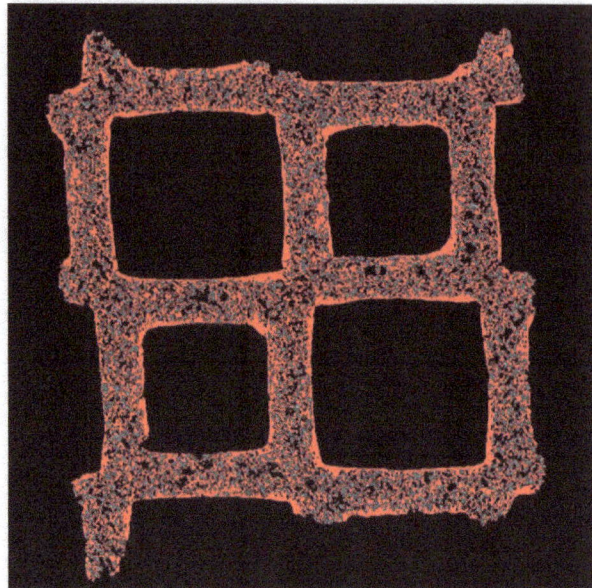

Reprinted from Ref. [15-3]. Pacific Northwest National Laboratory.

Gao et al. [15-2, 15-3] also used high-resolution X-ray CT to look at cross-sections of the DPF substrate and determine where the SCR washcoat ended up. In one example, shown in **Figure 15.8**, the catalyst ended up entirely within the porous filter walls. In another example, the catalytic coating was thicker, and started to coat the channel walls, as seen in **Figure 15.9**.

Even when an SDPF is implemented, it may be necessary to have an SCR system placed after it in the EAS. This configuration was evaluated by Guo et al. [15-16], as part of an evaluation of SDPF performance with a 3.0-L V-6 LDDE, and has been considered by several others as they have looked to implement SDPF systems in production. Georgiadis et al. [15-17] evaluated SDPF performance as part of a novel EAS configuration and control scheme.

15.3 Combined Three-Way Catalyst and Gasoline Particulate Filter

As discussed in Chapter 11, GPFs were first launched in a mass-production application by Daimler in early 2014 [15-18]. Starting in 2017, the number of GPF applications increased rapidly because of the PN standard in the EU's Euro 6d-TEMP standard for light-duty vehicles (LDV), as described in Section 3.2.3. The real driving emission (RDE) standards, described in Section 4.1.10, also provide a strong regulatory push for better EAS performance. By mid-2018, one GPF manufacturer, Corning, supplied 1 million GPF substrates for the European market, which indicates that within approximately one year the technology reached a market penetration of at least 10% of gasoline-fueled spark-ignition (SI) engines in LDV. GPF may be adopted for some port fuel injected (PFI) engines, even though LDVs with PFI engines are not subject to PN or PM emission standards that necessitate a GPF for compliance.

Most early GPF applications have included an uncoated GPF positioned downstream of a TWC unit. As the GPF technology has matured, GPF substrates have also had a TWC washcoat applied to combine the functionality in one unit, as described by Richter et al. [15-19], Ogyu et al. [15-20], and Xia et al. [15-21]. This combination of TWC and GPF is sometimes called a four-way catalyst or a three-way catalyst filter (TWCF).

As discussed in Chapter 2, when direct-injection (DI) SI engines generate particulates, they tend to make a very large number of very small particles; that is, the PN is high even though the PM is low. Thus, the GPF is needed to reduce the PN to below the regulatory threshold. As discussed in Chapter 11, the particles generated are trapped in the GPF, but the exhaust gas temperature typically stays hot enough to quickly oxidize any trapped particulates. Thus, a GPF does not need to be actively regenerated in the way that a DPF does. In addition, as discussed in Chapter 9, TWCs are designed to survive the hotter exhaust temperatures of an SI engine. Thus, the combination of TWC and GPF brings together two functions that are compatible in one device [15-19, 15-21].

As with the SDPF, described in Section 15.2, one of the issues with the combined TWC–GPF is adding the TWC washcoat to the GPF substrate. The washcoat adheres best when it permeates the wall, but it must not block too many of the channels through the porous filter wall or else it will significantly increase the pressure drop of the combined unit.

15.4 Combined Three-Way Catalyst and Hydrocarbon Trap

Another potential combination that is currently in development is to add HC trap functionality to a TWC to improve the low-temperature performance during a cold start. The HC trap is described in Section 14.2. In addition to some patents on this combined system approach [15-22, 15-23], Chang et al. [15-24] studied this combination. One of the challenges they noted is that HC traps often start releasing HC before the TWC has reached its light-off temperature of approximately 250°C, which limits the HC conversion benefit of the HC trap. Future internal combustion engines are expected to be more efficient, which means that in future engines the exhaust gas temperature entering the EAS will be cooler on average. For their advanced HC trap, Chang et al. [15-24] used an extruded zeolite substrate with 300 cpsi (cell/in.^{2}; 47 cells/cm^2). Their catalyst loading of 4 g/in.^{3} (0.24 g/cm^3) was approximately double the normal HC trap catalyst loading of approximately 2 g/in.^{3} (0.12 g/cm^3) of zeolite washcoat on a ceramic substrate. Comparisons of the HC removal performance of the conventional and advanced HC traps with a low-temperature TWC, shown in **Figure 15.10**, indicate that either trap improves the low-temperature HC emissions. Over the 30-s interval of the test, the inlet conditions were approximately 1,700 ppm HC and 80°C gas temperature. In these conditions, the conventional HC trap collects approximately 70% of the HC fed to it, and the advanced HC trap, 85%.

FIGURE 15.10 Comparison of HC removal performance over 30 s with inlet gas temperature of 80°C and inlet HC concentration of 1,700 ppm. (Figure 6 in [15-24].)

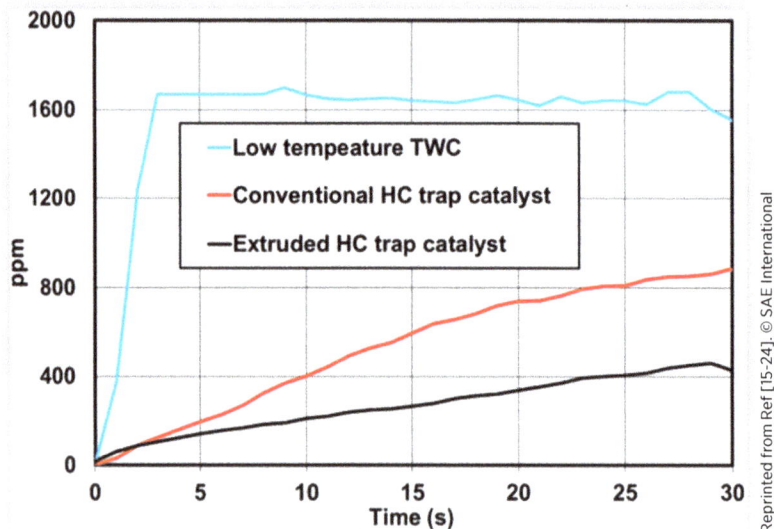

FIGURE 15.11 Comparison of HC release during temperature ramp from 80°C to 650°C. (Figure 7 in [15-24].)

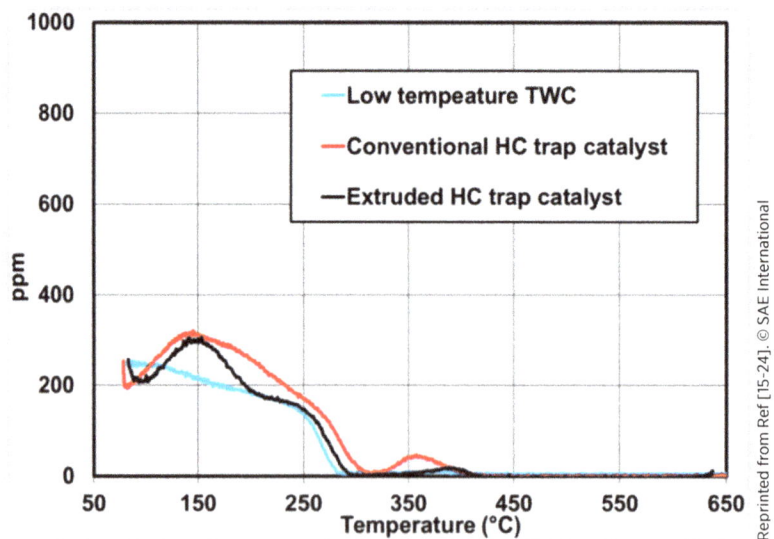

On the subsequent warm-up of these materials, there is little difference in the HC release between the conventional and advanced HC traps, as shown in **Figure 15.11**. The HC traps have a release peak at approximately 150°C that is not present with the low-temperature TWC. On-vehicle tests suggest that that the HC trap can reduce non-methane hydrocarbons (NMHC) emissions over Bag 1 of the FTP-75 cycle by 5 mg/mi., from 22 mg/mi. down to 17 mg/mi. The HC trap has no effect on NOx emissions.

For the combined TWC and HC trap, the best approach is to layer the washcoats so that the HC trap function is under the TWC. Although the additional diffusion barrier of the TWC washcoat can impair the ability of the HC molecules to reach the HC trap layer underneath, when the HC species are released from the trap they pass through the TWC on their way back to the bulk gas flow in the monolith channel. This route will give the HC species the opportunity to be converted before they re-enter the exhaust gas.

As described in Section 14.2.2, Wiebenga et al. [15-25] studied several mixtures of Pd and rhodium (Rh) on one of two substrates, described in Table 14.2. With a cordierite substrate, they coated a TWC layer over an HC trap layer, and Pd was included in both washcoat layers. With an extruded zeolite substrate, the zeolite was the HC trap, and it was coated with a TWC. In this case, only the TWC layer had Pd. Rh was only placed in the TWC washcoat layer regardless of the substrate. Wiebenga et al. [15-25] evaluated the trapping ability of the combined system in bench reactor tests using several HC species, as discussed in Section 14.2.2.

15.5 Passive Selective Catalytic Reduction System

As discussed in Chapter 12, one of the challenges of urea-SCR is that it requires packaging a urea-water solution (UWS) tank and dosing system. Another significant challenge is that the urea-SCR system needs the exhaust gas to be at least 200°C, and preferably at least 250°C, to ensure that the urea decomposition and NOx reduction reactions proceed. This minimum exhaust gas temperature also mitigates deposits and other side reactions that impair NOx reduction.

An approach to manage these challenges is called passive SCR [15-26]. Here, a catalytic converter is placed upstream of an SCR system in the tailpipe. The upstream catalytic converter can be either an LNT or a TWC, and its functions in the combined system include the following:

- Remove NOx from the exhaust during fuel-lean or stoichiometric conditions
- Generate NH_3 for use on the downstream SCR during fuel-rich conditions

The NH_3 generated during the fuel-rich operating period is stored in the downstream SCR system and used later during stoichiometric or fuel-lean exhaust gas conditions to convert NOx to N_2.

The main advantage of passive SCR is that the upstream catalytic converter replaces the UWS storage tank and dosing system on the vehicle. Instead, the fuel already carried on board and its combustion by-products are the reducing agents used in the EAS, which simplifies the logistics for the application. NH_3 generation requires an extended or "deep" rich pulse, where λ is < 1, in the upstream EAS unit, which can create issues with excess CO or HC emissions.

15.5.1 Lean NOx Trap and Selective Catalytic Reduction in Series

One passive SCR configuration uses an LNT to generate NH_3 during the fuel-rich part of its operating cycle. The NH_3 is stored in a downstream SCR system and used later

FIGURE 15.12 Concepts for LNT–SCR passive SCR system configurations.

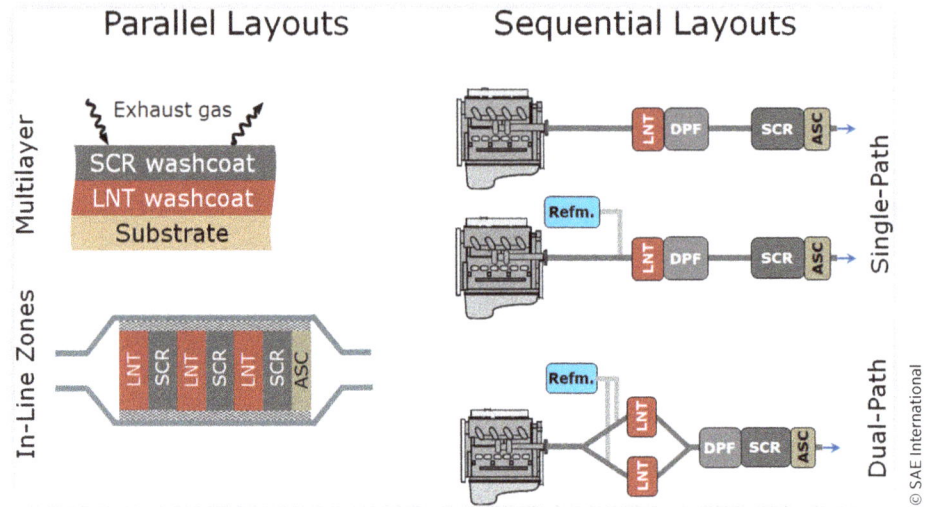

during fuel-lean operation to convert NOx to N_2. This configuration has been considered by several research teams since the early 2000s [15-26, 15-27, 15-28, 15-29, 15-30, 15-31, 15-32, 15-33, 15-34, 15-35, 15-36, 15-37].

As described in Chapter 10, the LNT uses the by-products of incomplete combustion, such as NMHC, hydrogen (H_2), or carbon monoxide (CO), as reducing agents to convert NOx into N_2 or NH_3. In particular, Section 10.1.3 discusses how the LNT generates NH_3.

The ammonia formation in an LNT depends on the air–fuel ratio (AFR) during and the duration of the regeneration process [15-33]. It occurs due to reaction of NOx with hydrocarbons under rich exhaust conditions, as seen on an empty LNT, or it can be formed due to the conversion of stored NOx during lean exhaust conditions. In particular, the use of hydrocarbon post-injection to assist LNT regeneration leads to a significant spike in ammonia formation. Therefore, there is potential for an upstream LNT to be used as the source of the ammonia reductant for SCR, replacing UWS dosing [15-33]. The combined LNT-SCR system is capable of effective operation over a wider temperature window than either catalyst alone. Note that the combined system uses either a sequential or parallel layout, as depicted in **Figure 15.12**.

Lambert [15-27] describes the advantages of a combining the LNT and SCR functions in the EAS. A combined system concept where the LNT and SCR functions are both placed within the same unit using axial zone coating or two monoliths in series is shown in **Figure 15.13**. Lambert and colleagues found that the combined system performance was comparable to that of a larger LNT system, at least with the laboratory samples available when they were doing their work.

FIGURE 15.13 Lean NOx trap and SCR combined system concept. (Figure 21 in [15-27].)

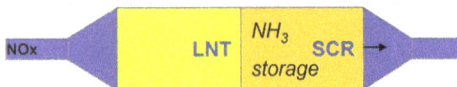

Reprinted from Ref. [15-27]. US Department of Energy

Around the same time, engineers at Eaton Corporation were working on a combined LNT–SCR system that included a reformer catalyst system upstream of the LNT [15-28, 15-29]. HC would be introduced into the exhaust gas by running the engine rich or by injecting fuel into the EAS. The reformer reacted these exhaust gas HC with water or oxygen gas (O_2) to form synthesis gas, which is a mixture of CO and H_2. The synthesis gas components are very effective reducing agents, and convert the NOx in the LNT into N_2 and NH_3. A prototype fuel reformer catalyst was tested with the LNT and SCR systems using exhaust from a multicylinder engine, as described by McCarthy and Holtgreven [15-29]. Their work included EAS aging tests, especially to assess the performance of the reformer, LNT, and SCR systems as the EAS was aged on-engine.

Wittka et al. [15-30, 15-31] investigated using an LNT with Pt and Rh catalyst to generate NH_3 for a downstream Cu/zeolite SCR system. Because the NH_3 generated in the LNT can be stored downstream, the lean period of the LNT was extended because any NOx that breaks through was converted in the downstream SCR. Likewise, a lower NOx conversion in the LNT will be mitigated by the NOx conversion in the SCR unit, and yield an acceptable overall NOx conversion in the EAS. These results have been supported by on-engine and in-vehicle tests [15-31].

Another approach is to combine the LNT and SCR functions into a single catalytic converter, perhaps by layering the catalyst washcoats within the monolith so that the SCR layer overlays the LNT layer, as described by several researchers [15-32, 15-34, 15-35]. Zheng et al. [15-32, 15-36], for example, ran the LNT in a fast cycle, as described in Section 10.1, in which 6 s of lean operation is followed by 1 s of rich operation. In this case, the LNT–SCR combination has significantly improved NOx conversion when the exhaust gas temperature is <250°C. Multi-zone catalyst architectures, with alternating LNT and SCR catalyst layers coated onto a single substrate, were found to yield significantly higher NOx conversion efficiencies than a series LNT–SCR combination [15-35].

If an SCR system is downstream of the LNT, the NH_3 released from the LNT will supplement what is dosed to reduce NOx in the SCR. In some EAS configurations, the LNT is the only source of NH_3 on the vehicle for the SCR reactions, which therefore means that an extended or deep rich cycle is needed to reduce the NO_2 all the way to NH_3. In this case, there is almost always a DPF in between the LNT and SCR unit, which will need to be uncatalyzed so that it does not oxidize the NH_3. For the combined system function, the LNT is run with a deeper rich cycle—either with a lower λ, extended duration, or both–to deliberately generate NH_3. This NH_3 is then adsorbed on the SCR catalyst downstream and used to remove any additional NOx during the lean cycle. Heavy-duty vehicles (HDVs) may include a urea dosing system, anyway, to ensure that there is always enough NH_3 available to convert the NOx to harmless species.

15.5.2 Three-Way Catalyst and Selective Catalytic Reduction in Series

One can also combine a TWC and SCR system to form a passive SCR system, just as one can combine an LNT and SCR, as described in Section 15.5.1. The TWC–SCR

FIGURE 15.14 Layout of PASS aftertreatment system [15-43].

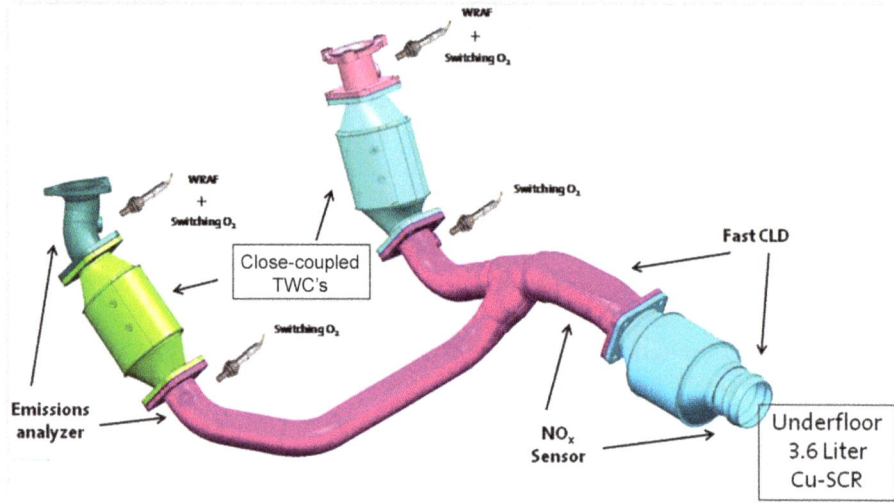

combination is more commonly applied to lean-burn SI engines, because lean-burn SI engines also operate with a stoichiometric AFR as the load increases. Therefore, the lean-burn SI engine can make use of the normal TWC functionality, described in Chapter 8, and support the function of the SCR. The TWC–SCR combination has been studied by several researchers [15-38, 15-39, 15-40, 15-41, 15-42] since around2009.

As with the LNT–SCR combination, in the TWC–SCR configuration the engine exhaust is held in a rich condition, with $\lambda < 1$, so that the TWC generates NH_3 from the net reducing exhaust. This NH_3 is stored on the SCR during the rich mode, and is then consumed during the following lean operating period by the NOx in the exhaust gas following the standard and fast reactions [15-39]. The challenge with this passive SCR system is managing the magnitude and duration of the rich period operation to minimize the fuel consumption penalty and the CO and HC emissions to ensure that they do not exceed the on-cycle or in-use limits. An additional challenge with using the TWC is ensuring that there is enough NOx available in the exhaust to generate the desired amount of NH_3, since the TWC does not store NOx as the LNT does.

As example of the TWC–SCR configuration is the Passive Ammonia SCR System (PASS) developed by General Motors for use in both lean and stoichiometric SI gasoline exhaust systems [15-43]. The PASS system is shown in **Figure 15.14**.

The effectiveness of the TWC–SCR system depends on the efficient and selective generation and use of NH_3 over the TWC and SCR catalysts [15-41]. Such a passive SCR approach can achieve > 99.5% NOx reduction efficiencies in lean gasoline engines [15-41]. Furthermore, the addition of a NOx storage component on the TWC can be used to increase the quantity of ammonia formation during rich operation as well as enable longer lean operating times [15-44].

References

15-1.   Girard, J.W., Cavataio, G., and Lambert, C.K., "The Influence of Ammonia Slip Catalysts on Ammonia, N2O and NOX Emissions for Diesel Engines," SAE Technical Paper 2007-01-1572, 2007, https://doi.org/10.4271/2007-01-1572.

15-2.   Stewart, M.L., Kamp, C.J., Gao, F., Wang, Y., and Engelhard, M.H., "Coating Distribution in a Commercial SCR Filter," *Emission Control Science and Technology* **4**, no. 4 (2018): 260–270, https://doi.org/10.1007/s40825-018-0097-3.

15-3.   Gao, F., Wang, Y., Engelhard, M., Crum, J., Stewart, M., and Kamp, C.J., "Coating Distribution in a Commercial SCR-Filter," *2017 CLEERS Workshop*, Ann Arbor, MI, 2017.

15-4.   Villamaina, R., Nova, I., Tronconi, E., Maunula, T., and Keenan, M., "The Effect of CH_4 on NH_3-SCR Over Metal-Promoted Zeolite Catalysts for Lean-Burn Natural Gas Vehicles," *Topics in Catalysis* **61**, no. 18-19 (2018): 1974–1982, https://doi.org/10.1007/s11244-018-1004-4.

15-5.   Olowojebutu, S. and Steffen, T., "A Review of the Literature on Modelling of Integrated SCR-in-DPF Systems," SAE Technical Paper 2017-01-0976, 2017, https://doi.org/10.4271/2017-01-0976.

15-6.   Patchett, J.A. and SantaMaria, M., "Integration of Ammonia SCR Catalysts with Oxidation Catalysts to Achieve NOx and Particulate Emissions Standards," *SAE Heavy Duty Diesel Emissions Control TOPTEC Symposium*, Gothenburg, 2003.

15-7.   Keenan, M., "Exhaust Emissions Control: 60 Years of Innovation and Development," SAE Technical Paper 2017-24-0120, 2017, https://doi.org/10.4271/2017-24-0120.

15-8.   Lee, J.H., Paratore, M.J., and Brown, D.B., "Evaluation of Cu-Based SCR/DPF Technology for Diesel Exhaust Emission Control," *SAE Int. J. Fuels Lubr.* **1**(1): 96–101, 2008, https://doi.org/10.4271/2008-01-0072.

15-9.   Oladipo, B., Bailey, O., Price, K., Balzan, N., and Kaul, S., "Simplification of Diesel Emission Control System Packaging Using SCR Coated on DPF," *US Dept. of Energy Diesel Engine Emissions and Energy Reductions Conference (DEER)*, Dearborn, MI, 2008.

15-10. Devarakonda, M.N., Stewart, M., Rappé, K.G., Male, J., and Herling, D., "Technical Challenges in the Integration of DPF and SCR Aftertreatment on a Single Substrate - Review from a Systems and Modeling Perspective," *12th CLEERS Workshop*, Dearborn, MI, 2009.

15-11. Rappe, K., "Combination and Integration of DPF-SCR Aftertreatment Technologies," Department of Energy Annual Merit Review, 2012.

15-12. Rappé, K.G., "Integrated Selective Catalytic Reduction-Diesel Particulate Filter Aftertreatment: Insights into Pressure Drop, NOx Conversion, and Passive Soot Oxidation Behavior," *Industrial & Engineering Chemistry Research* **53**, no. 45 (2014): 17547–17557, https://doi.org/10.1021/ie502832f.

15-13. Chilumukuru, K., Gupta, A., Ruth, M., Cunningham, M., Kothand araman, G., Cumaranatunge, L., and Hess, H., "Aftertreatment Architecture and Control Methodologies for Future Light Duty Diesel Emission Eegulations," *SAE Int. J. Engines* **10**, no. 4 (2017): 1580–1587, https://doi.org/10.4271/2017-01-0911.

15-14. Cumaranatunge, L., "A Study of the Soot Burning Efficiency of an SCRF Catalyst vs. a CSF during Active Regeneration Events," *2016 CLEERS Workshop*, Ann Arbor, MI, 2016.

15-15. Sharp, C., Webb, C.C., Neely, G., Carter, M., Yoon, S., and Henry, C., "Achieving Ultra Low NOX Emissions Levels with a 2017 Heavy-Duty On-Highway TC Diesel Engine and an Advanced Technology Emissions System - Thermal

Management Strategies," *SAE Int. J. Engines* 10(4): 1697–1712, 2017, https://doi.org/10.4271/2017-01-0954.

15-16.   Guo, G., Warner, J., Cavataio, G., Dobson, D., Badillo, E., and Lambert, C., "The Development of Advanced Urea-SCR Systems for Tier 2 Bin 5 and beyond Diesel Vehicles," SAE Technical Paper 2010-01-1183, 2010, https://doi.org/10.4271/2010-01-1183.

15-17.   Georgiadis, E., Kudo, T., Herrmann, O., Uchiyama, K., and Hagen, J., "Real Driving Emission Efficiency Potential of SDPF Systems without an Ammonia Slip Catalyst," SAE Technical Paper 2017-01-0913, 2017, https://doi.org/10.4271/2017-01-0913.

15-18.   Lanzerath, P., Wunsch, R., and Schön, C., "The First Series-Production Particulate Filter for Mercedes-Benz Gasoline Engines," *17. Internationales Stuttgarter Symposium*, Stuttgart, Springer, 2017.

15-19.   Richter, J.M., Klingmann, R., Spiess, S., and Wong, K.-F., "Application of Catalyzed Gasoline Particulate Filters to GDI Vehicles," *SAE Int. J. Engines* 5, no. 3 (2012): 1361-1370, https://doi.org/10.4271/2012-01-1244.

15-20.   Ogyu, K., Ogasawara, T., Nagatsu, Y., Yamamoto, Y., Higuchi, T., and Ohno, K., "Feasibility Study on the Filter Design of Re-Crystallized SiC-GPF for TWC Coating Application," SAE Technical Paper 2015-01-1011, 2015, https://doi.org/10.4271/2015-01-1011.

15-21.   Xia, W., Zheng, Y., He, X., Yang, D., Shao, H., Remias, J., Roos, J., and Wang, Y., "Catalyzed Gasoline Particulate Filter (GPF) Performance: Effect of Driving Cycle, Fuel, Catalyst Coating," SAE Technical Paper 2017-01-2366, 2017, https://doi.org/10.4271/2017-01-2366.

15-22.   Kachi, N. and Nishizawa, K., "Catalytic Converter with Multilayered Catalyst System," US20010006934A1, Jul. 16, 2001.

15-23.   Lupescu, J.A. and Jen, H., "Combined Hydrocarbon Trap and Catalyst," US20130287659A1, Oct. 31, 2013.

15-24.   Chang, H.-L., Chen, H.-Y., Koo, K., Rieck, J., and Blakeman, P., "Gasoline Cold Start Concept (gCSC™) Technology for Low Temperature Emission Control," *SAE Int. J. Fuels Lubr.* 7(2): 480–488, 2014, https://doi.org/10.4271/2014-01-1509.

15-25.   Wiebenga, M.H., Oh, S.H., and Qi, G., "Cold-Start Emission Reduction Potential and Limitations of Commercial Passive Hydrocarbon Adsorbers," *Emission Control Science and Technology* 3, no. 1 (2016): 47–58, https://doi.org/10.1007/s40825-016-0052-0.

15-26.   Hackenberg, S. and Ranalli, M., "Ammonia on a LNT: Avoid the Formation or Take Advantage of It," SAE Technical Paper 2007-01-1239, 2007, https://doi.org/10.4271/2007-01-1239.

15-27.   Lambert, C., "Advanced CIDI Emission Control System Development," Report No. 898797, 2006, https://doi.org/10.2172/898797.

15-28.   McCarthy, J.J. and Taylor, W., "LNT + SCR Aftertreatment for Medium-Heavy Duty Applications: A Systems Approach," *DEER*, Detroit, MI, 2007.

15-29.   McCarthy, J. and Holtgreven, J., "Advanced NOx Aftertreatment System Performance Following 150 LNT Desulfation Events," SAE Technical Paper 2008-01-1541, 2008, https://doi.org/10.4271/2008-01-1541.

15-30.   Wittka, T., Holderbaum, B., Maunula, T., and Weissner, M., "Development and Demonstration of LNT+SCR System for Passenger Car Diesel Applications," *SAE Int. J. Engines* 7(3): 1269–1279, 2014, https://doi.org/10.4271/2014-01-1537.

15-31.   Wittka, T., Holderbaum, B., Dittmann, P., and Pischinger, S., "Experimental Investigation of Combined LNT+ SCR Diesel Exhaust Aftertreatment," *Emission Control*

Science and Technology **1**, no. 2 (2015): 167–182, https://doi.org/10.1007/s40825-015-0012-0.

15-32. Zheng, Y., Li, M., Luss, D., and Harold, M., "Fast Cycling NSR + SCR Over Dual-Layer Catalysts," *2015 CLEERS Workshop*, 2015.

15-33. Westerberg, B. and Fridell, E., "A Transient FTIR Study of Species Formed during NOx Storage in the Pt/BaO/Al$_2$O$_3$ System," *Journal of Molecular Catalysis A: Chemical* **165**, no. 1-2 (2001): 249–263, https://doi.org/10.1016/S1381-1169(00)00431-3.

15-34. Liu, Y., Harold, M.P., and Luss, D., "Coupled NOx Storage and Reduction and Selective Catalytic Reduction Using Dual-Layer Monolithic Catalysts," *Applied Catalysis B: Environmental* **121** (2012): 239–251, https://doi.org/10.1016/j.apcatb.2012.04.013.

15-35. Theis, J.R., Dearth, M., and McCabe, R., "LNT+ SCR Catalyst Systems Optimized for NOx Conversion on Diesel Applications," SAE Technical Paper 2011-01-0305, 2011, https://doi.org/10.4271/2011-01-0305.

15-36. Zheng, Y., Li, M., Harold, M., and Luss, D., "Enhanced Low-Temperature NOx Conversion by High-Frequency Hydrocarbon Pulsing on a Dual Layer LNT-SCR Catalyst," *SAE Int. J. Engines* **8**(3): 1117–1125, 2015, https://doi.org/10.4271/2015-01-0984.

15-37. Zheng, Y., Li, M., Wang, D., Harold, M.P., and Luss, D., "Rapid Propylene Pulsing for Enhanced Low Temperature NOx Conversion on Combined LNT-SCR Catalysts," *Catalysis Today* **267** (2016): 192-201, https://doi.org/10.1016/j.cattod.2015.10.029.

15-38. Li, W., Perry, K.L., Narayanaswamy, K., Kim, C.H., and Najt, P., "Passive Ammonia SCR System for Lean-burn SIDI Engines," *SAE Int. J. Fuels Lubr.* **3**(1): 99–106, 2010, https://doi.org/10.4271/2010-01-0366.

15-39. Prikhodko, V.Y., Parks, J.E., Pihl, J.A., and Toops, T.J., "Ammonia Generation over TWC for Passive SCR NO$_x$ Control for Lean Gasoline Engines," *SAE Int. J. Engines* **7**(3): 1235–1243, 2014, https://doi.org/10.4271/2014-01-1505.

15-40. Theis, J., Kim, J., and Cavataio, G., "Passive TWC+SCR Systems for Satisfying Tier 2 Bin2 Emission Stand ards on Lean-Burn Gasoline Engines," *2015 CLEERS Workshop*, Dearborn, MI, 2015.

15-41. Prikhodko, V.Y., Parks, J.E., Pihl, J.A., and Toops, T.J., "Ammonia Generation and Utilization in a Passive SCR (TWC+SCR) System on Lean Gasoline Engine," *SAE Int. J. Engines* **9**(2): 1289–1295, 2016, https://doi.org/10.4271/2016-01-0934.

15-42. Chen, P., Lin, Q., Prikhodko, V.Y., and Parks, J.E., "Non-Uniform Cylinder-to-Cylinder Combustion for Ammonia Generation in a New Passive SCR System," *2018 Annual American Control Conference (ACC)*, IEEE, 2018.

15-43. Guralp, O., Qi, G., Li, W., and Najt, P., "Experimental Study of NO$_x$ Reduction by Passive Ammonia-SCR for Stoichiometric SIDI Engines," SAE Technical Paper 2011-01-0307, 2011, https://doi.org/10.4271/2011-01-0307.

15-44. Prikhodko, V., Pihl, J., Toops, T., and Parks, J., "Effects of NO$_x$ Storage Component on Ammonia Formation in TWC for Passive SCR NO$_x$ Control in Lean Gasoline Engines," SAE Technical Paper 2018-01-0946, 2018, https://doi.org/10.4271/2018-01-0946.

16

Alternative Fuels

If you're implying that I switched out that rot-gut excuse for alternative fuel with my all natural sustainable organic bio-fuel, [...] you're dead wrong man... [points to Sarge] It was him.

—Filmore (a cartoon microbus), "Cars 2"

For the purposes of this chapter, "alternative fuel" applies to any fuel for internal combustion engines (ICEs) that is not gasoline or diesel fuel distilled from crude oil. The term can include fossil fuels, such as natural gas, although in most cases it refers to fuels that are generated from plants or other renewable feedstocks. There are some fuel blends that are already approved for general use, such as some blends of ethanol in gasoline or of biodiesel in distillate diesel. For mobile applications, the most convenient alternative fuels are gases or liquids.

One advantage of alternative fuels is that they generate less net carbon dioxide (CO_2) than traditional distillate fuels do when burned. For example, ethanol (C_2H_5OH) derived from an agricultural source is approximately zero net carbon. Alternative fuels may also emit less CO_2 per unit of energy obtained from the fuel, as with methane (CH_4), which is the primary constituent of natural gas. In most liquid fuels, the ratio of hydrogen (H) to carbon (C) in the fuel is about 1.8:1, whereas in natural gas it is just under 4:1. Thus, a greater proportion of the energy generated from natural gas comes from the formation of water (H_2O) than from CO_2, compared with traditional fuels. Another advantage is that alternative fuels may be produced entirely locally, and thus reduce a country's dependence on energy imports.

In general, alternative fuels are not drop-in replacements for the distillate fuels they are displacing. Thus, an engine may need to be designed to support flexibility in fuel source, and especially to operate in a way that takes advantage of the unique fuel properties.

After reading this chapter, one will better understand why alternative fuels are of interest to the automotive engineering community. One will also get an overview of which fuels have been seriously considered as options for use in ICEs, their properties, and their influence on emissions.

16.1 **Motivation for Alternative Fuels**

The economy thrives on transportation of people and goods, and this transportation requires energy. Petroleum-based hydrocarbons are a stable and dense form of energy for use on a vehicle or similar application. Distillate hydrocarbon fuels are used in passenger car, commercial vehicle, and off-road applications worldwide. Fuels represent a large proportion of global petroleum production.

Crude oil is a fungible commodity in that there is a global market for the production, processing, and use of crude oil for fuels that is somewhat independent of source or market. Nevertheless, as a commodity of immense global strategic importance, world events can affect the pricing and vice versa. This inherent volatility in crude oil pricing affects households and businesses worldwide. Thus, for example, to help mitigate this volatility in crude oil pricing, the US Department of Energy (DOE) "is supporting research to greatly improve the fuel efficiency and reduce the emissions produced by both light and heavy-duty vehicles." [16-1]

Alternative fuels can have many advantages regarding impacts on the environment and society. Most notably, the benefits include sustainability, fuel security, and economic benefits to the regions that produce the fuel feedstocks.

Alternative fuels can also benefit society and the environment. Favoring the use of local feedstocks can help improve energy security, reduce transportation costs, and mitigate the risk of future supply shocks. For example, pulp and paper industry waste in Finland and Sweden can be used as a feedstock to make dimethyl ether (DME, $H_3C-O-CH_3$). Likewise, one of the benefits of ethanol production in Brazil and the US is that in both places local feedstocks are used to make the fuel.

16.2 **Common Issues with Alternative Fuels**

A fundamental issue with most alternative fuels is that the fuel itself often costs more than the distillate fuel it is trying to replace, be that gasoline or diesel, especially on a unit energy basis. In addition, many of the fuels described in this chapter require some level of modification to the engine calibration, if not hardware, to take the best advantage of the fuel's properties.

Many of the alternative fuels described in this chapter are made from agricultural feedstocks. Thus the production of fuels may compete directly with the production of food. There has been considerable interest in using agricultural by-products such as corn stover or low-input crops, such as switch grass to provide feedstocks for biofuels production. Unfortunately, there is not yet a cost-effective approach to convert the cellulose present in plants into fuel precursor chemicals at the production volumes needed for fuels production, which is on the order of millions of barrels* per day in the US alone. Another issue with crop-derived fuels is the transportation cost of getting the feedstocks to a fuel production plant as most of them cannot be transported via pipelines like crude oil can.

---

* One oil barrel equals 42 US gallons or 159 liters.

The consumption of water in the course of producing the alternative fuels is an increasingly critical issue and may affect local production in some areas of the world. Agriculturally derived feedstocks require water to grow the plants, of course, but fuel production itself consumes water. For example, water and hydrocarbon (HC) or carbohydrate fuel feedstocks can be reacted to make synthesis gas (a.k.a. syngas, a mix of H_2 and CO) as a starting point to making fuels.

Alternative fuels present several common issues. A key area of concern is that many fuels are not drop-in replacements, because they have physical properties different from the fuel they are replacing. These properties may include the following:

- Boiling point
- Flash point
- Density
- Viscosity
- Lubricity
- Octane number or knock resistance for spark-ignition (SI) engines
- Cetane number or ignition characteristics for compression-ignition (CI) engines
- Energy density on a mass or volumetric basis

The alternatives may interact differently with common materials exposed to liquid fuel on the engine. This is especially a challenge if the fuel negatively affects fuel injection system* (FIS) components by facilitating deposits or wear.

For example, biodiesel fuels, described in Section 16.10, start to turn cloudy or gel or turn waxy as they get cold, but at temperatures higher than distillate diesel does. Another example is that ethanol, described further in Section 16.6, is not compatible with some materials found in SI engine fuel system components, such as some rubber hose compounds. Also, because alcohols are miscible in water, over time they tend to increase the water content of fuel blends with alcohol, which can affect the corrosiveness of the fuel.

Key fuel properties for alternative fuels are listed in **Table 16.1** [16-2, 16-3, 16-4, 16-5, 16-6, 16-7, 16-8, 16-9], including the research octane number (RON) for gasoline alternatives and cetane number for diesel alternatives, respectively.

16.3 **Natural Gas**

Natural gas is most often associated with methane (CH_4), its largest constituent, but it is in fact a mixture of gases, whose composition varies dependent upon the source. Typical compositions are 75%–90% methane, 5%–20% ethane (C_2H_6), 3%–15% propane (C_3H_8) and trace amounts of butane (C_4H_{10}) and heavier hydrocarbons. The structures of methane and ethane are shown in **Figure 16.1**; propane and butane are described further in Section 16.4. Because natural gas is a blend of compounds that may liquefy under pressure, it is possible for the heavier compounds (butane or heavier) to accumulate in the tank over time as the lighter ones (methane and ethane) are used first.

---
* The fuel injection system (FIS) is also known as the fuel injection equipment (FIE).

TABLE 16.1 Selected fuel properties of common alternative fuels [16-2, 16-3, 16-4, 16-5, 16-6, 16-7, 16-8, 16-9].

Fuel	Lower heating value, LHV (MJ/kg)	Density (kg/m³)	RON	Cetane number
Distillate gasoline	40.2–41.9	720-775	92–95	—
Distillate diesel	42.9–43.1	820-845	15–25	40-51
Natural gas*	40.3–49.1[†]	0.73-0.83[†]	130	
LPG*	46.1	2.25 (540 liq.)	94–112	8
Propane*	46.3	2.0 (510 liq.)	112	—
Butane*	45.6	2.7 (580 liq.)	94	8
DME*	28.8	2.05 (667 liq.)	—	55
Ethanol	26.9	790	110	—
1-Butanol	33.1	810	94–96[‡]	—
2-Butanol	33	806	101[‡]	—
Isobutanol	33	802	113[‡]	—
Fischer-Tropsch diesel, low-T synthesis	43.9	784	0–20	≥80
Fischer-Tropsch diesel, high-T synthesis	43.2	780	0–60	50-60
FAME biodiesel (RME)	36.5–37.2	880	—	47-65
Renewable diesel	43.1	780	—	≥70

* Fuel is a gas at typical ambient conditions. Second density value is for liquefied fuel.

[†] LHV and density vary strongly with natural gas composition, especially methane content.

[‡] Blending RON value.

FIGURE 16.1   Major constituents of natural gas are methane (a) and ethane (b) [16-10, 16-11].

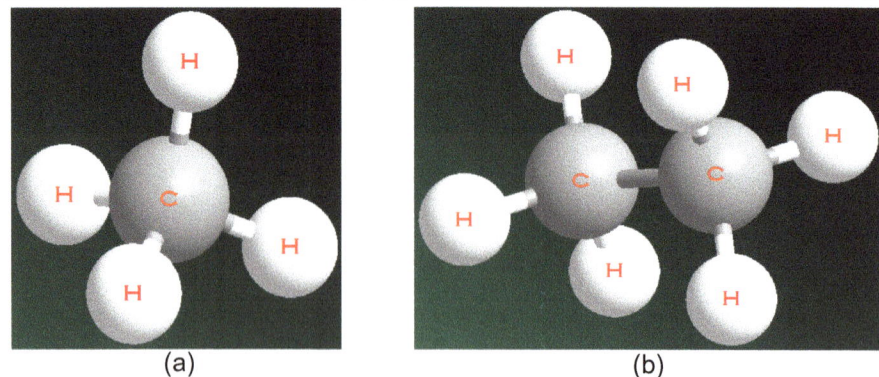

(a)   (b)

National Institutes of Standards and Technology

Natural gas is an abundant and relatively low-cost resource in the US, especially in light of new discoveries of shale gas reserves and novel extraction methods that have been commercialized since the late 2000s. These developments in natural gas production technology have dramatically increased supply of natural gas in the US, making the price more competitive against diesel fuel when crude oil prices are higher. This benefit may not be available in all markets, as outside US the price of natural gas is typically pegged to the price of crude oil.

Compressed natural gas (CNG) and liquefied natural gas (LNG) are the most commonly used natural gas fuels for internal combustion engines (ICE) in the transportation sector. Both options can take advantage of the well-established natural gas distribution network via pipelines. The pipeline gas needs to be processed into CNG or LNG and distributed to fueling stations. Also, the vehicle fueling infrastructure is limited. CNG is more commonly used because it is easier to handle than the cryogenic LNG, even though CNG has a lower volumetric energy density than LNG does. The fuel tank for either CNG or LNG is heavier than the equivalent tank for gasoline or diesel fuel. Also, the fuel system needs to be changed for use with natural gas instead of a liquid fuel. These conversions significantly increase the initial cost of the vehicle, which is an impediment to broader adoption.

For stationary applications, the engine can be fueled directly from the natural gas utility feed to the facility, assuming the facility-delivered flow capacity is sufficient for the engine. There are some engines at oil or gas production or at landfill sites that use natural gas straight from the well, but this feed usually contains high levels of sulfur, CO_2, or other diluents and is a poor fuel requiring additional engine modifications for practical use.

There are several commercially available ICEs that use natural gas as a fuel. The majority are used for stationary power generation or back-up power applications, but vehicles or ICEs for vehicles are also available from several manufacturers that span the range from light-duty vehicles (LDV) through to heavy-duty on-road applications, including buses and vocational vehicles. Natural gas is not a good fuel for CI engines, as the cetane number of methane is not high enough to support CI combustion on its own. Thus, natural gas engines typically use one of two options in their operation.

First, the engine can run a fuel-lean mixture of natural gas and air where the natural gas is injected into the ports to create a homogeneous blend. Then, a diesel fuel injector in the cylinder sprays fuel to initiate combustion once the natural gas–air blend has been compressed. This approach retains the efficiency benefits of lean combustion, but retains the expensive diesel FIS for use as the ignition system. Also, the exhaust aftertreatment system (EAS) needs to be able to reduce NOx in net oxidizing conditions, which means using a lean NOx trap or selective catalytic reduction system.

Second, the engine can be adapted to burn a stoichiometric mixture of natural gas and air using a spark plug to ignite the mixture in the cylinder. This approach has the advantage of being able to use a cheaper three-way catalyst (TWC) instead of the typical EAS needed for lean combustion. But the spark plug structure may limit the peak firing pressure (PFP) in the cylinder, and thus limit the engine power for a given displacement.

Natural gas can produce soot, even though there are no carbon-carbon bonds that serve as precursors. Particulates production is much lower than with a diesel-fueled CI engine, though. Nitrogen oxides (NOx) and carbon monoxide (CO) are potentially issues, depending on the combustion mode used in the ICE.

A common challenge is that not all of the methane in the fuel is burned in the engine, and so it enters the EAS. Unfortunately, the catalytic combustion of methane to carbon dioxide (CO_2) and water (H_2O) is difficult, and requires special catalysts for the purpose and relatively high exhaust temperatures [16-12, 16-13, 16-14, 16-15]. Cargnello et al. [16-16] have investigated a palladium (Pd) catalyst coated in ceria (CeO_2) that appears to support complete catalytic combustion of methane at 400°C, and platinum (Pt) is also common. These methane oxidation catalysts are susceptible to poisoning by sulfur oxides (SOx), especially when water is present [16-17].

16.4 **Propane and Butane**

Propane (C_3H_8) and butane (C_4H_{10}) can also be used as fuels in ICE. Their structures are shown in **Figure 16.2**, which assumes that the butane is n-butane and not isobutane (a.k.a. 2-methylpropane). Blends of propane and butane are typically called LPG, which can range from about 100% propane to 100% butane depending on the location and season. For example, a summer LPG blend typically contains more butane, whereas a winter LPG blend contains more propane. Note that the fuel properties, such as RON, do not vary linearly with composition [16-18].

LPG blends are gases at typical ambient conditions, but condense to a liquid when pressurized. Thus, the fuel can be stored on a vehicle in a pressure vessel as a denser liquid that expands into a gas when used to fuel the engine.

LPG is typically used in SI engines that were converted from using gasoline, and so tend to run in a stoichiometric mixture. HC emissions from LPG-fueled engines tend to be lower than those from gasoline-fueled engines, making them attractive for some applications, although NOx and CO emissions will be comparable. LPG-fueled engines are common in some non-road applications such as material handling vehicles. The main drawback is that propane and butane are both fossil fuels. The LPG fueling network for vehicles is also sparse [16-21]. Also, although the prices in the US are comparable to gasoline, these fuels are expensive compared to other fuel options such as natural gas that also require engine conversion work.

FIGURE 16.2   Major constituents of liquefied petroleum gas (LPG) are (a) propane and (b) n-butane [16-19, 16-20].

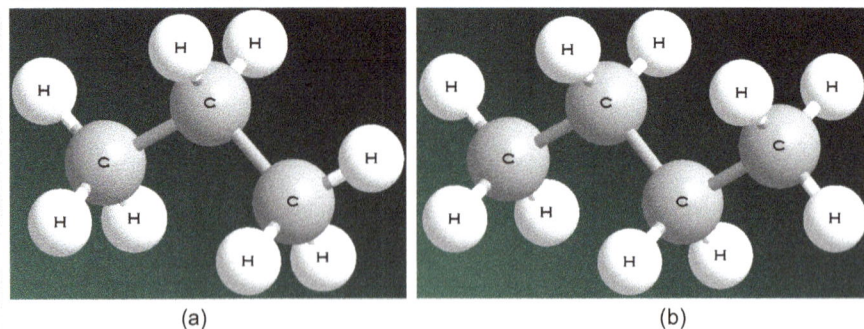

(a)   (b)

National Institutes of Standards and Technology

16.5 **Dimethyl Ether**

Dimethyl ether (DME, or $H_3C-O-CH_3$) is the simplest ether. It comprises two methyl groups (-CH_3) that are bonded to a central oxygen atom (O), as shown in **Figure 16.3**. DME is a gas at typical ambient temperatures and pressures.Its chemical structure is similar to methanol (CH_3OH) in that DME contains oxygen and no carbon-carbon bonds, but its vapor pressure characteristics are more like those of propane. DME's structure therefore reduces the possibility of forming PM emissions during combustion. Unlike methanol, DME has a high cetane number, and performs well as a fuel in CI engines [16-4].

DME can be produced from methane, or by a condensation reaction with methanol that removes the resulting water, as follows:

$$2\,CH_3OH = CH_3OCH_3 + H_2O. \tag{16.1}$$

Aguayo et al. [16-23] have studied an alternate synthesis pathway using a novel catalyst. This pathway reacts carbon dioxide (CO_2) and syngas, a mixture of H_2 and CO, in a single step to make DME.

DME has been studied by researchers at several institutions since the 1990s for its suitability in heavy-duty diesel engines (HDDE), including Chalmers University [16-24], Volvo Trucks [16-25], and AVL [16-26]. More recently, Park and Lee [16-27] have reviewed the literature to assess DME's suitability as an alternative to diesel fuel. When used in a CI engine, DME reduces PM emissions because there are no carbon–carbon bonds to support particulate precursors [16-4]. Because of this feature, a CI engine may run with a high rate of EGR to suppress NOx, as described in Chapter 2, Emissions Formation.

FIGURE 16.3   Structure of dimethyl ether (DME) [16-22].

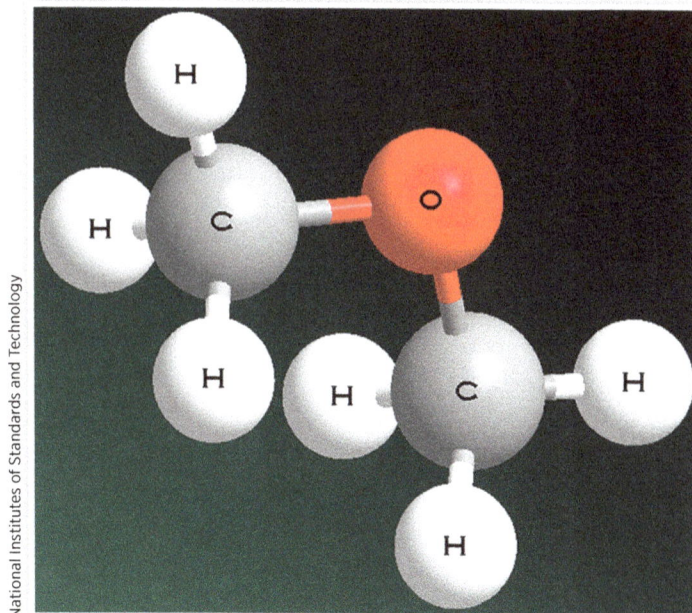

National Institutes of Standards and Technology

The main challenge with DME is that the CI engine and its fuel system need to be converted for use with the fuel [16-28], as DME is a gas at normal ambient conditions. Another significant challenge is that DME, like other ethers, is not stable against oxidation, and has a risk of forming dimethyl peroxide, a strong oxidizer that creates flammability risk. DME also has poor lubricity, which has a negative effect on the FIS durability. The energy density of DME is lower than distillate diesel fuel, as shown in Table 16.1. Thus, DME shows promise as an alternative to diesel, but the commercial incentives are not yet sufficient to displace distillate diesel fuel with DME.

16.6 **Ethanol**

Ethanol (C_2H_5OH) is one of the most widely utilized liquid biofuels. It may be combined with gasoline to create an ethanol blend fuel, or used in its pure form. Ethanol, molecular structure shown in **Figure 16.4**, is most commonly produced by fermenting carbohydrates obtained from natural sugars, starches, or other plant biomass from corn (maize), sugar cane, sugar beets, switchgrass, or fast-growing woody biomass. The US and Brazil are the two major producers of bioethanol presently. In the US, corn is the major primary crop used for ethanol production while in Brazil, sugar cane is the dominant crop source.

There are several advantages to using ethanol as an ICE fuel, both from an emissions and performance standpoint. The key advantage is that ethanol has a RON of 110, as shown in Table 16.1, which is significantly higher than the 92 RON for typical pump gasoline in the US or the 95 RON used in the European Union (EU) and other markets. Currently in the US, E10 blends of 10 vol% ethanol in gasoline are commonly available,

FIGURE 16.4   Structure of ethanol [16-29].

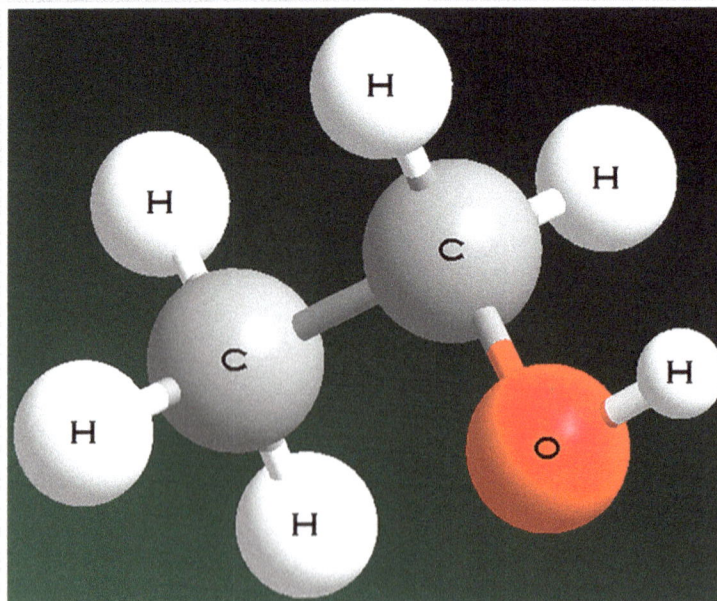

National Institutes of Standards and Technology

and E85 blends with 51 vol% to 83 vol% ethanol can also be found [16-4]. In addition, the US Environmental Protection Agency (EPA) has approved E15 blends for LDV from MY2002 to the present.

For ethanol that is derived from grains, such as corn, the net process is slightly carbon negative thanks to decades of process optimization. Ethanol made from sugar, such as in Brazil, is even more favorable as the sugars can be fermented directly without preprocessing. Nevertheless, as discussed earlier in Section 16.2, water use is a major challenge, especially in growing the plants which are used to make ethanol.

A performance disadvantage is that the volumetric energy content of ethanol is about 25% lower than that for distillate gasoline, as shown in Table 16.1, and most engine calibrations are not developed to take advantage of a higher ethanol content when it is available. Thus, the volumetric fuel consumption of ethanol is higher than that for gasoline, or the expected range from a full tank of fuel is lower. Aldehydes such as formaldehyde (CH_2O) and acetaldehyde (CH_3CHO) can be formed during combustion. As an oxygenated fuel, ethanol mitigates HC and CO formation, but can boost or suppress NOx depending on the combustion system and how it takes advantage of ethanol's relatively high heat of vaporization.

Other issues with ethanol include its materials compatibility with ICE fuel system components that are exposed to the liquid fuel. For example, increasing the ethanol content above 20 vol% in a fuel can cause many rubber compounds to shrink relative to their volume with E0 or E10 fuels. Also, ethanol is hygroscopic, meaning that it pulls water into it from the environment over time. This water content can make ethanol fuel blends more corrosive. In addition, ethanol has to be transported by rail car or truck tanker, since it is not compatible with the mix of refinery products that are distributed via pipelines.

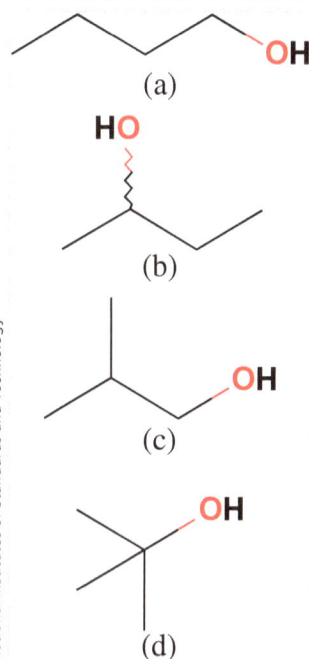

FIGURE 16.5 Structures of (a) 1-butanol, (b) 2-butanol, (c) isobutanol, and (d) tert-butanol [16-33, 16-34, 16-35, 16-36].

Overall, ethanol has promise as an alternative to gasoline, with the advantage of many years of use in the field and an understanding of common issues in industry. As with many of the other alternative fuels discussed in this chapter, it is not a drop-in replacement, but instead requires some modifications to the engine and calibration to get the most benefit.

16.7 **Butanols**

Butanol (C_4H_9OH) is another biofuel that is produced by fermentation from sugars or starches, like ethanol [16-30, 16-31, 16-32]. The structures of the various butanol isomers—1-butanol (a.k.a. n-butanol), 2-butanol, isobutanol (a.k.a. 2-methyl-1-pro-panol), and *tert*-butanol (a.k.a. 2-methyl-2-propanol)—are shown in **Figure 16.5**. Note that 2-butanol is a "handed" molecule, with R and S enantiomers defined by the arrangement of the methyl (–CH_3), ethyl (–C_2H_5), and hydroxyl (–OH) functional groups around the central carbon (C) atom.

The main production pathway for renewable butanol that has been developed to date uses ABE (acetone, butanol, ethanol) bacteria instead of yeast in the fermentation systems. Production challenges include butanol yield and efficiently separating the butanol product from the by-products. The main butanol isomer produced by

National Institutes of Standards and Technology

this fermentation process is 1-butanol, although 2-butanol has the more favorable RON value, as shown in Table 16.1. Ndaba et al. [16-31] identify some alternative production pathways that start with ethanol, for example.

The most common bacterium used for ABE fermentation is *Clostridium acetobutylicum*, although related clostridia bacteria can also be used [16-37]. This bacterial fermentation approach has the advantage that it can use the same process plant hardware as ethanol production does, which minimizes the capital investment. A related butanol fermentation process has been developed at GEVO and at Butamax, a joint venture of BP and DuPont, that produces isobutanol instead of 1-butanol.

With its four carbon atoms, butanol is closer than ethanol in nature to the distillate gasoline it can replace. For example, it is less hygroscopic than ethanol. Butanol-gasoline blends can also be transported by refinery pipeline, which has the potential to ease and improve fuel blending and to reduce transportation costs. In addition, because the proportion of oxygen in each butanol molecule is lower than the proportion of oxygen in ethanol, 16% butanol in gasoline is the equivalent of an E10 blend. Thus, more butanol can be used within existing regulatory frameworks in the US.

As an oxygenated fuel, butanol tends to suppress HC and CO emissions, but can boost NOx in the engine. As a lighter and more volatile molecule, butanol should be less likely to produce particulates in direct injection (DI) SI engines.

16.8 Synthetic Gasoline

The most common route to making synthetic gasoline is to use the ExxonMobil methanol to gasoline (MTG) process. This process starts with a hydrocarbon or carbohydrate feedstock that can be used to make synthesis gas by steam reforming or similar process,

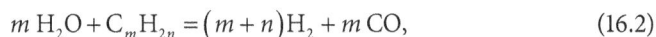

$$m\,H_2O + C_mH_{2n} = (m+n)H_2 + m\,CO, \tag{16.2}$$

where C_mH_{2n} is a generic HC input. For example, with $m = 1$ and $n = 2$, the base HC input is CH_4 and the process in Eq. 16.2 is steam methane reforming (SMR). The H_2 content of the synthesis gas can be increased by reacting the CO with more water in the water–gas shift (WGS) reaction over a catalyst,

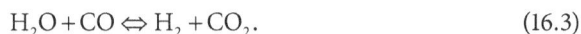

$$H_2O + CO \Leftrightarrow H_2 + CO_2. \tag{16.3}$$

The synthesis gas is then used to make methanol,

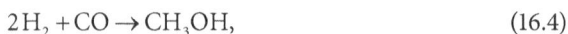

$$2H_2 + CO \rightarrow CH_3OH, \tag{16.4}$$

and the methanol is then reacted together over a catalyst to form multi-carbon molecules that are suitable for gasoline [16-38]. As a drop-in replacement for gasoline, the emissions from an ICE using synthetic gasoline are comparable to the baseline case.

Bosch has discussed the production of E-Fuels [16-39] as a sustainable zero-carbon option. E-Fuels are made by reacting H_2 with CO_2 to form fuels. To be sustainable, the H_2 is produced by electrolysis from water and renewable electricity sources, although at the scales needed for fuels production water could become a limiting resource. The CO_2 can come from the atmosphere or a more concentrated source. The H_2 and CO_2 form water and CO via the WGS reaction in Eq. 16.3, and the CO can be mixed with more H_2 to make a syngas blend. The syngas can be used to form methanol per Eq. 16.4, and the methanol in turn can be used to make gasoline using the ExxonMobil MTG process. Alternately, the syngas can be used in the Fischer–Tropsch process described in Section 16.9.

16.9 Fischer-Tropsch Fuels

The Fischer-Tropsch (FT) process is a method for making middle distillate fuels—kerosene and diesel fuel—from syngas. The FT process was originally developed in Germany in 1920s, but it has seen limited large-scale use because the overall chemical process is both capital- and water-intensive.

The first step is to reform a feedstock with steam to form syngas [16-28] following the generic reaction in Eq. 16.2. Historically, the most common feedstocks for FT fuels are natural gas or coal, but any feedstock that can be converted to syngas can be used. The syngas is then fed to a reactor with an iron (Fe) or cobalt (Co) catalyst that builds up straight-chain alkanes (C_nH_{2n+2}). Given enough time, these alkanes end up approximately in the same molecular range as middle distillate kerosene or diesel fuel components. Once the alkanes are built, they can be further processed in the refinery, for example, cracked into smaller molecules or hydrotreated, to form appropriate fuel components.

The main attraction to FT fuels is that they are a drop-in replacement for distillate fuels, as the products of the FT process are alkanes. This means that the aromatics content of the resulting fuel is negligible, which helps reduce particulate emissions. Also, the sulfur content of the resulting FT fuel is negligible because of how the fuel is made [16-28]. Other emissions are not significantly affected by the use of FT fuels, as they are a drop-in replacement and generally burn similarly to the base fuels they replace. The alkanes, being paraffins, can start forming waxy compounds at higher temperatures than distillate diesel does.

The primary challenge with the FT process is that it is expensive, and therefore requires cheap feedstocks, cheap energy inputs, and high distillate fuel prices to be commercially feasible. Once a plant is up and running, the main challenge is operating the FT process reactor to get the best product yield from the feedstock while minimizing the selectivity for byproducts such as CH_4 or CO_2. Also, because the products are alkanes only, the preference is to blend FT fuels with distillate fuels so that there are some olefins (alkenes) in the fuel, since they provide benefits to fuel properties such as lubricity and wax point.

16.10 **Biodiesel**

Biodiesel fuels are made by reacting a fat or oil with a short-chain alcohol such as methanol, ethanol, or 1-propanol (C_3H_7OH) to form a fatty-acid ester. The usual method for making biodiesel from fats or oils is base-catalyzed transesterification with an alcohol, typically methanol. This process can occur at low temperature with a conversion greater than 98%.

Of the alcohols, methanol is the most commonly used, and the resulting biodiesel is called fatty-acid methyl ester (FAME) biodiesel. Fats that have been used to make biodiesel include beef tallow, chicken fat, and coconut oil; oils include corn, jatropha, palm, rapeseed, and soybean oils. Soybean oil methyl ester biodiesel (SME) is the most commonly produced biodiesel in US, whereas rapeseed oil methyl ester (RME) biodiesel is more common in Europe. In general, each market uses the fat or oil that is most readily available.

When prepared, the biodiesel is 100% (neat) biodiesel, and is called B100. The neat biodiesel is usually blended with a distillate diesel blendstock to form a fuel that can be sold at the pump. The most common blends authorized for sale are B5, B7, and B20, which are 5%, 7%, and 20% biodiesel in distillate fuel, respectively.

Soybean oil contains a mixture of fatty acids, primarily linoleic acid (cis-9,cis-12-octadecadienoic acid, $C_{18}H_{32}O_2$) and oleic acid (cis-9-octadecenoic acid, $C_{18}H_{34}O_2$), with significant amounts of palmitic acid (n-hexadecanoic acid, $C_{16}H_{32}O_2$) and α-linolenic acid (cis,cis,cis-9,12,15-octadecatrienoic acid, $C_{18}H_{30}O_2$) [16-40, 16-41, 16-42, 16-43, 16-44]. The structures for these four major constituents are shown in the respective parts of **Figure 16.6**. SME is made from reacting these fatty acids with methanol to form a blend of methyl esters.

As evaluated by a gas chromatograph–mass spectrometer (GCMS) system, 100% SME has a very different molecular profile than conventional diesel fuel does. Distillate diesel includes a wide variety of molecules ranging from those with 11 carbon atoms to those with 25. In contrast, SME only has the two GCMS peaks at 16 and 18 carbons.

Biodiesel combustion reduces particulate emissions and increases NOx emissions and the soluble organic fraction (SOF) [16-45]. Its effect on particulate properties and EAS devices has only lately been studied. The effects of the combustion of B100 and biodiesel blends on particulates are of interest, especially since the biodiesel fuel components are oxygenated. The effects on particulate structure and reactivity are of particular interest.

Reviews on the use of biodiesel [16-46, 16-47] have consistently shown that using an oxygenated fuel like FAME biodiesel significantly reduces particulates emissions. Biodiesel can also reduce CO and non-methane hydrocarbon (NMHC) emissions, albeit at the expense of increasing NOx emissions. Biodiesel does not seem to significantly affect fuel consumption beyond the decrease in lower heating value (LHV) of the fuel, as shown in Table 16.1.

Investigating the effect of biodiesel blends on the regeneration of particulate filters [16-45] showed that biodiesel blending increased the SOF content of the particulates, which made available additional reactive HC to be oxidized in the diesel particulate filter (DPF). Biodiesel fuel was also shown to affect the carbon structure of the

FIGURE 16.6   Fatty acid constituents of soybean oil include (a) linoleic acid, (b) oleic acid, (c) palmitic acid, and (d) α-linolenic acid [16-41, 16-42, 16-43, 16-44].

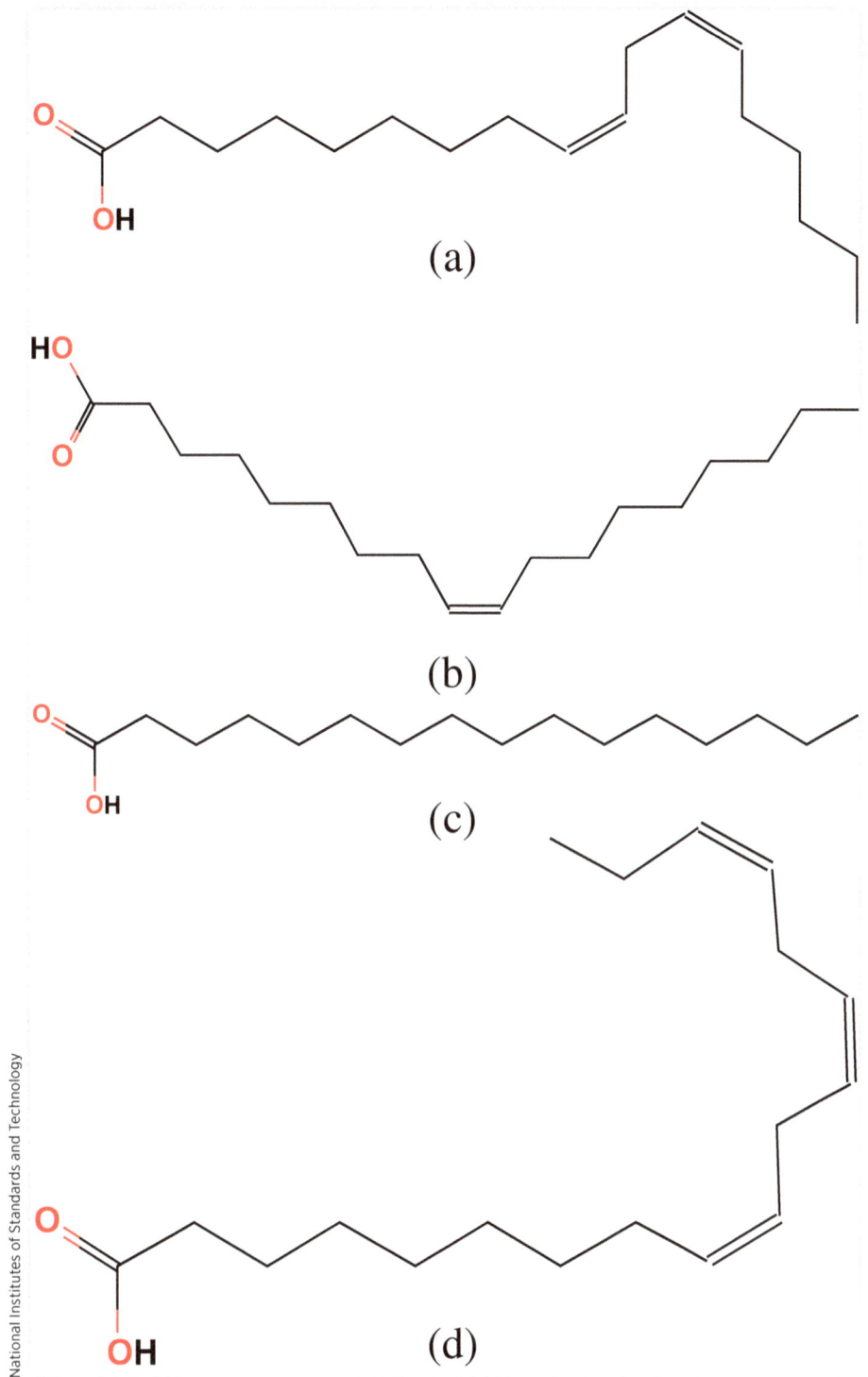

particulate, producing more amorphous carbon. Increasing the biodiesel content of the fuel reduces the temperature needed to trigger a DPF regeneration event, likely because there is more NOx in the exhaust and because the particulates have more SOF and a changed nanostructure.

An experimental study by Lahane and Subramanian [16-48] varied the biodiesel content of various fuels, ranging from B5 to B25, along with B50 and B100 blends. Their results suggest that fuel blends up to B15 may offer the best benefit in PM emissions reduction for the least penalty in NOx emissions. Likewise, Can et al. [16-49] added up to 15% EGR when using an SME B20 blend, and were able to decrease both NOx and PM emissions by up to 55% and 15%, respectively, at high loads. They also observed a slight fuel consumption penalty from the fuel blend and the EGR.

Regarding other physical properties, the lubricity of FAME biodiesel fuels is generally good, which is favorable. However, the fuels have a higher cloud point than distillate diesel does, where the cloud point is the temperature at which the fuel starts to look cloudy as it begins to turn into a gel. This cloud point issue is why the State of Minnesota (US) requires B5 in the winter and B20 in the summer, and exempts No. 1 diesel fuel from a minimum biodiesel content requirement [16-50]. FAME biodiesel fuels also need to be stored carefully, as any carbon–carbon double bonds in the fatty acids can be oxidized, making the fuel rancid.

16.11 Renewable Diesel

Renewable diesel is a hydrocarbon fuel that comes from renewable biomass feedstocks, but with constituents that are otherwise chemically similar to distillate diesel fuels [16-4]. There are several pathways to renewable diesel production that are currently in development, including from pyrolysis, hydrotreating, and hydrocracking [16-51, 16-52]. Of these process options, hydrotreating is the most common. Hydrotreating means reacting the input molecules, such as the fats shown in Figure 16.6, with enough H_2 that the oxygen is removed along with some or most of the carbon-carbon double bonds in the molecule. The resulting alkanes and olefins (alkenes) are indistinguishable from the distillate fuel components in the refinery. This means that the engine-out emissions, other engine operating characteristics, and effects on the EAS will be similar to what is seen with distillate fuels.

Neste renewable diesel is hydrotreated vegetable oil which is produced in a dedicated plant [16-9]. The Neste process makes branched alkanes because after reacting the fatty acids with hydrogen to strip out the functional groups, the molecules are isomerized into a branched form. Isomerization means rearranging the molecule's structure without changing the numbers of atoms within the molecule, such as the isomers of butanol shown in Figure 16.5.

The Neste renewable diesel product has a cetane number of over 70 and inherently low sulfur because of the feedstock [16-9]. The density is slightly lower than for distillate diesel, as shown in Table 16.1, so the volumetric energy is slightly lower than diesel even though the mass-specific energy content is the same.

Other companies co-process the renewable diesel fuel, meaning that the biomass inputs, such as fats and oils, are introduced to the main refinery process stream upstream of the hydrotreating unit. Phillips 66—the refinery operations of the former ConocoPhillips—uses hydrotreating in this manner to make renewable diesel, as do Sinclair Oil Corporation and Marathon Oil Corporation [16-53]. The biomass feedstocks are typically not >5% of the total process stream in the refinery, meaning that the product from the refinery will only have about 5% renewable diesel content.

For example, in 2007, ConocoPhillips and Tyson Foods, Inc., jointly announced that Tyson would provide animal fats as a feedstock to ConocoPhillips for use in renewable diesel production [16-54]. This relationship ran its course, and was subsequently dissolved. More recently, Phillips 66 has announced partnerships with Ryze Renewables and Renewable Energy Group, Inc., (REG) to build production facilities for renewable diesel products [16-55, 16-56], where the REG plant would use REG's BioSynfining® technology.

References

16-1. US Department of Energy, "Fuel Efficiency," https://www.energy.gov/eere/vehicles/fuel-efficiency. Accessed 1 Aug. 2019.

16-2. Advanced Motor Fuels, "Fuel Information – Butanol – Properties," https://www.iea-amf.org/content/fuel_information/butanol/properties. Accessed 22 Jul. 2019.

16-3. Advanced Motor Fuels, "Liquefied Petroleum Gas, LPG," https://www.iea-amf.org/content/fuel_information/lpg. Accessed 2 Aug. 2019.

16-4. Alternative Fuels Data Center, "Alternative Fuels and Advanced Vehicles," https://afdc.energy.gov/fuels/. Accessed 11 Jul. 2019.

16-5. Alleman, T.L., Eudy, L., Miyasato, M., Oshinuga, A., Allison, S., Corcoran, T., Chatterjee, S., Jacobs, T., Cherrillo, R.A., Clark, R., Virrels, I., Nine, R., Wayne, S., and Lansing, R., "Fuel Property, Emission Test, and Operability Results from a Fleet of Class 6 Vehicles Operating on Gas-To-Liquid Fuel and Catalyzed Diesel Particle Filters," SAE Technical Paper 2004-01-2959, 2004, https://doi.org/10.4271/2004-01-2959.

16-6. Dietsche, K.-H. and Reif, K., *Automotive Handbook*, 10th ed (Karlsruhe, Germany: Robert Bosch, GmbH, 2018).

16-7. Hashimoto, K., Ohta, H., Hirasawa, T., Arai, M., and Tamura, M., "Evaluation of Ignition Quality of LPG with Cetane Number Improver," SAE Technical Paper 2002-01-0870, 2002, https://doi.org/10.4271/2002-01-0870.

16-8. National Energy Technology Laboratory, "Fischer-Tropsch Synthesis," http://www.netl.doe.gov/research/coal/energy-systems/gasification/gasifipedia/ftsynthesis. Accessed 22 Jul. 2019.

16-9. Oyj, N., *Neste Renewable Diesel Hand book* (2016), 1–55.

16-10. National Institutes of Stand ards and Technology, "Listing of Experimental Geometry Data for CH_4 (Methane)," https://cccbdb.nist.gov/expgeom2.asp?casno=74828&charge=0. Accessed 18 Jul. 2019.

16-11. National Institutes of Stand ards and Technology, "Listing of Experimental Geometry Data for C_2H_6 (Ethane)," https://cccbdb.nist.gov/expgeom2.asp?casno=74840&charge=0. Accessed 18 Jul. 2019.

16-12. Lampert, J.K., Kazi, M.S., and Farrauto, R.J., "Palladium Catalyst Performance for Methane Emissions Abatement from Lean Burn Natural Gas Vehicles," *Applied Catalysis B: Environmental* **14**, no. 3-4 (1997): 211-223, https://doi.org/10.1016/S0926-3373(97)00024-6.

16-13. Choudhary, T.V., Banerjee, S., and Choudhary, V.R., "Catalysts for Combustion of Methane and Lower Alkanes," *Applied Catalysis A: General* **234**, no. 1–2 (2002): 1–23, https://doi.org/10.1016/S0926-860X(02)00231-4.

16-14. Okumura, K., Shinohara, E., and Niwa, M., "Pd Loaded on High Silica Beta Support Active for the Total Oxidation of Diluted Methane in the Presence of Water Vapor," *Catalysis Today* **117**, no. 4 (2006): 577–583, https://doi.org/10.1016/j.cattod.2006.06.008.

16-15. Auvray, X., Lindholm, A., Milh, M., and Olsson, L., "The Addition of Alkali and Alkaline Earth Metals to Pd/Al2O3 to Promote Methane Combustion. Effect of Pd and Ca Loading," *Catalysis Today* **299** (2018): 212–218, https://doi.org/10.1016/j.cattod.2017.05.066.

16-16. Cargnello, M., Delgado Jaén, J.J., Hernández Garrido, J.C., Bakhmutsky, K., Montini, T., Calvino Gámez, J.J., Gorte, R.J., and Fornasiero, P., "Exceptional Activity for Methane Combustion over Modular Pd@CeO2 Subunits on Functionalized Al2O3," *Science* **337**, no. 6095 (2012): 713–717, https://doi.org/10.1126/science.1222887.

16-17. Mowery, D.L., Graboski, M.S., Ohno, T.R., and McCormick, R.L., "Deactivation of PdO-Al2O3 Oxidation Catalyst in Lean-Burn Natural Gas Engine Exhaust: Aged Catalyst Characterization and Studies of Poisoning by H2O and SO2," *Applied Catalysis B: Environmental* **21**, no. 3 (1999): 157–169, https://doi.org/10.1016/S0926-3373(99)00017-X.

16-18. Morganti, K.J., Foong, T.M., Brear, M.J., da Silva, G., Yang, Y., and Dryer, F.L., "The Research and Motor Octane Numbers of Liquefied Petroleum Gas (LPG)," *Fuel* **108** (2013): 797–811, https://doi.org/10.1016/j.fuel.2013.01.072.

16-19. National Institutes of Standards and Technology, "Listing of Experimental Geometry Data for C3H8 (Propane)," https://cccbdb.nist.gov/expgeom2.asp?casno=74986. Accessed 18 Jul. 2019.

16-20. National Institutes of Standards and Technology, "Listing of Experimental Geometry Data for CH3CH2CH2CH3 (Butane)," https://cccbdb.nist.gov/expgeom2.asp?casno=106978&charge=0. Accessed 18 Jul. 2019.

16-21. US Department of Energy, "Propane Fueling Stations Alternative Fuels Data Center," https://afdc.energy.gov/fuels/propane_stations.html. Accessed 18 Jul. 2019.

16-22. National Institutes of Standards and Technology, "Listing of Experimental Geometry Data for CH3OCH3 (Dimethyl Ether)," https://cccbdb.nist.gov/expgeom2.asp?casno=115106&charge=0. Accessed 18 Jul. 2019.

16-23. Aguayo, A.T., Ereña, J., Sierra, I., Olazar, M., and Bilbao, J., "Deactivation and Regeneration of Hybrid Catalysts in the Single-Step Synthesis of Dimethyl Ether from Syngas and CO2," *Catalysis Today* **106**, no. 1–4 (2005): 265-270, https://doi.org/10.1016/j.cattod.2005.07.144.

16-24. Golovitchev, V.I., Nordin, N., and Chomiak, J., "Neat Dimethyl Ether: Is It Really Diesel Fuel of Promise?," SAE Technical Paper 982537, 1998, https://doi.org/10.4271/982537.

16-25. Hansen, K.F., Nielsen, L., Hansen, J.B., Mikkelsen, S.-E., Landälv, H., Ristola, T., and Vielwerth, K., "Demonstration of a DME (Dimethyl Ether) Fuelled City Bus," SAE Technical Paper 2000-01-2005, 2000, https://doi.org/10.4271/2000-01-2005.

16-26. Teng, H., McCandless, J.C., and Schneyer, J.B., "Thermochemical Characteristics of Dimethyl Ether - An Alternative Fuel for Compression-Ignition Engines," SAE Technical Paper 2001-01-0154, 2001, https://doi.org/10.4271/2001-01-0154.

16-27. Park, S.H. and Lee, C.S., "Applicability of Dimethyl Ether (DME) in a Compression Ignition Engine as an Alternative Fuel," *Energy Conversion and Management* **86** (2014): 848–863.

16-28. Khair, M.K. and Majewski, W.A. *Diesel Emissions and Their Control* (Warrandale, PA: SAE International, 2006).

16-29. National Institutes of Stand ards and Technology, "Ethanol," NIST Chemistry WebBook, SRD 69 https://webbook.nist.gov/cgi/cbook.cgi?Name=ethanol&Units=SI. Accessed 18 Jul. 2019.

16-30. Harvey, B.G. and Meylemans, H.A., "The Role of Butanol in the Development of Sustainable Fuel Technologies," *Journal of Chemical Technology & Biotechnology* **86**, no. 1 (2011): 2–9, https://doi.org/10.1002/jctb.2540.

16-31. Ndaba, B., Chiyanzu, I., and Marx, S., "n-Butanol Derived from Biochemical and Chemical Routes: A Review," *Biotechnology Reports* **8** (2015): 1-9, https://doi.org/10.1016/j.btre.2015.08.001.

16-32. da Silva Trindade, W.R. and dos Santos, R.G., "Review on the Characteristics of Butanol, Its Production and Use as Fuel in Internal Combustion Engines," *Renewable and Sustainable Energy Reviews* **69** (2017): 642-651, https://doi.org/10.1016/j.rser.2016.11.213.

16-33. National Institutes of Stand ards and Technology, "1-Butanol," NIST Chemistry WebBook, SRD 69 https://webbook.nist.gov/cgi/cbook.cgi?Name=butanol&Units=SI. Accessed 18 Jul. 2019.

16-34. National Institutes of Stand ards and Technology, "2-Butanol," NIST Chemistry WebBook, SRD 69 https://webbook.nist.gov/cgi/cbook.cgi?ID=C15892236&Units=SI. Accessed 18 Jul. 2019.

16-35. National Institutes of Stand ards and Technology, "1-Propanol, 2-methyl-," NIST Chemistry WebBook, SRD 69 https://webbook.nist.gov/cgi/cbook.cgi?Name=isobutanol&Units=SI. Accessed 18 Jul. 2019.

16-36. National Institutes of Stand ards and Technology, "2-Propanol, 2-methyl-," NIST Chemistry WebBook, SRD 69 https://webbook.nist.gov/cgi/cbook.cgi?Name=2-methyl-2-propanol&Units=SI. Accessed 18 Jul. 2019.

16-37. Jang, Y.-S., Malaviya, A., Cho, C., Lee, J., and Lee, S.Y., "Butanol Production from Renewable Biomass by Clostridia," *Bioresource Technology* **123** (2012): 653–663, https://doi.org/10.1016/j.biortech.2012.07.104.

16-38. National Energy Technology Laboratory, "Conversion of Methanol to Gasoline," https://www.netl.doe.gov/research/coal/energy-systems/gasification/gasifipedia/methanol-to-gasoline. Accessed 22 Jul. 2019.

16-39. Robert Bosch GmbH, "Synthetic Fuels," https://www.bosch.com/stories/synthetic-fuels/. Accessed 17 Jul. 2019.

16-40. Ivanov, D.S., Lević, J.D., and Sredanović, S.A., "Fatty Acid Composition of Various Soybean Products," *Food and Feed Research* **37**, no. 2 (2010): 65–70.

16-41. National Institutes of Stand ards and Technology, "9,12-Octadecadienoic Acid (Z,Z)-," NIST Chemistry WebBook, SRD 69 https://webbook.nist.gov/cgi/cbook.cgi?Name=linoleic+acid+&Units=SI. Accessed 22 Jul. 2019.

16-42. National Institutes of Stand ards and Technology, "Oleic Acid," NIST Chemistry WebBook, SRD 69 https://webbook.nist.gov/cgi/cbook.cgi?ID=C112801&Units=SI. Accessed 22 Jul. 2019.

16-43. National Institutes of Stand ards and Technology, "n-Hexadecanoic Acid," NIST Chemistry WebBook, SRD 69 https://webbook.nist.gov/cgi/cbook.cgi?Name=palmitic+acid&Units=SI. Accessed 22 Jul. 2019.

16-44. National Institutes of Stand ards and Technology, "9,12,15-Octadecatrienoic acid, (Z,Z,Z)-," NIST Chemistry WebBook, SRD 69 https://webbook.nist.gov/cgi/cbook.cgi?ID=C463401&Units=SI. Accessed 22 Jul. 2019.

16-45. Boehman, A.L., Song, J., and Alam, M., "Impact of Biodiesel Blending on Diesel Soot and the Regeneration of Particulate Filters," *Energy & Fuels* **19**, no. 5 (2005): 1857–1864, https://doi.org/10.1021/ef0500585.

16-46. Rakopoulos, C.D., Hountalas, D.T., Zannis, T.C., and Levendis, Y.A., "Operational and Environmental Evaluation of Diesel Engines Burning Oxygen-Enriched Intake Air or Oxygen-Enriched Fuels: A Review," SAE Technical Paper 2004-01-2924, 2004, https://doi.org/10.4271/2004-01-2924.

16-47. Lapuerta, M., Armas, O., and Rodriguez-Fernand ez, J., "Effect of Biodiesel Fuels on Diesel Engine Emissions," *Progress in Energy and Combustion Science* **34**, no. 2 (2008): 198-223, https://doi.org/10.1016/j.pecs.2007.07.001.

16-48. Lahane, S. and Subramanian, K., "Effect of Different Percentages of Biodiesel-Diesel Blends on Injection, Spray, Combustion, Performance, and Emission Characteristics of a Diesel Engine," *Fuel* **139** (2015): 537–545.

16-49. Can, Ö., Öztürk, E., Solmaz, H., Aksoy, F., Çinar, C., and Yücesu, H.S., "Combined Effects of Soybean Biodiesel Fuel Addition and EGR Application on the Combustion and Exhaust Emissions in a Diesel Engine," *Applied Thermal Engineering* **95** (2016): 115–124, https://doi.org/10.1016/j.applthermaleng.2015.11.056.

16-50. Minnesota Department of Agriculture, "Minnesota Biodiesel," https://www.mda.state.mn.us/environment-sustainability/minnesota-biodiesel. Accessed 22 Jul. 2019.

16-51. Jones, S.B., Valkenburt, C., Walton, C.W., Elliott, D.C., Holladay, J.E., Stevens, D.J., Kinchin, C., and Czernik, S., "Production of Gasoline and Diesel from Biomass via Fast Pyrolysis, Hydrotreating and Hydrocracking: A Design Case," Pacific Northwest National Lab.(PNNL), Richland , WA (United States) Report No. PNNL-18284, 2009.

16-52. Jones, S.B. and Snowden-Swan, L.J., "Production of Gasoline and Diesel from Biomass via Fast Pyrolysis, Hydrotreating and Hydrocracking: 2012 State of Technology and projections to 2017," Pacific Northwest National Lab.(PNNL), Richland , WA (United States) Report No. PNNL-22684, 2013.

16-53. Kotrba, R., "The Complex Dynamics of Coprocessing," *Biodiesel Magazine*, 2018.

16-54. White, A., "ConocoPhillips, Tyson Foods to Make Renewable Diesel," *Oil & Gas Journal*, 2007.

16-55. Phillips 66, "Ryze Renewables Partners with Phillips 66 to Build Next-Gen Renewable Diesel Fuel Plants in Nevada," 2018.

16-56. Phillips 66, "Phillips 66 and Renewable Energy Group Announce Plans for Large-Scale Renewable Diesel Facility on West Coast," 2018.

about the authors

Dr. John J. Kasab, P.E., is the Senior Technical Specialist for emissions controls at AVL Powertrain Engineering, Inc., in Plymouth, Michigan, where he has been since 2016. In this role, he focuses on criteria pollutants and exhaust aftertreatment systems and on fuel economy and greenhouse gas emissions. The applications of interest include both light-duty and heavy-duty vehicles and engines. In addition, he supports activities with reformers and fuel-cell systems engineering. He leads proposal development and project teams working in these areas and coordinates activities across AVL's sites globally.

Previously, he spent almost 13 years as a consulting engineer at Ricardo, Inc., where he proposed and led programs on emissions control technology for both criteria pollutants and greenhouse gas emissions, alternative fuels and lubricants, and reformer and fuel cell systems engineering. Prior to working at Ricardo, he also founded and managed Kasab Research, LLC, a small consultancy, and worked as a development engineer at Scania CV AB in Södertälje, Sweden, on DPF and SCR systems for Euro V commercial vehicle emissions.

He received B.S.Ch.E. and M.S.Ch.E. from Washington University in St. Louis. His Ph.D. in chemical engineering from the University of Wisconsin-Madison was on modeling chemical reactors under the direction of Professor Warren E. Stewart. He is also a registered Professional Engineer in the State of Wisconsin. He is a member of SAE International, the American Institute of Chemical Engineering, and is a life member of the Society of Women Engineers.

Dr. Andrea Strzelec is the Program Director for the Masters of Engineering in Engine Systems at the University of Wisconsin-Madison. Previously, she held faculty positions in Mechanical Engineering at Texas A&M University and Mississippi State University. She was the PI of the Combustion & Reaction Characterization Laboratory, where her sustainable energy research interests were devoted to heterogeneous reaction kinetics and characterization with specific focus on automotive combustion and emissions aftertreatment, low-temperature catalysis, particulate filtration, pyrolysis, and remediation of hydrocarbon contamination. She has taught classes in thermodynamics, heat transfer, internal combustion engines, combustion science, and automotive engineering. Before her academic career, she held postdoctoral fellowships at Oak Ridge and Pacific Northwest National Laboratories. She received her interdisciplinary Combustion Engineering Ph.D. from the University of Wisconsin-Madison, working through the Engine Research Center under the direction of Professor David E. Foster and Drs. C. Stuart Daw and Todd J. Toops of Oak Ridge National Laboratory. Her work has been published in SAE as well as top journals, including *Topics in Catalysis, Energy & Fuels, Fuel,* and *International Journal of Engine Research.*

Dr. Strzelec is the Chair of the SAE Exhaust Aftertreatment and Emissions Committee, Associate Editor of the Journal of Emissions Control Science & Technology, a recipient of the 2020 SAE Lloyd L. Withrow Distinguished Speaker Award, 2016 SAE Forrest R. McFarland Service Award, 2015 SAE Ralph R. Teetor Educational Award, and the recipient of the Texas A&M Mechanical Engineering 2015 Brittian Undergraduate Teaching Award. She is an active participant of the DOE CLEERS group, organizer for the Catalyst Systems session for ASME, member of ACS, AIChE, and SWE.

index

www.ingramcontent.com/pod-product-compliance
Lightning Source LLC
Chambersburg PA
CBHW051703210326

41597CB00032B/5353